RC Active Filter Design Handbook

THE WILEY ELECTRICAL AND ELECTRONICS TECHNOLOGY HANDBOOK SERIES

The Wiley Electrical and Electronics Technology Handbook Series provides technicians and engineers with up-to-date information on a wide range of topics in a rapidly evolving field. Every chapter in each volume is written by an authority on the subject, and takes a practical, how-to-do-it approach with numerous worked-out examples and step-by-step calculations. These procedures provide productive approaches to many problems encountered in electronics and electrical technology.

TOPICS OF FORTHCOMING HANDBOOKS

Op Amps
Digital Circuits
Microprocessors and Microcomputers
Communications
Electrical Machines
Automatic Control Systems
System Troubleshooting
Electrical and Process Measurements
Semiconductor Memories
Microwave Measurements
Robotics
Fiberoptics

RC ACTIVE FILTER DESIGN HANDBOOK

F. W. Stephenson
Virginia Polytechnic Institute and State University

JOHN WILEY & SONS
New York • Chichester • Brisbane • Toronto • Singapore

Library of Congress Cataloging in Publication Data:

Stephenson, F. W. (Frederick William)
RC active filter design handbook.

(The Wiley electrical and electronics technology handbook series)
Includes bibliographies and indexes.
1. Electric filters, Active. 2. Electric filters, Resistance-capacitance. I. Title. II. Title: R.C. active filter design handbook. III. Series.

TK7872. F5S75 1985 621.3815'32 84-23707
ISBN 0-471-86151-0

Printed in the United States of America

10 9 8 7 6 5 4 3 2 1

CONTRIBUTING AUTHORS

Baez-Lopez, David	Instituto Nacional de Astrofisica
Davis, Artice M.	San José State University
Geiger, R. L.	Texas A & M University
Huelsman, L. P.	University of Arizona
Lee, M. R.	Sheffield City Polytechnic
Mikhael, W. B.	West Virginia University
Nelin, Bert D.	Siemens Communications Systems, Inc.
Reineck, K. M.	University of Cape Town
Sánchez-Sinencio, E.	Instituto Nacional de Astrofisica
Sedra, Adel S.	University of Toronto
Steer, R. W., Jr.	Frequency Devices, Inc.
Stirk, G.	General Electric
Wierzba, Gregory M.	Michigan State University

PREFACE

Two significant technological advances have made RC active filters a practical proposition in many applications. These are the availability of low-cost, high-quality operational amplifiers and the precise tuning afforded by laser trimming of resistors. As a result of these advances, active filters are now widely used; consequently, the student, technician, and nonspecialist engineer alike need to understand the rudiments of filter design.

The purpose of the RC ACTIVE FILTER DESIGN HANDBOOK is to provide comprehensive, up-to-date and authoritative information on those filter structures which have proven themselves in practice. The main thrust of the book is aimed at providing a readily understandable manual for use by *electronics engineering technicians, students* and *electrical engineers who are not specialists in filter design.*

Practical aspects are strongly emphasized in the text, and mathematics is kept to a minimum. Whenever possible, formulas are presented as an integral part of the design process.

This handbook consists of 13 chapters. Chapter 1 introduces the active filter in the general context of filter design methods, indicating areas of application and fabrication procedures. Current trends and developments are discussed in relation to the quest for totally integrated structures. Chapter 2 outlines the approximation problem and presents essential information on a selection of commonly used functions. Frequency transformations are considered so that any practical set of specifications can be translated into a low-pass form. Chapter 3 considers the second-order function, the basic component of many active filter design procedures.

Chapters 4 and 5 introduce the active building blocks and relatively simple methods for analyzing networks that contain them. The design methods proposed in Chapters 6 to 9 are all well established. They represent structures that, within the limits presented, provide low sensitivity and readily reproduced performance.

Chapter 10 allows the reader to undertake the design of filters of an order greater than two. Two broad approaches are discussed; they represent the most commonly adopted techniques. Chapter 11 provides useful practical information related to active compensation, power supply decoupling, limitations of amplifier performance, and component selection.

In a modern text on active filters, it is essential that the reader be made aware of the relative ease with which both hand calculators and personal computers can be incorporated into the design process. The design equations presented in the handbook are readily programmed into calculator routines, and a more powerful, general approach is possible with the use of a personal computer. To illustrate this, Chapter 12 is devoted to the development of a design program for use with an IBM PC.

No handbook on active filters is complete without a comprehensive source of design data. Chapter 13 has, therefore, been compiled to provide a meaningful source of information for the practical designer.

This text instructs by means of worked examples. Practical filter designs, from *specification* to *realization*, are featured in Chapters 6 through 10.

The handbook's chapters have been written by leading designers from universities and industry. I wish to express my gratitude to all of the authors for their contributions to this text. In addition, I am indebted to Art Seidman, the Series Editor, for inviting me to edit this volume. While occasionally experiencing the frustrations of a novice editor, I have nonetheless enjoyed the challenge immensely. I am also indebted to Hank Stewart and his colleagues at John Wiley, who have been a constant source of encouragement throughout this project. I would also like to extend my appreciation to the reviewers Peter Aronhine, M. F. Doug DeMaw, Russell Heiserman, Forrest M. Mims III, Mac Van Valkenburg, Joseph Vellely, and Paul Wojnowiak for their detailed and constructive comments and suggestions.

Sincere thanks are due to my wife, Sally, and daughter, Sarah, who once again have suffered life with an author and editor.

Finally, I wish to acknowledge Virginia Polytechnic Institute and State University for the provision of facilities throughout the four years of manuscript preparation and to thank Mrs. Pat Manning and Mrs. Sandy Crigger of the Electrical Engineering Department for typing parts of the manuscript.

F. W. Stephenson

CONTENTS

CHAPTER 1

Introduction to Active Filters

Adel S. Sedra

University of Toronto

1.1 INTRODUCTION

This book is concerned with the design of a special type of electric wave filter: the RC-active filter. Such filters employ operational amplifiers (op amps) together with resistors and capacitors to realize a wide variety of filtering functions. The resulting networks are fabricated using either discrete components or, when production quantities justify it, thin or thick film technology.

RC-active filters find extensive use in the audio-frequency range (up to 20 kHz) and, with some effort in the choice of appropriate op amp type and circuit configuration, at frequencies up to 100 kHz or even higher. In the audio-frequency range, active filters are used in almost any electronic system that needs some sort of signal filtering. In particular, they are extensively employed in communication and instrumentation systems.

This introductory chapter describes some of the basic concepts that underlie the filter design process in general and RC-active filter design in particular. It describes in some detail the method used for specifying the transmission characteristics of filters. The two phases of filter design, approximation and realization, are delineated, and a historical evolution of RC-active filters is synopsized. This is followed with a brief description of filter realization technologies to help place the RC-active filter in proper perspective. Finally, the active filter design process is outlined and its various steps are related to the corresponding chapters of this book.

Figure 1.1 Black-box representation of a filter network.

1.2 FILTER TRANSMISSION

The electric wave filter is a circuit that processes an input signal (usually voltage) in a predetermined manner to produce an output signal (usually voltage). The filter networks with which we are concerned here are specified in terms of their operation in the frequency domain. To be specific, we are interested in *frequency selective* circuits whose function is to separate the different frequency components of input signals—*passing* some components and *stopping* others. Ideally, all such circuits are linear; thus, no new frequency components should appear at their outputs.

Figure 1.1 shows a black-box representation of a filter network. Both the input and output signals are shown as functions of frequency ω. (Note that ω denotes the angular frequency in rad/s and is related to the frequency f in hertz by $\omega = 2\pi f$). If a single-frequency sinusoid is applied at the input, then the output will be a sinusoid of the same frequency but with a different amplitude and shifted in phase. The filter *transmission* can therefore be measured in terms of the *transfer function* $T(\omega)$:

$$T(\omega) \equiv \frac{V_o(\omega)}{V_i(\omega)}$$

where $T(\omega)$ has a magnitude $|T(\omega)|$ and phase $\phi(\omega)$, that is,

$$T(\omega) = |T(\omega)|e^{j\phi(\omega)} \tag{1.1}$$

Thus, for an input sinusoid with a frequency ω_1 and amplitude V_1, the output sinusoid will have an amplitude $|T(\omega_1)|V_1$ and will be shifted in phase by an angle $\phi(\omega_1)$.

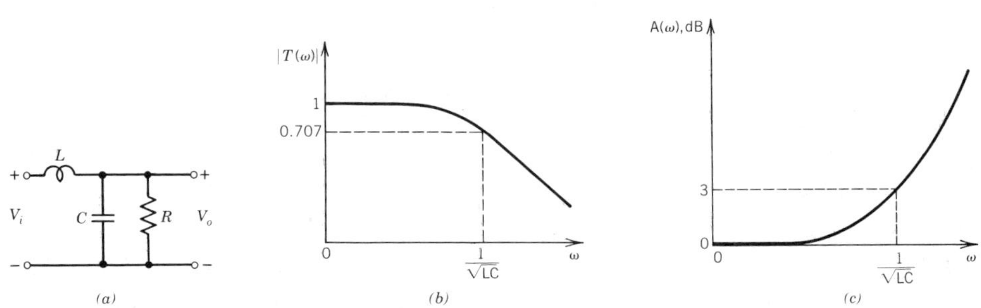

Figure 1.2 (*a*) *A simple filter circuit;* (*b*) *its magnitude of transmission for the special case when* $R\sqrt{C/L} = 1/\sqrt{2}$; *and* (*c*) *the attenuation function* $A(\omega)$ *for this particular case.*

Although in most filter applications only the magnitude of transmission, $|T(\omega)|$, is of interest, there are situations in which the phase response $\phi(\omega)$ is also specified. We elaborate on this point shortly.

The concept of magnitude response may be illustrated by considering the simple filter circuit of Fig. 1.2*a*. Using straightforward analysis, we obtain the voltage transfer function $T(\omega)$ as follows:

$$T(\omega) \equiv \frac{V_o(\omega)}{V_i(\omega)} = \frac{(1/j\omega L)}{(1/j\omega L) + [j\omega C + (1/R)]}$$

$$= \frac{1}{(1 - \omega^2 LC) + j(\omega L/R)}$$

Thus

$$|T(\omega)| = \frac{1}{\sqrt{(1 - \omega^2 LC)^2 + (\omega L/R)^2}}$$

and

$$\phi(\omega) = -\tan^{-1}\left(\frac{\omega L/R}{1 - \omega^2 LC}\right)$$

Figure 1.2*b* shows a sketch of the magnitude of transmission of this filter for the case $R\sqrt{C/L} = 1/\sqrt{2}$. The reader is urged to verify that for this particular choice of component values, the magnitude of transmission drops from 1 to $1/\sqrt{2}$ at $\omega = 1/\sqrt{LC}$, as indicated in the figure.

The magnitude of filter transmission, $|T(\omega)|$, is usually expressed in *decibels*,

$$G(\omega) = 20 \log_{10}|T(\omega)| \qquad \text{dB}$$

where $G(\omega)$ is the filter *gain* in decibels. Alternatively, some filter designers prefer to deal with the filter *loss* or *attenuation* $A(\omega)$,

$$A(\omega) = -20 \log_{10}|T(\omega)| \qquad \text{dB}$$

Referring to Fig. 1.2*b*, we note that the transmission of the filter shown is 0 dB at $\omega = 0$ (dc) and drops to -3 dB at $\omega = 1/\sqrt{LC}$. Alternatively, we can say that this filter provides 3 dB of attenuation at $\omega = 1/\sqrt{LC}$. Also, while the transmission or gain of this particular filter drops (*falls off*) monotonically (ripple-free), its attenuation or loss increases. The attenuation function $A(\omega)$ for this filter is shown in Fig. 1.2*c*.

1.3 FILTER SPECIFICATION

As mentioned above, a filter is a two-port network whose purpose is to *pass* signals whose frequency spectrum lies within a specified range and *stop* signals whose frequency spectrum falls outside this range. Thus, ideally a filter has a frequency band over which its attenuation is required to be 0 dB (the filter *passband*) and a

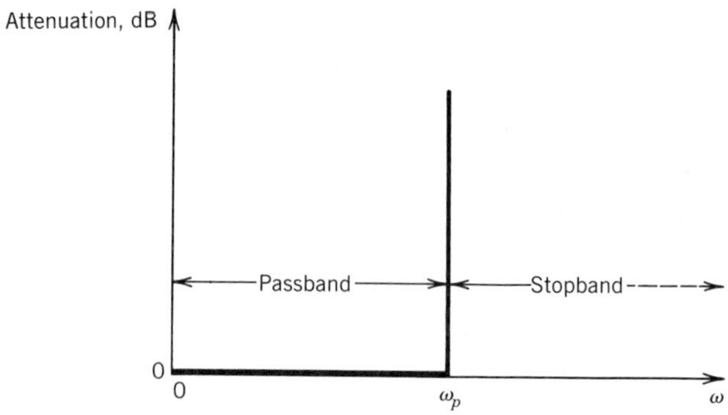

(a)

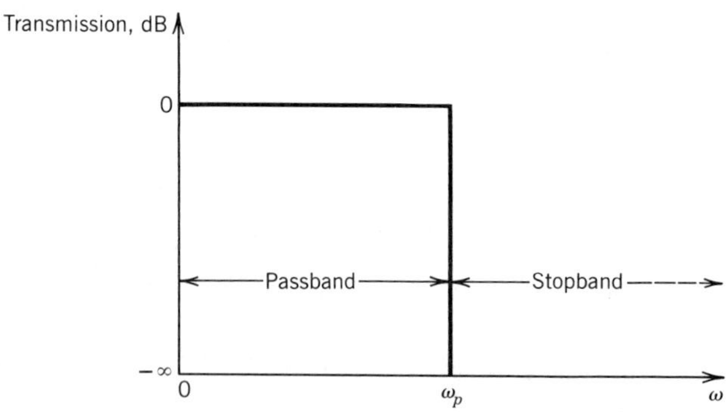

(b)

Figure 1.3 (a) Attenuation characteristics of the ideal low-pass filter; (b) transmission characteristics of the ideal low-pass filter.

frequency band over which the attenuation is required to be infinite (the filter *stopband*). As an example, Fig. 1.3 shows the attenuation and transmission characteristics of an ideal *low-pass* filter. Note that this filter is classified as low pass because it passes low-frequency signals while stopping high-frequency signals.

The ideal low-pass filter characteristics shown in Fig. 1.3 cannot be realized using a circuit with a finite number of components. It follows that one has to find a more realistic way of specifying the filter attenuation characteristics. Toward that end we first note that there is no need for insisting on 0 dB of attenuation over the filter passband. It makes little difference to the operation of the filter as a frequency selective device whether the passband attenuation is 0 dB or, say, 6 dB as is usually the case in passive *LC* filters. Of course, active filters are capable of providing gain and can be designed to do so. From the point of view of designing the filter, however, it is convenient to normalize the filter specifications so that the passband transmission is unity (0 dB).

Although the exact value of passband transmission is of little significance, the constancy of the filter transmission over the passband is very important. To appreciate this, consider an input signal composed of two sinusoidal components of different frequencies, both within the passband. In order that the output signal be an exact replica of the input signal, the two components must experience the same magnitude of transmission and must be delayed by equal times. The first condition gives rise to the requirement of constant passband transmission. The second condition, namely, equal time delay at all passband frequencies, implies the need for a phase function $\phi(\omega)$ that is linearly related to ω. This linear phase property, however, is not always required.

Physical networks are not capable of providing constant transmission over a frequency interval. It follows that the filter user must accept some variation in passband transmission. Fortunately, the filter design process is such that the designer can a priori specify an upper limit on the allowable variation in passband transmission. This is indicated in Fig. 1.4 where $A_{\max}$ denotes the maximum deviation (from the ideal 0 dB) allowed in passband attenuation.

Although physical networks can have infinite attenuation at distinct frequencies, they cannot do so over a band of frequencies. It follows that the filter user must accept a certain amount of transmission in the filter stopband. Fortunately, the filter design process is such that the designer can specify a priori an upper limit on the amount of this transmission or, equivalently, a minimum amount of required stopband attenuation. This is indicated in Fig. 1.4 by $A_{\min}$, the minimum acceptable stopband attenuation.

To complete our discussion of the method for specifying filters, note that in sketching the specifications in Fig. 1.4 we have taken account of the fact that filter attenuation cannot increase abruptly at the edge of the passband. We have therefore introduced a *transition band*, from ω_p to ω_s, over which the attenuation rises from $A_{\max}$ to $A_{\min}$. The narrower the transition band, the more selective the filter becomes. Unfortunately, however, we pay for this increase in selectivity because the resulting network becomes more complex and hence more expensive.

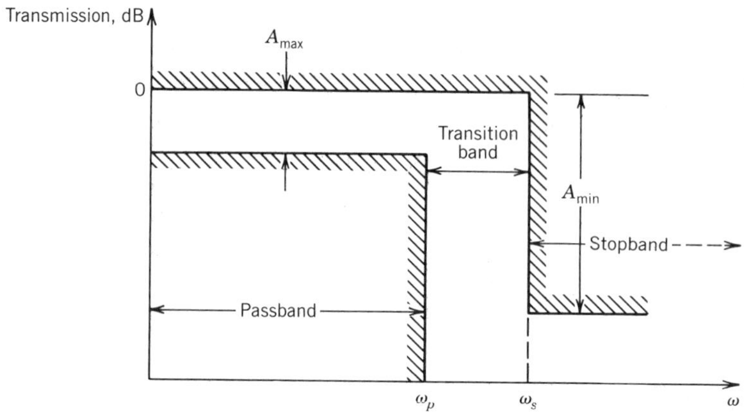

Figure 1.4 Method for specifying the transmission characteristics of a low-pass filter.

In general, the tighter the bounds the filter user imposes—that is, lower A_{max}, higher A_{min}, and narrower transition band or equivalently lower selectivity ratio (ω_s/ω_p)—the more complex and expensive the final circuit must be. Also, if phase linearity is an additional requirement, the design process is further complicated.

1.4 FILTER TYPES

Figure 1.5 shows transmission specifications for the four major filter types: *low-pass* (*LP*), *high-pass* (*HP*), *bandpass* (*BP*), and *bandstop* (*BS*). Note that while the bandpass filter has two stopbands and one passband, the opposite is true for the bandstop filter. Although the attenuation in the two stopbands of the bandpass filter are shown equal, this is not always the case.

In addition to the four filter types shown in Fig. 1.5, the need sometimes arises for a filter having constant transmission over the entire frequency range of interest, ideally for $0 \leq \omega \leq \infty$ but having specified phase characteristics. Such a network finds application in pulse transmission systems where, to preserve the shape of the pulse signals, phase linearity is needed. The phase characteristics required of the filter are those that would make the phase function of the overall system linear. Such a filter is known as a *delay equalizer* because its specification and design is usually in terms of the derivative of phase, known as the *group delay* function* $\tau(\omega)$,

$$\tau(\omega) = \frac{-d\phi(\omega)}{d\omega} \tag{1.2}$$

Note that linear phase implies constant group delay.

* Group delay is also known as *envelope delay*, since it is the time delay experienced by the envelope of an amplitude modulated signal as it passes through a bandpass filter.

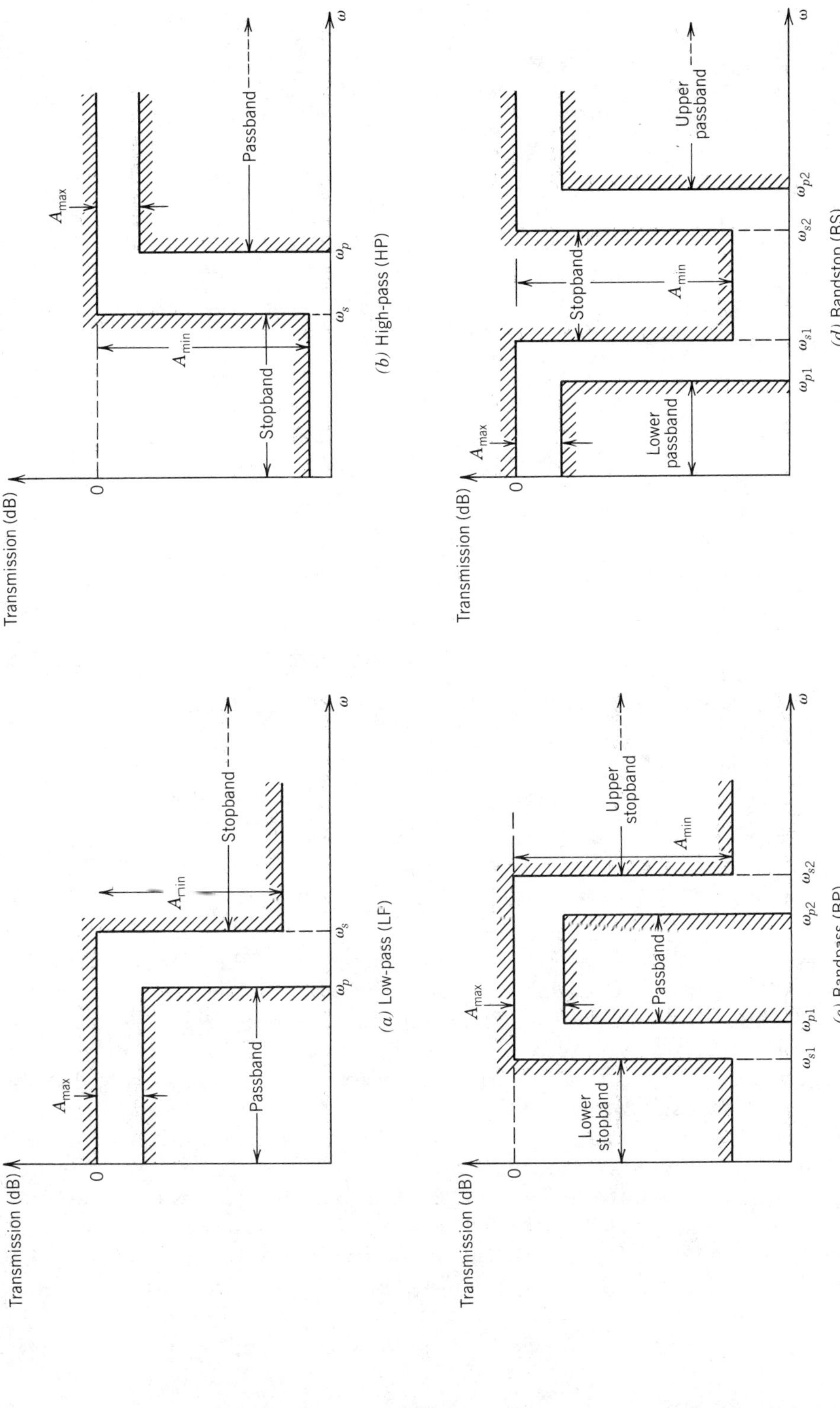

Figure 1.5 The four major filter types.

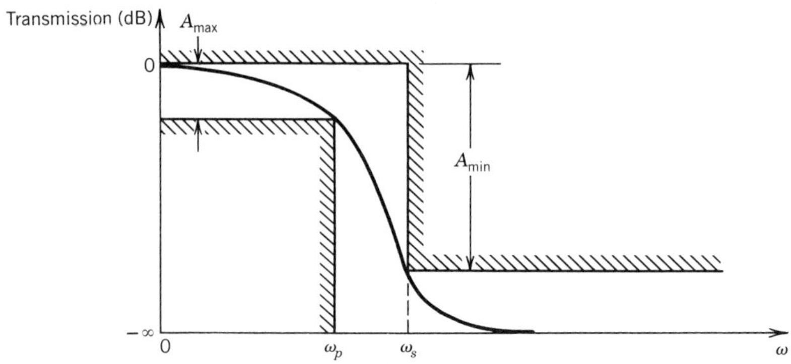

(a) Monotonic response.

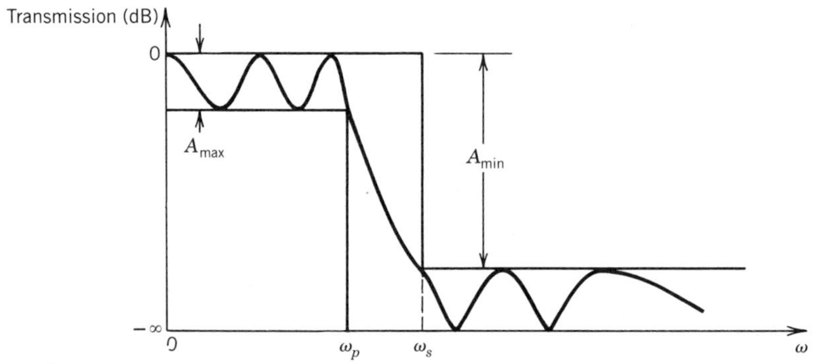

(b) Equiripple response.

Figure 1.6 Two different transmission functions that meet identical low-pass specifications.

1.5 THE FILTER DESIGN PROCESS

Given a set of transmission specifications such as one of the four shown in Fig. 1.5, the first step in the design consists of finding a suitable transfer function $T(\omega)$ whose corresponding attenuation function $A(\omega) = -20 \log_{10}|T(\omega)|$ meets the given specifications. Figure 1.6 shows sketches of two different transmission functions that meet identical low-pass transmission specifications. Note that the transmission function depicted in Fig. 1.6*a* decreases monotonically and attains the maximum allowable deviation in passband attenuation (A_{max}) at the passband edge (ω_p) and the minimum permissible stopband attenuation (A_{min}) at the stopband edge (ω_s). In contrast, the function depicted in Fig. 1.6*b* exhibits ripples in both the passband and the stopband. Chapter 2 shows that transfer functions of the type in Fig. 1.6*b* provide more efficient approximations to the given transmission specifications than those of the type shown in Fig. 1.6*a*. In other words, by allowing ripples in both the passband and stopband attenuation, one obtains a transfer function $T(\omega)$ of lower order, and the resulting filter circuit is simpler.

The problem of finding a suitable transfer function $T(\omega)$ is known as *filter approximation.* Chapter 2 is devoted to this topic. Here it suffices to say that a considerable amount of design data in the form of filter tables and computer programs are currently available to help in the solution of the filter approximation problem.

Once the approximation problem is solved and a suitable transfer function is obtained, the filter designer must turn his attention to the second phase of the design process: filter realization. Filter realization is the process of finding a circuit whose voltage transfer function $V_o(\omega)/V_i(\omega)$ is equal to the function $T(\omega)$ obtained from approximation. Before addressing the realization problem and the various technologies available for filter realization, we briefly discuss the filter transfer function.

1.6 THE FILTER TRANSFER FUNCTION

Filter approximation is most conveniently performed using the Laplace complex frequency variable s. Thus the result of the approximation process is a function $T(s)$ from which we can obtain the filter transmission by substituting $s = j\omega$. Moreover, most realization methods operate directly on $T(s)$ rather than $T(j\omega)$. Therefore, it is useful at this stage to consider briefly the form that the function $T(s)$ takes. Specifically, $T(s)$ can be written as the ratio of two polynomials as follows:

$$T(s) = \frac{a_m s^m + a_{m-1} s^{m-1} + \cdots + a_1 s + a_0}{s^n + b_{n-1} s^{n-1} + \cdots + b_1 s + b_0} \tag{1.3}$$

Hence n is the *order* of the filter and for stability reasons (i.e., in order that the final circuit be stable) the degree of the numerator polynomial must be less than or equal to that of the denominator polynomial, that is, $m \leq n$. The numerator and denominator coefficients, $a_0, a_1, \ldots, a_m$ and $b_0, b_1, \ldots, b_{n-1}$, are real numbers. The numerator and denominator polynomials can be factored and $T(s)$ expressed in the form

$$T(s) = \frac{a_m (s - z_1)(s - z_2) \cdots (s - z_m)}{(s - p_1)(s - p_2) \cdots (s - p_n)} \tag{1.4}$$

Here the numerator roots, $z_1, z_2, \ldots, z_m$, are called the transfer function *zeros* or *transmission zeros*, and the denominator roots $p_1, p_2, \ldots, p_n$ are called the transfer function *poles* or the *natural modes.* Each zero or pole can be either a real or a complex number. Complex zeros and poles, however, must occur in conjugate pairs. That is, if $-1 + j2$ happens to be a zero, then $-1 - j2$ also must be a zero.

Although active-RC circuits are capable of realizing transfer functions whose zeros have any arbitrary values, the zeros are usually pure imaginary, that is, of

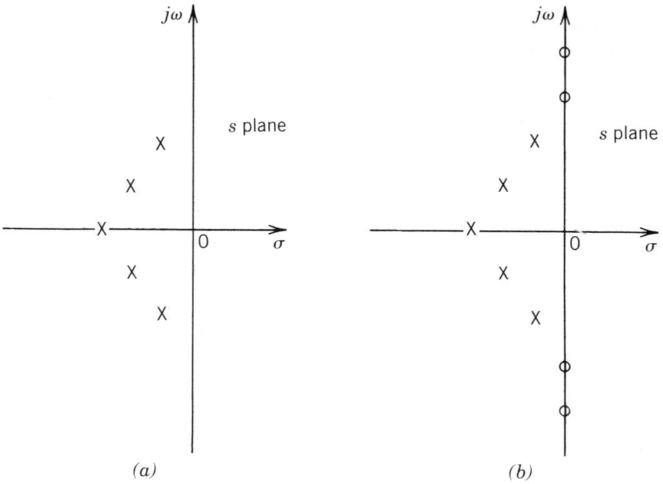

Figure 1.7 Pole-zero patterns corresponding to the two low-pass attenuation functions depicted in Fig. 1.6.

the form $\pm j\omega_Z$. Recalling that for physical frequencies $s = j\omega$, we see that a pure-imaginary zero implies that the transmission will be exactly zero when $\omega = \omega_Z$. It follows that to obtain the maximum attenuation effect from the transfer function zeros, they should be pure imaginary and with values in the filter stopband. To illustrate, Fig. 1.7*b* shows a sketch of the complex *s* plane with the poles (indicated by *X*s) and the zeros (indicated by 0s) of the low-pass filter function of Fig. 1.6*b*. Note that this function has five poles, and thus the filter is of fifth order ($n = 5$). We have two pairs of complex conjugate poles and one real-axis pole. It is very important to observe that here, and indeed for any filter, all of the poles must have negative real parts. An alternative way of stating this is: All of the poles must be in the left half of the *s*-plane. This restriction on pole location follows from the requirement that the filter network must be stable. That is, the filter must not generate signals on its own.

Returning to the example in Fig. 1.7*b*, we see that there are two pairs of imaginary-axis zeros. The fifth zero is at $s = \infty$ (i.e., at $\omega = \infty$). It follows that, for this filter, the degree of the numerator of the transfer function, *m*, is 4. In general, the number of transmission zeros at infinity is $(n - m)$.

The $j\omega$-axis zeros result in the filter transmission being zero, or equivalently the filter loss or attenuation being infinite, at ω equal to the frequency of the zero. This is evident from the sketch of the transmission function shown in Fig. 1.6*b* where we note that the attenuation becomes infinite at two finite values of ω. We also note that the attenuation becomes infinite at $\omega = \infty$, which is a consequence of the transmission zero at infinity.

As another example Fig. 1.7*a* shows the pole-zero pattern corresponding to the transmission function of Fig. 1.6*a*. Here we note that there are no transmission zeros at finite values of ω; all five transmission zeros are at $s = \infty$ ($\omega = \infty$). This is borne out by the shape of the transmission characteristic in Fig. 1.6*a*, which decreases monotonically as $\omega \to \infty$ and does not reach infinity for any finite value of ω. The transfer function of this filter must have $m = 0$; that is, the numerator is simply a constant. Such a filter is known as an *all-pole* filter.

EXAMPLE 1.1 Determine the voltage transfer function and the pole-zero locations for the circuit of Fig. 1.2*a*

PROCEDURE

The transfer function can be written using the voltage divider rule as

$$T(s) \equiv \frac{V_o(s)}{V_i(s)} = \frac{(1/sL)}{(1/sL) + [sC + (1/R)]}$$

$$= \frac{(1/LC)}{s^2 + s(1/CR) + (1/LC)}$$

Thus the filter is of second order with two zeros at $s = \infty$ and a pair of poles at

$$s = -\frac{1}{2CR} \pm \sqrt{\left(\frac{1}{2CR}\right)^2 - \frac{1}{LC}} \tag{1.5}$$

For the special case considered in Figure 1.2(b), $R\sqrt{C/L} = 1/\sqrt{2}$, that is, $R = 1/\sqrt{2C/L}$. Substituting in Eq. (1.5) results in

$$s = -\frac{1}{\sqrt{2LC}}(1 \pm j)$$

In concluding this section, we must point out that almost all of the filters of interest in this book have complex poles. The exceptions, of course, are odd-order (i.e., n odd) filters for which one pole must be real. Furthermore, the more selective the filter response, the closer the poles must be to the $j\omega$-axis. This point is especially important in the context of the active filter design problem, as will be explained in Section 1.10.

1.7 HISTORY OF FILTER REALIZATION TECHNOLOGIES

As mentioned earlier, the object of realization is to find a physical network whose voltage transfer function $V_o(s)/V_i(s)$ is equal to the function $T(s)$ obtained from approximation. The filter realization process is obviously dependent on the technology that one plans to use for constructing the filter. In this book we are interested in only one realization technology: RC-active circuits. Nevertheless, it is important to place this technology in proper perspective relative to the variety of technologies currently available for filter realization.

The concept of the electric wave filter was introduced in 1915 by Campbell in the United States and independently by Wagner in Germany. The earliest filters were realized as passive networks employing inductors, capacitors, and, occasionally, resistors. These early filters were designed using an approximate method known as *image parameter* design.

The modern method for LC filter design, known as *insertion-loss synthesis*, was introduced in the 1930s by Darlington in the United States and Cauer in Germany. However, because of its reliance on extensive, numerically sensitive computations, insertion-loss synthesis was not widely used until the 1950s when digital computers became available. Since then the design, analysis, and fabrication of LC filter networks has been perfected. A considerable number of design aids in the form of design manuals, tables, and computer programs have been produced, and the resulting LC filters perform remarkably well.

In spite of the fact that LC filters have served the electronics and communications industry extremely well, they do have disadvantages. Chief among these is the need for bulky, heavy inductors in low-frequency filter applications. With the invention of the transistor in the early 1950s, interest was generated in the attempt to achieve frequency-selective responses with circuits composed of active elements (transistors), resistors and capacitors: The search for inductorless filters had begun. That the theoretical answer to this question is positive was quickly discovered. It was to be many years, however, before active-RC filters would become a practical reality. At this point it is particularly interesting to note that a paper published in 1954 proposing a variety of vacuum tube–RC circuits to realize second-order filter functions has since become a classic. A good number of the circuit configurations that Sallen and Key proposed in that paper are still in use today (with the vacuum tubes replaced with op amps, of course). They are examined in Chapter 6.

The introduction of the integrated circuit (IC) op amp in the mid-1960s, and specifically the inexpensive, versatile, and durable 741-type in 1967, proved to be another turning point in the history of active filters. In the decade that followed, hundreds of papers were written on the design of op amp–RC filters. During the 1970s such filters became a practical reality and virtually replaced LC filters in audio-frequency applications. Furthermore, because their design and construction is much simpler than that of the LC types, filter design became accessible to the nonspecialist. Currently, op amp–RC filters are being employed in a wide variety of electronic systems.

In concluding this section, it should be mentioned that passive LC networks remain a viable proposition for filter design at frequencies higher than about 100 kHz. There are, of course, other technologies; these are noted in the next section.

1.8 OTHER TECHNOLOGIES FOR INDUCTORLESS FILTERS

In addition to the RC-active circuit there are a number of other technologies for the realization of inductorless filters. In particular, we should mention crystal filters, mechanical filters, microwave filters, surface acoustic wave (SAW) filters, digital filters, and, most recently, switched-capacitor filters. Books dealing with each one of these subjects are listed at the end of this chapter.

Crystal, mechanical, microwave, and SAW filters are used in particular and special applications. From this point of view none of these technologies can be considered as a serious alternative to RC-active filters. Also, in spite of recent excellent books on these topics, they remain highly specialized and by-and-large out of reach of the nonspecialist.

A digital filter is, in effect, a computer that implements a filtering algorithm on signal samples that have been converted to digital form (numbers) using an analog-to-digital converter (ADC). The output of the digital filter is a set of numbers from which an analog signal can be reconstructed using a digital-to-analog converter (DAC) followed by a sample-and-hold circuit and a simple smoothing filter. A *continuous-time* filter is also required at the input to limit the bandwidth of the input signal, thus avoiding aliasing.*

Research in digital filtering techniques has been very active for about two decades. For about the last 15 years there have been annual predictions that digital filters have become a practical reality and that the end of analog filters is in sight. This, of course, has not occurred and, in the opinion of this writer, will not happen. To be sure, digital signal processing is currently being extensively employed in nonreal-time applications and some real-time applications in which the signal is already in digital form. Furthermore, the very recent (1983) appearance of single-chip *digital signal processors* (DSPs) will definitely increase the range of application of digital filtering. It does not appear likely, however, that digital filters will replace analog filters as simple, inexpensive, self-contained system building blocks. This statement is particularly true in light of the recent emergence of the switched-capacitor filter.

The introduction of the *switched-capacitor* (SC) filter in 1977 and the subsequent extensive research and development work have quickly established this filter technology as a viable one for the production of very large quantity, low-frequency, fully integrated analog filters. The basic idea of the SC filter is the use of switched capacitors to replace the resistors in active-RC filter structures. Thus the basic components of an SC filter are op amps, capacitors, and analog switches. All these components can be fabricated using either the NMOS or CMOS technology, thus making it possible for the first time to fabricate a fully integrated analog filter. Furthermore, since these same technologies are employed in the fabrication of digital ICs, one now has the ability to design sophisticated analog and digital functions on the same IC chip.

SC filter design is, to a large extent, based on the methods that have been evolved for the design of RC-active filters, which are studied in detail in this book. It should be noted, however, that although SC filters are analog circuits (because the signal amplitude is not quantitized), they are discrete-time (or sampled-data) systems due to their switching nature.

* *Aliasing*, or foldover distortion, occurs if the sampling rate is not fast enough. In such cases the frequency characteristics of a continuous signal and its sampled sequence are no longer linearly related. Hence the continuous-time signal cannot be constructed from the sampled sequence. The effect can be overcome by increasing the sampling rate and/or bandlimiting the presampled signal.

1.9 FABRICATION TECHNOLOGIES

Currently, there are three possible fabrication technologies for RC-active filters: discrete, thick-film, and thin-film circuits. In all three technologies monolithic IC op amps are employed. Discrete circuits are assembled using resistors, capacitors, and IC op amps on a printed circuit board. Typically, metal film resistors and polystyrene capacitors are employed. The op amps used are usually encapsulated in dual-in-line packages (DIPs). Tuning may be performed by including one or more potentiometers in the circuit. Such potentiometers can then be replaced by suitable fixed-value resistors. Discrete circuits are simple to assemble and provide the most economical approach to fabricating small quantities of RC-active filters.

Thick-film circuits consist of resistive, conductive, or dielectric inks fused onto the surface of a ceramic substrate. These inks form the thick-film resistors, conductors, and capacitors. Typically, the layers deposited are 10 to 50 μm in thickness. The op amps used are in chip form and are directly bonded onto the substrate. In some cases discrete capacitors are used and are bonded to the substrate. Tuning can be accomplished by laser trimming or sandblasting the resistors. The processing steps for fabricating thick-film circuits are relatively simple. The resulting circuits are small, light, and reliable. This technology is economical for medium-quality filters that have to be produced in quantities of 10,000 to 100,000 per year.

Thin-film circuits consist of resistors and capacitors deposited on a ceramic substrate using evaporation, electroplating, or sputtering. To achieve the desired geometry and the very small thicknesses of 0.005 to 1 μm, photolithographic and chemical etching techniques are employed. Typically, the resistors are metal nitride and can be tuned very accurately by laser trimming. The capacitors are usually tantalum and the op amps used are in chip form and are bonded onto the substrate. This technology for fabricating RC-active filters has been refined to a high degree and has been very successful in producing small units that are accurate and reliable. The resulting filters can be packaged in standard DIPs.

1.10 ACTIVE FILTER DESIGN

As already mentioned, to obtain a selective frequency response, the natural modes of the filter must be complex and located close to the $j\omega$-axis in the s plane. This of course is possible with LCR networks. In contrast, RC networks have the property that their natural modes are restricted to lie on the negative real axis. Thus it is not possible to obtain highly selective frequency responses using RC networks alone.

The basic idea underlying the design of active-RC filters is the use of the active element (op amp) together with the RC network in a feedback configuration in a way that results in the overall network having complex conjugate natural modes that can be located as close to the $j\omega$-axis as dictated by the selectivity requirement of the filter. In other words, feedback, either negative or positive or a combination of the two polarities, is employed to move the natural modes of the RC network from their negative real axis locations to the desired complex conjugate locations. That feedback is capable of moving natural modes is known from the study of feedback theory.

Although it is theoretically possible to realize a filter transfer function of any order using one op amp together with an appropriate RC network, such realizations have been shown to be impractical. Specifically, such realizations result in circuits that are very *sensitive* to small variations in the values of the resistors and capacitors and to the inevitable nonidealities of the op amp such as its finite gain and its gain roll-off with frequency. A sensitive filter realization is one that shows large changes in the filter transmission as a result of small changes in component values. The component value changes may simply be the original, known tolerances. Although these can be corrected by tuning the filter, changes in environmental and/or operating conditions could cause the transfer function of a sensitive realization to change beyond acceptable limits.

A sensitive active filter realization is also susceptible to the filter breaking into oscillation. This occurs if the component-value changes cause the natural modes of the circuit to move into the right-half of the s-plane. Oscillation is a problem that cannot exist in a passive filter but is quite possible in an active filter that is not carefully designed.

As a result of a great deal of research work into the sensitivity properties of active-RC networks, a number of methods have been developed for the design of low-sensitivity filter networks. In this book we study the two most popular methods: cascade design and design based on the simulation of LC ladder networks.

Cascade Design

The cascade design method is based on the fact that it is possible to realize second-order filter functions using simple, low-sensitivity circuits. A second-order filter function can be written as the ratio of two quadratic polynomials,

$$T(s) = \frac{a_2 s^2 + a_1 s + a_0}{s^2 + b_1 s + b_0} \tag{1.6}$$

and thus is known as a biquadratic function or simply a *biquad*. Chapter 3 is devoted to the study of such functions.

Circuits for the realization of second-order filter functions are known as biquadratic circuits or again simply as biquads. After a study of op amps in Chapter 4 and analysis techniques in Chapter 5, the succeeding chapters (Chapters 6 to 9) deal with a variety of biquad circuits. If a second-order filter is desired, then one of these biquads circuits can be used. More generally, cascading a number of biquad circuits realizes any given even-order transfer function. To be more specific, a given even-order transfer function $T(s)$ can always be written as the product of second-order transfer functions,

$$T(s) = T_1(s)T_2(s) \cdots T_{n/2}(s)$$

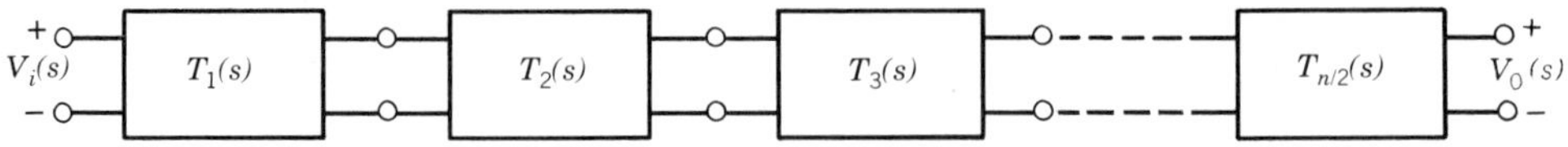

Figure 1.8 *Cascading (n/2) biquad circuits having low output impedances results in an overall transfer function* $V_0(s)/V_i(s) = T_1(s)T_2(s) \cdots T_{n/2}(s)$.

where n is the filter order. Each of the biquadratic transfer functions in the product can be realized using an appropriate circuit from the selection given in Chapters 6 to 9. Since (as will be shown) all these circuits have low-impedance output terminals, the circuits can be cascaded, as shown in Fig. 1.8, with no change resulting in their individual transfer functions. It follows that the overall transfer function of the cascade network is equal to the product of the individual transfer functions, which is the original function $T(s)$.

Odd-order functions can be dealt with in the same manner except that one of the sections in the cascade must have a first-order transfer function. The general form of such functions is

$$T(s) = \frac{a_1 s + a_0}{s + b_0}$$

Such functions are easily realized using an op amp together with an RC network, or even just an RC network. This completes our introduction to the cascade method of design. More will be said in Chapter 10.

Designs Based on LC Ladder Simulation

Although the cascade method of active filter design is very simple and results in excellent realizations, it is not suitable for the realization of filters that have very tight specifications. Cascade realizations of such filters are usually too sensitive to component changes to be practical for volume production.

For tightly specified filters a design based on the simulation of an LC ladder prototype is recommended. That is, first the filter function is realized using an LC network, usually in the form of a ladder terminated in resistors on both sides, such as the one shown in Fig. 1.9. These *doubly terminated* LC ladder networks can be designed so that over the filter passband, maximum power is drawn from

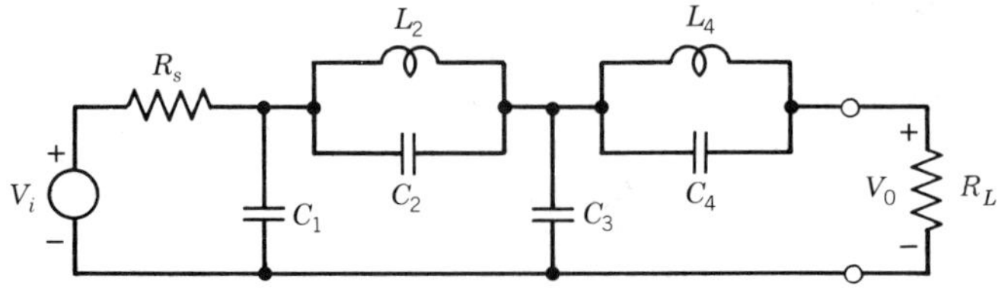

Figure 1.9 *A doubly terminated LC ladder network realization of a fifth-order low-pass filter. This network is capable of providing the transmission characteristics shown in Fig. 1.6b.*

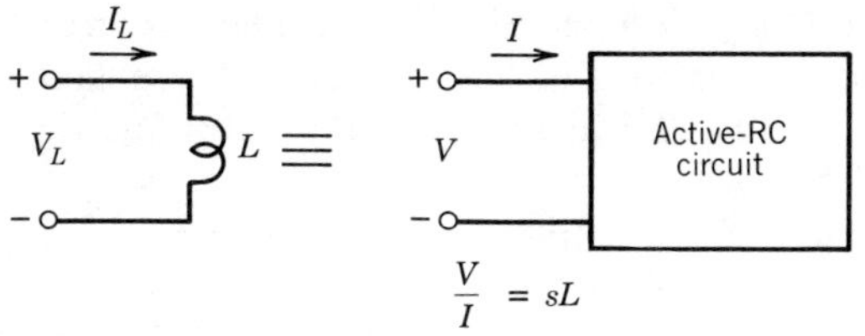

Figure 1.10 Inductance simulation.

the source. In 1966 H. J. Orchard showed that an LC network designed for maximum power transfer possesses extremely low sensitivities to variations in the values of its components. It follows that an active-RC filter, which is designed based on such a low-sensitivity LC prototype, will exhibit low sensitivities to its component values. This is the idea behind the design methods that are based on LC ladder simulation.

There are basically two approaches to the simulation of LC networks: component simulation and operational simulation. In the component simulation method, each inductor in the LC prototype is replaced by an active-RC network whose input impedance is inductive (see Fig. 1.10). This technique and some of its variants will be discussed in detail in Chapter 10.

In the operational simulation method the resulting active-RC circuit provides an analog computer simulation of the current-voltage relationships of the LC prototype network. To be specific, consider an inductance L in the LC network. It is characterized by

$$I_L = \frac{V_L}{sL}$$

The simulating active-RC circuit must contain an integrator whose output voltage V_2 may be expressed in terms of its input voltage V_1 as

$$V_2 = -\frac{V_1}{s\tau}$$

By selecting the time constant τ to be proportional to the value of the inductance L, this integrator simulates the operation of the inductance with V_1 corresponding to V_L and V_2 corresponding to I_L (see Fig. 1.11). Similarly, active-RC integrators

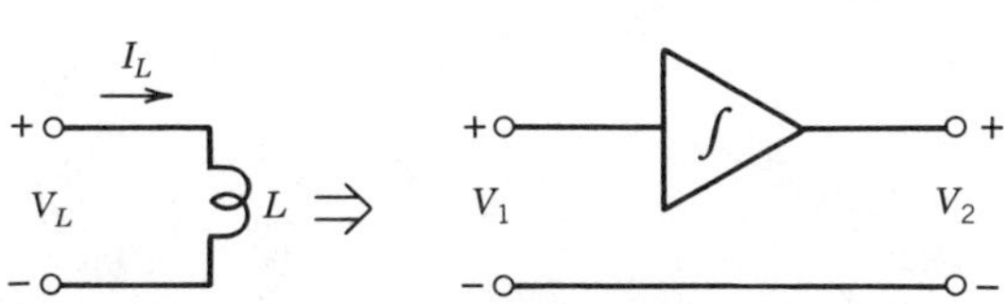

Figure 1.11 To simulate the operation of an inductance the RC-active circuit uses an integrator with a time constant proportional to the inductance L. The voltage V_1 is the analog of V_L and voltage V_2 is the analog of I_L.

can be used to simulate the current-voltage characteristic of the capacitors in the LC network. Finally, the loop and node equations of the LC network are simulated using weighted summers in the active-RC circuit. The operational simulation method is discussed in Chapter 10.

The design of an active-RC filter is not complete without a study of the limitations imposed on its performance by the inevitable tolerances and other nonidealities of its components. Such a study enables the designer to estimate a priori the expected deviations in the filter transmission. Chapter 11 provides a detailed study of the practical limitations of RC-active filters.

1.11 REFERENCES

[1] A. S. Sedra and P. O. Brackett, *Filter Theory and Design: Active and Passive*, Portland, Oregon: Matrix Publishers, 1978. An advanced text and reference book dealing with the approximation and realization of passive and active filters.

[2] M. E. Van Valkerburg, *Analog Filter Design*, New York: Holt, Rinehart and Winston, 1982. An excellent introduction to the filter design area; it includes material on switched-capacitor filters.

[3] G. Daryanani, *Principles of Active Networks Synthesis and Design*, New York: Wiley, 1976.

[4] M. S. Ghausi and K. R. Laker, *Active Filter Design*, Englewood Cliffs, N.J.: Prentice-Hall, 1981. Includes an excellent chapter on switched-capacitor filters.

[5] R. Schaumann, M. Soderstrand, and K. Laker, eds., *Modern Active Filter Design*, IEEE Press, New York, 1981. Contains an excellent collection of papers on active-RC and switched-capacitor filter design.

[6] G. C. Temes and S. K. Mitra, eds., *Modern Filter Theory and Design*, New York: Wiley, 1973. Includes material on a variety of realization technologies.

[7] R. A. Johnson, *Mechanical Filters in Electronics*, New York: Wiley-Interscience, 1982. A modern text on this alternative filter technology.

[8] E. Christian, *LC Filters*, New York: Wiley-Interscience, 1983. A modern text on practical LC filter design.

[9] A. Antoniou, *Digital Filters*, New York: McGraw-Hill, 1980. A modern text on this alternative filter technology.

CHAPTER 2

Approximation

Gregory M. Wierzba
Michigan State University

2.1 INTRODUCTION

A filter is a circuit in which a signal is passed from the input to the output in a band of frequencies and is blocked or stopped at the output for all other frequencies. These bands of frequencies are called the passband and the stopband, respectively. Figure 2.1 depicts the behavior of the magnitude of an ideal low-pass filter versus the frequency. For this filter, the gain for all frequencies below ω_p is one, that is, it passes all frequencies from zero to ω_p. The gain for all frequencies above ω_p is zero; that is, it blocks all frequencies greater than ω_p.

In any real circuit there is some maximum allowable loss in the passband and some minimum acceptable attenuation in the stopband. Figure 2.2 illustrates how Fig. 2.1 is modified to take these losses into account.

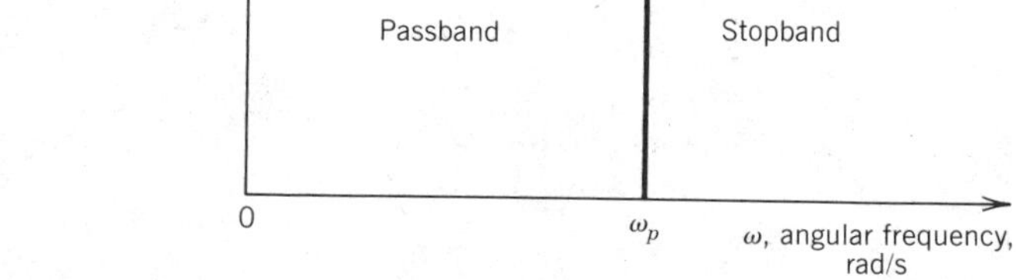

Figure 2.1 Ideal low-pass filter.

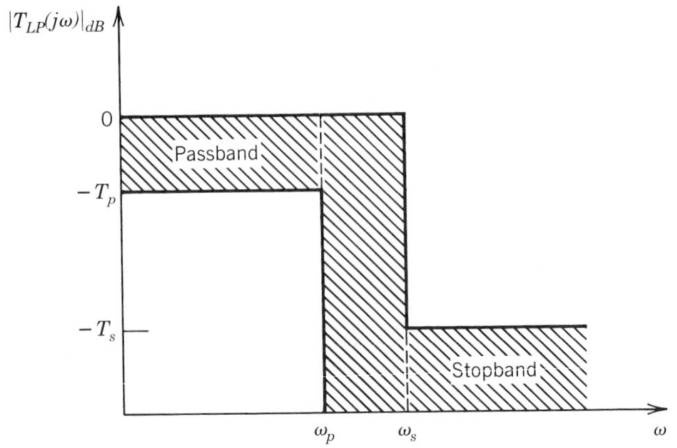

Figure 2.2 Acceptable bounds of a low-pass filter.

The specifications for a filter are usually given in terms of the losses in the passband at ω_p and in the stopband at ω_s. The approximation problem is to find a function whose characteristics lie within the cross-hatched regions of Fig. 2.2.

2.2 MAGNITUDE APPROXIMATION TECHNIQUES

The following transfer functions exhibit a behavior like that described in Fig. 2.1 and 2.2:

$$T(s) = \frac{a}{s + a}, \quad \frac{b}{s^2 + as + b}, \quad \frac{c}{s^3 + as^2 + bs + c}, \quad \ldots$$

For these equations, the magnitude of $T(s = j\omega)$ approaches one (0 dB) as ω approaches zero and the magnitude approaches zero as ω approaches infinity. The plot of $|T(j\omega)|$ versus frequency rolls off at $-6n$ dB/octave, where n is the order of $T(s)$; that is, the highest power of s in $T(s)$.

EXAMPLE 2.1 Determine the lowest-order low-pass transfer function that satisfies the filter specifications given and determine the constants of $T(s)$:

$$f_p = 20 \text{ kHz} \qquad T_p = 3 \text{ dB}$$

$$f_s = 80 \text{ kHz} \qquad T_s = 10 \text{ dB}$$

PROCEDURE

1 At $\omega_p = (2\pi)20$ K, $|T(j\omega)| = -3$ dB $= 0.707$ and at $\omega_s = (2\pi)80$ K, $|T(j\omega)| = -10$ dB $=$ 0.316. In two octaves, the curve needs to roll off -7 dB or -3.5 dB/octave. Therefore, select $n = 1$.

2 $T(s) = \dfrac{a}{s + a}, \qquad |T(j\omega)|^2 = \dfrac{a^2}{\omega^2 + a^2}$

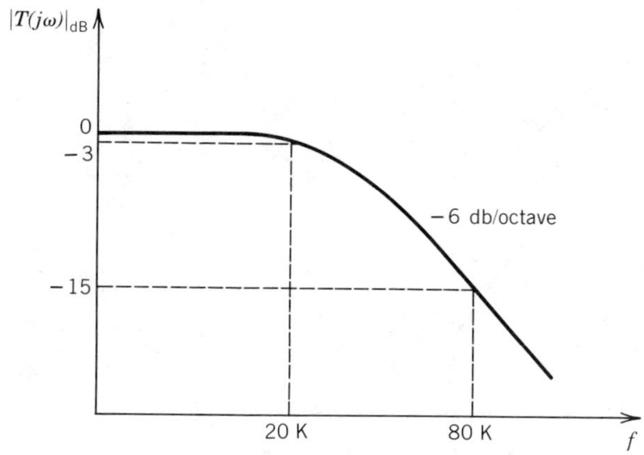

Figure 2.3 Selected approximation for Problem 2.1.

At $\omega = (2\pi)20$ K, $|T(j\omega)| = 0.707$. Solving for a, we find that $a = 125{,}625$. Hence

$$T(s) = \frac{125{,}625}{s + 125{,}625}$$

The results are plotted in Fig. 2.3.

This method of approximation is suitable for low-order transfer functions, but as n increases it becomes difficult to solve for the constants in $T(s)$. In the following sections, some well-known approximations are presented along with a simple means for determining n.

2.3 BUTTERWORTH APPROXIMATION

Consider the following low-pass transfer function magnitude:

$$|T(j\omega)| = \frac{1}{\sqrt{1 + \varepsilon^2(\omega/\omega_p)^{2n}}} \tag{2.1}$$

where ε is a constant and n is the order of $T(s)$. The response of this transfer function is illustrated in Fig. 2.4. For $\varepsilon = 1$, $|T(j\omega)|$ is down 3 dB from its dc level at $\omega = \omega_p$.

The magnitude response $|T(j\omega)|$ has the property that its first $(2n - 1)$ derivatives are zero at $\omega = 0$. This causes the magnitude response to be as flat as possible at $\omega = 0$. For this reason, this transfer function is sometimes referred to as having maximally flat magnitude [1] and was first studied in regard to filters by S. Butterworth [2].

It is convenient to normalize $|T(j\omega)|$ by replacing ω/ω_p by Ω, that is, for $\omega = \omega_p$, $\Omega = 1$. Thus

$$|T(j\Omega)| = \frac{1}{\sqrt{1 + \varepsilon^2\Omega^{2n}}} \tag{2.2}$$

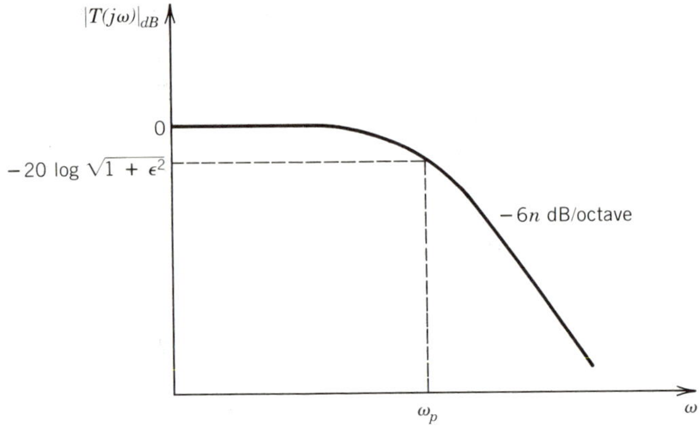

Figure 2.4 Butterworth approximation.

Table 2.1 lists some Butterworth polynomials, $D(s)$, for $\varepsilon = 1$, where $T(s) = 1/D(s)$. An alternative listing, presenting the root locations, is included in Chapter 13 as Table 13.1. $|T(j\Omega)|$ is plotted in Fig. 2.5.

To select the appropriate Butterworth low-pass transfer function, we need to specify ω_p, T_p, ω_s, and T_s. Knowing ω_p and T_p, we can solve Eq. 2.1 for ε at $\omega = \omega_p$.

In order to use Table 2.1 for selecting the Butterworth polynomials, we need to normalize our specifications. Having done this, we can solve Eq. 2.2 for n with

Table 2.1 **Butterworth polynomials**

n	$D(s)$
1	$s + 1$
2	$s^2 + 1.41421s + 1$
3	$(s + 1)(s^2 + s + 1)$
4	$(s^2 + 0.76537s + 1)(s^2 + 1.84776s + 1)$
5	$(s + 1)(s^2 + 0.61803s + 1)(s^2 + 1.61803s + 1)$
6	$(s^2 + 0.51764s + 1)(s^2 + 1.41421s + 1)$ $(s^2 + 1.93186s + 1)$
7	$(s + 1)(s^2 + 0.44504s + 1)(s^2 + 1.24698s + 1)$ $(s^2 + 1.80194s + 1)$
8	$(s^2 + 0.39018s + 1)(s^2 + 1.11114s + 1)$ $(s^2 + 1.66294s + 1)(s^2 + 1.96157s + 1)$
9	$(s + 1)(s^2 + 0.3473s + 1)(s^2 + s + 1)$ $(s^2 + 1.53209s + 1)(s^2 + 1.87939s + 1)$
10	$(s^2 + 0.31287s + 1)(s^2 + 0.90798s + 1)$ $(s^2 + 1.41421s + 1)(s^2 + 1.78201s + 1)$ $(s^2 + 1.97538s + 1)$

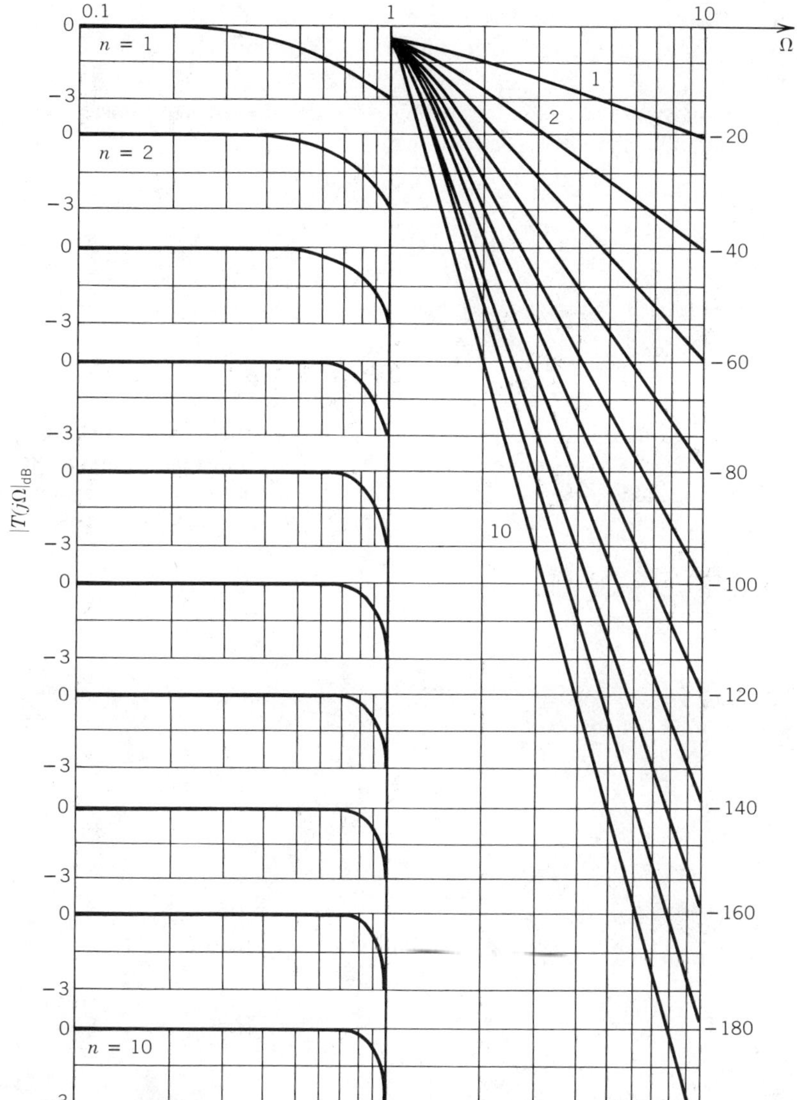

Figure 2.5 Butterworth magnitude response.

$\varepsilon = 1$. To simplify the design process, Eq. 2.2 has been solved pictorially in Chapter 13 (Fig. 13.10). Note here that $T_p \equiv A_{\max}$ and $T_s \equiv A_{\min}$.

Because of our specifications, it may turn out that n is not an integer. If this is the case and we always round up to the next highest integer, then our specifications will be always satisfied.

Now that we have specified n, we can form the normalized $T(s)$ from Table 2.1. The denormalization of $T(s)$ is done in two steps. First, the normalized

($\varepsilon = 1$) T_p is shifted from $\Omega = 1$ to $\omega = \omega_p$, and second, the specified T_p is shifted to $\omega = \omega_p$. This procedure is illustrated in the following example.

EXAMPLE 2.2 Determine the expression for the lowest-order Butterworth transfer function that can satisfy the following filter specifications:

$$\omega_p = 1000 \text{ rad/s} \qquad T_p = 1 \text{ dB } (= A_{\max})$$

$$\omega_s = 6000 \text{ rad/s} \qquad T_s = 20 \text{ dB } (= A_{\min})$$

PROCEDURE

1 Determine ε by evaluating the following formula:

$$\varepsilon = \sqrt{10^{T_p/10} - 1}$$
$$= \sqrt{10^{1/10} - 1} = 0.5088$$

2 Determine n by using the nomograph given in Fig. 13.9. To use this nomograph, locate the points $A_{\max}$, $A_{\min}$, $\Omega_s = \omega_s/\omega_p$. Connect these points as shown in Fig. 2.6 and locate the point x. This is the value of n needed to satisfy the filter specifications. If x does not intersect a curve, select the next higher curve for the value of n. Therefore, for this specification $n = 2$.

3 Form the normalized $T(s)$ from Table 2.1.

$$T(s) = \frac{1}{s^2 + 1.414s + 1}$$

4 To denormalize the response replace $j\Omega$ by $j\omega/\omega_p$; that is, replace s by s/ω_p. This shifts the -3 dB point from $\Omega = 1$ to $\omega = \omega_p$.

$$T_1(s) = T(s)|_{s \to s/1000} = \frac{1}{(s/1000)^2 + (1.414/1000)s + 1}$$

5 To move the -1 dB point to $\omega = \omega_p$, replace s by $\varepsilon^{1/n}s$.

$$T_{LP}(s) = T_1(s)|_{s \to \varepsilon^{1/2}s} = T_1(s)|_{s \to 0.7133s}$$
$$= \frac{1.965(10^6)}{s^2 + 1982s + 1.965(10^6)}$$

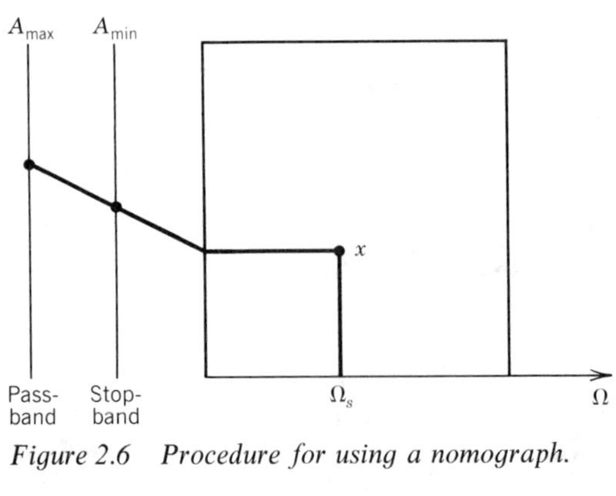

Figure 2.6 Procedure for using a nomograph.

6 From our design, $T_{LP}(j\omega)$ has a magnitude of -1 dB at $\omega = \omega_p = 1000$ rad/s. Here $|T_{LP}(j6000)|$ is found by evaluating the following magnitude expression:

$$|T_{LP}(j6000)| = \frac{1.965(10^6)}{\{[1.965(10^6) - (6000)^2]^2 + [(1982)(6000)]^2\}^{1/2}} = 5.451(10^{-2}) = -25.27 \text{ dB}$$

Thus the specifications have been satisfied.

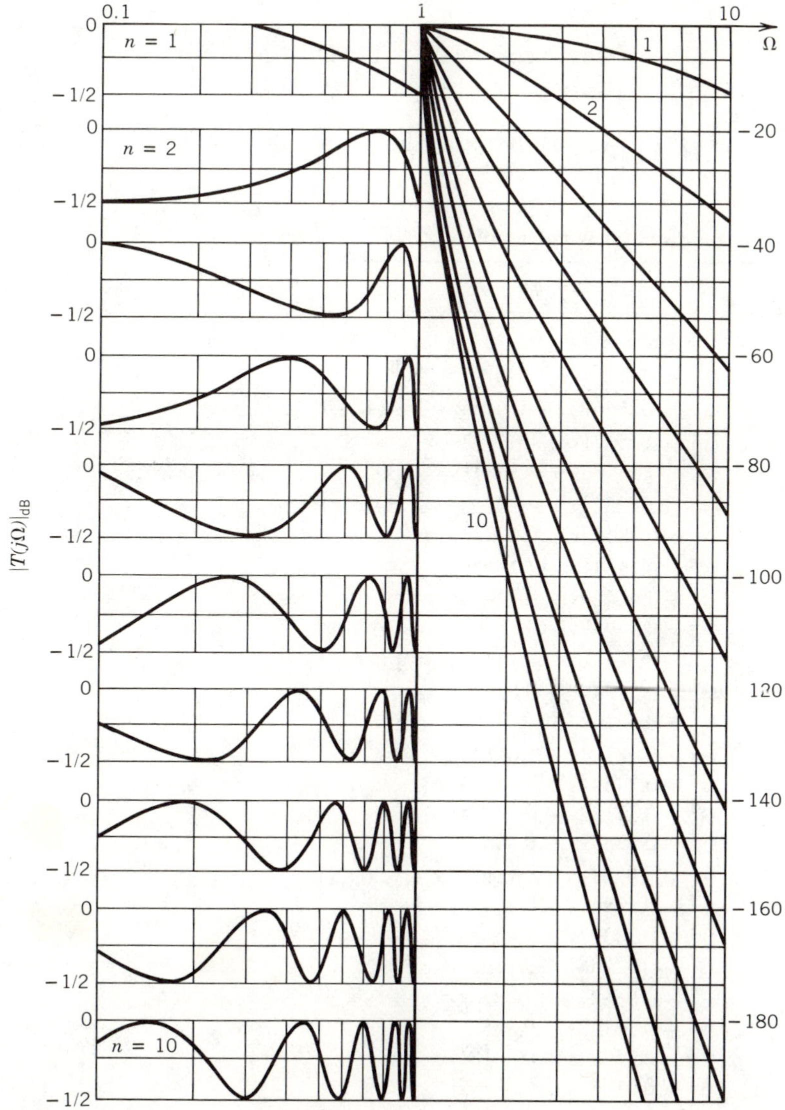

Figure 2.7 The 0.5 dB ripple Chebyshev magnitude response ($\varepsilon = 0.3493$).

2.4 CHEBYSHEV APPROXIMATION

As we have seen in Section 2.3, the Butterworth approximation has the advantage of maximally flat magnitude at $\omega = 0$. As ω approaches ω_p, the approximation to a flat passband gets progressively worse.

Consider the following low-pass transfer function magnitude

$$|T(j\Omega)| = \frac{1}{\sqrt{1 + \varepsilon^2 C_n^2(\Omega)}} \tag{2.3}$$

where

$$C_n(\Omega) = \cos(n \cos^{-1} \Omega) \qquad \text{for } |\Omega| \le 1$$
$$= \cosh(n \cosh^{-1} \Omega) \qquad \text{for } |\Omega| \ge 1$$

Figure 2.7 illustrates $|T(j\Omega)|$ versus frequency for $\varepsilon = 0.3493$. At $\Omega = 1$, $|T(j\Omega)| = -20 \log \sqrt{1 + \varepsilon^2} = -0.5$ dB for $\varepsilon = 0.3493$.

Table 2.2 **0.5 dB ripple Chebyshev polynomials**

n	N	$D(s)$
1	2.86278	$s + 2.86278$
2	1.43139	$s^2 + 1.42562s + 1.5162$
3	0.7157	$(s + 0.62646)(s^2 + 0.62646s + 1.14245)$
4	0.35785	$(s^2 + 0.35071s + 1.06352)$ $(s^2 + 0.84668s + 0.35641)$
5	0.17892	$(s + 0.36232)(s^2 + 0.22393s + 1.03578)$ $(s^2 + 0.58625s + 0.47677)$
6	0.089463	$(s^2 + 0.1553s + 1.02302)$ $(s^2 + 0.42429s + 0.59001)$ $(s^2 + 0.57959s + 0.157)$
7	0.044731	$(s + 0.25617)(s^2 + 0.11401s + 1.01611)$ $(s^2 + 0.31944s + 0.67688)$ $(s^2 + 0.4616s + 0.25388)$
8	0.022365	$(s^2 + 0.08724s + 1.01193)$ $(s^2 + 0.24844s + 0.74133)$ $(s^2 + 0.37182s + 0.35865)$ $(s^2 + 0.43859s + 0.08805)$
9	0.011183	$(s + 0.19841)(s^2 + 0.06891s + 1.00921)$ $(s^2 + 0.19841s + 0.78936)$ $(s^2 + 0.30397s + 0.45254)$ $(s^2 + 0.37288s + 0.15634)$
10	0.0055914	$(s^2 + 0.0558s + 1.00734)$ $(s^2 + 0.16193s + 0.8257)$ $(s^2 + 0.25222s + 0.53181)$ $(s^2 + 0.31781s + 0.23791)$ $(s^2 + 0.3523s + 0.05628)$

Table 2.3 **1 dB ripple Chebyshev polynomials**

n	N	$D(s)$
1	1.96523	$s + 1.96523$
2	0.98261	$s^2 + 1.09773s + 1.10251$
3	0.4913	$(s + 0.49417)(s^2 + 0.49417s + 0.9942)$
4	0.24566	$(s^2 + 0.67374s + 0.2794)$ $(s^2 + 0.27907s + 0.98651)$
5	0.12283	$(s + 0.28949)(s^2 + 0.46841s + 0.4293)$ $(s^2 + 0.17892s + 0.98832)$
6	0.061415	$(s^2 + 0.12436s + 0.99073)$ $(s^2 + 0.33976s + 0.55772)$ $(s^2 + 0.46413s + 0.12471)$
7	0.030706	$(s + 0.20541)(s^2 + 0.09142s + 0.99268)$ $(s^2 + 0.25615s + 0.65346)$ $(s^2 + 0.37014s + 0.23045)$
8	0.015353	$(s^2 + 0.07002s + 0.99414)$ $(s^2 + 0.19939s + 0.72354)$ $(s^2 + 0.29841s + 0.34086)$ $(s^2 + 0.352s + 0.07026)$
9	0.0076764	$(s + 0.15933)(s^2 + 0.05533s + 0.99523)$ $(s^2 + 0.15933s + 0.77539)$ $(s^2 + 0.24411s + 0.43856)$ $(s^2 + 0.29944s + 0.14236)$
10	0.0038382	$(s^2 + 0.04483s + 0.99606)$ $(s^2 + 0.1301s + 0.81442)$ $(s^2 + 0.20263s + 0.52053)$ $(s^2 + 0.25533s + 0.22664)$ $(s^2 + 0.28304s + 0.045)$

This transfer function has the property that passband losses are rippled between two limits. It is sometimes referred to as an equal-ripple approximation. This function was first studied in regard to designing steam engines by P. L. Chebyshev [3] and was first applied to filters by W. Cauer [4].

The advantage of the Chebyshev approximation over the Butterworth approximation is the increased attenuation in the stopband for the same-order transfer function even though both tend toward the same high-frequency asymptote. However, this increased attenuation is achieved at the expense of inferior passband behavior.

The curves shown in Fig. 2.7 are usually referred to as a 0.5 dB ripple approximation. Table 2.2 lists the 0.5 dB ripple Chebyshev polynomials, $D(s)$, where $T(s) = N/D(s)$. Other Chebyshev approximations are listed in Tables 2.3 to 2.5. The 1 dB ripple frequency response curve is given in Fig. 2.8. A more extensive listing is presented in Tables 13.2 to 13.7.

Table 2.4 **2 dB ripple Chebyshev polynomials**

n	N	$D(s)$
1	1.30756	$s + 1.30756$
2	0.65378	$s^2 + 0.80382s + 0.82306$
3	0.32689	$(s + 0.36891)(s^2 + 0.36891s + 0.8861)$
4	0.16345	$(s^2 + 0.20977s + 0.92868)(s^2 + 0.50644s + 0.22157)$
5	0.081723	$(s + 0.21831)(s^2 + 0.13492s + 0.95217)$ $(s^2 + 0.35323s + 0.39315)$
6	0.040863	$(s^2 + 0.09395s + 0.96595)$ $(s^2 + 0.25667s + 0.53294)$ $(s^2 + 0.35061s + 0.09993)$
7	0.020431	$(s + 0.15534)(s^2 + 0.06913s + 0.97461)$ $(s^2 + 0.19371s + 0.63539)$ $(s^2 + 0.27991s + 0.21239)$
8	0.010215	$(s^2 + 0.05298s + 0.98038)$ $(s^2 + 0.15089s + 0.70978)$ $(s^2 + 0.22582s + 0.3271)$ $(s^2 + 0.26637s + 0.0565)$
9	0.0051077	$(s + 0.12063)(s^2 + 0.04189s + 0.9844)$ $(s^2 + 0.12063s + 0.76455)$ $(s^2 + 0.18482s + 0.42773)$ $(s^2 + 0.22671s + 0.13153)$
10	0.002554	$(s^2 + 0.03395s + 0.9873)$ $(s^2 + 0.09853s + 0.80567)$ $(s^2 + 0.15347s + 0.51178)$ $(s^2 + 0.19338s + 0.21788)$ $(s^2 + 0.21436s + 0.03625)$

To select the appropriate ripple, Chebyshev low-pass transfer function, we need to specify ω_p, T_p, ω_s, and T_s. The ripple needed is T_p. If T_p is not one of the tables given, select the nearest case.* Then, if the next smaller case is chosen, the resulting design will exceed the specifications given.

The order of n can be determined by solving Eq. 2.3. To simplify the design process, however, a nomograph is given in Chapter 13 (Fig. 13.10). Note here that $T_p \equiv A_{\max}$ and $T_s \equiv A_{\min}$.

Having selected n, we form $T(s)$ from the appropriate table. The denormalization of $T(s)$ is achieved in one step by shifting the ripple specification from $\Omega = 1$ to $\omega = \omega_p$.

This procedure is illustrated in the following example.

* If this is not satisfactory, the exact root locations of $D(s)$ may be evaluated with the aid of formulas presented in [7].

Table 2.5 **3 dB ripple Chebyshev polynomials**

n	N	$D(s)$
1	1	$s + 1$
2	0.5	$s^2 + 0.64359s + 0.70711$
3	0.25	$(s + 0.29804)(s^2 + 0.29804s + 0.83883)$
4	0.125	$(s^2 + 0.17001s + 0.9029)$ $(s^2 + 0.41044s + 0.19579)$
5	0.0625	$(s + 0.17719)(s^2 + 0.10951s + 0.9359)$ $(s^2 + 0.2867s + 0.37689)$
6	0.031249	$(s^2 + 0.07631s + 0.95475)$ $(s^2 + 0.20849s + 0.52173)$ $(s^2 + 0.2848s + 0.08872)$
7	0.015624	$(s + 0.12624)(s^2 + 0.05618s + 0.96642)$ $(s^2 + 0.15742s + 0.6272)$ $(s^2 + 0.22748s + 0.20419)$
8	0.0078131	$(s^2 + 0.04307s + 0.97413)$ $(s^2 + 0.12266s + 0.70353)$ $(s^2 + 0.18358s + 0.32085)$ $(s^2 + 0.21655s + 0.05025)$
9	0.0039064	$(s + 0.09809)(s^2 + 0.03407s + 0.97947)$ $(s^2 + 0.09809s + 0.75962)$ $(s^2 + 0.15028s + 0.4228)$ $(s^2 + 0.18434s + 0.1266)$
10	0.0019532	$(s^2 + 0.02761s + 0.98332)$ $(s^2 + 0.08013s + 0.80168)$ $(s^2 + 0.12481s + 0.50779)$ $(s^2 + 0.15727s + 0.2139)$ $(s^2 + 0.17433s + 0.03226)$

EXAMPLE 2.3 Determine the expression for the lowest-order Chebyshev transfer function that can satisfy the filter specifications given in Example 2.2.

PROCEDURE

1 Determine the ripple needed from T_p. Therefore, a 1 dB ripple approximation is needed.

2 Determine n by using the nomograph given in Fig. 13.10. Therefore, $n = 2$.

3 Form the normalized $T(s)$ from Table 2.3:

$$T(s) = \frac{0.9826}{s^2 + 1.098s + 1.103}$$

4 To denormalize the response, replace s by s/ω_p. This shifts the -1 dB point from $\Omega = 1$ to $\omega = \omega_p$.

$$T_{\text{LP}}(s) = T(s)|_{s \to s/1000} = \frac{0.9826(10^6)}{s^2 + 1098s + 1.103(10^6)}$$

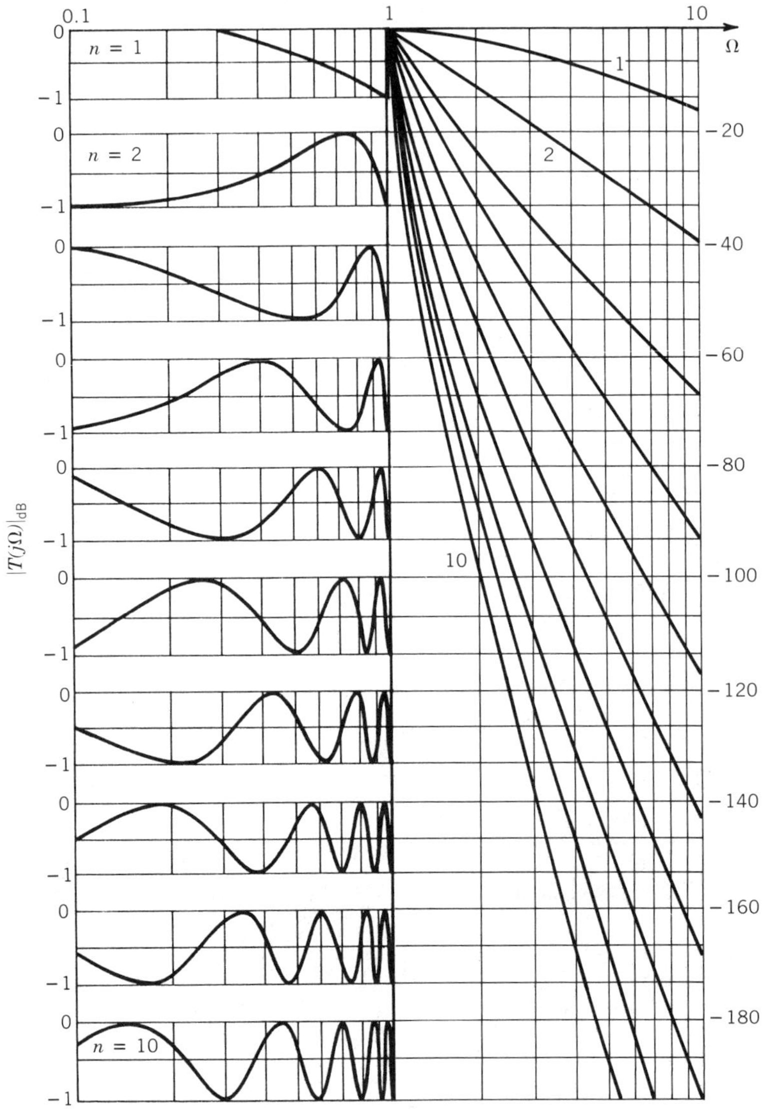

Figure 2.8 The 1 dB ripple Chebyshev magnitude response ($\varepsilon = 0.5088$).

5 From our design, $T_{LP}(j\omega)$ has a magnitude of -1 dB at $\omega = \omega_p = 1000$ rad/s. Then $|T_{LP}(j6000)|$ is found by evaluating the following magnitude expression:

$$|T_{LP}(j6000)| = \frac{0.9826(10^6)}{\{[1.103(10^6) - (6000)^2]^2 + [(1098)(6000)]^2\}^{1/2}}$$

$$= 2.767(10^{-2})$$

$$= -31.16 \text{ dB}$$

Therefore, the attenuation for the Chebyshev approximation is greater than the Butterworth approximation at ω_s for the same-order transfer function.

Unfortunately, a set of Chebyshev polynomials is needed for each ripple specification and thus is not as flexible as the Butterworth polynomials. Other Chebyshev approximations can be found with the aid of the programs provided in Chapter 12.

2.5 CAUER APPROXIMATION

As we have seen in Section 2.4, ripple in the passband results in increased attenuation in the stopband at ω_s. As the frequency increases, the attenuation increases beyond that specified for T_s. This overdesign can be remedied by allowing equal ripple in the stopband as well as in the passband.

Consider the following low-pass transfer function magnitude [5]:

$$|T(j\Omega)| = \frac{1}{\sqrt{1 + R_n^2(\Omega)}}$$

where

$$R_n(\Omega) = K_e \frac{\prod^{n/2} (\Omega_i^2\Omega_s - \Omega^2)}{\prod^{n/2} (\Omega_s - \Omega_i^2\Omega^2)} \qquad \text{for } n \text{ even}$$

$$= K_0 \frac{\Omega \prod^{(n-1)/2} (\Omega_i^2\Omega_s - \Omega^2)}{\prod^{(n-1)/2} (\Omega_s - \Omega_i^2\Omega^2)} \qquad \text{for } n \text{ odd}$$

Figure 2.9 illustrates $|T(j\Omega)|$ versus frequency for $n = 4$.

The determination of $R_n(\Omega)$ requires the use of elliptic functions, and $|T(j\Omega)|$ is sometimes referred to as an elliptic approximation. This function was first studied in regard to filters by W. Cauer [4].

The advantage of the Cauer approximation over the Chebyshev and Butterworth approximations is increased attenuation at ω_s for the same order transfer function.

Table 2.6 [5] lists the Cauer polynomials, $N(s)$ and $D(s)$, for a passband ripple of 1 dB, where $T(s) = N(s)/D(s)$. A more extensive listing is presented in Tables 13.9 to 13.11.

To select the appropriate Cauer low-pass transfer function, we need to specify ω_p, T_p, ω_s and T_s. The passband ripple needed is T_p.

The order of n is simply found by using the nomograph given in Fig. 13.11, where $A_{max} \equiv T_p$ and $A_{min} \equiv T_s$. Having selected n, $T(s)$ is formed from the table using Ω_s. If Ω_s is not listed in the table, then one should select the nearest case. However, if the next smaller case is chosen and T_s is adequate, then the resulting design will always satisfy the specifications given. If T_s is not adequate, then the next order should be selected.

The denormalization of $T(s)$ is accomplished in one step by shifting the passband ripple specification from $\Omega = 1$ to $\omega = \omega_p$. This procedure is illustrated in the following example.

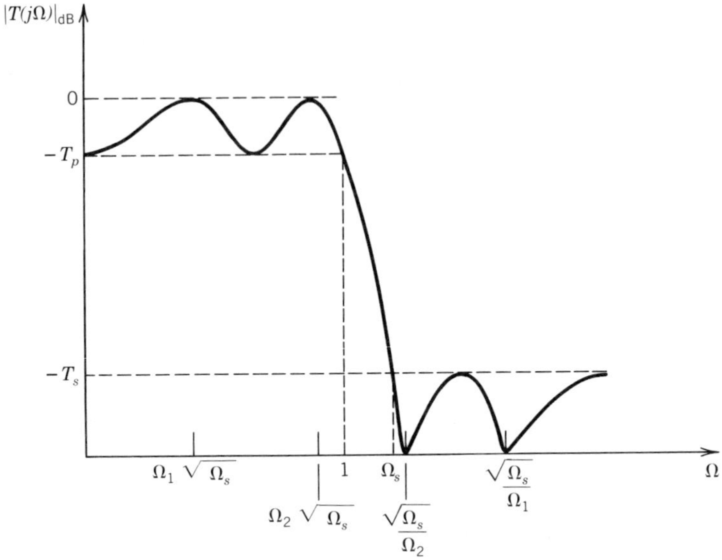

Figure 2.9 Cauer magnitude response for n = 4.

EXAMPLE 2.4 Determine the expression for the lowest-order transfer function that can satisfy the following filter specifications:

$$\omega_p = 400 \text{ rad/s} \qquad T_p = 1.1 \text{ dB } (= A_{\max})$$

$$\omega_s = 800 \text{ rad/s} \qquad T_s = 30 \text{ dB } (= A_{\min})$$

PROCEDURE

1 Determine the ripple needed in the passband from T_p. Therefore, a 1 dB ripple approximation will satisfy this specification.

2 Calculate $\Omega_s = \omega_s/\omega_p$. Thus $\Omega_s = 2$.

3 Determine n from the nomograph given in Fig. 13.11. Therefore $n = 3$.

4 Form the normalized T(s) from Table 2.6.

$$T(s) = \frac{(0.1059)(s^2 + 5.153)}{(s + 0.54)(s^2 + 0.4341s + 1.011)}$$

5 From Table 2.6, for $n = 3$ and $\Omega_s = 2$, $T_s = 34.454$ dB.

6 To denormalize the response, replace s by s/ω_p. This shifts the -1 dB point from $\Omega = 1$ to $\omega = \omega_p$.

$$T_{LP}(s) = T(s)\big|_{s \to s/400}$$
$$= \frac{(42.36)[s^2 + 8.245(10^5)]}{(s + 216)[s^2 + 173.6s + 1.617(10^5)]}$$

A set of Cauer rational functions is needed for each passband and stopband ripple specification. Additional sets of Cauer approximations can be found in Tables 13.9 to 13.11.

Table 2.6 Cauer polynomials for $T_p = 1$ dB

T_s (dB)	n	$N(s)$	$D(s)$
		(a) $\Omega_s = 1.5$	
11.194	2	$(0.27562)(s^2 + 3.92705)$	$s^2 + 0.87942s + 1.21443$
25.176	3	$(0.21562)(s^2 + 2.80601)$	$(s + 0.59102)(s^2 + 0.3754s + 1.02371)$
39.518	4	$(0.01057)(s^2 + 2.53555)(s^2 + 12.09931)$	$(s^2 + 0.72998s + 0.364281)(s^2 + 0.20882s + 0.99881)$
53.875	5	$(0.01318)(s^2 + 2.42552)(s^2 + 5.43765)$	$(s + 0.33785)(s^2 + 0.45775s + 0.51707)$ $(s^2 + 0.13308s + 0.99496)$
		(b) $\Omega_s = 2$	
17.095	2	$(0.13971)(s^2 + 7.4641)$	$s^2 + 0.99894s + 1.17008$
34.454	3	$(0.10589)(s^2 + 5.15321)$	$(s + 0.53996)(s^2 + 0.43407s + 1.01059)$
51.906	4	$(0.002539)(s^2 + 4.59326)(s^2 + 24.2272)$	$(s^2 + 0.70255s + 0.3192)(s^2 + 0.24296s + 0.99323)$
69.360	5	$(0.003169)(s^2 + 4.36495)(s^2 + 10.56773)$	$(s + 0.3126)(s^2 + 0.46468s + 0.47187)$ $(s^2 + 0.15525s + 0.9919)$

2.6 THOMSON APPROXIMATION

In the previous sections we considered the frequency response of various approximations. Figure 2.10 illustrates the time response of the Butterworth low-pass transfer functions of Table 2.1 to a unit step input. Likewise, the step responses of the 1 dB ripple Chebyshev low-pass transfer functions are shown in Fig. 2.11. As can be seen, the Butterworth functions have better time responses than the Chebyshev functions. This is partially due to the fact that time is inversely proportional to frequency.

Consider the following low-pass transfer function:

$$T(s) = \frac{D_n(0)}{D_n(s)}$$

where

$$D_0(s) = 1$$

$$D_1(s) = s + 1$$

$$D_2(s) = s^2 + 3s + 3$$

$$\vdots$$

$$D_n(s) = (2n - 1)D_{n-1}(s) + s^2 D_{n-2}(s)$$

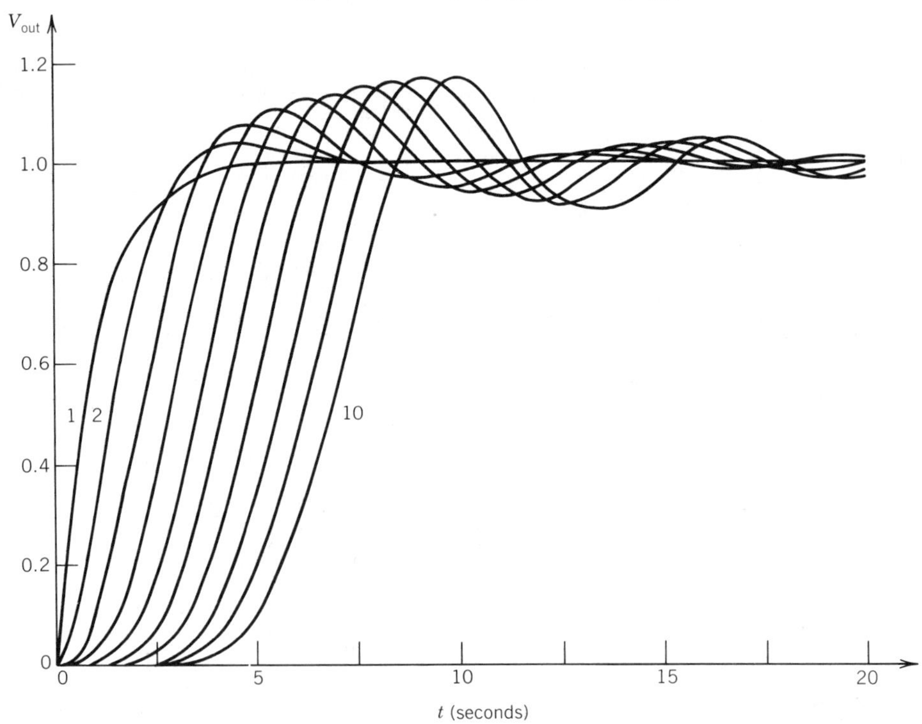

Figure 2.10 Butterworth step response.

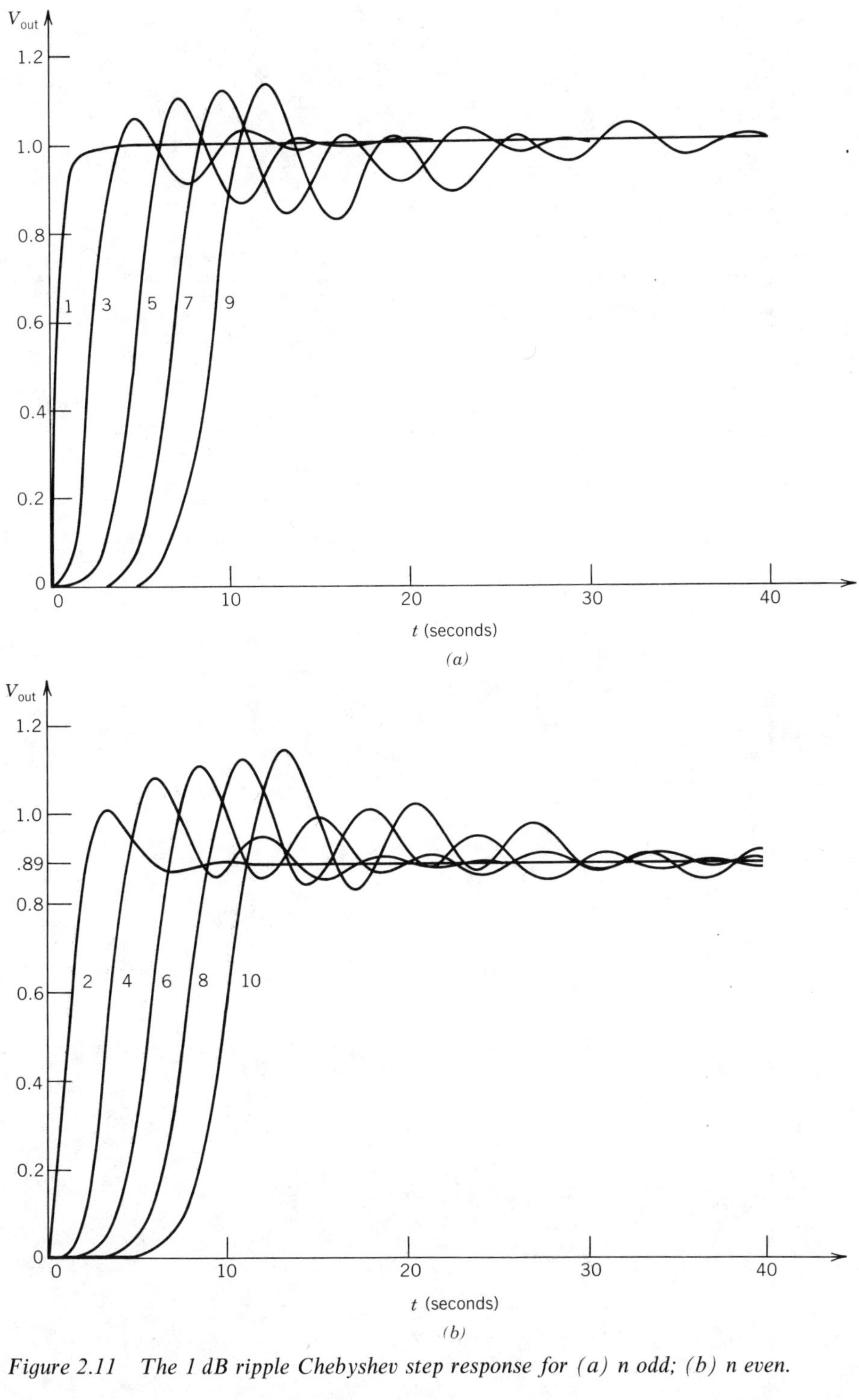

Figure 2.11 The 1 dB ripple Chebyshev step response for (a) n odd; (b) n even.

Table 2.7 **Thomson polynomials**

n	$\Omega_{3\text{ dB}}$	t_d	$\frac{t_d\Omega_{3\text{ dB}}}{2.2}$	N	$D(s)$
1	1.000	0.693	0.347	1	$s + 1$
2	1.362	0.900	0.557	3	$s^2 + 3s + 3$
3	1.756	0.958	0.841	15	$(s + 2.32219)(s^2 + 3.67781s + 6.45943)$
4	2.115	0.979	0.941	105	$(s^2 + 5.79242s + 9.14013)(s^2 + 4.20758s + 11.4878)$
5	2.427	0.989	1.091	945	$(s + 3.64674)(s^2 + 6.70391s + 14.27248)$ $(s^2 + 4.64935s + 18.15631)$
6	2.703	0.994	1.221	10,395	$(s^2 + 5.03186s + 26.51403)(s^2 + 7.47142s + 20.85282)$ $(s^2 + 8.49672s + 18.80113)$
7	2.952	0.997	1.338	135,135	$(s + 4.97179)(s^2 + 5.37135s + 36.59679)$ $(s^2 + 8.14028s + 28.93655)(s^2 + 9.51658s + 25.66644)$
8	3.179	0.998	1.442	2,027,025	$(s^2 + 5.67797s + 48.43202)(s^2 + 8.73658s + 38.56925)$ $(s^2 + 10.40968s + 33.93474)(s^2 + 11.17577s + 31.97723)$
9	3.391	0.999	1.540	34,459,434	$(s + 6.29702)(s^2 + 5.95852s + 62.04144)$ $(s^2 + 9.27688s + 49.7885)(s^2 + 11.20994s + 43.64665)$ $(s^2 + 12.25874s + 40.58927)$
10	3.591	0.999	1.631	654,729,157	$(s^2 + 6.21783s + 77.4427)(s^2 + 9.77244s + 62.62559)$ $(s^2 + 11.93506s + 54.83916)(s^2 + 13.84409s + 48.66755)$ $(s^2 + 13.23058s + 50.58236)$

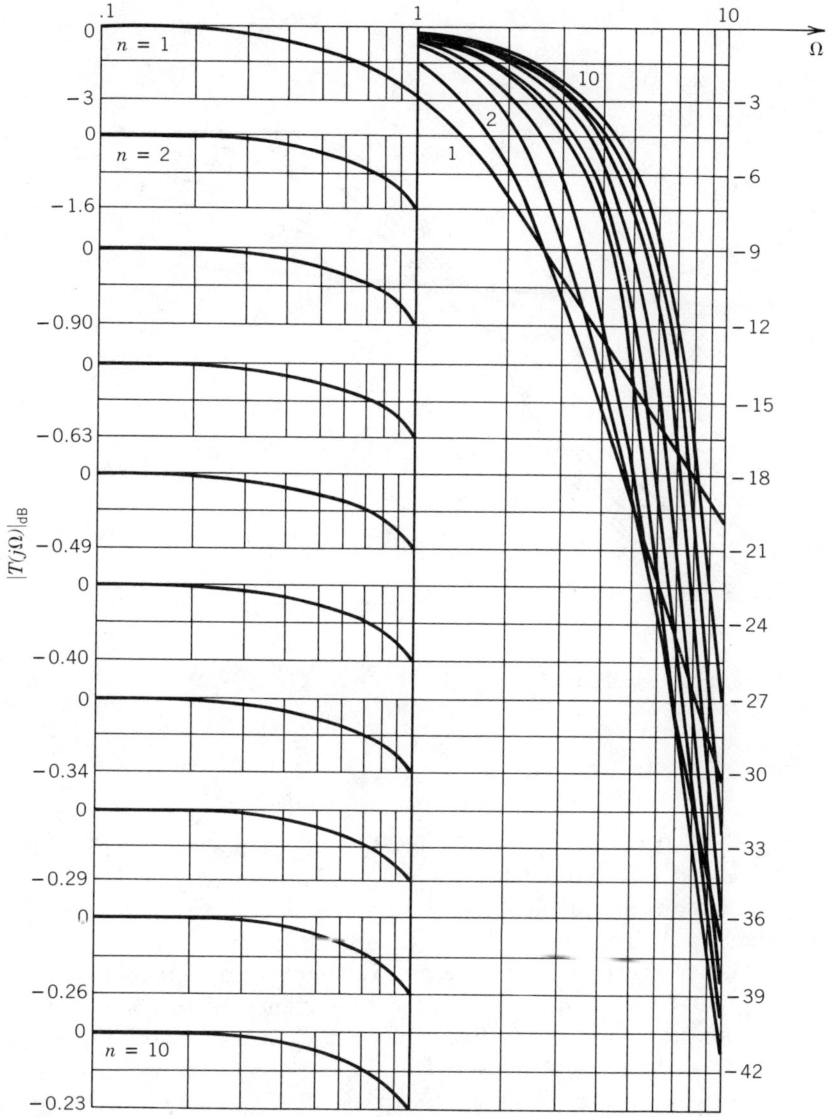

Figure 2.12 Thomson magnitude response.

Table 2.7 lists the Thomson polynomials, $D(s)$, where $T(s) = N/D(s)$, the normalized delay time,* t_d, for the step response, and the normalized 3 dB bandwidth, $\Omega_{3\text{ dB}}$, in rad/s. (An alternative listing, presenting the root locations, is included in Chapter 13 as Table 13.8.) The magnitude of $T(j\Omega)$ is plotted in Fig. 2.12 and the step

* The time required to reach 50% of the final value.

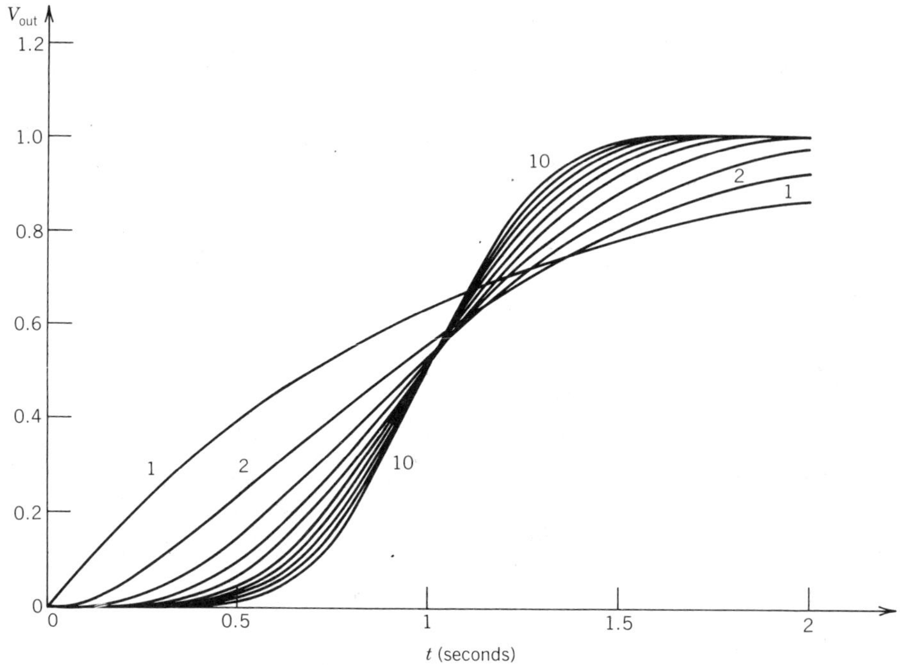

Figure 2.13 Thomson step response.

response in Fig. 2.13. It turns out that the product of the normalized rise time,* t_r, and the normalized 3 dB bandwidth is equal to approximately 2.2.

The $D_n(s)$ functions are called the Bessel polynomials and were first studied in regard to filters by W. Thomson [6]. The advantages of the Thomson approximation over the other approximations are no overshoot† and increased rise time for increased order. These advantages are a consequence of the fact that the Thomson approximation has a nearly linear phase shift in the passband. Because of the improved time response, these filters find use in pulse transmission systems.

To select the appropriate Thomson low-pass transfer function, we need to specify the desired rise time, $\hat{t}_r$, and the desired delay time, $\hat{t}_d$. The ratio of the desired time delay to the desired rise time is approximately equal to the normalized time delay times the normalized 3 dB bandwidth divided by 2.2.

Using this information, we can select n from Table 2.7 and thus form the normalized $T(s)$. To achieve the desired rise time, we need to shift the time scale

* The time required to rise from 10 to 90% of the final value.

† The difference between the peak value and the final value.

in Fig. 2.13 to the appropriate value. This is done by frequency scaling $T(s)$ by the ratio of the normalized rise time to the desired rise time.

This procedure is illustrated in the following example.

EXAMPLE 2.5 Design a low-pass filter with a rise time $\hat{t}_r$ of 3 ms and a time delay, $\hat{t}_d$, of 1.5 ms.

PROCEDURE

1 Evaluate the following formula:

$$\frac{t_d \Omega_{3\ \mathrm{dB}}}{2.2} = \frac{\hat{t}_d}{\hat{t}_r} = 0.5$$

2 Using $\hat{t}_d/\hat{t}_r$, select n from Table 2.7. Thus $n = 2$.

3 Form the normalized $T(s)$.

$$T(s) = \frac{3}{s^2 + 3s + 3}$$

4 The normalized rise time for $T(s)$ shown in Fig. 2.13 is

$$t_r = \frac{2.2}{\Omega_{3\ \mathrm{dB}}}$$
$$= \frac{2.2}{1.362}$$
$$= 1.615$$

5 Determine the frequency scale factor, ω_p, by evaluating the following formula:

$$\omega_p = \frac{t_r}{\hat{t}_r}$$
$$= 538.4$$

6 To denormalize the response replace s by s/ω_p. This shifts the step response time scale from t seconds to $t \times 3\ \mathrm{ms}/t_r$; that is, 1 s becomes 1.858 ms.

$$T_{\mathrm{LP}}(s) = T(s)_{s \to s/538.4}$$
$$= \frac{8.696(10^5)}{s^2 + 1.615(10^3)s + 8.696(10^5)}$$

7 The denormalized time delay is found by evaluating the following formula:

$$\hat{t}_d = \frac{t_d \hat{t}_r}{t_r}$$
$$= \frac{(0.9)(3)(10^{-3})}{1.615}$$
$$= 1.668\ \mathrm{ms}$$

Thus our transfer function has a risetime of 3 ms and a delay time of 1.668 ms.

If a wide pulse is applied to the network, then the response to the falling edge of the pulse can be found by using the mirror image of Fig. 2.13 and shifting the scale up by 1 V.

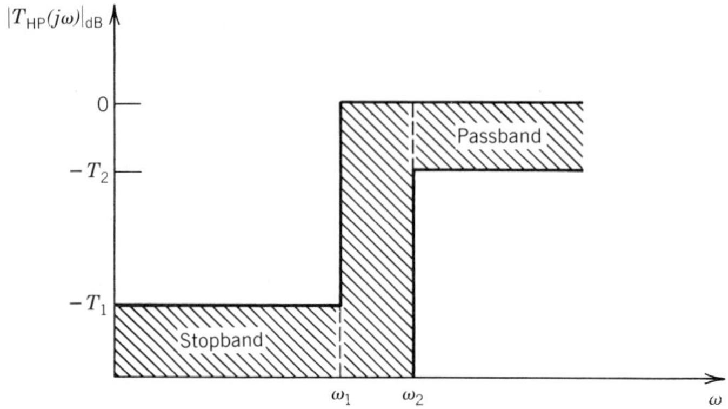

Figure 2.14 Acceptable bounds of a high-pass filter.

2.7 FREQUENCY TRANSFORMATIONS

The approximations in the previous sections dealt with selecting low-pass transfer functions. Instead of repeating this process for high-pass, bandpass, and band-reject transfer functions, it is easier to transform the high-pass, bandpass, and band-reject specifications into an equivalent low-pass specification, find the low-pass transfer function, and transform the result back into the desired transfer function [7].

High-Pass Functions

Figure 2.14 depicts the acceptable bounds of behavior of a high-pass filter.

EXAMPLE 2.6 Determine the expression for the lowest-order high-pass transfer function given in terms of T_1, T_2, ω_1, and ω_2.

PROCEDURE

1 Determine the equivalent low-pass specification with the following formulas:

$$T_p = T_2 \qquad \omega_p = \frac{1}{\omega_2}$$

$$T_s = T_1 \qquad \omega_s = \frac{1}{\omega_1}$$

2 Find $T_{LP}(s)$ using the specifications in 1 and the desired approximation.

3 The high-pass transfer function is formed by replacing s with $1/s$ in $T_{LP}(s)$.

$$T_{HP}(s) = T_{LP}(s)\big|_{s \to 1/s}$$

Bandpass Functions

Figure 2.15 depicts the acceptable bounds of behavior for a symmetrical bandpass filter, where

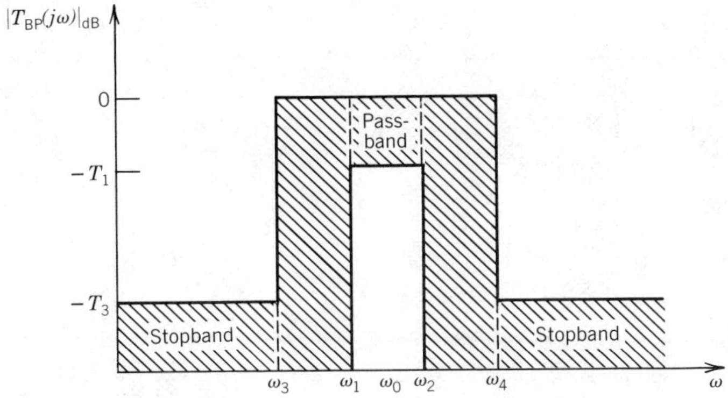

Figure 2.15 Acceptable bounds of a symmetrical bandpass filter.

$$\omega_o = \sqrt{\omega_1 \omega_2} = \sqrt{\omega_3 \omega_4}$$

EXAMPLE 2.7 Determine the expression for the lowest-order bandpass transfer function given in terms of T_1, T_3, ω_1, ω_2, ω_3, ω_4, and where

$$\omega_o = \sqrt{\omega_1 \omega_2} = \sqrt{\omega_3 \omega_4}$$

PROCEDURE

1 Determine the equivalent low-pass specification with the following formulas:

$$T_p = T_1 \qquad \omega_p = \omega_2 - \omega_1$$
$$T_s = T_3 \qquad \omega_s = \omega_4 - \omega_3$$

2 Find $T_{LP}(s)$ using the specifications in step 1 and the desired approximation.

3 The bandpass transfer function is formed by replacing s with $(s^2 + \omega_0^2)/s$.

$$T_{BP}(s) = T_{LP}(s)|_{s \to (s^2 + \omega_0^2)/s}$$

Band-Reject Functions

Figure 2.16 depicts the acceptable bounds of behavior for a symmetrical band-reject (or bandstop) filter, where

$$\omega_o = \sqrt{\omega_1 \omega_2} = \sqrt{\omega_3 \omega_4}$$

EXAMPLE 2.8 Determine the expression for the lowest-order band-reject transfer function given in terms of T_1, T_3, ω_1, ω_2, ω_3, ω_4, and where

$$\omega_o = \sqrt{\omega_1 \omega_2} = \sqrt{\omega_3 \omega_4}$$

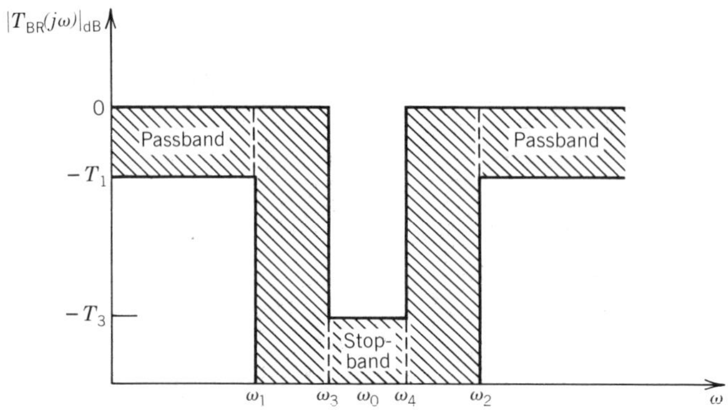

Figure 2.16 Acceptable bounds of a symmetrical band-reject filter.

PROCEDURE

1 Determine the equivalent low-pass specification with the following formulas:

$$T_p = T_1 \qquad \omega_p = \frac{1}{\omega_2 - \omega_1}$$

$$T_s = T_3 \qquad \omega_s = \frac{1}{\omega_4 - \omega_3}$$

2 Find $T_{LP}(s)$ using the specifications in item 1 and the desired approximation.

3 The band-reject transfer function is formed by replacing s with $s/(s^2 + \omega_0^2)$.

$$T_{BR}(s) = T_{LP}(s)|_{s \to s/(s^2 + \omega_0^2)}$$

2.8 SUMMARY

This chapter has presented information on the more popular approximating functions—namely, Butterworth, Chebyshev, Cauer, and Thomson. The relative merits of these functions have been described; root locations enable the selection of appropriate transfer functions. The chapter concludes by illustrating the technique for translating practical specifications to and from a normalized low-pass characteristic.

2.9 REFERENCES

[1] V. D. Landon, "Cascade Amplifiers with Maximal Flatness," *RCA Review*, vol. 5, pp. 347–362, 481–497, 1941.

[2] S. Butterworth, "On the Theory of Filter Amplifiers," *Exp. Wireless and Wireless Eng.*, vol. 7, pp. 536–541, 1930.

[3] P. L. Chebyshev, Théorie des Mécanismes Connus sous le Nom de Parallelogrammes, *Oeuvres*, vol. 1, St. Petersburg, 1899.

[4] W. Cauer, *Synthesis of Linear Communication Networks*, New York: McGraw-Hill, 1958.

[5] L. P. Huelsman and P. E. Allen, *Introduction to the Theory and Design of Active Filters*, New York: McGraw-Hill, 1980.

[6] W. E. Thomson, "Delay Networks Having Maximally Flat Frequency Characteristics," *Proc. IEE*, part 3, vol. 96, pp. 487–490, November, 1949.

[7] P. Bowron and F. W. Stephenson, *Active Filters for Communications and Instrumentation*, London: McGraw-Hill, 1979.

[8] E. Christian and E. Eisenman, *Filter Design Tables and Graphs*, New York: Wiley, 1966.

[9] R. W. Daniels, *Approximation Methods for Electronic Filter Design*, New York: McGraw-Hill, 1974.

CHAPTER 3

The Second-Order Function

K. M. Reineck
University of Cape Town
South Africa

3.1 INTRODUCTION

The second-order function plays a significant role in the design of active filters. Many of the structures discussed in later chapters are based upon the biquadratic transfer function and form the basis for cascade synthesis. These simpler networks provide a sound pedagogic understanding of circuit behavior.

This chapter is therefore concerned with forming a technical foundation upon which later discussions may be based. It begins by considering the concept of poles and zeros and their combination to form special cases (low-pass, high-pass, bandpass, bandstop, and all-pass) of the general biquadratic function. The properties of these functions are studied in some detail so that their critical parameters associated with gain and phase responses are well understood.

The realization of the basic filter forms by means of formal RC synthesis is discussed so as to provide a basis for understanding some of the passive structures utilized in active RC filters. Finally, the concept of denormalization is considered.

3.2 THE POLYNOMIAL FUNCTION

As a basis for the description and realization of networks, certain fundamentals of general network theory must first be introduced. As shown in Chapter 1, the design begins with the approximation of the desired transfer characteristics.

A system or network can be represented by a black box with an input and output pair of terminals as in Fig. 3.1. There are, of course, systems that can have

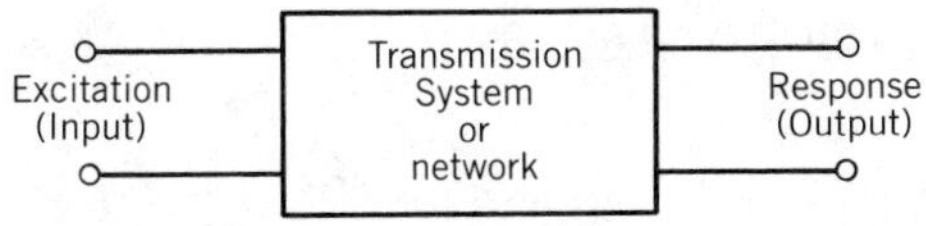

Figure 3.1 System or network representation.

more than one input and one output. The network is made up of physical elements such as resistors, capacitors, and energy sources. When energy sources (e.g., operational amplifiers) are included, the network is considered to be an *active network*. To understand the function of a complex system, it is first necessary to consider the input and output relationships. All physical elements are nonlinear to a greater or lesser degree. However, in the majority of cases, the nonlinear nature of the network element is not too pronounced at low signal levels. In these circumstances the linear approximation generally yields results that are in close agreement with the analytically predicted behavior.

In network theory we distinguish between *network analysis*, which is concerned with determining the response (output) given the excitation (input), and *network synthesis*, where the network is designed to yield given response and excitation characteristics.

Synthesis generally presents more difficulty than network analysis. Although the input/output relationship of a network is unique, it can be satisfied by more than one network. As an example, consider the two networks in Fig. 3.2, which show the same input–output relationship (input = current; output = voltage or $Z = V/I$). The impedance seen looking into the two terminals is the same for both circuits. In differential equation form the circuits may be described as:

$$\frac{d^2V}{dt^2} + 2\frac{dV}{dt} = \frac{d^2I}{dt^2} + 4\frac{dI}{dt} + 3I \tag{3.1}$$

Time Function and Spectra

Current and voltage are normally real functions of time. This means that for real values of t, the function $k(t)$ is also real. Of equal importance is the description of the spectrum $K(s)$, which is obtained from the time function by application of the one-sided Laplace transformation. A time function $k(t)$ defined for the time interval $0 \to \infty$ is uniquely assigned to a function $K(s)$:

$$K(s) = \mathscr{L}[k(t)] \tag{3.2}$$

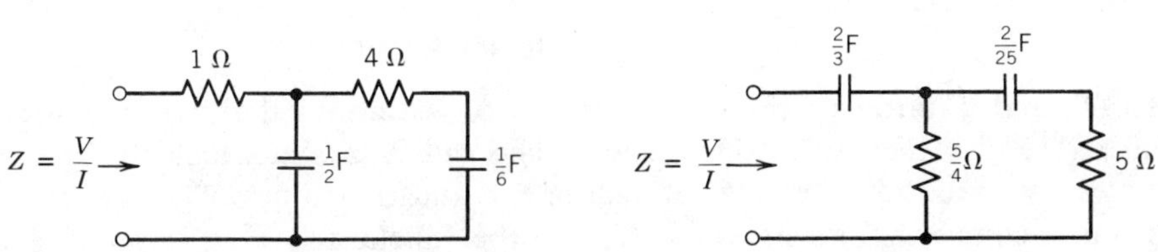

Figure 3.2 Networks with equivalent excitation/response relationships.

where $s = \sigma + j\omega$, which is the complex frequency. The spectrum function $K(s)$ can be obtained as:

$$K(s) = a(s) + jb(s) = \mathscr{L}[k(t)] = \int_{t=0^-}^{\infty} k(t)e^{-st}\,dt \tag{3.3}$$

With the use of s as the Laplace-transform variable, the impedance of a capacitance is simply given by $Z = 1/sC$. Setting $s = j\omega$ yields the conventional impedances as functions of frequency, for example, $1/j\omega C$. Thus ω in the text will be considered as a real variable. In contrast, the appearance of s in an equation indicates that the properties as a function of the complex frequency variable s are considered. Equation 3.1 describes the networks of Fig. 3.2 in differential equation form. From the discussion in this section it is clear that a transformation from the time domain to the frequency domain using the Laplace transform is possible. The result is a polynomial function, which also describes the input/output relationship of a network. For the circuits shown in Fig. 3.2 this polynomial function is known as the driving point impedance that refers excitation (input) and response (output) to the same pair of terminals. Hence:

$$Z(s) = \frac{V}{I} = \frac{s^2 + 4s + 3}{s^2 + 2s} \tag{3.4}$$

The driving point impedance is seen to be a rational function consisting of a numerator and a denominator polynomial. By converting to the factored form, Eq. 3.4 becomes

$$Z(s) = \frac{(s + 1)(s + 3)}{s(s + 2)} \tag{3.5}$$

Poles, Zeros, and the Frequency Domain Response

Consider some arbitrary function of a variable s such as $T(s)$. As s varies in magnitude, $T(s)$ goes through gyrations of one sort or another. For certain values of s the function $T(s)$ may even become zero. For example, $T(s) = s - 1$ becomes zero for $s = 1$. These particular values of s are called the *zeros* of $T(s)$.

For some other values of s the function may go to infinity. For example, $T(s) = 1/(s + 1)$ goes to infinity for $s = -1$. Such specific values of s are called the singularities or *poles*.

Ordinarily, a function $T(s)$ will have poles and zeros for certain values of s. If the complex variable is represented by $s = \sigma + j\omega$, where σ and ω are the real and imaginary parts of s, respectively, the function of the complex variable is

$$T(s) = T(\sigma + j\omega) = T_1(\sigma, \omega) + jT_2(\sigma, \omega) \tag{3.6}$$

where T_1 and T_2 are equations of a surface located above the two dimensional s-plane. (See also the end of this section.) The values of s at which the function becomes zero are the *zeros* of the function (z_i), and the values of s at which the function becomes infinite are the *poles* of the function (p_i). It is possible for multiple-order poles or zeros to exist at the same point on the s-plane. Zeros and poles of a function $T(s)$ can be plotted on the s-plane. In Eq. 3.5 the factored

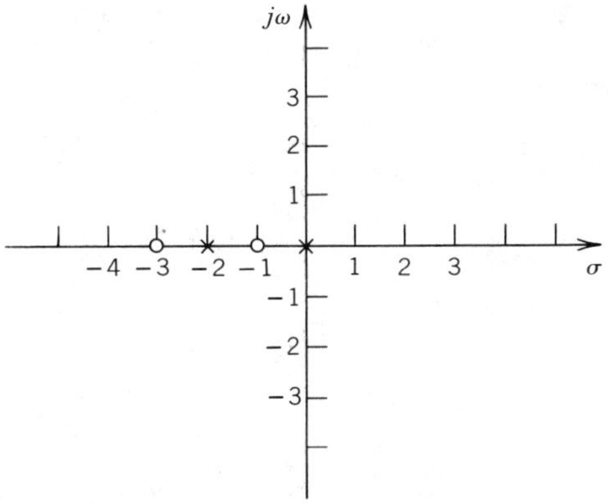

Figure 3.3 *Pole-zero plot of Eq. 3.5.*

form of a driving point impedance was given for which the poles and zeros can now be identified.

If we designate a zero with a circle and a pole by a cross, poles and zeros may be plotted in the complex plane as in Fig. 3.3. Note that for purely passive RC networks the poles are located on the negative side of the real axis, whereas zeros may be located in the left as well as the right half of the s-plane but not on the positive real axis. However, for active RC networks complex pole locations in the left half of the s-plane (LHP) may be generated. For the purpose of determining the pole-zero locations, consider the following transfer function with a zero at $z = -1$.

$$T(s) = \frac{H(s+1)}{s^2 + \sqrt{2}s + 1} \tag{3.7}$$

The two poles are located at

$$P_1 = -\frac{\sqrt{2}}{2} + j\sqrt{\frac{2}{4}} = -0.707 + j0.707$$

$$P_2 = -\frac{\sqrt{2}}{2} - j\sqrt{\frac{2}{4}} = -0.707 - j0.707$$

The pole-zero plot is given in Fig. 3.4.

If the function $T(s)$ is to be evaluated for all real frequencies, we can view a point $s = j\omega$ as moving up the $j\omega$-axis from 0 to ∞. For each position of $s = j\omega$ the vector M_i or m_i has a different magnitude and phase. The overall magnitude is obtained by taking the length of the numerator vector divided by the product of the lengths of the two denominator vectors. The overall phase angle is given by the numerator vector angle minus the two angles of the denominator vectors. Thus

$$T(j\omega) = \frac{M_1}{m_1 m_2} e^{j(\beta_1 - \alpha_1 - \alpha_2)} \tag{3.8}$$

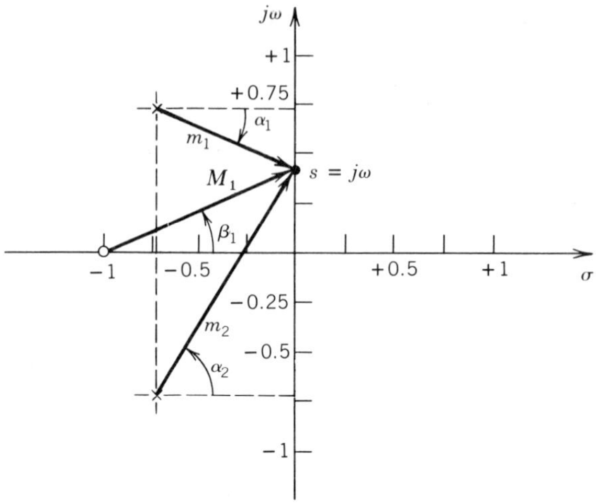

Figure 3.4 *Pole-zero plot with s-plane vectors.*

It should be noted that the s-plane vectors do not contain any information about the constant multiplier, also known as the gain constant H of Eq. 3.7. In cases where the constant is specified, it is an easy matter to multiply by it. However, in other cases the constant may be known only after realization of the network.

EXAMPLE 3.1 Obtain the pole-zero plot for a Butterworth second-order bandpass approximation with a passband width of $B = \omega_2 - \omega_1 = 4 - 3 = 1$ rad/s and $\omega_0 = \sqrt{12} = 3.464$ rad/s.

PROCEDURE

1 Obtain the first-order LP Butterworth function from Table 2.1.

$$T(s) = \frac{H}{s + 1}$$

2 The frequency transformation to bandpass form is obtained by replacing s by

$$s = \frac{s^2 + \omega_0^2}{Bs} = \frac{s^2 + 12}{s}$$

in the LP function (see Section 2.7). Thus

$$T(s) = \frac{Hs}{s^2 + s + 12}$$

is the second-order bandpass function with the zeros and poles located at

$$z_1 = 0$$

$$p_1 = -\frac{1}{2} + j\sqrt{\frac{47}{4}}$$

$$p_2 = -\frac{1}{2} - j\sqrt{\frac{47}{4}}$$

3 The pole-zero plot is shown in Fig. 3.5*a*.

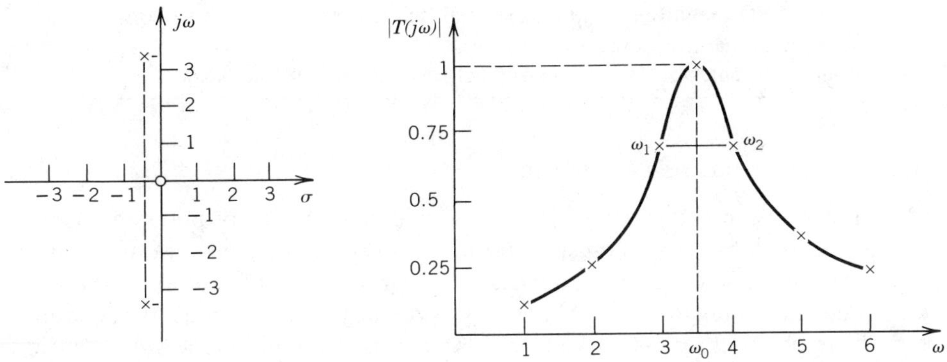

(a) *(b)*

ω (*rad/s*)	M_1 (*Units*)	m_1 (*Units*)	m_2 (*Units*)	$\lvert T(j\omega)\rvert$ (*Units*)
1	1	2.48	4.45	0.091
2	2	1.51	5.45	0.243
3	3	0.66	6.45	0.707
3.43	3.43	0.50	6.87	1.000
4	4	0.76	7.44	0.707
5	5	1.65	8.44	0.359
6	6	2.62	9.44	0.243

(c)

(d)

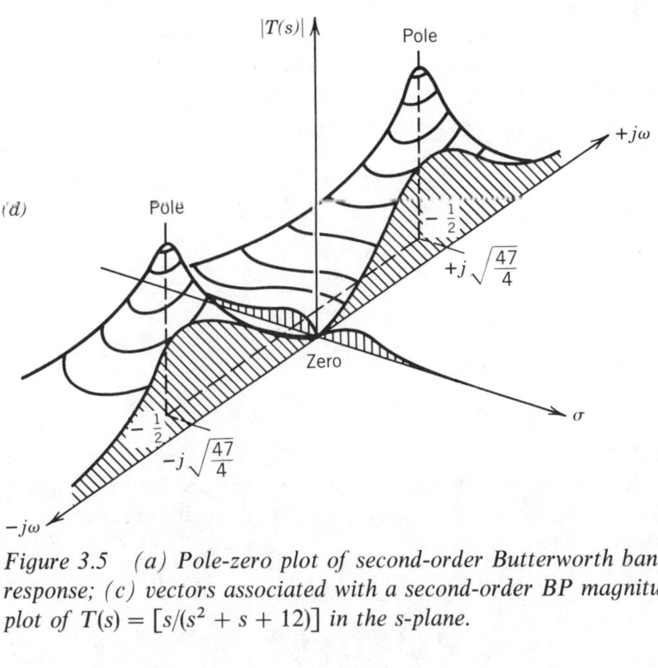

Figure 3.5 (a) Pole-zero plot of second-order Butterworth bandpass; (b) magnitude response; (c) vectors associated with a second-order BP magnitude response; (d) contour plot of $T(s) = [s/(s^2 + s + 12)]$ *in the s-plane.*

4 Evaluation of the magnitude values at several frequencies allows for the construction of the frequency domain response shown in Fig. 3.5*b*.
5 A contour representation of the pole-zero plot is depicted in Fig. 3.5*d*.

Time Domain or Impulse Response

It has already been shown that network functions are ratios of polynomials in s. The various coefficients of s must be positive and real, since they are functions of the network parameters. A consequence of this property is that complex poles and zeros occur in conjugate pairs. This fact was already observed with the functions used above. Restricting networks to be stable must be considered as another important network property. This means that a bounded input excitation to the circuit must yield a bounded response.

Passive networks are stable by their very nature because they contain no energy sources. Active networks, however, do contain such energy sources, which, together with the input excitation, could produce an unbounded output. Such unstable networks are of no use as practical filters.

By considering the response of a network to an impulse or step function, it is possible to determine the stability of the circuit in a convenient manner. The impulse response is obtained by taking the inverse Laplace transform of the partial fraction expansion of the network function. For a network function with a simple pole, the impulse response is

$$K(t) = \mathscr{L}^{-1} \frac{H}{s - p_1} = He^{p_1 t} \tag{3.9}$$

If p_1 is positive, the impulse response is seen to increase exponentially with time corresponding to an unstable circuit, whereas for p_1 negative or at the origin, the circuit is stable (Fig. 3.6). Other characteristic pole positions, together with their associated time domain responses, are also given in Fig. 3.6.

EXAMPLE 3.2 Check if the following network functions are stable:

1 $T(s) = \dfrac{s}{s^2 - 2s + 3}$

2 $T(s) = \dfrac{s - 2}{s^2 + 9}$

PROCEDURE

1 The function has a zero at $z = 0$ and two poles at $p_1 = 1 + j\sqrt{2}$ and $p_2 = 1 - j\sqrt{2}$. This function cannot be realized as a stable network because the poles are in the right half of the s-plane. Furthermore, by inspection it is unstable, since all coefficients are not of the same sign (Routh–Hurwitz criterion).

2 There is a zero at $z_1 = 2$ and two poles at $p_1 = j3$ and $p_2 = -j3$. This function can be

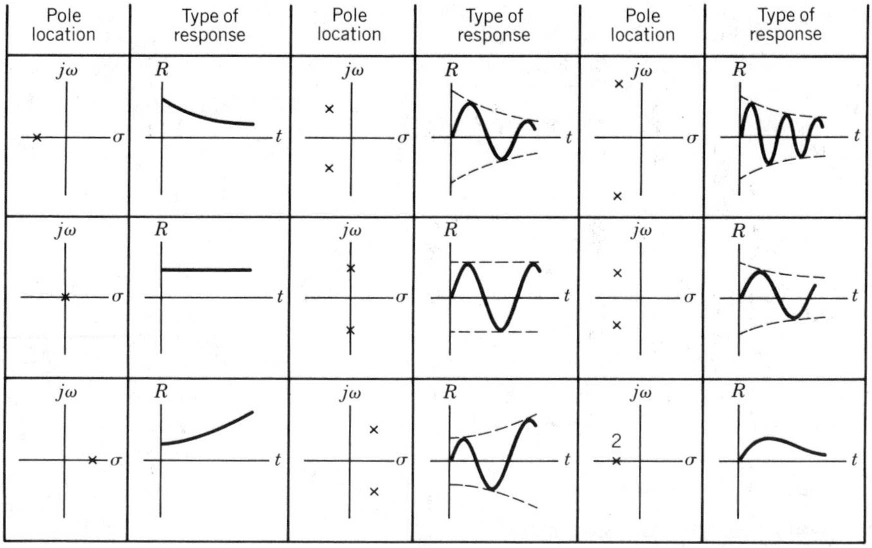

Figure 3.6 Various first- and second-order time-domain responses as a function of the pole locations of the network function.

realized as a stable network because the poles are on the $j\omega$-axis. The zero in the right half-plane does not violate any of the stability requirements.

3.3 MAXIMUM GAIN CONSIDERATIONS

The maximum realizable gain of an RC two-port network has been studied extensively by Fialkow and Gerst [1, 2], while Paige and Kuh [3] have derived the upper bound of the realizable gain constant H for general RC ladder networks. As indicated earlier, the transfer function generally consists of a numerator polynomial $N(s)$, a denominator polynomial $D(s)$, and a multiplier constant H known as the gain constant.

When considering what can be obtained from passive networks alone without amplifiers or similar devices, the description of the gain constant H is just as important as is $N(s)$ and $D(s)$. In most textbooks on network theory, H is either totally ignored or just left to chance.

The problem of the upper bound of realizable gain of general RC ladders can be attacked by considering the poles and zeros of a function. The transfer function is of the form:

$$T(s) = \frac{V_o}{V_i} = H \frac{\prod_{i=1}^{m} (s + z_i)}{\prod_{i=1}^{n} (s + p_i)} \tag{3.10}$$

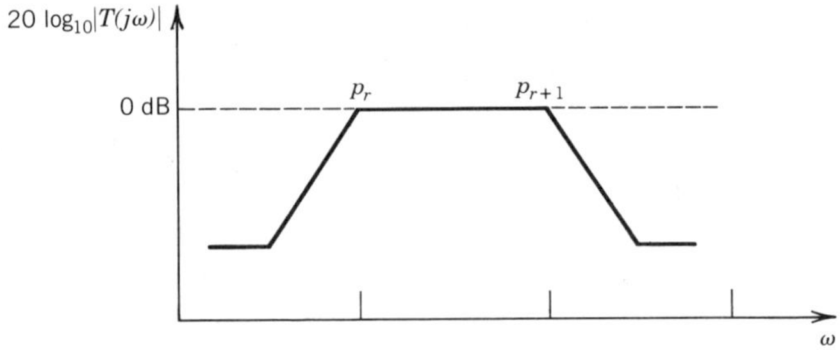

Figure 3.7 General asymptotic plot.

For RC networks, some of the well-known realizability conditions are:

1 $m \leq n$.

2 z_i must be real and nonnegative.

3 p_i must be real, positive, and distinct.

A value for the upper bound of the gain constant H_{max} may be obtained by forming asymptotic sketches of the magnitude, also called Bode plots. For functions with poles and zeros on the negative real axis, such plots can easily be derived from the corner frequencies. The slopes of the asymptotes are simply ± 6 dB/octave. The gain constant H controls the level, but not the shape, of the plot.

In the case of a bandpass-type asymptotic plot, the maximum asymptote is represented by $(p_r - p_{r+1})$, where r refers to the corner frequencies of the bandpass function. The upper bound H_{max} for an RC ladder is fixed by the plot, where the maximum asymptote is a tangent to the 0 dB line [3]. A plot for a bandpass case is shown in Fig. 3.7. A passive RC ladder network can at most have unity gain. This means that the input signal passes through the network without being attenuated. In terms of an output-to-input voltage ratio, this amounts to unity; that is, $V_o/V_i = 1$ or, in decibels, $20 \log_{10} 1 = 0$ dB.

In order to determine the numerical value of H_{max} the following cases can be distinguished.

1 $r + 1 \leq m \leq n$

$$H_{max} = \frac{\prod_{r+1}^{n} p_i}{\prod_{r+1}^{m} z_i} \tag{3.11}$$

2 $r = 0$ (low-pass)

$$H_{max} = \frac{\prod_{1}^{n} p_i}{\prod_{1}^{m} z_i} \tag{3.12}$$

3 $m < r + 1 < n$

$$H_{max} = \prod_{r+1}^{n} p_i \tag{3.13}$$

4 $r = m = n$ (high-pass)

$$H_{max} = 1 \tag{3.14}$$

Case 1 represents a general case that includes the bandpass. Case 3 is a situation where the slopes of the asymptotes above p_{r+1} are always negative because transmission zeros above p_{r+1} are all at infinity. Generally, it can be stated that for the low-pass and high-pass cases H_{max} can be realized. This is not so for the bandpass situation where H_{max} may only be approached arbitrarily closely.

EXAMPLE 3.3 Determine H_{max} for the following low-pass transfer function:

$$T(s) = H\frac{(s + 2.5)(s + 4)}{(s + 0.5)(s + 2)}$$

Draw the asymptotic plot of $T(s)$.

PROCEDURE

1 According to the formula given in Eq. 3.12:

$$H_{max} = \frac{0.5 \times 2}{2.5 \times 4} = \frac{1}{10}$$

2 For this gain constant the asymptotic plot will start on the 0 dB line as shown in Fig. 3.8. All corner frequencies are seen to lie on the 0 dB line when the factored form of $T(s)$ is used.

EXAMPLE 3.4 Consider the following high-pass transfer function:

$$T(s) = H\frac{s^2}{(s + 1)(s + 2)}$$

and determine H_{max}.

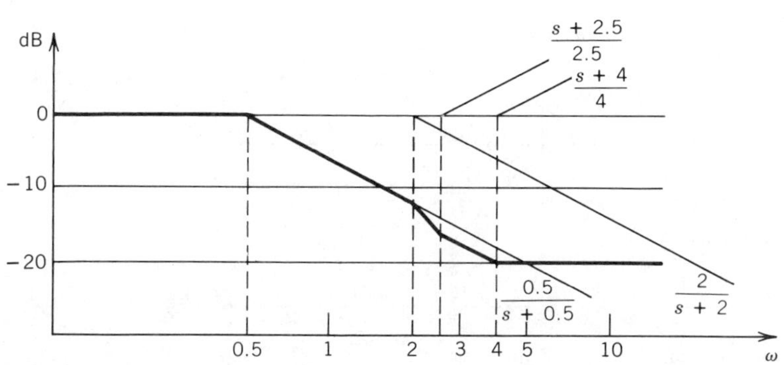

Figure 3.8 Asymptotic plot of biquadratic LP function.

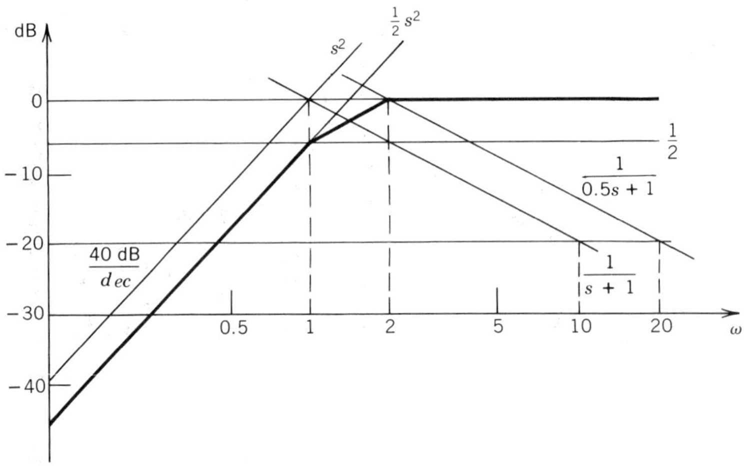

Figure 3.9 HP function asymptotic plot.

PROCEDURE

1 According to Eq. 3.14

$$H_{max} = 1$$

2 For this gain constant the high-frequency asymptote will be on the 0 dB line (Fig. 3.9). The bold line shows the complete Bode plot.

EXAMPLE 3.5 Determine H_{max} for the following bandpass transfer function:

$$T(s) = H\frac{(s+1)(s+4)}{(s+2)(s+3)} = H\frac{2}{3}\frac{(s+1)(0.25s+1)}{(0.5s+1)\left(\frac{s}{3}+1\right)}$$

PROCEDURE

1 Draw the Bode plot with corner frequencies located on the 0 dB line (Fig. 3.10).

2 Identify the maximum asymptote represented by $p_r - p_{r+1}$ (①). Here we have poles $p_1 - p_2$, which cause the asymptotic line to lie above the 0 dB line. However, for an RC

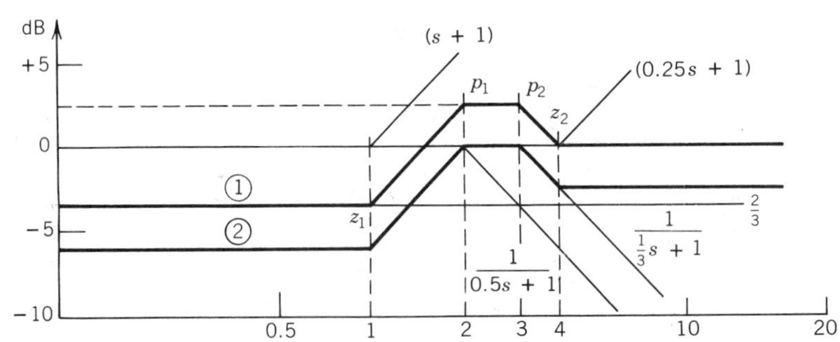

Figure 3.10 Asymptotic plot of BP-type function.

network, maximum gain only allows for the asymptotic line to lie on the 0 dB level.

3 Choose H_{max} so that the $p_1 - p_2$ asymptote lies on the 0 dB line. According to Eq. 3.11 the maximum gain constant becomes

$$H_{max} = \frac{p_2}{z_2} = \frac{3}{4}$$

where $r = 1$.

4 Thus the complete transfer function with H_{max} included becomes

$$T(s) = \frac{3}{4}\frac{(s+1)(s+4)}{(s+2)(s+3)}$$

5 The final Bode plot for $H_{max} = \frac{3}{4}$ is shown as the bold trace ② in Fig. 3.10. As can be seen, $H_{max} = \frac{3}{4}$ has shifted the maximum asymptote onto the zero dB line.

It has been shown that H_{max} can be explicitly determined in terms of the poles and zeros of the given transfer function. In view of considerations such as noise and distortion, the maximization of the gain constant must be regarded as important.

3.4 TYPES OF SECOND-ORDER TRANSFER FUNCTIONS

The second-order function is important because of its relative simplicity. In active filters it represents a very useful building block for which the mathematical properties are readily obtainable and whose physical characteristics are easily measurable and adjustable.

A circuit realized from a given network function will be frequency-dependent. It will filter select frequencies; therefore, filters can be classified according to the functions they perform. The various filter types will now be specified according to their transfer functions. The frequency response plot makes use of semilogarithmic coordinates. The gain or magnitude of the function is usually given in decibels while the frequency axis or abscissa has a logarithmic scale.

Low-Pass Filters

Characteristically, a low-pass (LP) filter passes frequencies from dc to some specified cutoff frequency ω_o whereas high frequencies are attenuated. A typical frequency response is shown in Fig. 3.11.

To meet the LP requirement, a filter characteristic has to stay outside the shaded region. In the passband a certain amount of ripple is allowed. This ripple appears as a peak and, for $Q > 1/\sqrt{2}$, occurs at

$$\omega = \omega_0\sqrt{1 - (1/2Q_p^2)} \tag{3.15}$$

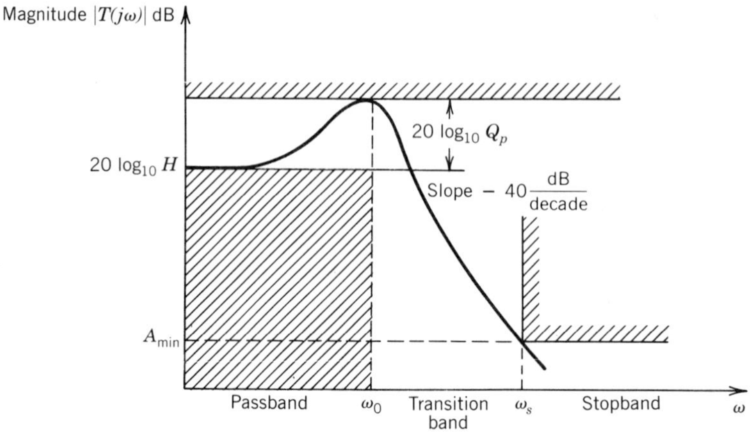

Figure 3.11 General LP frequency response.

Frequencies above the stopband cutoff frequency ω_s are required to have at least $A_{\min}$ dB attenuation. With the parameters ω_o, ω_s, Q_p, and $A_{\min}$ the LP filter requirements of Fig. 3.11 are completely described. The second-order LP transfer function gain characteristic is

$$T_{\text{LP}}(s) = \frac{H}{s^2 + (\omega_o/Q_p)s + \omega_o^2} \tag{3.16}$$

$T(s)$ has two zeros at $s = \infty$. To find the high-frequency behavior of this function, it is necessary to consider $|T(j\omega)|$ for $\omega \gg \omega_o$. Generally,

$$|T(j\omega)| = \frac{H}{|-\omega^2 + \omega_o^2 + j\omega(\omega_o/Q_p)|} \tag{3.17}$$

At high frequencies, however,

$$|T(j\omega)| = \frac{H}{\omega^2} \tag{3.18}$$

The phase response of this function is governed to a large extent by the Q factor. For $Q > 5$ the phase undergoes a rapid shift of π radians about ω_o. At $\omega = \omega_o$ the phase is $\phi(\omega_o) = -\pi/2$.

EXAMPLE 3.6 Use $\omega = 10$ and 100 rad/s to determine the slope of attenuation at high frequencies for the LP function of Eq. 3.16.

PROCEDURE

1 For $\omega = 10$ we get $20 \log_{10} \frac{1}{100} = -40$ dB.

2 For $\omega = 100$ the attenuation will be $20 \log_{10} \frac{1}{100^2} = -80$ dB.

3 It can readily be seen that the rate of attenuation of the second-order LP function is -40 dB/decade.

EXAMPLE 3.7 Determine the peak gain for a low-pass function, assuming that $Q_p > 1/\sqrt{2}$.

PROCEDURE

1 The frequency at which the magnitude peaks was given in Eq. 3.15. Inserting this value into the magnitude function yields the peak gain as

$$
\begin{aligned}
|T(j\omega)|_{\max} &= \left|\frac{H}{-\omega^2 + \omega_o^2 + j\omega(\omega_o/Q_p)}\right| \\
&= \left|\frac{H}{-\omega_o^2[1-(1/2Q_p^2)] + \omega_o^2 + j\omega_o\sqrt{1-(1/2Q_p^2)}(\omega_o/Q_p)}\right| \\
&= \left|\frac{H}{(\omega_o^2/2Q_p^2) + j(\omega_o^2/Q_p)\sqrt{1-(1/2Q_p^2)}}\right| \\
&= \frac{H}{\sqrt{(\omega_o^4/4Q_p^4) + (\omega_o^4/Q_p^2) - (\omega_o^4/2Q_p^4)}} \\
&= \frac{H}{\sqrt{(\omega_o^4/Q_p^2) - (\omega_o^4/4Q_p^4)}} \\
&= \frac{HQ_p}{\omega_o^2\sqrt{1-(1/4Q_p^2)}} \simeq \frac{HQ_p}{\omega_0^2} \qquad \text{for } 4Q_p^2 \gg 1
\end{aligned}
$$

EXAMPLE 3.8 Plot the phase of the following low-pass function:

$$T(s) = \frac{H}{s^2 + (\omega_o/Q_p)s + \omega_o^2}$$

with $H = \omega_o^2$, $Q_p = 2$, and $\omega_o = 1$ rad/s.

PROCEDURE

1 Write the transfer function as

$$T(s) = \frac{\omega_o^2}{s^2 + (\omega_o/2)s + \omega_o^2}$$

2 Rewrite the transfer function in terms of the frequency variable $j\omega$.

$$T(j\omega) = \frac{1}{[1-(\omega^2/\omega_o^2)] + j(\omega/2\omega_o)}$$

3 The phase function is identified as

$$\phi(\omega) = -\tan^{-1}\frac{\omega/2\omega_o}{1-(\omega^2/\omega_o^2)}$$

4 Plot the phase response as a function of frequency (see Fig. 3.12).

High-Pass Filters

A high-pass (HP) filter passes frequencies above a specified cutoff frequency ω_o while low frequencies are attenuated. The HP requirement is shown in Fig. 3.13.

(a)

ω rad/s	$\phi(\omega)$
0.2	$-5.94°$
0.4	$-13.39°$
0.6	$-25.11°$
0.8	$-48.01°$
1.0	$-90°$
1.2	$-126.25°$
1.4	$-143.90°$
1.6	$-152.85°$
2.0	$-161.56°$

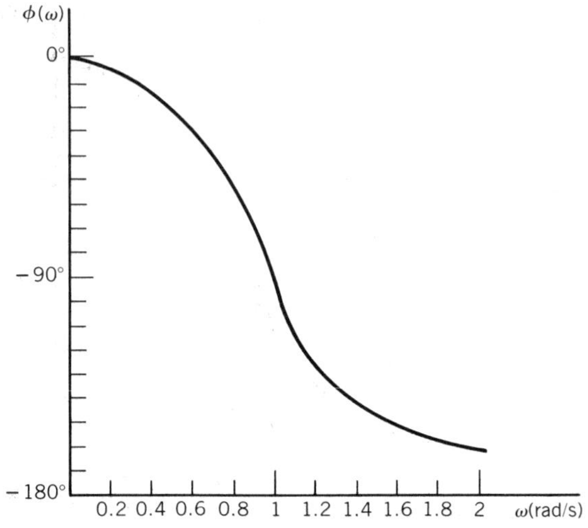

Figure 3.12 (a) Phase as a function of frequency; (b) phase response of second-order LP function.

As in the LP case the parameters ω_o, ω_s, Q_p and A_{min} completely characterize the HP filter requirements. Again a certain amount of ripple may occur in the passband, provided that $Q_p > 1/\sqrt{2}$. The maximum gain of the function in this case will occur at the frequency

$$\omega = \omega_o \frac{1}{\sqrt{1 - (1/2Q_p^2)}} \tag{3.19}$$

for which the peak gain will be

$$|T(j\omega)|_{\max} = \frac{HQ_p}{\sqrt{1 - (1/4Q_p^2)}} \tag{3.20}$$

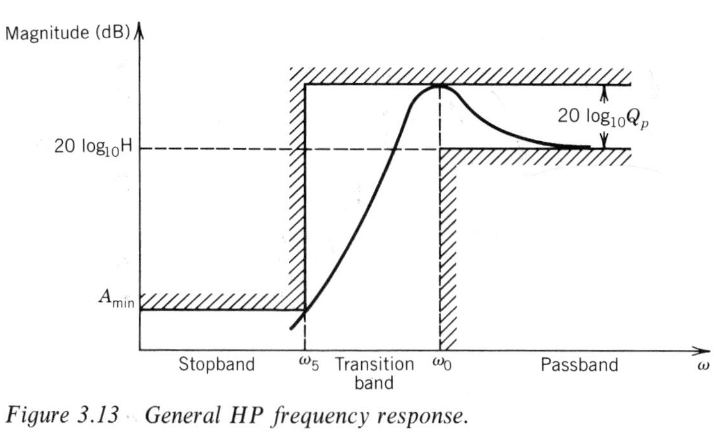

Figure 3.13 General HP frequency response.

For $Q < 1/\sqrt{2}$ no peaking will occur and maximum gain of the function at high frequencies will be $|T(j\omega)|_{\omega \to \infty} = H$.

The general second-order HP transfer function is of the form

$$T_{\mathrm{HP}}(s) = \frac{Hs^2}{s^2 + (\omega_o/Q_p)s + \omega_o^2} \tag{3.21}$$

$H(s)$ has two zeros at $s = 0$, and the two poles of the function will be situated in the left half of the s-plane. The magnitude response is given by

$$|T(j\omega)| = \left| \frac{H(-\omega^2)}{(-\omega^2 + \omega_o^2) + j\omega(\omega_o/Q_p)} \right| \tag{3.22}$$

For low frequencies ($\omega \ll \omega_o$) the approximation reduces to

$$|T(j\omega)|_{\omega \to 0} = H\frac{\omega^2}{\omega_o^2} \tag{3.23}$$

The phase response of this function is again governed by the Q factor. For $Q > 5$ the phase undergoes a rapid shift of nearly π radians about ω_o. The phase shift is π radians at low frequencies, 0 radians at high frequencies, and, at $\omega = \omega_o$, the phase is $\phi(\omega_o) = \pi/2$ radians.

EXAMPLE 3.9 Determine the slope of the HP response for $\omega \ll \omega_o$ using the HP function equation (3.21).

PROCEDURE

1 The cutoff frequency ω_o in normalized form is approximately $\omega_o = 1$ rad/s so that values of $\omega = 0.01$ and 0.1 rad/s could be chosen to represent a low-frequency decade.

2 For $\omega = 0.01$ the attenuation will be $20 \log_{10} 10^{-4} = -80$ dB.

3 For $\omega = 0.1$ we get $20 \log_{10} 10^{-2} = -40$ dB.

4 As can be seen, the rate of rise will be 40 dB/decade. This is equivalent to 12 dB/octave.

Bandpass Filters

Bandpass (BP) filters pass frequencies in a certain frequency band while attenuating frequencies on either side of this band. The passband extends from a lower cutoff frequency ω_1 to an upper cutoff frequency ω_2. Bandpass filters of order greater than two may again exhibit a certain amount of ripple in the passband. A typical second-order gain response is shown in Fig. 3.14.

As can be seen, the two stopbands extend from dc to ω_3 and ω_4 to infinity. The general second-order BP transfer function is given by:

$$T(s) = \frac{Hs}{s^2 + (\omega_o/Q_p)s + \omega_o^2} \tag{3.24}$$

High- and low-frequency slopes of a BP filter are 20 dB/decade. The zero at the origin provides for infinite attenuation at dc. The width of the passband B is

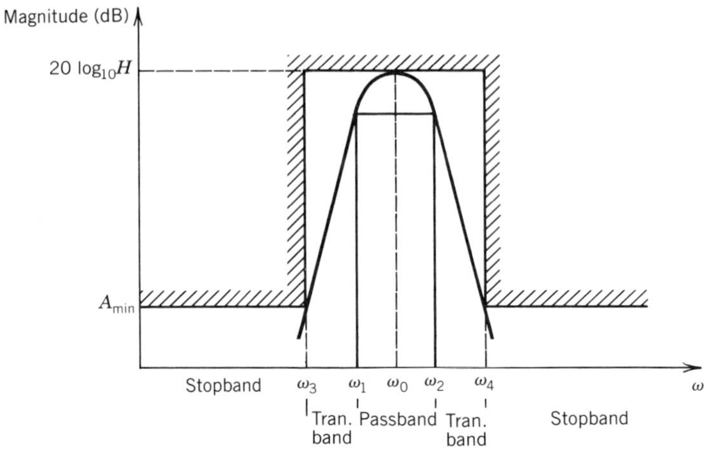

Figure 3.14 General BP frequency response.

given by

$$B = \omega_2 - \omega_1 = \frac{\omega_o}{Q_p} \tag{3.25}$$

Regardless of the value of Q_p, the magnitude curve peaks exactly at $\omega = \omega_o$. The phase at ω_o is 0 rad while at $\omega \ll 1$ the phase is $\pi/2$, and at $\omega \gg 1$ it will be $-\pi/2$. For high Q_p a rapid phase shift of π radians will occur about ω_o.

Bandstop Filters

Bandstop (BS) filters are used to attenuate a band of frequencies, while passing frequencies on either side of the reject-band. Thus the passband extends from dc to a lower cutoff frequency ω_1 and from an upper cutoff frequency ω_2 to infinity.

A typical BS filter response is shown in Fig. 3.15.

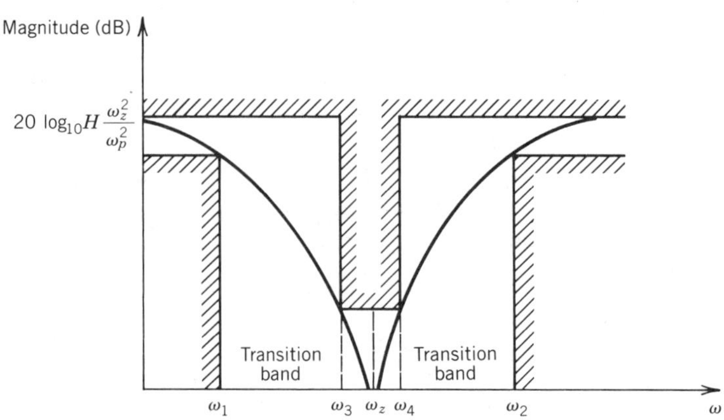

Figure 3.15 Symmetrical BS frequency response.

The second-order function of the BS gain characteristic is

$$T_{BS}(s) = \frac{H(s^2 + \omega_z^2)}{s^2 + (\omega_p/Q_p)s + \omega_p^2} \tag{3.26}$$

Here ω_z represents the frequency of zero transmission. The stopband represents the rejected frequencies ω_3 to ω_4 and is given by

$$\omega_4 - \omega_3 = \frac{\omega_z}{Q_p} \tag{3.27}$$

A symmetric gain response on a log frequency scale is obtained if $\omega_p = \omega_z$. If $H = 1$, the magnitude will be unity for both low and high frequencies. Since Q_p controls the sharpness of the notch, higher Q_ps result in narrower transition bands.

The phase of $T(j\omega)$ at higher frequencies is the negative of the phase at the lower frequencies below ω_p. For $\omega \gg 1$ the phase is practically zero. The phase jumps π radians at ω_p (phase discontinuity). For the frequencies $\omega \leq \omega_z$ the phase will be in the range 0 to $-\pi/2$ radians while, for $\omega \geq \omega_z$, it will extend from $\pi/2$ to 0 rad.

The bandwidth, defined as the band of frequencies over which the magnitude is down $1/\sqrt{2}$ (3 dB) or more, is given as

$$B = \frac{\omega_p}{Q_p} \tag{3.28}$$

For $\omega_z > \omega_p$, a low-pass notch will be obtained, whereas for $\omega_p > \omega_z$ it will be a high-pass notch response (Fig. 3.16).

Gain Equalizers

Gain equalizers are used to shape the gain or magnitude response as a function of frequency. With a gain equalizer the shape of the magnitude response is not characterized by a passband or stopband as was the case with the filter types

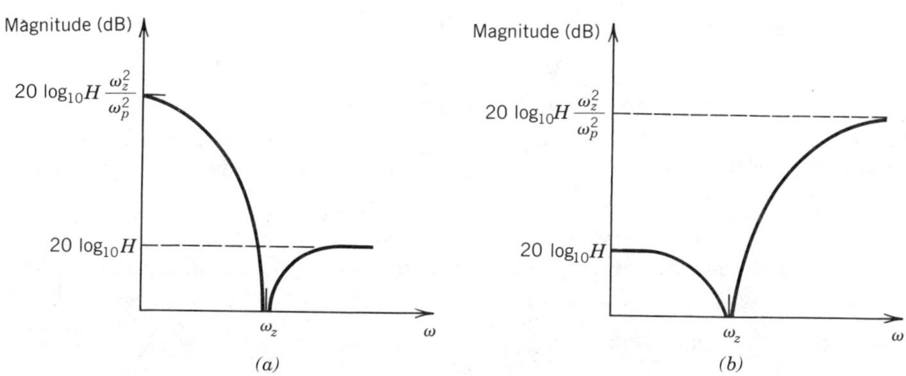

Figure 3.16 (a) Second-order low-pass notch response; (b) second-order high-pass notch response.

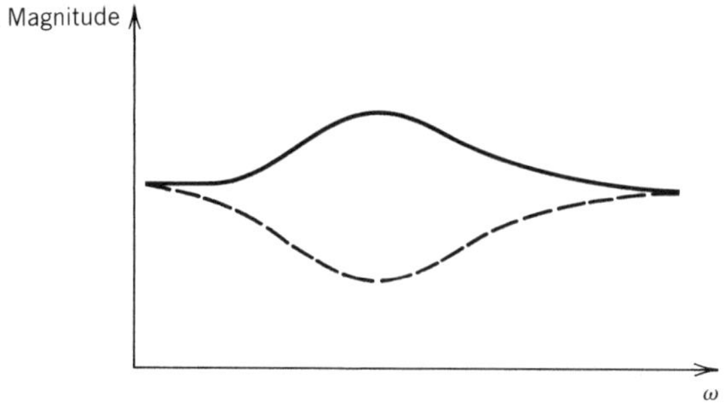

Figure 3.17 Gain-equalizer frequency response.

discussed so far. Therefore, any magnitude response that cannot be classified as one of the four standard filters (LP, HP, BP, and BS) will be considered a gain equalizer.

The gain equalizer response may take the form of a hump or dip as shown in Fig. 3.17, which is an emphasis or de-emphasis, respectively, of a band of frequencies. This emphasis or de-emphasis may also be made to occur at the high or low end of a frequency spectrum.

The general transfer function of a second-order gain equalizer is a biquadratic function

$$T_{\mathrm{GE}}(s) = \frac{H[s^2 + (\omega_z/Q_z)s + \omega_z^2]}{s^2 + (\omega_p/Q_p)s + \omega_p^2} \tag{3.29}$$

The subscripts z and p designate the zeros and poles, respectively.

EXAMPLE 3.10 Determine the gain response of the following biquadratic function:

$$T(s) = H\frac{s^2 + 12.5s + 1}{s^2 + 2.5s + 1} \qquad \text{with } H = 1$$

PROCEDURE

1 First consider the function at low and high frequencies:

$$s = j\omega = 0 \qquad \text{and} \qquad s = j\omega \to \infty$$

For both frequencies $|T(j\omega)| = 1$ or $20 \log_{10}|T(j\omega)| = 0$ dB.

2 However, for $s = j\omega_z = j\omega_p = 1$, the magnitude is $|T(j\omega)| = 12.5/2.5 = 5$ or $20 \log_{10} 5 = 13.979$ dB. The function, therefore, exhibits a hump at the characteristic frequency of $\omega_z = \omega_p = 1$.

3 The quality factor of this hump or frequency selectivity* is equal to $Q = \sqrt{Q_p^2 - 2Q_z^2} = 0.3836$. In a gain equalizer of this type $Q_p > Q_z$; therefore, the root cannot become negative.

4 The -3 dB bandwidth is given by $B = \omega_z/Q = 2.6068$ rad/s.

5 The frequency response of the gain equalizer is shown in Fig. 3.18.

* The cutoff rate or "skirt" steepness of a filter is also referred to as the filter selectivity. This selectivity is strongly related to the Q of the transfer function.

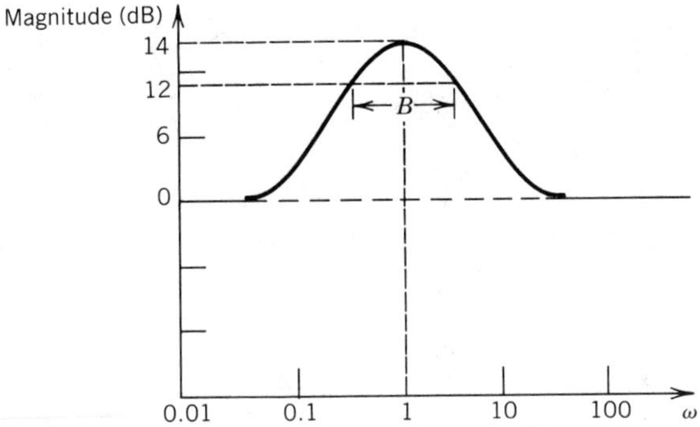

Figure 3.18 Gain-equalizer response.

EXAMPLE 3.11 Determine the gain response of the following biquadratic function:

$$T(s) = H\frac{s^2 + 0.5s + 1}{s^2 + 2.5s + 1} \qquad \text{with } H = 1$$

PROCEDURE

1 If we consider the function at the low- and high-frequency ends, $s = j\omega = 0$ and $s = j\omega \to \infty$, it is noted that in both cases $|T(j\omega)| = 1$ as in Example 3.10.

2 However, for $s = j\omega_z = j\omega_p = 1$ the magnitude is $|T(j\omega)| = 0.5/2.5 = 0.2$ or $20 \log_{10} 0.2 = -13.979$ dB. The function is seen to exhibit a dip at the characteristic frequency of $\omega_z = 1$.

3 The frequency selectivity of this gain equalizer is given by

$$Q = \frac{Q_p Q_z}{\sqrt{Q_z^2 - 2Q_p^2}} = 0.417$$

For this type of gain equalizer $Q_z > Q_p$; therefore, the expression for Q remains real.

4 The bandwidth at the -3 dB points with $\omega_z = \omega_p = 1$ is

$$B = \frac{\omega_z}{Q} = 2.398 \text{ rad/s}$$

5 A plot of the gain response is given in Fig. 3.19.

All-Pass Filters (Time-Delay Equalization Filters)

The filter characteristics of the previous sections were mainly concerned with the gain characteristic or magnitude response $|T(j\omega)|$. However, as indicated by the transfer function for steady-state frequencies $s = j\omega$,

$$T(j\omega) = |T(j\omega)|e^{j\phi(\omega)} \tag{3.30}$$

a phase response also exists. In frequency selective filters, as discussed in the previous sections, phase is important but is not an overriding characteristic. When considering audio applications, the phase behavior may justifiably be ignored because the human ear is insensitive to phase changes. In digital transmissions,

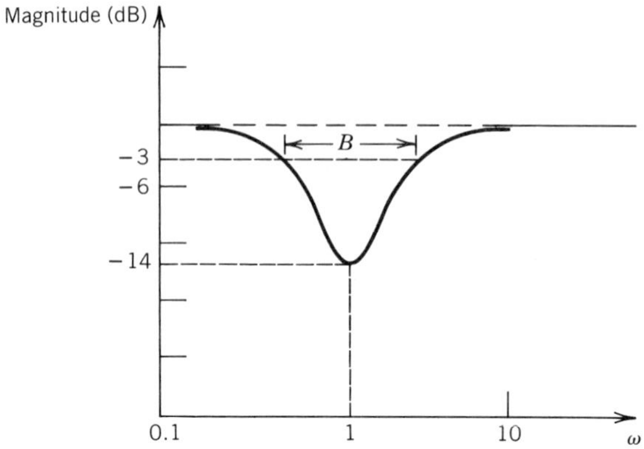

Figure 3.19 *Gain-equalizer response.*

however, distortions in the wave shape due to phase changes may become significant. With an all-pass filter the magnitude response remains constant for all frequencies while the phase response should be linear if no phase distortion is to be present. As such, an all-pass filter can be used to linearize the phase of a given system. The phase response is generally given as

$$\phi(\omega) = -\tau_o\omega \tag{3.31}$$

where τ_o is a constant representing a certain time delay by which a pulse may be retarded between the input and output terminals of an all-pass filter.

The time delay in the case of linear phase is defined as the negative of the slope of the phase response:

$$\tau(\omega) = -\frac{d\phi(\omega)}{d\omega} = \tau_o \tag{3.32}$$

According to Eq. 3.31 the delay must be flat for all frequencies; that is, the phase must be linear to achieve distortionless digital transmission. If gain-shaping (frequency-selective) filters are used in such transmissions, the delay characteristic will not be flat; therefore, a need for correction exists. Such delay correction is referred to as delay equalization, and all-pass filters may be used to achieve this. Thus the purpose of an all-pass filter (or delay equalizer) is to introduce the necessary delay shape to make the total delay as flat as possible. Although delay equalization linearizes the delay characteristic, it nevertheless increases the total delay of the system.

A second-order delay equalizer (all-pass filter) characteristic is described by the function

$$T_{AP}(s) = H\frac{s^2 - (\omega_p/Q_p)s + \omega_p^2}{s^2 + (\omega_p/Q_p)s + \omega_p^2} \tag{3.33}$$

The maximum gain constant will be $H_{\max} = 1$. The complex poles and zeros are symmetrical about the $j\omega$-axis. The gain is

$$20 \log_{10} \left. \frac{s^2 - (\omega_p/Q_p)s + \omega_p^2}{s^2 + (\omega_p/Q_p)s + \omega_p^2} \right|_{s=j\omega} = 10 \log_{10} \left[(\omega_p^2 - \omega^2)^2 + \left(\frac{\omega_p}{Q_p} \omega \right)^2 \right] - 10 \log_{10} \left[(\omega_p^2 - \omega^2)^2 + \left(\frac{\omega_p}{Q_p} \omega \right)^2 \right] = 0 \text{ dB} \tag{3.34}$$

that is, equal transmission for all frequencies and hence its designation as an all-pass function. The delay of the function depends largely on the coefficients ω_p/Q_p and ω_p^2 and is given by:

$$\text{delay} = \frac{-d\phi(\omega)}{d\omega} = 2 \frac{(\omega_p/Q_p)(\omega_p^2 + \omega^2)}{(\omega_p^2 - \omega^2)^2 + (\omega_p^2/Q_p^2)\omega^2} \tag{3.35}$$

EXAMPLE 3.12 Determine the magnitude and phase response of a second-order all-pass filter.

PROCEDURE

1 The second-order all-pass function is of the form

$$T(s) = H \frac{s^2 - (\omega_p/Q_p)s + \omega_p^2}{s^2 + (\omega_p/Q_p)s + \omega_p^2} \qquad \text{where } H_{\max} = 1$$

2 Setting $s = j\omega$, we obtain the response to real frequencies as

$$T(j\omega) = \frac{(\omega_p^2 - \omega^2) - j\omega(\omega_p/Q_p)}{(\omega_p^2 - \omega^2) + j\omega(\omega_p/Q_p)}$$

3 The magnitude is given by

$$|T(j\omega)| = \frac{\sqrt{(\omega_p^2 - \omega^2)^2 + \omega^2(\omega_p^2/Q_p^2)}}{\sqrt{(\omega_p^2 - \omega^2)^2 + \omega^2(\omega_p^2/Q_p^2)}} = 1$$

4 The phase is given by

$$\phi(\omega) = -2 \tan^{-1} \left| \frac{\omega/\omega_p Q_p}{1 - (\omega/\omega_p)^2} \right| = -2 \tan^{-1} \left| \frac{\Omega/Q_p}{1 - \Omega^2} \right|$$

The ratio $\omega/\omega_p = \Omega$ is also known as the normalized frequency. If we plot the phase versus the normalized frequency with Q_p as the parameter, the all-pass phase responses of Fig 3.20 are obtained.

5 To determine the point of inflection, the second derivative must be set equal to zero, which gives

$$\Omega = \sqrt{\sqrt{4 - (1/Q_p^2)} - 1}$$

It can be seen that for $Q_p \gg 1$ the point of inflection occurs at $\Omega = 1$. Note that for $Q_p < \frac{1}{2}$ the curve bends in the same way for all values of Ω, whereas for high Q_p values the bend is first clockwise and then anticlockwise, which means that the second derivative of the phase changes sign.

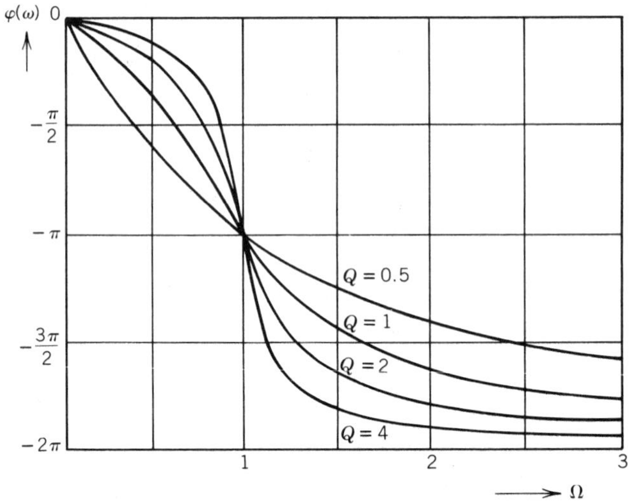

Figure 3.20 Second-order all-pass function phase response.

For $Q_p \gg 1$ the point of inflection practically occurs at $\Omega = 1$. It should be noted that all-pass functions provide a phase lag.

The essential features of the various transfer function types discussed are summarized in Table 3.1.

3.5 SIGNIFICANCE OF TRANSFER FUNCTION COEFFICIENTS

Often some confusion exists regarding the definition of the quality factor or frequency selectivity Q. In the text Q has been considered as being the overall Q-factor of a specific network function, whereas Q_z and Q_p in the function characterize the location of poles or zeros with respect to the real and imaginary axis of the s-plane.

The quality factors for the zeros or the poles of a transfer function are defined via the coefficients of the function. If a general biquadratic function is given as

$$T(s) = H \frac{s^2 + a_1 s + a_0}{s^2 + b_1 s + b_0} \tag{3.36}$$

then Q_z and Q_p are defined as

$$Q_z = \frac{\omega_z}{a_1} \tag{3.37}$$

$$Q_p = \frac{\omega_p}{b_1} \tag{3.38}$$

where $\omega_z = \sqrt{a_0}$ and $\omega_p = \sqrt{b_0}$ are the undamped natural frequencies of the numerator and denominator polynomials, respectively. The significance of H has already been discussed in Section 3.3. It suffices to say that for passive RC synthesis a maximum value of H can be predetermined, whereas for active RC realizations H may be set to any specific gain level. The frequency selectivity or quality factor Q of the complete transfer function is a practical and also easily measurable term, defined as

$$Q = \frac{\omega_o}{\omega_2 - \omega_1} \tag{3.39}$$

where ω_2 and ω_1 are the upper and lower cutoff frequencies at which the output signal level is $1/\sqrt{2}$ or 3 dB lower than the maximum transmission at ω_o (see Fig. 3.14).

The factors Q_z and Q_p are quite general and apply to any second-order network while the frequency selectivity Q, because of its practical definition, is only sensibly applicable to bandpass, bandstop, and gain-equalizer networks. However, high-Q low-pass and high-pass responses also fit the general Q definition, since their peak resembles that of a bandpass.

3.6 SYNTHESIS OF RC BUILDING BLOCKS

The problem of determining the response of a given network to an excitation function is solved through analysis of the network. There is always a unique solution to this type of problem. However, the case where both the excitation and response functions are given and the network has to be found is termed *synthesis*. Here, if a solution exists, it will not be unique. In fact, a vast number of possible solutions may exist from which a specific choice must be made. The number of possible solutions may be reduced by the stipulation of certain optimum criteria. Examples of such criteria are sensitivity, component spread, number of elements, and so on.

The class of the network may be another consideration. Since it has already been predetermined to use RC networks, the realizability requirements for such passive circuits must be stated briefly. Network functions of all passive circuits and all stable active circuits must:

1 Be rational functions in s with real coefficients.

2 Not have poles in the right half of the s-plane.

3 Not have multiple poles on the $j\omega$-axis.

The realization of a network proceeds by deriving a driving point impedance or admittance that will have a frequency response as described by its driving point function. As already discussed in Section 3.2, the driving point impedance is described by the ratio of voltage and current at a particular port or two-terminal pair of a network. The reciprocal of a driving point impedance is a driving point admittance. An RC building block may generally be represented by a two-port as shown in Fig. 3.21. For this arrangement the *open*-circuit impedance parameter

Table 3.1 **Characteristics of general second-order network functions**

Classification	*Transfer function*		*Frequency of peak gain* or trough loss†*
Low-pass	$T(s) = \dfrac{H}{s^2 + (\omega_0/Q_p)s + \omega_0^2}$		$\omega^* = \omega_0\sqrt{1 - (1/2Q_p^2)}$ for $Q_p > \dfrac{1}{\sqrt{2}}$
High-pass	$T(s) = \dfrac{Hs^2}{s^2 + (\omega_0/Q_p)s + \omega_0^2}$		$\omega^* = \omega_0\dfrac{1}{\sqrt{1 - (1/2Q_p^2)}}$ for $Q_p > \dfrac{1}{\sqrt{2}}$
Bandpass	$T(s) = \dfrac{Hs}{s^2 + (\omega_0/Q_p)s + \omega_0^2}$		$\omega^* = \omega_0$
Band-reject (symmetrical)	$T(s) = \dfrac{H(s^2 + \omega_z^2)}{s^2 + (\omega_p/Q_p)s + \omega_p^2}$	$\omega_z = \omega_p$	$\omega^\dagger = \omega_p$
Band-reject (LP)	$T(s) = \dfrac{H(s^2 + \omega_z^2)}{s^2 + (\omega_p/Q_p)s + \omega_p^2}$	$\omega_z^2 > \omega_p^2$	$\omega^\dagger = \omega_z$
Band-reject (HP)	$T(s) = \dfrac{H(s^2 + \omega_z^2)}{s^2 + (\omega_p/Q_p)s + \omega_p^2}$	$\omega_z^2 < \omega_p^2$	$\omega^\dagger = \omega_z$
Gain equalizer $\dfrac{\omega_z}{Q_z} > \dfrac{\omega_p}{Q_p}$	$T(s) = \dfrac{H[s^2 + (\omega_z/Q_z)s + \omega_z^2]}{s^2 + (\omega_p/Q_p)s + \omega_p^2}$	$\omega_z = \omega_p$	—
Loss equalizer $\dfrac{\omega_p}{Q_p} > \dfrac{\omega_z}{Q_z}$	$T(s) = \dfrac{H[s^2 + (\omega_z/Q_z)s + \omega_z^2]}{s^2 + (\omega_p/Q_p)s + \omega_p^2}$	$\omega_z = \omega_p$	—
All-pass	$T(s) = \dfrac{H[s^2 - (\omega_p/Q_p)s + \omega_p^2]}{s^2 + (\omega_p/Q_p)s + \omega_p^2}$		—

H_{max} is defined as that value of the gain constant H which will allow for the realized passive RC network to have maximum realizable gain.

Maximum Gain Constant H_{max}	*Peak gain (max) or trough loss (min)* *High/low frequency passband gain**	*Phase Response*
$H_{max} = \omega_0^2$	$\lvert T(j\omega)\rvert_{max} = \dfrac{HQ_p}{\omega_0^2\sqrt{1-(1/4Q_p^2)}}$ for $Q_p > \dfrac{1}{\sqrt{2}}$ $\lvert T(j\omega)\rvert^* = \dfrac{H}{\omega_0^2}$	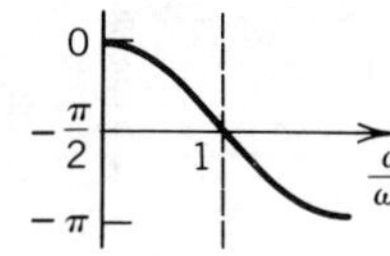
$H_{max} = 1$	$\lvert T(j\omega)\rvert_{max} = \dfrac{HQ_p}{\sqrt{1-(1/4Q_p^2)}}$ for $Q_p > \dfrac{1}{\sqrt{2}}$ $\lvert T(j\omega)\rvert^* = H$	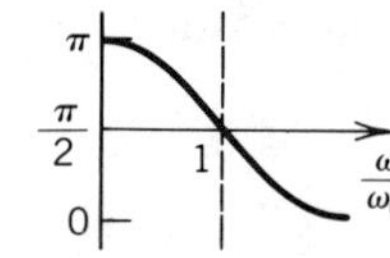
$H_{max} = \dfrac{\omega_0}{Q_p}$	$\lvert T(j\omega)\rvert_{max} = \dfrac{HQ_p}{\omega_0}$	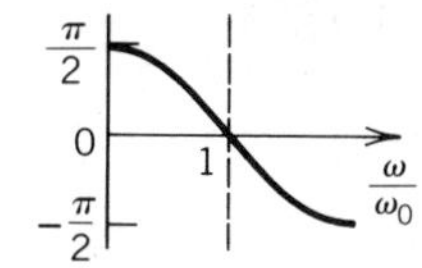
$H_{max} = 1$	$\lvert T(j\omega)\rvert^* = H$	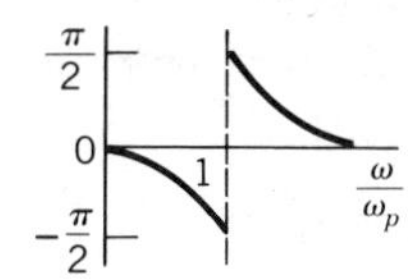
$H_{max} = \dfrac{\omega_p^2}{\omega_z^2}$	$\lvert T(j\omega)\rvert^* = \dfrac{H\omega_z^2}{\omega_p^2}$	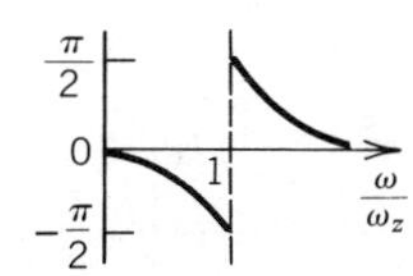
$H_{max} = 1$	$\lvert T(j\omega)\rvert^* = H$	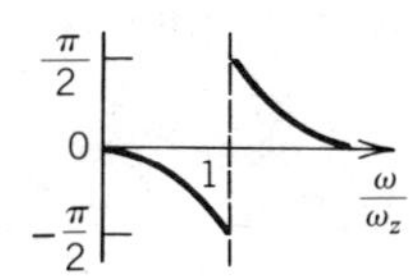
$H_{max} = \dfrac{\omega_p Q_z}{\omega_z Q_p}$	$\lvert T(j\omega)\rvert_{max} = \dfrac{H\omega_z Q_p}{\omega_p Q_z}$ $\lvert T(j\omega)\rvert^* = H$	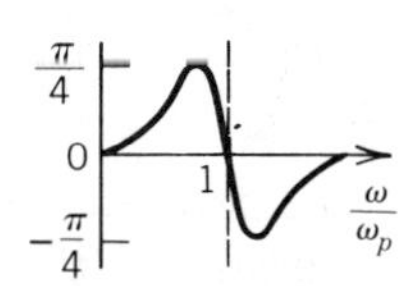
$H_{max} = 1$	$\lvert T(j\omega)\rvert_{min} = \dfrac{H\omega_2 Q_p}{\omega_p Q_z}$ $\lvert T(j\omega)\rvert^* = H$	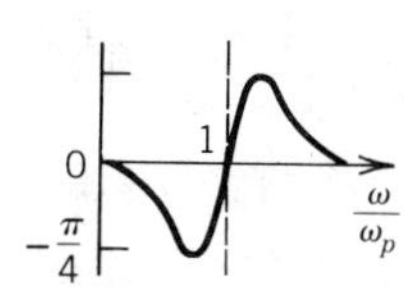
$H_{max} = 1$	$\lvert T(j\omega)\rvert^* = H$	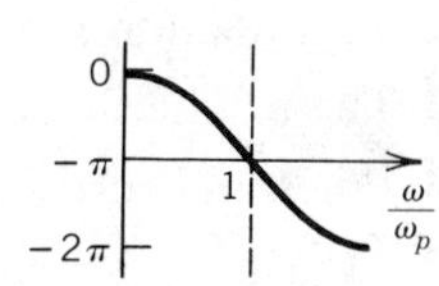

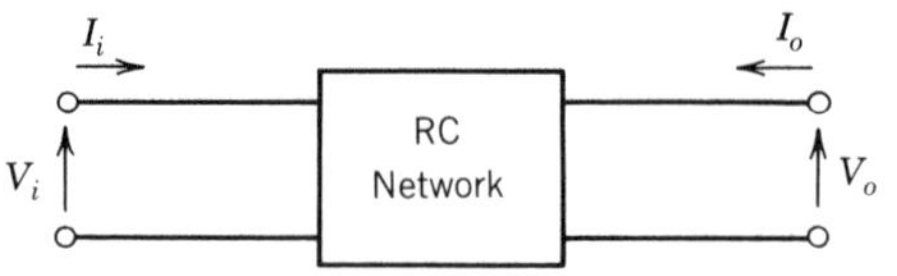

Figure 3.21 Two-port with defined current and voltage directions.

equations are

$$V_i = z_{11} I_i + z_{12} I_o \tag{3.40}$$

$$V_o = z_{21} I_i + z_{22} I_o \tag{3.41}$$

with

$$z_{11} = \left.\frac{V_i}{I_i}\right|_{I_o=0} \qquad z_{12} = \left.\frac{V_i}{I_o}\right|_{I_i=0} \qquad z_{21} = \left.\frac{V_o}{I_i}\right|_{I_o=0} \qquad \text{and} \qquad z_{22} = \left.\frac{V_o}{I_o}\right|_{I_i=0}$$

Note that both z_{11} and z_{22} represent driving point impedances. For the same circuit arrangement the *short*-circuit admittance parameter equations are

$$I_i = y_{11} V_i + y_{12} V_o \tag{3.42}$$

$$I_o = y_{21} V_i + y_{22} V_o \tag{3.43}$$

with

$$y_{11} = \left.\frac{I_i}{V_i}\right|_{V_o=0} \qquad y_{12} = \left.\frac{I_i}{V_o}\right|_{V_i=0} \qquad y_{21} = \left.\frac{I_o}{V_i}\right|_{V_o=0} \qquad \text{and} \qquad y_{22} = \left.\frac{I_o}{V_o}\right|_{V_i=0}$$

Here y_{11} and y_{22} are driving point admittances.

By using the two-port parameter conversion table (Table 5.1) it is possible to consider parameter interrelationships. Thus the open-circuit voltage transfer function can be described as

$$T(s) = \frac{V_o}{V_i} = \frac{z_{21}}{z_{11}} = \frac{-y_{21}}{y_{22}} \tag{3.44}$$

Hence both a transfer impedance (or admittance) and a driving point impedance (or admittance) is associated with each transfer function.

For passive RC driving point functions, additional realizability properties apply.

1 Numerator and denominator polynomial degrees must not differ by more than one.
2 The function must have a nonnegative real part for all $s = j\omega$.
3 The residues of $Z_{RC}(s)[Y_{RC}(s)]$ are real and positive (negative).
4 Poles and zeros must lie on the negative real axis and alternate.
5 At $\omega = 0$, the driving point impedance (admittance) function is either a positive constant or has a pole (zero).

Table 3.2. **Some simple functions that may be used for synthesis by inspection.**

Impedance form	*Admittance form*	*Circuit*
R	$\frac{1}{R}$	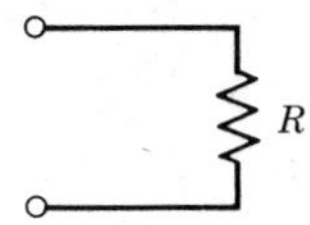
$\frac{1}{sC}$	sC	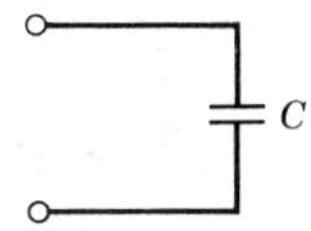
$R + \frac{1}{sC} = \frac{s + (1/CR)}{(1/R)s}$	$\frac{sC/R}{(1/R) + sC} = \frac{(1/R)s}{s + (1/CR)}$	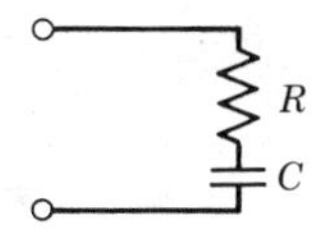
$\frac{1}{sC + (1/R)} = \frac{1/C}{s + (1/CR)}$	$\frac{1}{R} + sC = \frac{s + (1/CR)}{1/C}$	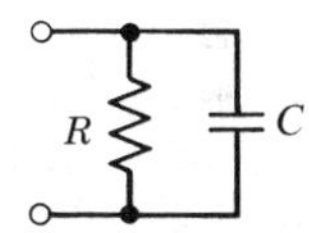

6 At $\omega = \infty$ driving point impedance (admittance) function is either a positive constant or has a zero (pole).

7 The slope of $[dZ_{\text{RC}}(\sigma)/d\sigma][dY_{\text{RC}}(\sigma)/d\sigma]$ is negative (positive).

Partial Fractions

Every rational polynomial function may be uniquely subdivided into a summation of partial fractions. These partial fractions are of relatively simple form, and each of them can be realized almost by inspection to represent simple networks as shown in Table 3.2.

Consider the following driving point impedance function:

$$Z(s) = \frac{N(s)}{D(s)} = \frac{s^2 + a_1 s + a_0}{s^2 + b_1 s + b_0} = \frac{s^2 + 4s + 3}{s^2 + 2.5s + 1} \tag{3.45}$$

where $N(s)$ is the numerator polynomial and $D(s)$ is the denominator polynomial. The roots of $N(s)$ of Eq. 3.45 are

$$z_1 = -\frac{a_1}{2} + \sqrt{\left(\frac{a_1}{2}\right)^2 - a_0} = -1 \quad \text{and} \quad z_2 = -\frac{a_1}{2} - \sqrt{\left(\frac{a_1}{2}\right)^2 - a_0} = -3$$

while the poles are at

$$p_1 = -\frac{b_1}{2} + \sqrt{\left(\frac{b_1}{2}\right)^2 - b_0} = -0.5 \quad \text{and} \quad p_2 = -\frac{b_1}{2} - \sqrt{\left(\frac{b_1}{2}\right)^2 - b_0} = -2$$

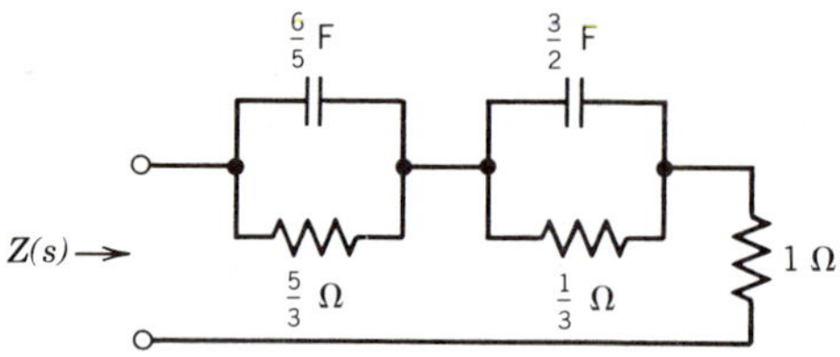

Figure 3.22 RC network realized by partial fraction expansion.

Rewriting Eq. 3.45 in factored form yields

$$Z(s) = \frac{(s - z_1)(s - z_2)}{(s - p_1)(s - p_2)} = \frac{(s + 1)(s + 3)}{(s + 0.5)(s + 2)} \tag{3.46}$$

Next the partial fraction expansion can be applied. As all poles are simple, the partial fractions will be of the form:

$$Z(s) = \frac{A}{s + 0.5} + \frac{B}{s + 2} + C \tag{3.47}$$

Partial fractions are the terms to the right of the equation. The coefficients A, B, and C are obtained by equating them to $N(s)$ so that

$$As + 2A + Bs + 0.5B + Cs^2 + 2.5Cs + C = s^2 + 4s + 3 \tag{3.48}$$

Arranging corresponding coefficients yields

1 $Cs^2 = s^2$ or $C = 1$

2 $As + Bs + 2.5Cs = 4s \rightarrow A = 1.5 - B$

3 $2A + 0.5B + C = 3$

so that all unknowns can be evaluated. With C known, B and A can be determined. From (3) and (1) and (2)

$$3 - 2B + 0.5B = 2 \rightarrow B = \tfrac{2}{3}$$

Thus

$$A = \tfrac{5}{6}$$

The partial fraction expansion of $Z(s)$ of Eq. 3.45 therefore yields

$$Z(s) = \frac{\frac{5}{6}}{s + 0.5} + \frac{\frac{2}{3}}{s + 2} + 1 \tag{3.49}$$

With the aid of Table 3.2 an RC network can be realized, resulting in the circuit of Fig. 3.22.

EXAMPLE 3.13 Use the driving point impedance given to synthesize an RC driving point impedance and admittance if

$$Z(s) = \frac{s + 2}{(s + 1)(s + 3)}$$

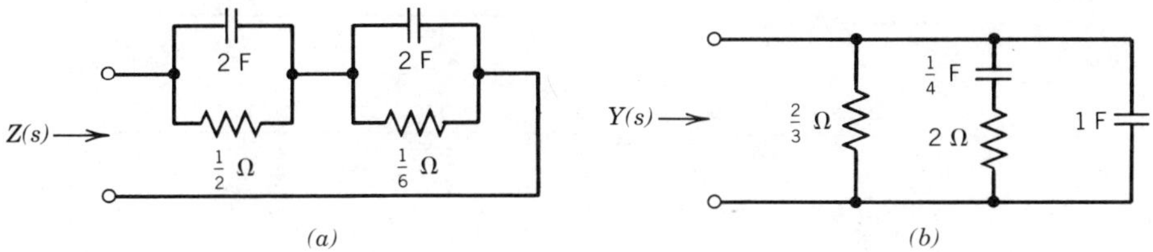

Figure 3.23 (a) Driving point impedance realization; (b) driving point admittance realization.

PROCEDURE

1 The poles and zeros alternate and lie on the negative real axis. An RC network can therefore be realized.

2 The impedance function yields the following partial fractions:

$$Z(s) = \frac{\frac{1}{2}}{s+1} + \frac{\frac{1}{2}}{s+3}$$

3 Using Table 3.2, we can find the circuits described by the two partial fraction terms. The complete circuit is shown in Fig. 3.23*a*.

4 The admittance function in partial fraction form is found to be

$$Y(s) = \frac{1}{Z(s)} = \frac{3}{2} + \frac{\frac{1}{2}s}{s+2} + s$$

Expanding $Y(s)/s$ into partial fractions yields the same type of terms as for $Z(s)$. This means that the expansion of $Y(s)$ has terms multiplied by s.

5 Again Table 3.2 is used to identify the circuit components, as shown in Fig. 3.23*b*. The circuits are identical in performance. Looking into the terminals of both circuits, we see $Z(s)$ as well as $Y(s)$.

RC Ladder Networks

It was shown in Eq. 3.44 that transfer functions are expressed in terms of impedance and admittance parameters. Whereas the numerator of a transfer function is generally associated with the transmission properties of the function, the denominator is seen to be associated with the driving point impedances or admittances. It is the latter that are used in the synthesis of the network; however, due regard must be given to the transmission requirements. This means that in the case of a low-pass function, the driving point impedance/admittance must realize a network exhibiting this low-pass characteristic. Provided that all realizability requirements for the synthesis of an RC network are fulfilled, the driving point must be seen to realize the transmission zeros stipulated in the original transfer function. Thus, for example, an all-pole low-pass function (as in Eq. 3.16) will have two transmission zeros at infinity:

$$T(s) = \frac{1}{s^2} = 0 \text{ for } s \to \infty$$

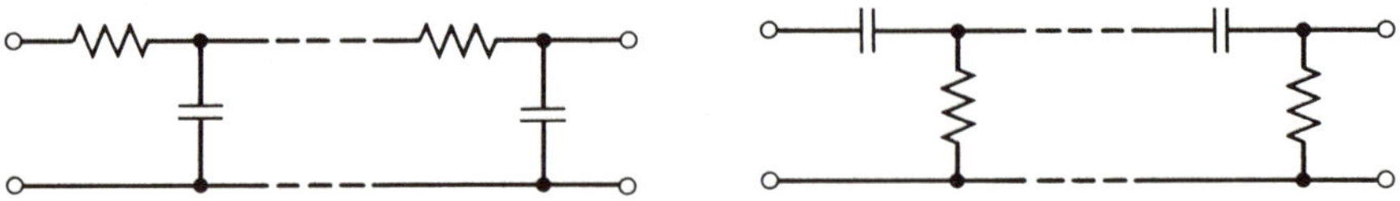

(a) (b)

Figure 3.24 (a) Realization of zeros of transmission at infinity; (b) realization of zeros of transmission at the origin.

In contrast, a bandpass function (as in Eq. 3.24) will have one transmission zero at the origin and one at infinity

$$\underset{s\to 0}{T(s)} = 0 \qquad \text{and} \qquad \underset{s\to\infty}{T(s)} = \frac{1}{s} = 0$$

The driving point function must realize these zeros of transmission.

With a ladder development, where the components are obtained successively in series and shunt arms of the network, the realization of zeros at infinity involves series and shunt components, as shown in Fig. 3.24*a*, whereas for transmission zeros at the origin the series and shunt components are as in Fig. 3.24*b*.

The ladder development due to Cauer [4] is also called the continued fraction expansion. In its implementation it makes use of the continued division process of a divide, invert, divide procedure. Polynomials arranged in descending order (from left to right) will allow for the realization of transmission zeros at infinity while an arrangement in ascending order realizes transmission zeros at the origin.

EXAMPLE 3.14 Synthesize a ladder network realizing series resistors and shunt capacitors.

$$Z(s) = \frac{(s+1)(s+3)}{s(s+2)} = \frac{s^2+4s+3}{s^2+2s}$$

PROCEDURE

1 Arrange polynomials in descending order and carry out a continued division process.

s^2+2s	s^2+4s+3 s^2+2s	1	Z division
$2s+3$	s^2+2s $s^2+\frac{3}{2}s$	$\frac{1}{2}s$	Y division
$\frac{1}{2}s$	$2s+3$ $2s$	4	Z division
3	$\frac{1}{2}s$ $\frac{1}{2}s$	$\frac{1}{6}s$	Y division
		0 = open circuit	

2 The first step is an impedance division resulting in a series resistance of 1 Ω.

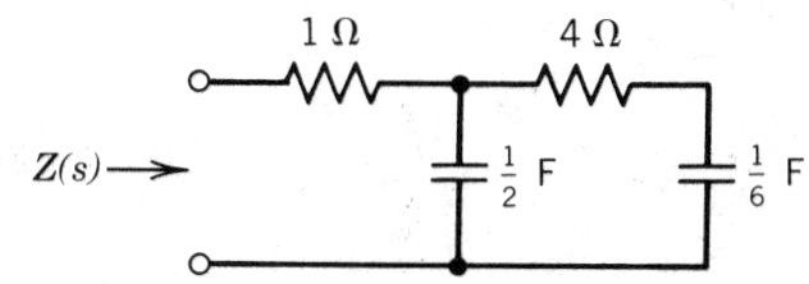

Figure 3.25 Network realization of transmission zeros at infinity.

3 The continued division process then inverts the function so that the next step is an admittance division, which results in a capacitance of $\frac{1}{2}$ F, and so on.

4 The last step is an admittance division, whereupon the ladder development is complete.

5 Ending with zero after the admittance division step means that the last component is followed by an open-circuit condition. Had the final division been that of an impedance, ending in zero, the component relating to this impedance would have been followed by a short-circuit condition.

6 The network realized by the above division process is shown in Fig. 3.25.

EXAMPLE 3.15 Repeat the previous example by expanding about the origin to give series capacitors and shunt resistors. The impedance function is again

$$Z(s) = \frac{(s+1)(s+3)}{s(s+2)}$$

PROCEDURE

1 The stipulated requirement calls for the realization of zeros at the origin. The polynomials must be arranged in ascending order; then the continued division process can proceed.

$2s + s^2 \;)\; 3 + 4s + s^2 \;(\; \frac{3}{2s}$ Z division

$\quad 3 + \frac{3}{2}s$

$\frac{5}{2}s + s^2 \;)\; 2s + s^2 \;(\; \frac{4}{5}$ Y division

$\quad 2s + \frac{4}{5}s^2$

$\frac{1}{5}s^2 \;)\; \frac{5}{2}s + s^2 \;(\; \frac{25}{2s}$ Z division

$\quad \frac{5}{2}s$

$s^2 \;)\; \frac{1}{5}s^2 \;(\; \frac{1}{5}$ Y division

$\quad \frac{1}{5}s^2$

0 = open circuit

2 Here the first step is an impedance division realizing a series capacitance, that is, $1/sC = 3/2s$ or $C = \frac{2}{3}$F.

3 The other steps follow with the second step being an admittance division, etc.

4 The final network is shown in Fig. 3.26.

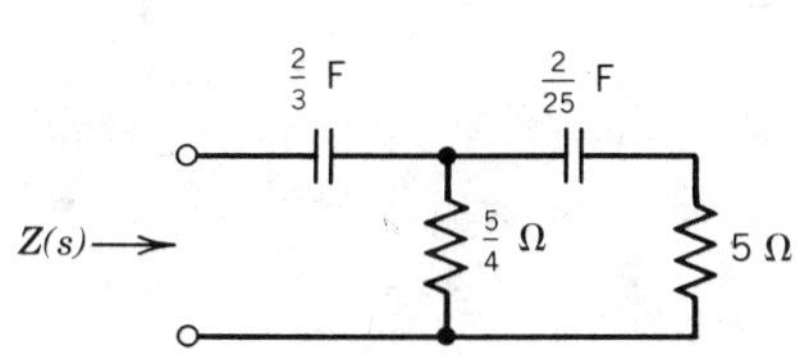

Figure 3.26 Network realization of transmission zeros at the origin.

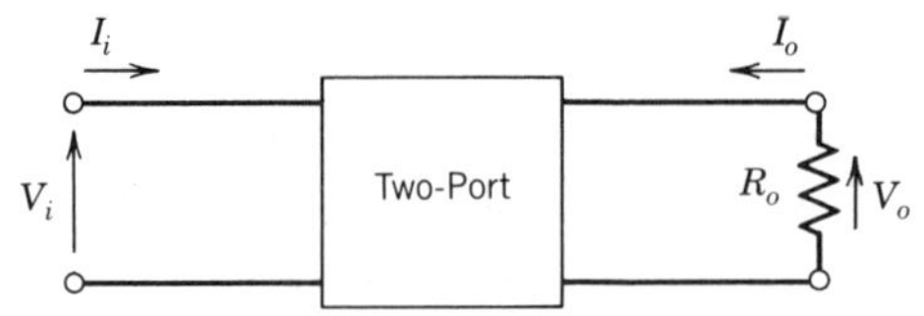

Figure 3.27 Two-port or four-terminal network representation.

Transfer Function Synthesis

As stated earlier, the transfer function of a realizable filter is a ratio of polynomials. Transfer functions relate information received at one port to information obtained at the other port. It is therefore obvious that various transfer functions will exist. For the purpose of arriving at a listing of these transfer functions, consider Fig. 3.27. Transfer functions of the response/excitation type are

$$T(s) = \frac{V_o}{V_i} = \frac{-y_{21}}{y_{22} + (1/R_o)} \tag{3.50}$$

$$A(s) = \frac{I_o}{I_i} = \frac{-z_{21}}{z_{22} + R_o} \tag{3.51}$$

$$Z(s) = \frac{V_o}{I_i} = \frac{R_o z_{21}}{z_{22} + R_o} \tag{3.52}$$

$$Y(s) = \frac{I_o}{V_i} = \frac{y_{21}}{y_{22}R_o + 1} \tag{3.53}$$

also the open-circuit voltage transfer function

$$T(s) = \left(\frac{V_o}{V_i}\right)_{R_o=\infty} = \frac{-y_{21}}{y_{22}} \tag{3.54}$$

the short-circuit current transfer function

$$A(s) = \left(\frac{I_o}{I_i}\right)_{R_o=0} = \frac{-z_{21}}{z_{22}} \tag{3.55}$$

the open-circuit transfer impedance

$$Z(s) = \left(\frac{V_o}{I_i}\right)_{R_o=\infty} = z_{21} \tag{3.56}$$

and the short-circuit transfer admittance

$$Y(s) = \left(\frac{I_o}{V_i}\right)_{R_o=0} = y_{21} \tag{3.57}$$

The special cases $R_o = \infty$ and $R_o = 0$ represent idealized situations. In the one case a very large load (e.g., a very high input impedance into an amplifier) is assumed; in the other case, a small input impedance (as for many transistor circuits) is envisaged.

Low-Pass Realizations

EXAMPLE 3.16 Synthesize the following low-pass transfer function as an open-circuit voltage transfer function:

$$T(s) = \frac{H}{s^2 + 4s + 3}$$

PROCEDURE

1 According to Eq. 3.12 (Section 3.3) $H_{max} = 3$. We now use Eq. 3.54:

$$T(s) = \left(\frac{V_o}{V_i}\right)_{R_o=\infty} = \frac{-y_{21}}{y_{22}}$$

2 In order for the maximum gain constant to be effectively realized by the network, it is necessary for H_{max} to be multiplied into the denominator polynomial of the above transfer function.

$$-y_{21} = \frac{1}{F(s)} \qquad y_{22} = \frac{\frac{1}{3}s^2 + \frac{4}{3}s + 1}{F(s)}$$

3 The admittances must be represented by a rational function (quotient of two polynomials); therefore, the introduction of an auxiliary polynomial $F(s)$ is necessary. The only restriction on $F(s)$ is that it must be chosen so that y_{22} is a realizable RC admittance, but otherwise its selection is arbitrary.

4 From the realizability conditions of Section 3.6 it will be remembered that, for driving point functions, the degrees of numerator and denominator polynomials may at most differ by order one. Also the poles and zeros must alternate so that a possible choice is $F(s) = s + 2$.

5 The admittance identifications are therefore

$$-y_{21} = \frac{1}{s+2} \qquad y_{22} = \frac{\frac{1}{3}(s+1)(s+3)}{s+2}$$

6 Because the voltage transfer function is low-pass in form, that is, two zeros of transmission at infinity, the polynomials of the driving point admittance must be arranged in descending order for the continued division process. Thus

$s+2$	$\frac{1}{3}s^2+\frac{4}{3}s+1$	$\frac{s}{3}$	*Y* division			
	$\frac{1}{3}s^2+\frac{2}{3}s$					
	$\frac{2}{3}s+1$	$s+2$	$\frac{3}{2}$	*Z* division		
		$s+\frac{3}{2}$				
		$\frac{1}{2}$	$\frac{2}{3}s+1$	$\frac{4}{3}s$	*Y* division	
			$\frac{2}{3}s$			
			1	$\frac{1}{2}$	$\frac{1}{2}$	*Z* division
				$\frac{1}{2}$		

0 = short circuit
or voltage source

7 Synthesis of the above low-pass voltage transfer function yields the network shown in Fig. 3.28.

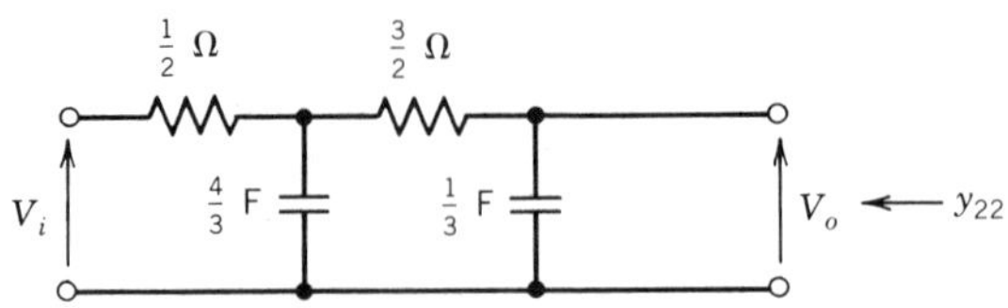

Figure 3.28 Second-order LP filter.

High-Pass Synthesis

EXAMPLE 3.17 Synthesize the following high-pass open-circuit voltage transfer function:

$$T(s) = \frac{Hs^2}{s^2 + 3s + 1}$$

PROCEDURE

1 According to Eq. 3.14, $H_{\max} = 1$ (see also Table 3.1).

2 Thus the transfer function can be written as

$$T(s) = \left(\frac{V_o}{V_i}\right)_{R_o=\infty} = \frac{-y_{21}}{y_{22}}$$

$$-y_{21} = \frac{s^2}{F(s)} \qquad y_{22} = \frac{s^2 + 3s + 1}{F(s)}$$

3 As was the case with the low-pass realization in Example 3.16, an auxiliary polynomial $F(s)$ is required. A possible choice is $F(s) = s + 1$ so that

$$-y_{21} = \frac{s^2}{s + 1} \qquad y_{22} = \frac{(s + 0.382)(s + 2.618)}{s + 1}$$

4 Synthesis of a high-pass function requires the realization of two transmission zeros at the origin. According to the discussion under "RC Ladder Networks," this requires that the polynomials of the driving point admittance y_{22} be arranged in ascending order.

5 The continued division process may then proceed.

$1 + s$	$1 + 3s + s^2$	1	Y division			
	$1 + s$					
	$2s + s^2$	$1 + s$	$\frac{1}{2s}$	Z division		
		$1 + \frac{1}{2}s$				
		$\frac{1}{2}s$	$2s + s^2$	4	Y division	
			$2s$			
			s^2	$\frac{1}{2}s$	$\frac{1}{2s}$	Z division
				$\frac{1}{2}s$		
				$0 =$ short circuit or voltage source		

6 The final high-pass filter is shown in Fig. 3.29.

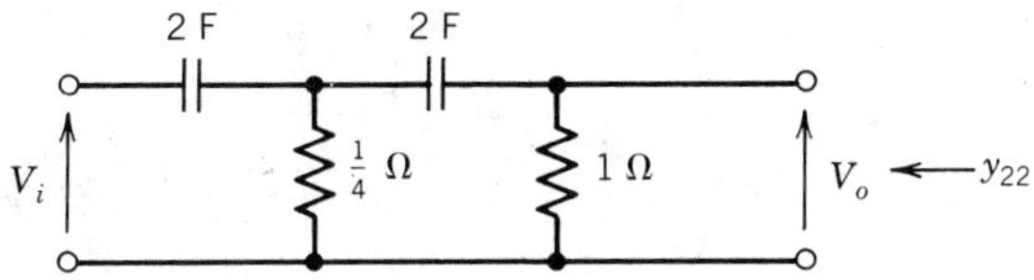

Figure 3.29 Second-order HP filter.

Bandpass Realizations

EXAMPLE 3.18 Obtain an RC realization of the following bandpass voltage transfer function:

$$T(s) = \frac{Hs}{s^2 + 4s + 2}$$

PROCEDURE

1 From Table 3.1 the maximum gain constant is found to be

$$H_{\max} = 4$$

2 In order for the maximum gain constant to be effectively realized by the network, it is necessary for $H_{\max}$ to be multiplied into the denominator polynomial of the above transfer function.

3 Following the procedure adopted in previous examples, we obtain

$$T(s) = \left(\frac{V_o}{V_i}\right)_{R_o=\infty} = \frac{-y_{21}}{y_{22}}$$

$$-y_{21} = \frac{s}{F(s)} \qquad y_{22} = \frac{\frac{1}{4}s^2 + s + \frac{1}{2}}{F(s)} = \frac{\frac{1}{4}(s + 0.586)(s + 3.414)}{F(s)}$$

4 A suitable auxiliary polynomial is $F(s) = s + 1$. Thus

$$-y_{21} = \frac{s}{s+1} \qquad y_{22} = \frac{\frac{1}{4}s^2 + s + \frac{1}{2}}{s+1}$$

5 In the case of this bandpass function, one zero of transmission at infinity and at the origin must be realized.

6 Let us consider removing the pole at infinity first.

Divisor	Dividend / subtracted	Quotient	
$s + 1$	$\frac{1}{4}s^2 + s + \frac{1}{2}$ $\frac{1}{4}s^2 + \frac{1}{4}s$	$\frac{1}{4}s$	Y division
$\frac{3}{4}s + \frac{1}{2}$	$s + 1$ $s + \frac{2}{3}$	$\frac{4}{3}$	Z division
$\frac{1}{3}$	$\frac{1}{2} + \frac{3}{4}s$ $\frac{1}{2}$	$\frac{3}{2}$	Y division
$\frac{3}{4}s$	$\frac{1}{3}$ $\frac{1}{3}$	$\frac{4}{9}\frac{1}{s}$	Z division
	0		

Here change to ascending order to realize transmission zero at the origin

0 = short circuit or voltage source

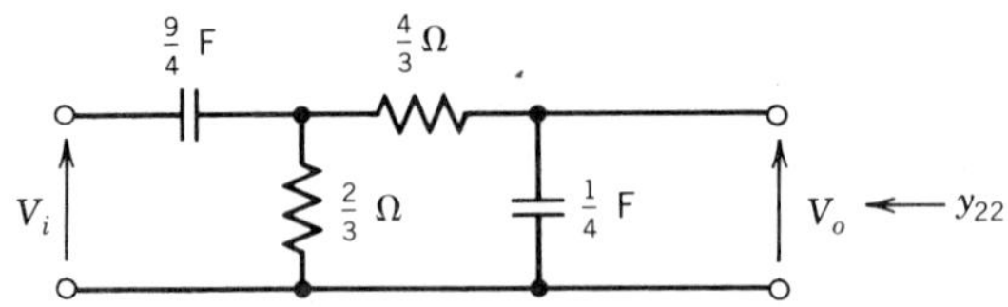

Figure 3.30 Second-order BP filter.

7 The final bandpass circuit is shown in Fig. 3.30.

8 Analysis of this circuit produces a voltage transfer function as

$$T(s) = \frac{s}{\frac{1}{3}s^2 + \frac{4}{3}s + \frac{2}{3}} = \frac{3s}{s^2 + 4s + 2}$$

9 Here $H = 3$ and not $H_{max} = 4$ as we had designed. It is clearly seen that maximum gain cannot be realized in this case, a fact that was pointed out in Section 3.3.

EXAMPLE 3.19 Consider the same bandpass in Example 3.18. Again use the maximum gain constant $H_{max} = 4$ but with an auxiliary polynomial as

$$F(s) = s + \tfrac{7}{10}$$

that is, where the constant term is now smaller than in the previous example.

PROCEDURE

1 The admittances are

$$-y_{21} = \frac{s}{s + \frac{1}{10}} \qquad y_{22} = \frac{\frac{1}{4}s^2 + s + \frac{1}{2}}{s + \frac{7}{10}}$$

2 The realization may proceed as follows.

$s + \frac{7}{10}$	$\frac{1}{4}s^2 + s + \frac{1}{2}$ $\frac{1}{4}s^2 + \frac{7}{40}s$	$\frac{1}{4}s$	Y division			
	$\frac{33}{40}s + \frac{1}{2}$	$s + \frac{7}{10}$ $s + \frac{20}{33}$	$\frac{40}{33}$	Z division		
		$\frac{31}{330}$	$\frac{1}{2} + \frac{33}{40}s$ $\frac{1}{2}$	$\frac{165}{31}$	Y division	
			$\frac{33}{40}s$	$\frac{31}{330}$ $\frac{31}{330}$	$\frac{124}{1089}\frac{1}{s}$	Z division
					0 = short circuit or voltage source	

Change to ascending order (see Example 3.18)

3 The final circuit is shown in Fig. 3.31.

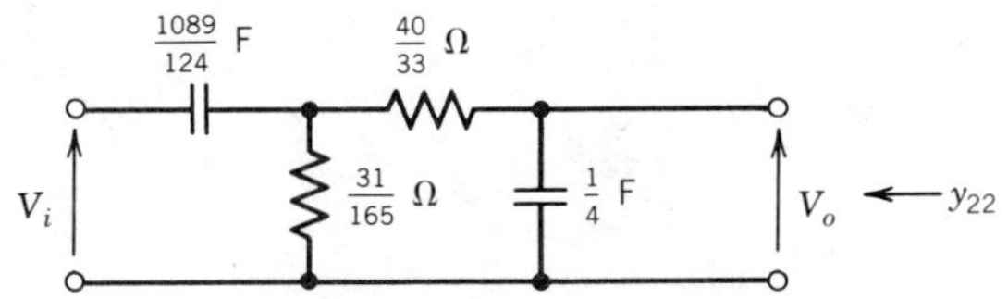

Figure 3.31 Alternative second-order BP filter.

4 Analysis of this filter gives the voltage transfer function as

$$T(s) = \frac{s}{\frac{10}{33}s^2 + \frac{40}{33}s + \frac{20}{33}} = \frac{3.3s}{s^2 + 4s + 2}$$

5 As can be seen, an improvement over the gain constant of Example 3.18 has been achieved. However, this has only been possible at the expense of the component spread. It can clearly be seen that some components are very large while others are relatively small.

Transmission Zero-Producing Networks (Bandstop Circuits)

The synthesis of a bandstop, low-pass notch, or high-pass notch filter RC network requires the realization of transmission zeros on the imaginary axis in the case of an infinite notch depth or in complex locations if the notch depth is to be at a specified level. Single RC ladder networks cannot produce this type of transmission zero. However, multiple transmission paths between input and output terminals can provide a solution to this problem. The use of parallel RC ladder networks to provide these multiple paths was first suggested by Guillemin [5]; a general synthesis procedure will be presented here. The starting point is a transfer function of the form

$$T(s) = H\frac{s^2 + a_1 s + a_0}{s^2 + b_1 s + b_0} \tag{3.58}$$

The significance of the gain constant H has already been discussed. The voltage transfer function of Eq. 3.58 can be expressed in terms of the network admittance parameters of a parallel arrangement of two ladder networks, here called the (a) and (b) ladders:

$$T(s) = -\frac{y_{21az} + y_{21b}}{y_{22az} + y_{22b}} \tag{3.59}$$

The general circuit configuration would be as shown in Fig. 3.32.

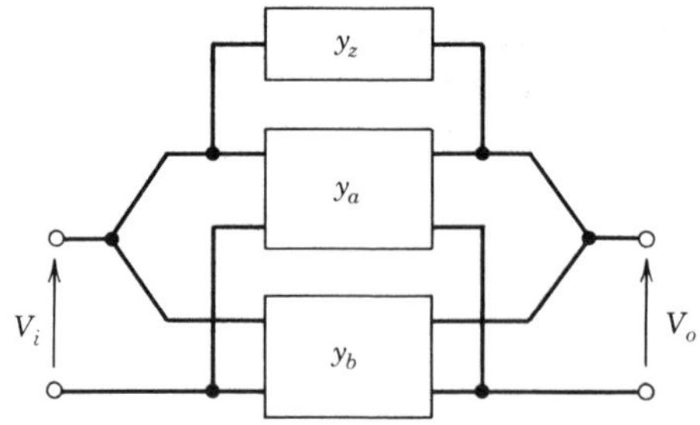

Figure 3.32 General parallel ladder network.

Consider the gain constant H as associated with the denominator polynomial. For the low-pass case $H_{max} = b_0/a_0$ so that Eq. 3.58 can be rewritten as

$$T(s) = \frac{s^2 + a_1 s + a_0}{(a_0/b_0)s^2 + (a_0/b_0)s + a_0} \tag{3.60}$$

As in our earlier ladder realizations, an auxiliary polynomial $F(s)$ is required, which may be one order less than the transfer function polynomial. According to a realizability condition the constant term of $F(s)$ must be larger than a_1. It is therefore convenient to write

$$F(s) = s + (\omega_m + a_1) \tag{3.61}$$

The transfer function of Eq. 3.60 can now be conveniently partitioned into two ladders, one representing a high-pass and the other a low-pass ladder.

$$y_{21az} = \frac{a_1 s + s^2}{(\omega_m + a_1) + s} \qquad y_{22az} = \frac{\xi s + s^2}{(\omega_m + a_1) + s} \tag{3.62}$$

$$y_{21b} = \frac{a_0}{s + (\omega_m + a_1)} \qquad y_{22b} = \frac{[a_0/b_0 - 1]s^2 + \eta s + a_0}{s + (\omega_m + a_1)} \tag{3.63}$$

The suffix z in the admittance parameters of the high-pass ladder equation (3.62) indicates its association with a certain bridging element y_z. Before proceeding with ladder synthesis y_z must be removed.

It can easily be shown that the coefficient of s in the numerator of Eq. 3.60 is solely determined by the bridging element multiplied by the constant term of the auxiliary polynomial. Thus

$$y_z = \frac{a_1 s}{\omega_m + a_1} \tag{3.64}$$

and removing y_z from both y_{21az} and y_{22az} of Eq. 3.62 yields

$$y_{21a} = \frac{[\omega_m/(\omega_m + a_1)]s^2}{(\omega_m + a_1) + s} \qquad y_{22a} = \frac{(\xi - a_1)s + [\omega_m/(\omega_m + a_1)s^2}{(\omega_m + a_1) + s} \tag{3.65}$$

Realization of the two parallel ladders may now proceed. The first division process for ladder (a) is an impedance division step realizing a series capacitance. For ladder (b), the first division step yields a shunt capacitance.

High-Pass Ladder Synthesis

$$
\begin{array}{l|l|l}
(\xi - a_1)s + \dfrac{\omega_m}{\omega_m + a_1} s^2 & (\omega_m + a_1) + s & \dfrac{\omega_m + a_1}{\xi - a_1}\dfrac{1}{s} \quad Z \text{ division} \\
 & (\omega_m + a_1) + \dfrac{\omega_m}{\xi - a_1} s & \\
\hline
\left(1 - \dfrac{\omega_m}{\xi - a_1}\right) s & (\xi - a_1)s + \dfrac{\omega_m}{\omega_m + a_1} s^2 & \dfrac{\xi - a_1}{1 - \omega_m/(\xi - a_1)} \quad Y \text{ division} \\
 & \quad \text{etc.} &
\end{array}
$$

Low-Pass Ladder Synthesis

$$
\begin{array}{l|l|l}
s + (\omega_m + a_1) & \left(\dfrac{a_0}{b_0} - 1\right) s^2 + \eta s + a_0 & \left(\dfrac{a_0}{b_0} - 1\right) s \quad Y \text{ division} \\
 & \left(\dfrac{a_0}{b_0} - 1\right) s^2 + (\omega_m + a_1)\left(\dfrac{a_0}{b_0} - 1\right) s & \\
\hline
\eta s - (\omega_m + a_1)\left(\dfrac{a_0}{b_0} - 1\right) s + a_0 & s + (\omega_m + a_1) & \dfrac{1}{\eta - (\omega_m + a_1)(a_0/b_0 - 1)} \quad Z\text{-division} \\
 & \quad \text{etc.} &
\end{array}
$$

For this synthesis the following realizability conditions emerge:

$$\xi > \omega_m + a_1 \tag{3.66}$$

$$\eta > \frac{a_0}{\omega_m + a_1} + (\omega_m + a_1)\left(\frac{a_0}{b_0} - 1\right) \tag{3.67}$$

The network realized by this synthesis procedure is depicted in Fig. 3.33.

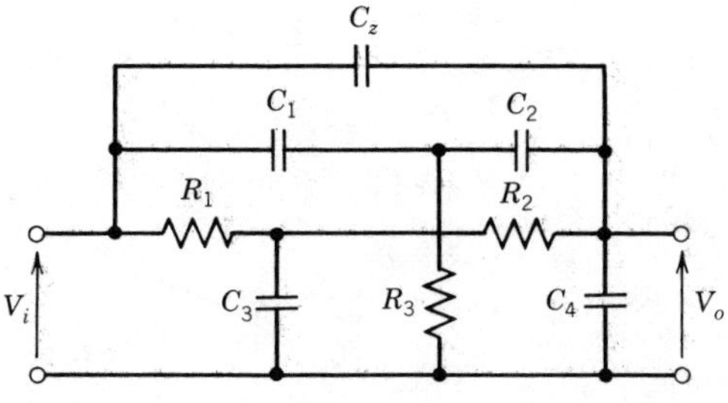

Figure 3.33 Basic RC zero section.

A listing of the general component expressions follows.

$$
\begin{aligned}
C_1 &= \frac{\omega_m/(\omega_m + a_1)}{1 - [\omega_m/(\xi - a_1)]} \\
C_2 &= \frac{\xi - a_1}{\omega_m + a_1} \\
C_3 &= \frac{\eta - (\omega_m + a_1)[a_0/b_0) - 1]}{(\omega_m + a_1) - a_0/\{\eta - (\omega_m + a_1)[(a_0/b_0) - 1]\}} \\
C_4 &= \left(\frac{a_0}{b_0} - 1\right) \\
R_1 &= \frac{(\omega_m - a_1) - a_0/\{\eta - (\omega_m + a_1)[(a_0/b_0) - 1]\}}{a_0} \\
R_2 &= \frac{1}{\eta - (\omega_m + a_1)[(a_0/b_0) - 1]} \\
R_3 &= \frac{1 - [\omega_m/(\xi - a_1)]}{\xi - a_1} \\
C_z &= \frac{a_1}{\omega_m + a_1}
\end{aligned}
\qquad (3.68)
$$

Bandstop Realization

EXAMPLE 3.20 Realize the following voltage transfer function:

$$T(s) = H\frac{s^2 + 1}{s^2 + 2.5s + 1}$$

PROCEDURE

1 Refer to the discussion on p. 81. It is noted that $a_1 = 0$ and $b_0 = a_0$; therefore, $C_z = C_4 = 0$. The circuit thus reduces to that of a standard twin-T.

2 Provided that the conditions of Eqs. 3.66 and 3.67 are met, the realization may proceed. Select $\omega_m = 1$ so that $F(s) = s + 1$; it must be ensured that $\xi > 1$ and $\eta > 1$. Since $\xi + \eta = 2.5$, one may select $\xi = 1.25$ and $\eta = 1.25$.

3 The components can now be determined from Eq. 3.68 as:

$$C_1 = \frac{1}{1 - (1/1.25)} = 5 \text{ F} \qquad R_1 = 1 - \frac{1}{1.25} = 0.2\ \Omega$$

$$C_2 = 1.25 \text{ F} \qquad R_2 = \frac{1}{1.25} = 0.8\ \Omega$$

$$C_2 = \frac{1.25}{1 - (1/1.25)} = 6.25 \text{ F} \qquad R_3 = \frac{1 - (1/1.25)}{1.25} = 0.16\ \Omega$$

4 The complete network is shown in Fig. 3.34.

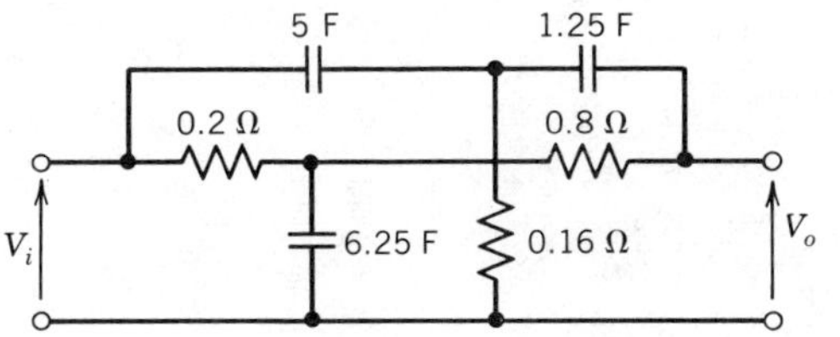

Figure 3.34 Band-reject filter.

EXAMPLE 3.21 Synthesize a low-pass notch filter having the voltage transfer function given by

$$T(s) = H\frac{s^2 + 10}{s^2 + 2.5s + 1}$$

PROCEDURE

1 As discussed earlier, we note that $a_1 = 0$ so that $C_z = 0$. Also $a_0 > b_0$ and therefore $H_{max} = b_0/a_0 = 0.1$. The synthesis produces a loaded twin-T.

2 Select $\omega_m = 1$ so that $F(s) = s + 1$. With this, $\xi > 1$ and $\eta > (10 + 9)$, but since $\xi + \eta = 25$ one may select $\xi = 3$ and $\eta = 22$.

3 The components can now be determined

$$C_1 = \frac{1}{1 - \frac{1}{3}} = 1.5 \text{ F} \qquad R_1 = \frac{1 - \frac{10}{13}}{10} = 0.02307\ \Omega$$

$$C_2 = 3 \text{ F} \qquad R_2 = \frac{1}{13} = 0.0769\ \Omega$$

$$C_3 = \frac{22 - 9}{1 - [10/(22 - 9)]} = 56.33 \text{ F} \qquad R_3 = \frac{1 - \frac{1}{3}}{3} = 0.2222\ \Omega$$

$$C_4 = 9 \text{ F}$$

4 The final circuit is shown in Fig. 3.35.

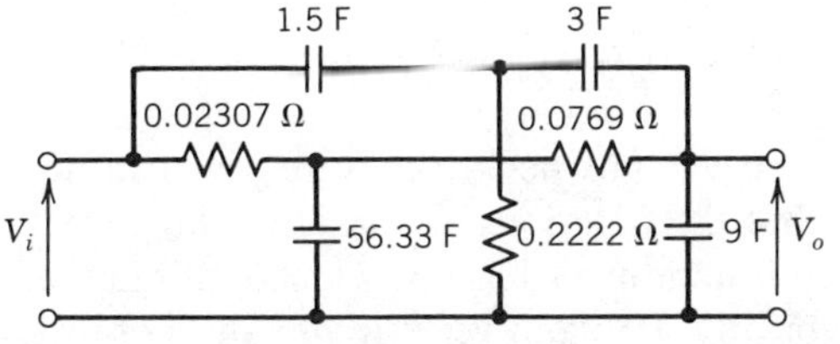

Figure 3.35 Low-pass notch filter.

Gain-Equalizer Realization

EXAMPLE 3.22 Derive a gain-equalizer circuit from the given voltage transfer function.

$$T(s) = H\frac{s^2 + 0.5s + 1}{s^2 + 2.5s + 1}$$

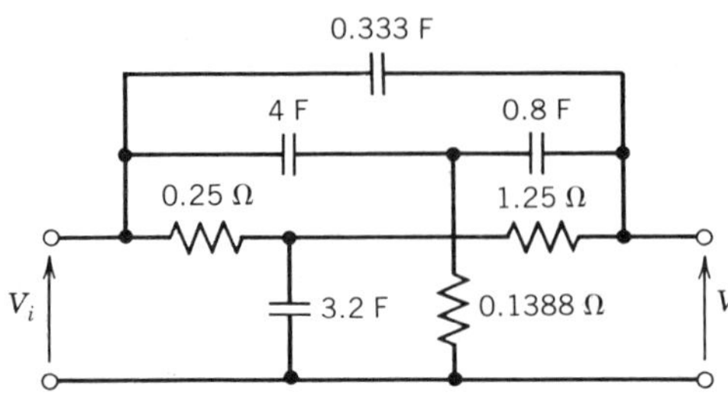

Figure 3.36 Gain-equalizer network.

PROCEDURE

1 Again refer to the discussion on p. 81. In this case $b_0 = a_0$ so that $C_4 = 0$ and $H_{max} = 1$. The synthesis produces a bridged twin-T circuit.

2 By choosing $\omega_m = 1$, the auxiliary polynomial is $F(s) = s + 1$; then $\xi > 1.5$ and $\eta > 1/1.5$ will produce a network. Because $\xi + \eta = 2.5$, one may select $\xi = 1.7$ and $\eta = 0.8$.

3 The component values are

$$C_1 = \frac{1/1.5}{1 - (1/1.2)} = 4 \text{ F} \qquad R_1 = 1.5 - \frac{1}{0.8} = 0.25\ \Omega$$

$$C_2 = \frac{1.2}{1.5} = 0.8 \text{ F} \qquad R_2 = \frac{1}{0.8} = 1.25\ \Omega$$

$$C_3 = \frac{0.8}{1.5 - (1/0.8)} = 3.2 \text{ F} \qquad R_3 = \frac{1 - (1/1.2)}{1.2} = 0.1388\ \Omega$$

$$C_z = \frac{0.5}{1.5} = 0.333 \text{ F}$$

4 The final circuit is shown in Fig. 3.36.

3.7 DENORMALIZATION

The functions used for the synthesis of the various networks were given in normalized form. Normalized form means that the functions are designed for a unity cutoff frequency. When synthesized, such functions yield networks containing unnaturally large values of capacitance (typically, 1 F) and very small values of resistance (typically a few ohms).

The process of converting such a network to contain more practical element values is called *scaling*. Many applications, however, also require a transformation in frequency. This means that the operation of the filter will be made to apply to a frequency band different from that of the normalized frequency band, where the cutoff frequency is unity. For the purpose of generality and ease of comparison, it is initially convenient to obtain the circuit elements in their normalized form.

Transformation to a different frequency range, as well as impedance scaling, is then applied as a final step.

Frequency Denormalization

To change the cutoff frequency of a normalized LP function from $\omega_o = \omega_n = 1$ rad/s to an LP function with a cutoff frequency of Ω_n rad/s is a process defined as denormalization. It requires the replacement of s_n by s/Ω_n where s_n is the normalized frequency variable; Ω_n is the frequency denormalization factor, normally the desired cutoff frequency in radians per second; and s is the actual operating frequency variable. To ensure that the impedance of a circuit element remains invariant, it is necessary that for a capacitance

$$\frac{1}{s_n C_n} = \frac{1}{sC} = \frac{1}{s_n \Omega_n C} \qquad \text{or} \qquad C = \frac{C_n}{\Omega_n} \tag{3.69}$$

For completeness, consider also the inductance element for which

$$s_n L_n = sL = s_n \Omega_n L \qquad \text{or} \qquad L = \frac{L_n}{\Omega_n} \tag{3.70}$$

The only other element remaining is the resistive component. Because resistors are independent of frequency, frequency scaling or denormalization cannot be applied.

Impedance Denormalization

In order that impedance values be at practical levels, it may be necessary to convert the impedance to just such a level. It is desirable for the impedance levels of all components of a network to be of the same order of magnitude. This helps to optimize component sensitivities and is seen to improve notch depth in the case of parallel ladder networks. It also assists in the minimization of parasitic and loading effects, particularly with active filters. The impedance denormalizing factor Z_n is a dimensionless constant by which the normalized impedance is multiplied to effect scaling or denormalization. Thus

$$\begin{aligned} R &= Z_n R_n \\ C &= \frac{C_n}{Z_n} \\ L &= Z_n L_n \end{aligned} \tag{3.71}$$

Frequency and Impedance Denormalization

Denormalization usually requires the application of frequency as well as impedance scaling; therefore, the two effects have to be combined. The denormalization of the network elements of capacitance, inductance, and resistance is therefore given as

$$R = Z_n R_n \tag{3.72}$$

$$C = \frac{C_n}{Z_n \Omega_n} \tag{3.73}$$

$$L = \frac{Z_n L_n}{\Omega_n} \tag{3.74}$$

where the suffix n refers to the normalized value, Ω_n is the frequency denormalization factor, and Z_n is the impedance denormalization factor.

EXAMPLE 3.23 Consider the previously designed network of Example 3.16. The normalized component values are

$$R_{1n} = \tfrac{1}{2}\ \Omega, \qquad R_{2n} = \tfrac{3}{2}\ \Omega, \qquad C_{1n} = \tfrac{4}{3}\ \text{F}, \qquad \text{and } C_{2n} = \tfrac{1}{3}\ \text{F}.$$

Apply impedance scaling to obtain $R_1 = 500\ \Omega$ and design for a cutoff frequency of $f_n = 1000$ Hz.

PROCEDURE

1 The impedance scaling requirement necessitates an impedance denormalizing constant of $Z_n = 1000$.

2 From $f_n = 1000$ Hz the frequency denormalizing constant becomes

$$\Omega_n = 2\pi\ 1000 = 6283\ \text{rad/s}$$

3 Denormalization of the above components yields

$$R_1 = Z_n R_{1n} = 1000 \times 0.5 = 500\ \Omega$$

$$R_2 = Z_n R_{2n} = 1000 \times 1.5 = 1500\ \Omega$$

$$C_1 = \frac{C_{1n}}{Z_n \Omega_n} = \frac{\frac{4}{3}}{1000 \times 6283} = 0.2122 \times 10^{-6}\ \text{F} = 0.2122\ \mu\text{F}$$

$$C_2 = \frac{C_{2n}}{Z_n \Omega_n} = \frac{\frac{1}{3}}{1000 \times 6283} = 0.05305 \times 10^{-6}\ \text{F} = 0.05305\ \mu\text{F}$$

As can be seen, the impedance magnitude of these denormalized components is of the same order.

3.8 SUMMARY

This chapter has described the second-order function in considerable detail. The early sections deal with the fundamental topics of poles and zeros, types of function, and their associated frequency responses. Later sections present design and synthesis procedures for the special passive structures, the ladder and parallel ladder, which are the most frequently encountered in active filters.

3.9 REFERENCES

[1] A. D. Fialkow, and J. Gerst, "The Transfer Function of an RC Ladder Network," *J. Math. and Phys.*, vol. 30, pp. 49–72, 1951.

[2] A. D. Fialkow, "Two Terminal-Pair Networks Containing Two Kinds of Elements Only," *Proc. 1952 Symp. on Modern Network Synthesis*, pp. 50–65.

[3] A. Paige, and E. S. Kuh, "Maximum Gain Realization of an RC Ladder Network," *IRE Trans. on Circuit Theory*, vol. CT 7, pp. 32–40, March 1960.

[4] L. Weinberg, *Network Analysis and Synthesis*, New York: McGraw-Hill, 1962.

[5] E. A. Guillemin, "Synthesis of RC Networks," *J. Math. and Phys.*, 28, pp. 22–42, 1949.

[6] M. E. Van Valkenburg, *Introduction to Modern Network Synthesis*, New York: Wiley, 1960.

[7] A. Budak, *Passive and Active Network Analysis and Synthesis*, Boston: Houghton Mifflin, 1974.

CHAPTER 4

Active Components

R. L. Geiger

Texas A&M University

4.1 INTRODUCTION

Active filters are generally designed by interconnecting basic functional building blocks with resistors and/or capacitors. Often several basic building blocks are combined to create larger functional building blocks. In this chapter the most basic building blocks are discussed. These include amplifiers, integrators, and generalized immittance converters such as synthetic inductors, negative resistors, and the frequency-dependent negative resistors (FDNR).

4.2 OPERATIONAL AMPLIFIERS

The term *active filter* implies the presence of active devices along with passive components. Active devices include vacuum tubes, bipolar transistors, field effect transistors, and operational amplifiers (op amps). As was stated in Chapter 1, active filters generally offer size, cost, and performance advantages over passive filters in the audio frequency range (20 Hz to 20 kHz).

Vacuum tubes were used extensively as the active devices until the early 1960s. The bipolar transistor replaced the vacuum tube in most applications at that time because it was smaller, more economical, and more reliable. The *op amp* has been the dominant active device in active filters since the early 1970s. It offers performance, size, and economic advantages when compared to filters designed with discrete transistors as well as major improvements in design simplicity. Bipolar transistors and vacuum tubes still offer some advantages at high frequencies and high power levels.

An operational amplifier is constructed by interconnecting a large number of transistors (typically 20 or more) on a single silicon substrate. Since the area of silicon required to make an op amp (around $\frac{1}{20}$ in. × $\frac{1}{20}$ in.) is small, hundreds of op amps are fabricated simultaneously on a single silicon wafer. After fabrication, the wafers are broken apart, and the operational amplifiers are typically mounted in the familiar 8-pin and 14-pin IC packages. Since the silicon area required for the op amp is so small, up to two op amps are available in some 8-pin packages and up to four in some 14-pin packages. The major practical limitation on the number of op amps placed in an IC package is the number of pins available on the package. Because of the extremely high volume demand and the small amount of silicon area required for an op amp, the price of popular op amps is quite low, comparable to that of single transistors and the individual passive components used in the filter. Several op amps are often used in the design of a single active filter.

Before discussing the electrical characteristics of operational amplifiers, it is instructive to consider an example that will shed some light on why operational amplifiers evolved. The circuit of Fig. 4.1 is recognized as a *feedback amplifier*, where it is assumed that (neglecting the voltage drop across R_S):

$$V_o = A(s)V_D \tag{4.1}$$

$$V_f = \beta V_o \tag{4.2}$$

and

$$V_D = V_i - V_f \tag{4.3}$$

independent of R_L and R_S. Routine analysis reveals that

$$\frac{V_o}{V_i} = \frac{A}{1 + A\beta} \tag{4.4}$$

It can be observed that if A is very large so that $A\beta \gg 1$, then

$$\frac{V_o}{V_i} \simeq \frac{A}{A\beta} \simeq \frac{1}{\beta} \tag{4.5}$$

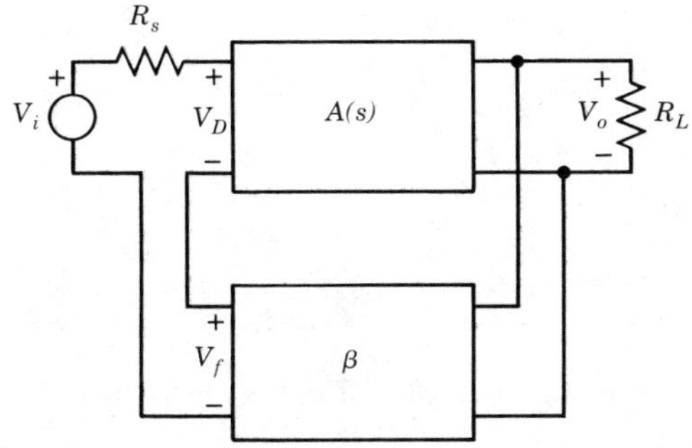

Figure 4.1 Feedback amplifier structure.

The feedback β network is often passive, containing only resistors and/or capacitors; as such the characteristics of β are accurately controllable. If, for example, the β network is a simple two-resistor voltage divider with gain $\beta = R_1/(R_1 + R_2)$, then for A large the gain of the feedback amplifier will be given by

$$\frac{V_o}{V_i} \simeq 1 + R_2/R_1 \tag{4.6}$$

Note that although A may not be accurately controllable, the fact that it is large makes V_o/V_i accurately controllable. Since different values of R_2/R_1 can be selected, a single high-gain amplifier can be used to design a multitude of precision amplifiers.

At this point, the desired characteristics of the A amplifier can be considered. Beyond the high gain (ideally infinite) the amplifier should have a very high input impedance (ideally infinite) and very low output impedance (ideally zero) so that the R_S and R_L do not affect Eqs. 4.1, 4.2, and 4.3, as required in the above analysis. The applications of high-gain amplifiers extend far beyond that for the simple feedback amplifier just considered and include all the active filters that are discussed in this and later chapters.

Ideal Characteristics

Two symbols that are commonly used for operational amplifiers are shown in Fig. 4.2. The symbol of Fig. 4.2*b* is used when it is necessary to explicitly indicate the biasing voltages, V^+ and V^-. Since the presence and value of the biasing voltages are often assumed, the symbol of Fig. 4.2*a* is more widely used. Although no ground terminal is explicitly shown, it is assumed that the nodal voltages, V_1, V_2, and V_o, are all with respect to ground and that they are related by the expression:

$$V_o = A(V_1 - V_2) \tag{4.7}$$

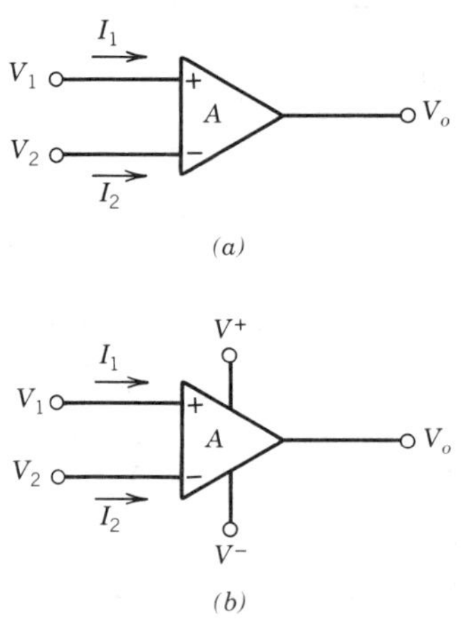

Figure 4.2 Op amp symbols: (a) Conventional; (b) with designation of V^+ and V^-.

where A is the voltage gain of the op amp. This relationship implies that an op amp is a difference amplifier.

An ideal op amp is one with infinite voltage gain A, (at all frequencies, including dc), infinite input impedance (at both the "+" terminal when the "−" terminal is grounded and at the "−" terminal when the "+" terminal is grounded), and zero output impedance. Although it might appear, to those not familiar with op amps, that the infinite voltage gain would be difficult to deal with, we shall see shortly that circuit analysis is actually much simpler when the op amps are assumed to be ideal.

At this point it should be emphasized that *the op amp is never used directly as a high-gain difference amplifier.* The voltage gain A is typically very large (often over 100,000) and varies considerably from device to device; as such it would cause the output of the op amp to *saturate** if even a very small dc voltage (e.g., 200 μV) were present at the input. The input offset voltage of the op amp itself (discussed later) generally is large enough to cause the open-loop op amp to saturate.

The input port of the ideal op amp has some interesting characteristics since it behaves in some ways as both an open and a short circuit. The open circuit behavior is a result of the assumption that the input impedance is infinite and consequently $I_1 = I_2 = 0$. The short-circuit behavior of the input port ($V_1 = V_2$) follows from the assumption that $A = \infty$. To see this, observe from Eq. 4.7 that if the differential input voltage, $V_1 - V_2$, were anything but zero, multiplication by A (which is assumed infinite) would result in an unacceptable infinite output voltage. A port at which the difference voltage is always zero and the input currents are also zero is termed a *null port.* If one of the terminals of a null port is at ground, a *virtual ground* is said to exist at the other terminal. It must be emphasized, however, that if a physical short circuit is placed across the input port of an op amp or if an open circuit to either the "+" or "−" terminals is created, the input port no longer behaves as a null port.

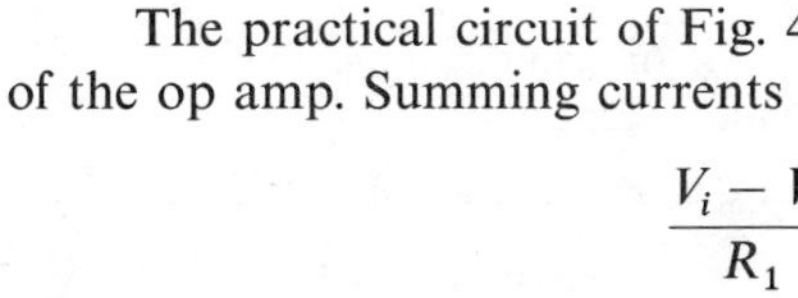

Figure 4.3 Practical inverting amplifier.

The practical circuit of Fig. 4.3 will now be analyzed using the ideal model of the op amp. Summing currents at the node labeled V_2, we obtain

$$\frac{V_i - V_2}{R_1} + \frac{V_o - V_2}{R_2} = I_2 \tag{4.8}$$

* The term *saturate* is often used to indicate that the op amp output voltage or output current is at a maximum or minimum value. When the op amp is saturated, further increases in the excitation will not result in the desired increase (or decrease) in the output and signal distortion will result. Voltage and current limitations are discussed later in this chapter.

From the ideal model of the op amp observe that $I_2 = 0$ and $V_2 - V_1 = 0$. Making these substitutions, we get

$$\frac{V_i}{R_1} + \frac{V_o}{R_2} = 0 \tag{4.9}$$

which is often expressed as

$$\frac{V_o}{V_i} = -\frac{R_2}{R_1} \tag{4.10}$$

This circuit thus serves as a basic inverting amplifier with gain that is determined solely by the ratio of two resistor values.

It may appear in this analysis that the assumption of infinite voltage gain for the ideal op amp was not used. Recall that it is this assumption that forced $V_1 = V_2$ and thus justified obtaining Eq. (4.9) from Eq. (4.8).

EXAMPLE 4.1 Design an amplifier with a voltage gain of -25 using a single operational amplifier. Assume that a dual ± 15 Vdc power supply is available.

PROCEDURE

1 The inverting amplifier of Fig. 4.3 can be used for this purpose. The values for R_1 and R_2 must be selected and the operational amplifier must be biased. It follows from Eq. (4.10) that R_1 and R_2 must satisfy the relationship

$$-25 = \frac{-R_2}{R_1}$$

2 A reasonable choice for the resistor R_2 is 47 kΩ (a standard value). The above equation can be solved for R_1 to obtain

$$R_1 = \frac{-47\ k\Omega}{-25} = 1.88\ \text{k}\Omega$$

3 The available dual ± 15 V power supply can be used directly for biasing the op amp. The completed circuit is shown in Fig. 4.4.

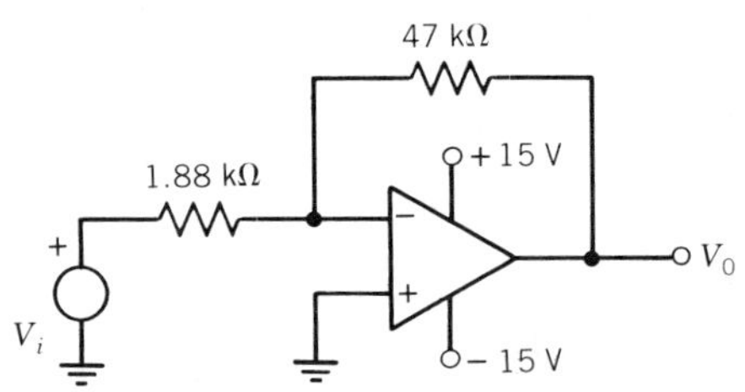

Figure 4.4 Inverting voltage amplifier with gain of −25.

Device Modeling

For many applications, the ideal model of the op amp discussed previously is adequate. At high frequencies and/or for high pole Qs (e.g., such as may be required for higher-order Chebyshev or Butterworth filters), the ideal model may not be adequate to obtain good agreement between theoretical and experimental results. This deviation is often attributable to a decrease in gain experienced by the op amp at high frequencies. In *internally compensated* op amps, this roll-off was intentionally introduced by the designer during compensation to maintain stability in low-gain feedback amplifier structures. Typically a single monolithic capacitor, internal to the op amp, has been used for this purpose. If this roll-off is included in the gain expression, the voltage gain of the op amp can be approximated by

$$A(s) = \frac{A_v \omega_a}{s + \omega_a} \simeq \frac{\omega_t}{s} \tag{4.11}$$

where A_v represents the dc gain and $-\omega_a$ denotes the dominant pole location in rad/s. For almost all applications, the simpler expression involving ω_t, which is equal by definition to $A_v \omega_a$, is adequate. The term ω_t is called the *gain bandwidth product* and has units of rad/s.

A plot of the gain of an operational amplifier as given by Eq. 4.11 is shown in Fig. 4.5. These plots are often included in the data sheets provided by the manufacturer for specific commercially available op amps. It is often more convenient to express the gain-bandwidth product in hertz rather than in radians per second. The symbol GB is descriptively used for this purpose and is related to ω_t by

$$\text{GB} = \frac{\omega_t}{2\pi} \tag{4.12}$$

An ideal op amp is one with $\text{GB} = \infty$. Practical op amps have gain-bandwidth products in the 300 kHz to 8 MHz range. The finite gain-bandwidth product is often the dominant factor that limits the frequency response and performance of active filters. Its effects will be discussed briefly in this chapter and in considerable detail in Chapter 11.

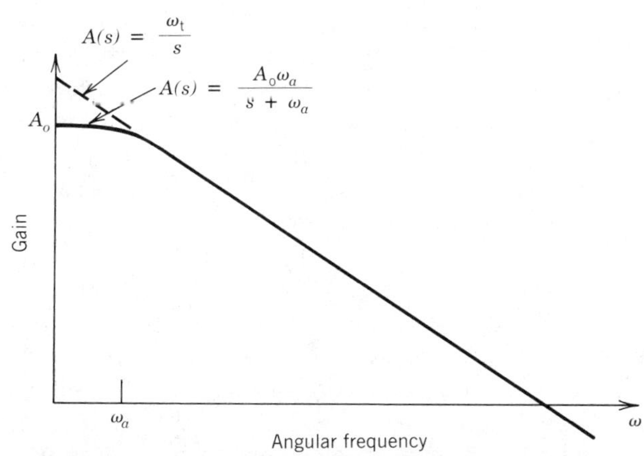

Figure 4.5 Frequency-dependent gain of internally compensated operational amplifier; logarithmic scales used for both axes.

Some op amps require external compensation. The user will be aware of these needs if the term *internally compensated* is not included in the manufacturer's data sheet describing the device. Compensation can often be achieved with a single external capacitor applied between two pins of the op amp that have been designated for this purpose by the manufacturer, although in some applications more complicated schemes are employed. While some improvements in filter performance are often possible with externally compensated op amps, the benefits are generally offset by the increased design complexity associated with the compensation along with the requirements for the additional passive components. Unless explictly stated, all op amps henceforth employed are assumed to be internally compensated.

Practical Considerations

A large number of parameters appear in the data sheets of operational amplifiers. The ones listed in Table 4.1 are often the most important. Typical values of these parameters for the popular 741 and 356 type op amps are also given. A brief discussion of each of these parameters follows.

V^+, V^- These voltages denote the dc voltages required to bias the op amp. Typically, the input voltages must be somewhat less than V^+ and somewhat larger than V^-. For normal operation of the 741, for example, the input voltage is restricted to lie between $V^- + 2\text{V}$ and $V^+ - 2\text{V}$. Often $+15$ V and -15 V are used for V^+ and V^-. Absolute maximum values for these voltages are given in the data sheets. Exceeding the absolute maxima will cause permanent damage to the op amp.

$V_{o,\max}$, $V_{o,\min}$ These are the maximum and minimum output voltages attainable. Attempts to exceed these values will be nondestructively met by output voltage clipping at $V_{o,\max}$ or $V_{o,\min}$. Manufacturers' data sheets typically specify how $V_{o,\max}$ and $V_{o,\min}$ relate to V^+ and V^-, respectively.

$I_{o,\max}$ This is the maximum output current the op amp can deliver (source or sink). Some op amps will be damaged if attempts to exceed this are made, although most op amps (including the 741 and 356) are short-circuit protected. Attempts to exceed the maximum output current in protected devices will result in an internal current limit and a resultant signal distortion. Increased current drive can be obtained by using external power boosting, which is discussed later.

V_{os} The voltage V_{os} is called the input offset voltage. It is a dc voltage that may be either positive or negative (only the worst case magnitude is typically given in data sheets) and can be modeled by a single dc voltage source in series with either the "+" or "−" terminal of the op amp as shown in Fig. 4.6. The average value of V_{os} taken over a large number of op amps should be zero. Offset null pins are available on some op amps (although in many applications they are not used) to allow nulling out of V_{os}. This parameter causes serious performance limitations to high-gain dc instrumentation amplifiers but can often be ignored in active filter applications.

Table 4.1 **Characteristics for 741 and 356 Op Amps**

Parameter	*Name*	741		356	
		Range	*Typical*	*Range*	*Typical*
V^+	Positive bias voltage	3 to 18V	15V	3 to 18 V	15 V
V^-	Negative bias voltage	-3 to -18 V	-15 V	-3 to -18 V	-15 V
$V_{o,\max}$	Maximum output voltage		$V^+ - 2$V		$V^+ - 2$ V
$V_{o,\min}$	Minimum output voltage		$V^- + 2$ V		$V^- + 2$ V
$I_{o,\max}$	Maximum output current (magnitude)		20 mA		20 mA
V_{os}	Offset voltage	Up to 6 mV	2 mV	Up to 10 mV	3 mV
GB	Gain-bandwidth product		1 MHz 6.28 M rad/s		5 MHz 31.4 M rad/s
A_v	DC open-loop voltage gain		100,000		200,000
I_{B1}, I_{B2}	Input bias currents		80 nA		30 pA
SR	Slew rate		0.5 V/μs		12 V/μs

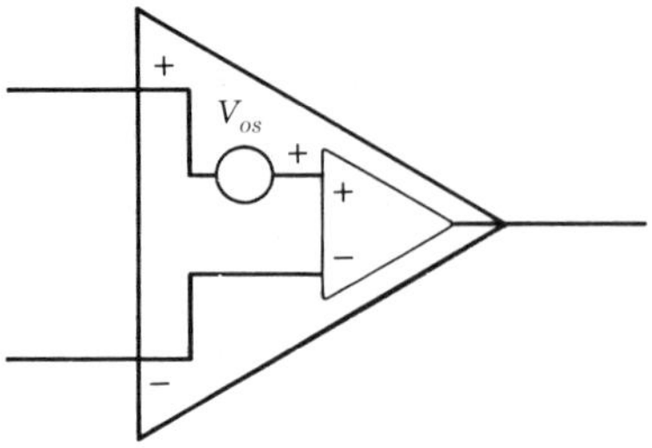

Figure 4.6 Offset voltage in operational amplifiers.

GB GB denotes the gain-bandwidth product of the op amp and was discussed in the previous section. It is generally either given in a tabular listing (typically in MHz) or determinable from the op amp gain curves (GB = frequency at which gain equals unity).

A_v A_v denotes the dc voltage gain of the op amp and is very large. The gain-bandwidth product rather than A_v itself is often of more interest.

I_{B1}, I_{B2} These are the bias currents of the op amp. They are dc currents and are used to bias the input terminals of the op amp. These currents will attain the required values provided that there is a dc path (a path containing only resistors, wires, or direct-coupled voltage sources) between both input terminals and either ground or an op amp output. Once the existence of the dc path has been established, bias current effects can be ignored in most filter applications.

SR SR is the *slew rate* of the op amp and indicates the maximum possible rate of change of the output voltage. If attempts are made to obtain an output that changes faster than the slew rate, the amplifier will be distorted and the rate of change will be equal to the slew rate.

With a desired square-wave output, the SR limitations can be noticed by observing the linear transitions (noninfinite slope) between the high and low states. At high frequencies, the SR distortion becomes very significant and can be observed as a triangular output rather than a square-wave output.

For a desired sinusoidal output of $V_o = V_m \sin 2\pi ft$, slew rate distortion will not occur provided that

$$2\pi f V_m \leq \text{SR} \tag{4.13}$$

For the sinusoidal output, the onset of SR distortion is observed as straight-line segments at the zero crossings and as a triangular output when the inequality in Eq. 4.13 is strongly violated. Slew rate distortion of both a sinusoid and a square-wave output is shown in Fig. 4.7.

EXAMPLE 4.2 The amplifier shown in Fig. 4.8 is used to amplify a 10 kHz sinusoid. Determine the maximum value of a sinusoidal input (in rms) that can be applied without SR distortion. The SR of the 741 is specified (see Table 4.1) as 0.5 V/μs.

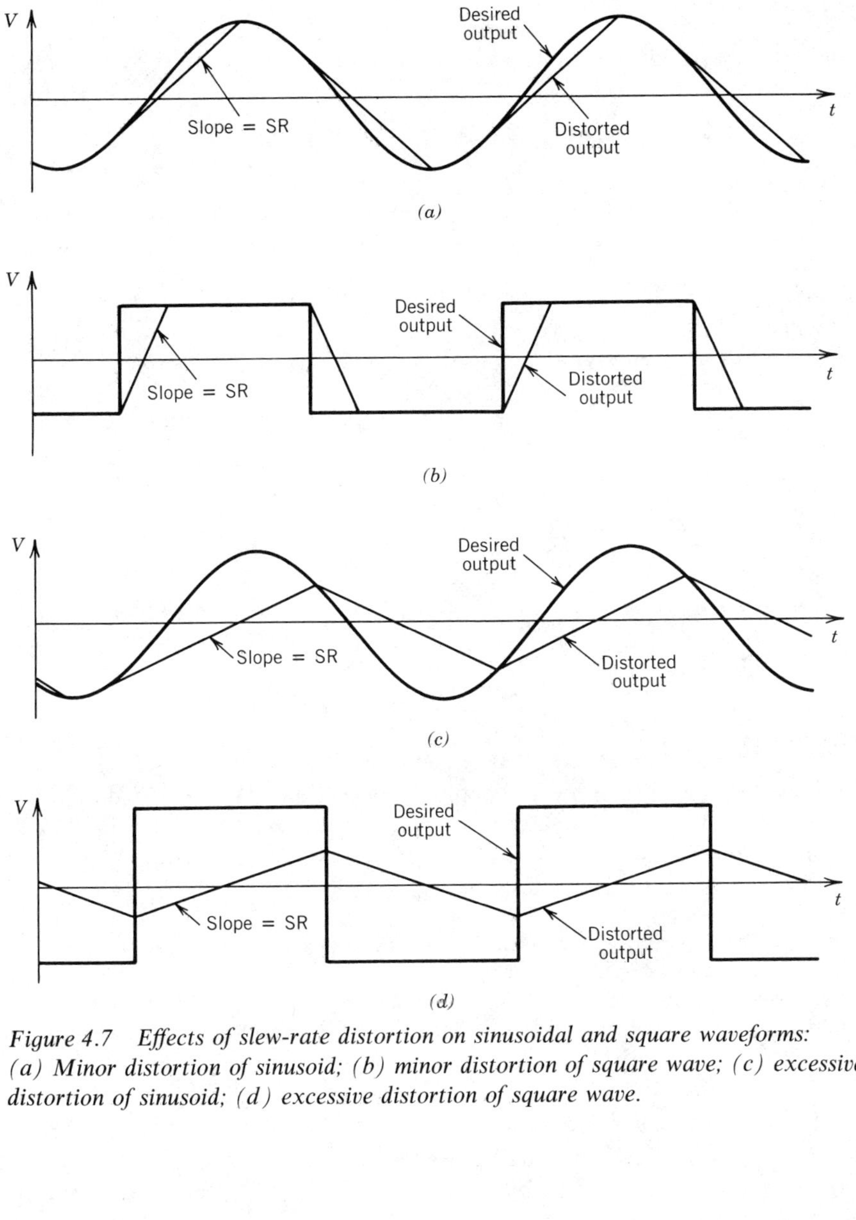

Figure 4.7 Effects of slew-rate distortion on sinusoidal and square waveforms: (a) Minor distortion of sinusoid; (b) minor distortion of square wave; (c) excessive distortion of sinusoid; (d) excessive distortion of square wave.

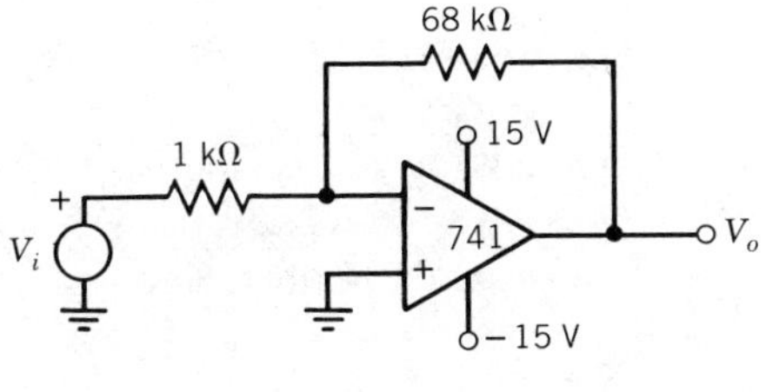

Figure 4.8 Example 4.2.

PROCEDURE

1 From Eq. 4.13, the maximum (peak) value of the output voltage at 10 kHz is given by

$$V_m = \frac{\text{SR}}{2\pi f} = \frac{(0.5\ V/\mu\text{s})}{2\pi 10^4\ \text{Hz}} = 7.96\ \text{V}$$

2 From Eq. 4.10, the gain of the amplifier is given by

$$\frac{V_o}{V_i} = \frac{-68\ \text{K}}{1\ \text{K}} = -68$$

3 Thus a sinusoidal input of 7.96 V/68 = 0.117 V (peak) is needed to produce the 7.96 V output.

4 The rms input voltage at 10 kHz must thus be less than $0.117/\sqrt{2} = 83$ mV.

EXAMPLE 4.3 Will the slew rate of the op amp affect the sinusoidal output of the amplifier shown in Fig. 4.9 for any values of V_m?

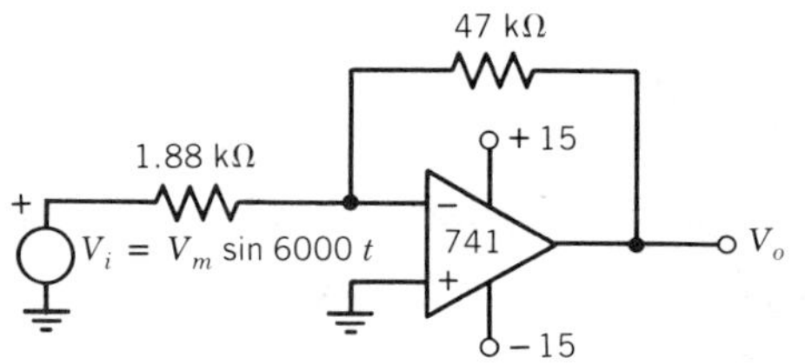

Figure 4.9 Example 4.3.

PROCEDURE

1 The slew rate of the 741 is 0.5 V/μs. The output voltage of the 741 is restricted to lie between V^- +2 V and $V^+ - 2\ V$, which in this case restricts the magnitude of V_o to 13 V.

2 At $\omega = 6000$ rad/s it follows from Eq. 4.13 that to prevent slew rate distortion, the peak value of the sinusoidal output must satisfy the expression

$$V_m \leq SR/2\pi f = \frac{0.5\ \text{V}/\mu\text{s}}{6000\ \text{rad/s}} = 83.3\ \text{V}$$

3 Since the power supplies limit the peak sinusoidal output to 13 V, which is considerably less than 83.3 V, it can be concluded that this amplifier will not be SR limited for any values of V_m.

Good filter designs will have outputs that are nearly independent of $I_{o,\max}$, $V_{o,\max}$, $V_{o,\min}$, V_{os}, I_{B1}, I_{B2}, GB, A_v, and the SR, since these are considered as parasitic parameters and vary considerably (±50%) from device to device and are affected by operating point, temperature, and so on.

Because good filter designs are nearly independent of the parasitic op amp parameters, it is generally adequate to refer to the manufacturer's data sheet for typical values of these parameters rather than experimentally measuring them. If a particular circuit design requires precise knowledge of any of these parameters, either an alternative approach or use of different operational amplifiers should be seriously considered.

Parameter	R_1	R_2	R_L	V_i
$V_{o,\max}$ $V_{o,\min}$	1 kΩ	10 kΩ	∞	$V_m \sin\{2\pi \times 10^2 \times t\}$
$I_{o,\max}$	1 kΩ	10 kΩ	Variable	$V_m \sin\{2\pi \times 10^3 \times t\}$
V_{os}	100	10 kΩ	∞	0
SR	1 kΩ	10 kΩ	∞	10 kHz square wave
GB	1 kΩ	10 kΩ	∞	$V_m \sin\{2\pi \times 10^4 \times t\}$

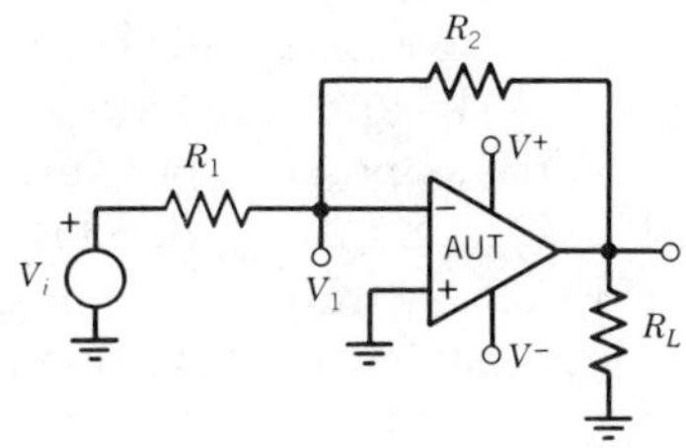

AUT denotes Amplifier Under Test.

Figure 4.10 Simple circuit for measuring basic op amp parameters.

Situations do arise, however, where it becomes necessary to measure some of the parasitic op amp parameters. Circuits for making these measurements can often be found in the manufacturer's application notes. A reasonably simple circuit that is useful for making such measurements is shown in Fig. 4.10. This can be used to measure $I_{o,\max}$, $V_{o,\max}$, $V_{o,\min}$, V_{os}, GB, and SR. These measurements should be made at the values of V^+ and V^- that will be used in a given application.

To measure $V_{o,\max}$ and $V_{o,\min}$, it is reasonable to select $R_2 = 10\text{ k}\Omega$, $R_1 = 1\ k\Omega$, $R_L = \infty$, and excite the network with a 1 kHz sinusoid. If the input amplitude is increased until clipping occurs at both the upper and lower peaks of the output waveform, the height of these peaks corresponds to $V_{o,\max}$ and $V_{o,\min}$, respectively.

$I_{o,\max}$ can be measured with the same basic circuit. To measure $I_{o,\max}$, the input amplitude is decreased with $R_L = \infty$ so that output clipping does not occur. The exact amplitude of the output sinusoid is not of major concern. By placing a potentiometer, R_L, on the output and decreasing its value, the output current will increase. If R_L is decreased so that clipping of the output occurs, then the height of the clipped peak divided by R_L is approximately equal to $I_{o,\max}$. (Actually, a small amount of output current flows through R_2, but it is quite small relative to $I_{o,\max}$ with $R_2 = 10\text{ k}\Omega$.)

To measure V_{os}, V_i is removed and the corresponding node of R_1 is connected to ground. The value of R_1 is decreased to 100 Ω. V_{os} will cause a dc voltage to appear on the output. V_{os} is determined by the equation

$$V_{os} = \frac{|V_o|}{101}$$

Note that the values of R_1 and R_2 were kept relatively small so that the bias currents, I_{B1} and I_{B2}, produce a negligible voltage at the output.

The slew rate can be easily determined by applying a square wave (10 kHz is reasonable with most op amps) at the input of sufficient amplitude to cause rather large output excursions. The magnitude of the slope of the square wave at the zero crossings is the slew rate. The magnitude of the slopes of the positive-going and negative-going waveforms are generally about the same.

The GB of internally compensated op amps can be measured from the circuit of Fig. 4.10 by applying a sinusoid to the input. The amplitude of the input should be as large as possible but small enough to avoid SR distortion (see Eq. 4.13 as well as $V_{o,\max}$ and $V_{o,\min}$ limitations). With a 10 kHz excitation, GB can be determined from the ratio of the magnitude of the sinusoid that appears at V_o to that which appears at the inverting input node to the op amp with the expression

$$\text{GB} = \frac{V_o}{V_1} \cdot 10 \text{ kHz}$$

For this measurement, the frequency was intentionally chosen high enough to make the amplitude of V_1 sufficiently large for easy measurement as well as to make the effects of the low-frequency dominant pole, ω_a (see Eq. 4.11), negligible. It was also chosen low enough to allow measurements with standard test equipment as well as to circumvent contributions due to additional parasitic poles that are present at higher frequencies.

The dc voltage gain, A_v, is somewhat more difficult to measure because it is so large, because offset voltage and bias current effects preclude direct open-loop measurements, and because of questions about the linearity of the device model and the effects of thermal feedback at low frequencies. Comments about measuring A_v as well as the bias currents of op amps can be found in [12]. The problems associated with measuring A_v are discussed by Solomon [13].

No discussion about measuring the bias currents I_{B1} and I_{B2} will be given, since the effects of these parameters are generally dominated by those imposed by V_{os} and GB. With FET input op amps (such as the 356) the bias currents are in the 30 pA range. Such small currents have negligible effects on most circuits. Bipolar input op amps (such as the 741) have bias currents over 1000 times as large, but even these currents typically play a minor role. As a rule of thumb, the designer can generally justify ignoring the bias currents for bipolar input op amps if all resistors connected to the "+" and "−" terminals of the op amp are smaller than 500 kΩ. Unless otherwise stated, the effects of bias currents will be neglected henceforth in this book.

4.3 ACTIVE BUILDING BLOCKS

Several basic active building blocks are considered in this section. Some of the building blocks are useful as stand-alone circuits. All can be combined to make more complicated structures.

Controlled Sources

The circuit of Fig. 4.3, which was shown to serve as an inverting amplifier, is an example of a voltage-controlled voltage source. This, along with several other useful controlled sources, is shown in Fig. 4.11. Expressions for the output voltage derived assuming ideal op amps are also given in Fig. 4.11. All of these expressions can be readily derived following the approach used in Section 4.2 for the inverting amplifier. Comments about each of these circuits follow. Practical limitations of the inverting and noninverting amplifier are discussed to present qualitative insight about the relative importance of some of the nonideal characteristics of op

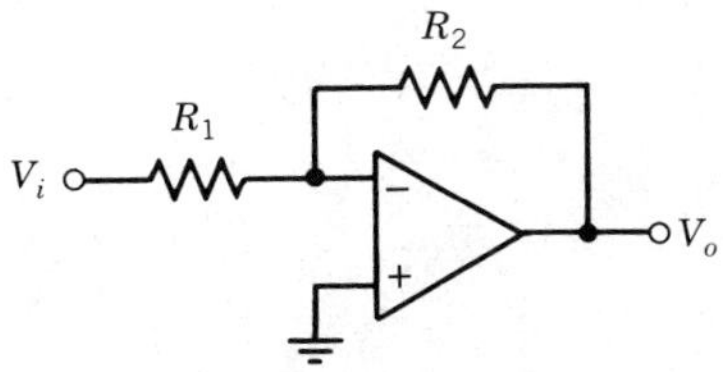

$$V_o = -\frac{R_2}{R_1} V_i$$

(*a*) Inverting amplifier.

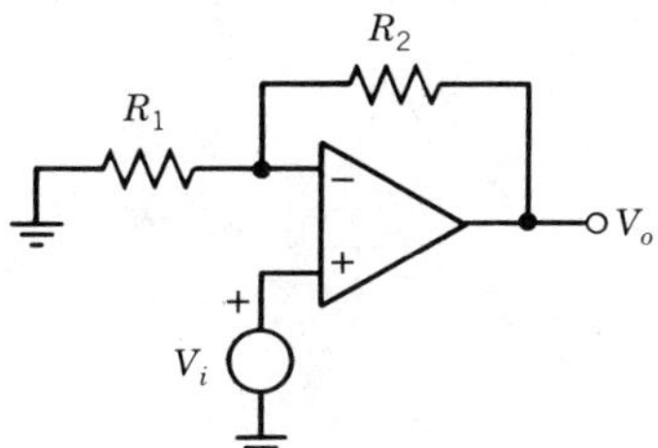

$$V_o = \left(1 + \frac{R_2}{R_1}\right) V_i$$

(*b*) Noninverting amplifier.

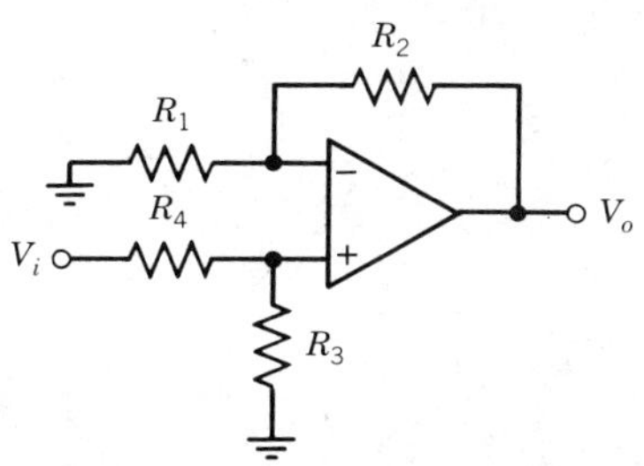

$$V_o = \left(1 + \frac{R_2}{R_1}\right)\left(\frac{R_3}{R_3 + R_4}\right) V_i$$

(*c*) Noninverting amplifier with gain reduction.

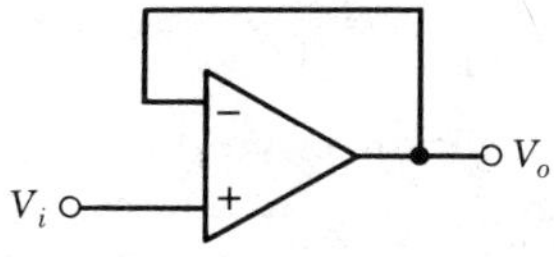

$$V_o = V_i$$

(*d*) Buffer amplifier.

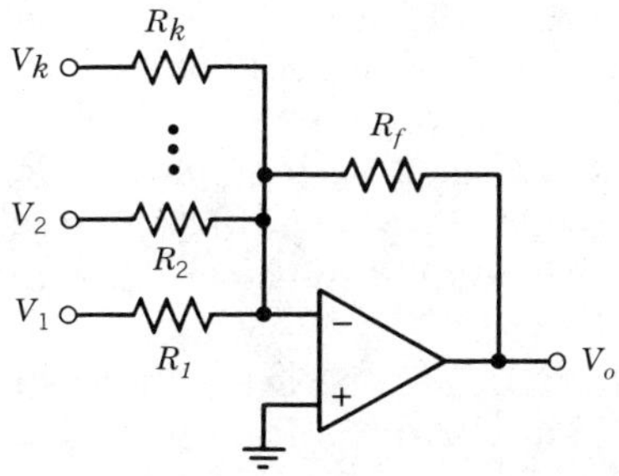

$$V_o = \frac{-R_f}{R_1} V_1 - \frac{R_f}{R_2} V_2 - \cdots - \frac{R_f}{R_k} V_k$$

(*e*) Summing amplifier.

Figure 4.11 Controlled sources.

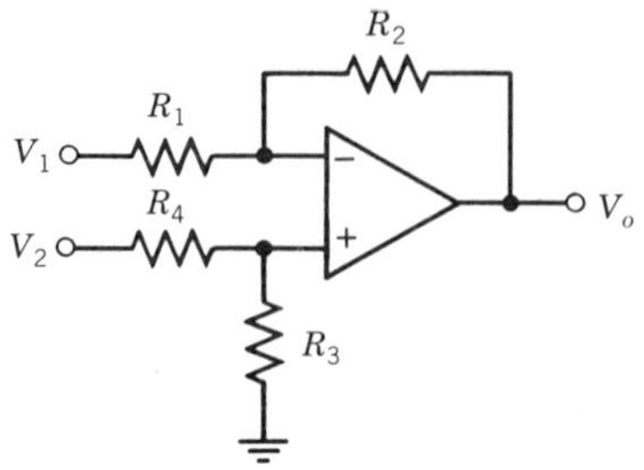

$$V_o = \left(\frac{R_3}{R_3 + R_4}\right)\left(1 + \frac{R_2}{R_1}\right)V_2 - \frac{R_2}{R_1}V_1$$

$$V_o \underset{\substack{R_1 = R_4 \\ R_2 = R_3}}{=} \frac{R_2}{R_1}(V_2 - V_1)$$

(*f*) Difference amplifier.

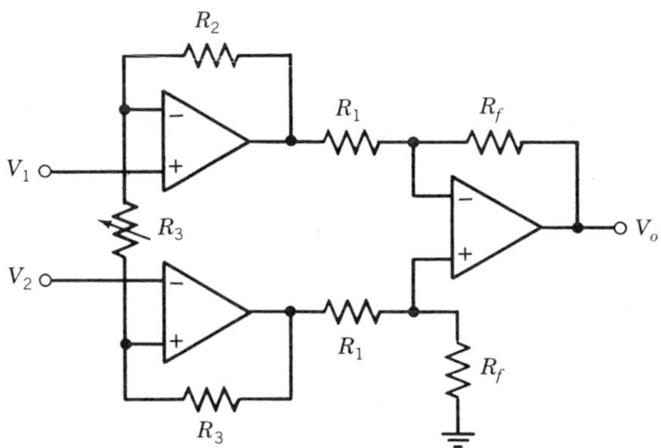

$$V_o = \frac{R_f}{R_1}\left(1 + \frac{2R_2}{R_3}\right)(V_2 - V_1)$$

(*g*) Instrumentation amplifier.

Figure 4.11(f) and (g) (continued)

amps. Although not specifically mentioned when discussing the remaining circuits, these characteristics also limit the performance of the latter structures.

Inverting Amplifier This amplifier is used extensively to obtain large or small gains. The ratio of the resistors, R_1 and R_2, determines the gain. The choice of the value of one of these resistors must be made by the designer. Large resistors tend to introduce noise in the circuit, whereas small resistors dissipate considerable power. In this and other circuits discussed in this chapter, a good rule of thumb is to select all resistors in the 1 kΩ to 100 kΩ range if design requirements permit. A discussion of some of the limitations of this circuit follows.

The input impedance is R_1 and may cause loading of the input source in some applications. The amplifier can be preceded by the unity gain buffer of Fig. 4.11*d* to increase the input impedance if required. The output impedance for all circuits in Fig. 4.11 is well under 1 Ω at low frequencies.

It can be shown that the gain rolls off at high frequencies because of the GB of the op amp. The 3 dB frequency (bandwidth) of this circuit is given by

$$f_{3\text{ dB}} = \frac{\text{GB}}{1 + (R_2/R_1)} \tag{4.14}$$

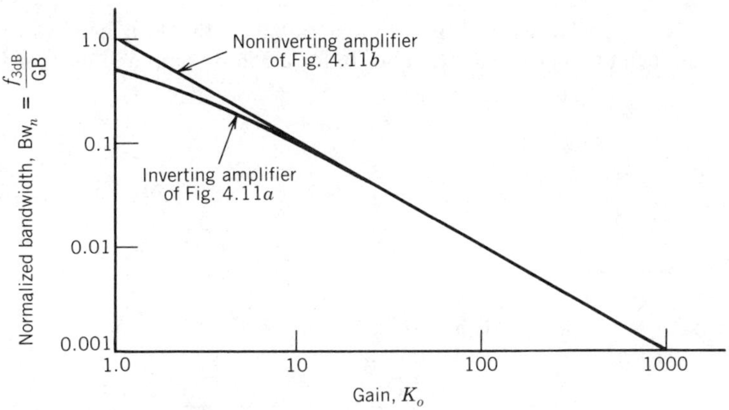

Figure 4.12 3-dB bandwidth of amplifiers of Fig. 4.11a and b.

From this equation a quick calculation of the 3 dB bandwidth can be made prior to design to determine if the amplifier will have sufficient bandwidth. The normalized 3 dB bandwidth, $BW_N = f_{3\text{ dB}}/GB$, is plotted versus the magnitude of the gain, $K_0 = R_2/R_1$, in Fig. 4.12. To find the 3 dB bandwidth of the amplifier, one must multiply the normalized bandwidth by the GB of the op amp used. Note from Eq. 4.14 that for large gains, $K_0 \gg 1$, the product $K_0 f_{3\text{ dB}}$ is approximately GB. For large-gain amplifiers, the bandwidth given in Eq. 4.14 is often unacceptably low. The bandwidth can be improved considerably by cascading several identical low-gain stages to obtain a large gain [1].

The offset voltage of the op amp introduces a dc offset at the output given by

$$V_{o,\text{offset}} = \left(1 + \frac{R_2}{R_1}\right) V_{os} \tag{4.15}$$

For low gains or sinusoidal excitations this factor is often negligible but may cause problems if R_2/R_1 is large and the excitation is dc. The effects can be reduced by using op amps with offset nulling capabilities.

Noninverting Amplifier The gain of the amplifer of Fig. 4.11*b* is always greater than or equal to one. The input impedance is very large (ideally infinite), since very little current flows into the "+" terminal of the op amp.

The 3 dB frequency of this circuit is also given by Eq. 4.14, but in this case the gain is $K_0 = 1 + R_2/R_1$ and the product of the gain and 3 dB bandwidth is constant and given by the expression

$$K_0 f_{3\text{ dB}} = GB \tag{4.16}$$

for all K_0. A plot of the bandwidth of this amplifier is included in Fig. 4.12. The effects of the offset voltage on this amplifier are the same as for the inverting amplifier and are also expressed in Eq. 4.15.

EXAMPLE 4.4 For the amplifier shown in Fig. 4.13, determine (a) the 3 dB bandwidth and (b) the effects of V_{os}. Then (c), repeat parts (a) and (b) for a 356 op amp and compare the results.

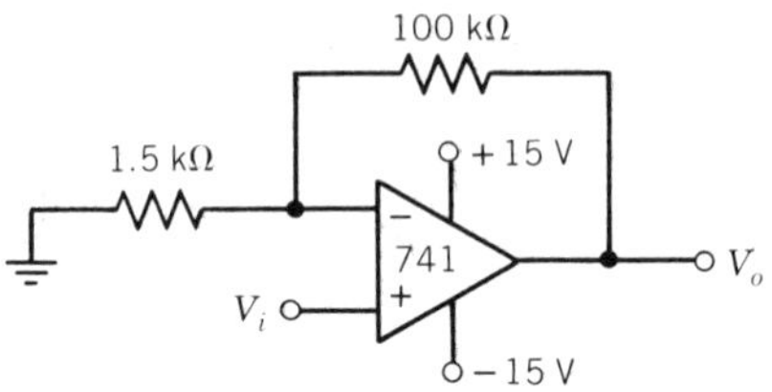

Figure 4.13 Example 4.4.

PROCEDURE

1 From Eq. 4.16 we see that the 3 dB bandwidth is given by

$$f_{3\,\text{dB}} = \frac{\text{GB}}{K_0}$$

where $K_0 = 1 + 100\ \text{K}/1.5\ \text{K} = 67.67$.

2 From Table 4.1, GB = 1 MHz, so

$$f_{3\,\text{dB}} = \frac{1\ \text{MHz}}{67.67} = 14.8\ \text{kHz}$$

3 Since the dc output due to the input offset voltage of the noninverting amplifier is identical to that of the inverting amplifier, it follows from Eq. 4.15 and Table 4.1 that the typical output offset magnitude is

$$V_{o,\text{offset}} = (67.67)(2\ \text{mV}) = 135\ \text{mV}$$

Thus the dc output voltage can be expected to be somewhere between −135 mV and +135 mV.

4 If we repeat the calculations for the 356 op amp, it follows that

$$f_{3\,\text{dB}} = 74\ \text{kHz}$$

and

$$V_{o,\text{offset}} = 203\ \text{mV}$$

5 Note that the 356 offers a significant improvement in amplifier bandwidth. The offset voltage effects with the 356 are somewhat worse.

Noninverting Amplifier with Gain Reduction Some applications require noninverting gains less than unity. A simple voltage attenuator could be used, but the output impedance would often be unacceptably high. The $R_3 \cdot R_4$ voltage divider in the circuit of Fig. 4.11*c* attenuates the signal applied to the "+" terminal and allows for gains less than 1. The input impedance of this circuit is $R_3 + R_4$.

Buffer Amplifier The buffer amplifier of Fig. 4.11*d* is a special case of the noninverting amplifier of Fig. 4.11*b* with $R_1 = \infty$ and $R_2 = 0$. Although it has the same gain as a wire, the input impedance is nearly infinite and the output impedance nearly zero. It is used extensively to increase the input impedance of more complicated structures.

Summing Amplifier The summing amplifier of Fig. 4.11*e* can be used to sum any number of signals. An inversion of all signals is evident from the output expression. A noninverting summer can be obtained by following this summer with the inverting amplifier of Fig. 4.11*a*. The resistor R_f can be used to simultaneously adjust the weight of all inputs. The resistors $R_1, \ldots R_k$ allow for independent weighting of the inputs $V_1, \ldots V_k$, respectively.

Difference Amplifier The resistors in the difference amplifier of Fig. 4.11*f* can be selected to have different weighting of the inverting and noninverting inputs, V_1 and V_2. By selecting $R_2 = R_3$ and $R_1 = R_4$, the weights are equal and the circuit becomes a true difference amplifier with gain R_2/R_1. Precise matching of R_2 to R_3 and R_1 to R_4 is required for proper operation of the difference amplifier.

The circuit is limited by the matching requirements, the noninfinite input impedance, and the inability to adjust the overall gain with a single resistor.

Instrumentation Amplifier The instrumentation amplifier of Fig. 4.11*g* serves the same basic function as the difference amplifier but offers some performance advantages obtained at the expense of a modest increase in component count. The input impedance facing V_1 and V_2 is nearly infinite and R_3 offers single-resistor adjustment of the overall difference gain. Precise matching of resistor values is required as was the case for the circuit of Fig. 4.11*f*.

Integrators

Several of the filter synthesis techniques discussed in later chapters use integrators as the basic functional building block. The evolution of such filters is quite natural since a linear filter can be thought of as a system that electrically solves a linear differential equation. The differential equation itself is expressed as sums of derivatives and/or integrals. Because of the inherent physical limitations of differentiator circuits, integrators are used in the implementation. Integrators are actually simple filters themselves and are used both as stand-alone systems and as a part of larger structures.

The gain of an integrator is given by

$$T_I(s) = \frac{I_o}{s} \tag{4.17}$$

where I_o is a real constant. If I_o is positive, the integrator is said to be noninverting; if negative, it is termed inverting.

One of the major limitations of the ideal integrator is the infinite dc gain that is present by definition. These infinite gain effects are generally compensated for in filter structures by existing feedback connections and are eliminated in stand-alone integrators by adding a small amount of loss as discussed later.

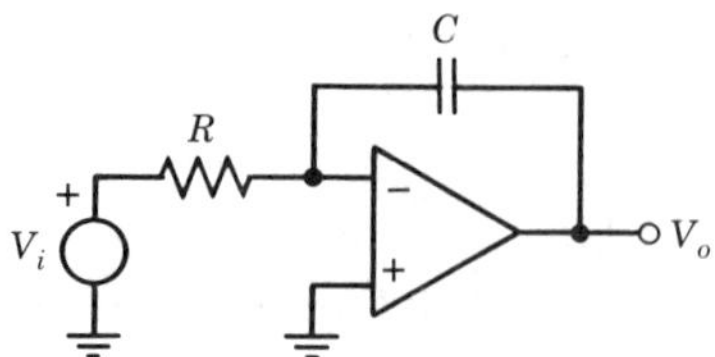

$$\frac{V_o}{V_i} = -\frac{1}{RCs}$$

(*a*) Simple integrator.

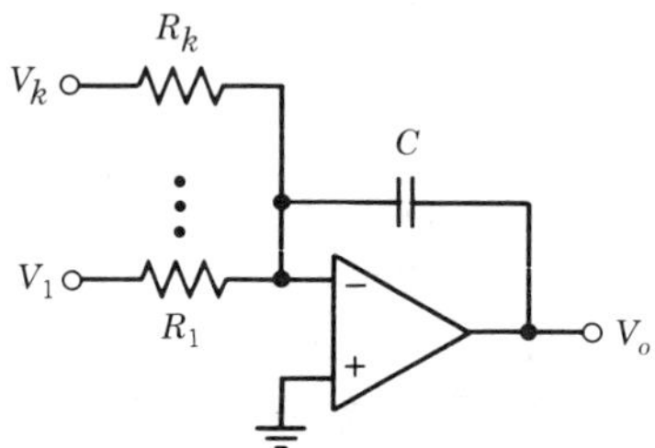

$$\frac{V_o}{V_i} = -\left(\frac{1}{R_1Cs} + \frac{1}{R_2Cs} + \cdots + \frac{1}{R_kCs}\right)$$

(*b*) Summing integrator.

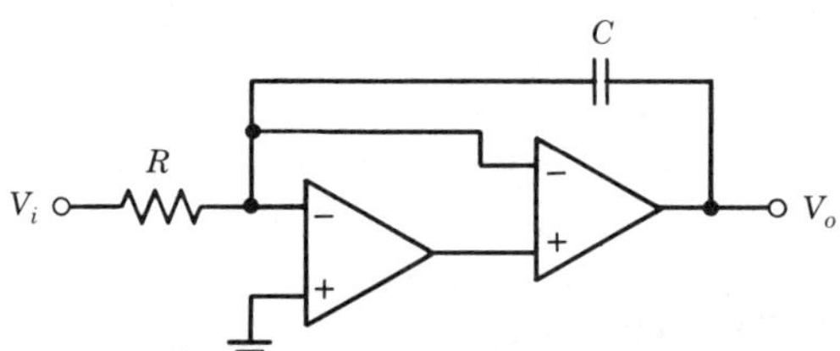

$$\frac{V_o}{V_i} = -\frac{1}{RCs}$$

(*c*) High-performance integrator.

Figure 4.14 Inverting integrators.

Several basic inverting integrators are shown in Fig. 4.14 along with the gain expressions obtainable by assuming ideal op amps. The noninverting counterparts are shown in Fig. 4.15. As can be seen from Figs. 4.14 and 4.15, the circuit topologies for the inverting integrators are less complex than for the noninverting case. Comments about each of these circuits follow.

Basic Inverting Integrator The most widely used inverting integrator is shown in Fig. 4.14*a*. The input impedance of this circuit is R. In the time domain, the output voltage of this integrator at any time t after a known time t_1 can be expressed as

$$V_o(t) = -\frac{1}{RC}\int_{t_1}^{t} V_i(\tau)\,d\tau + V_o(t_1) \tag{4.18}$$

In the frequency domain the gain is given by

$$\frac{V_o}{V_i} = -\frac{1}{RCs} \tag{4.19}$$

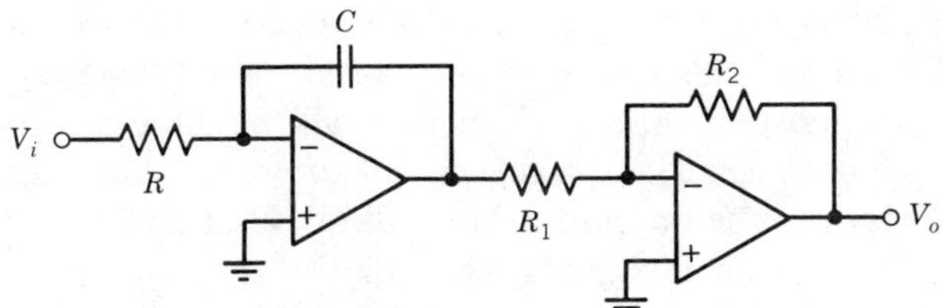

$$\frac{V_o}{V_i} = \left(\frac{R_2}{R_1}\right)\left(\frac{1}{RCs}\right)$$

(*a*) Basic noninverting.

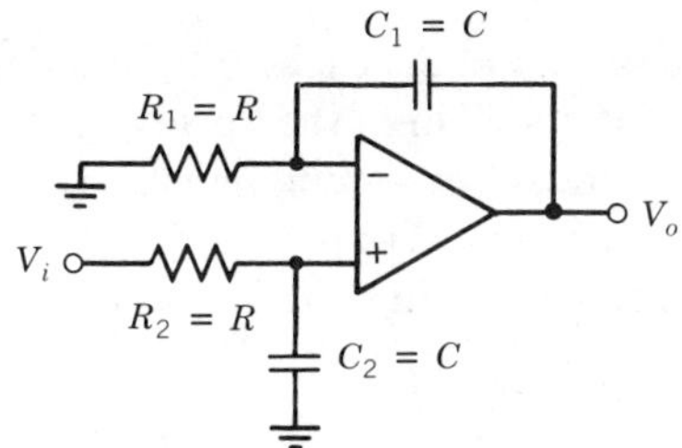

$$\frac{V_o}{V_i} = \frac{1}{RCs}$$

(*b*) Balanced time constant.

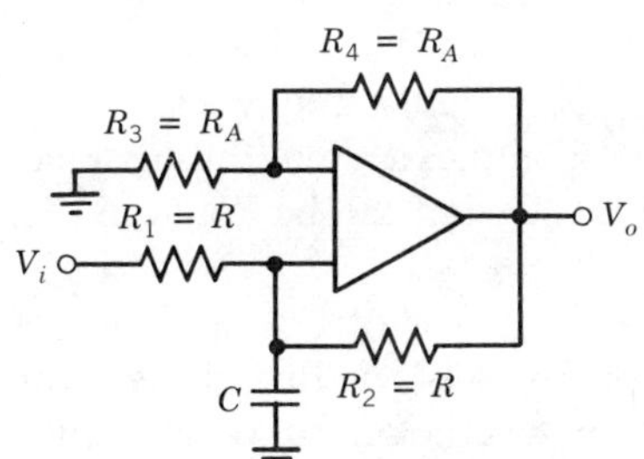

$$\frac{V_o}{V_i} = \frac{2}{RCs}$$

(*c*) Matched resistor noninverting.

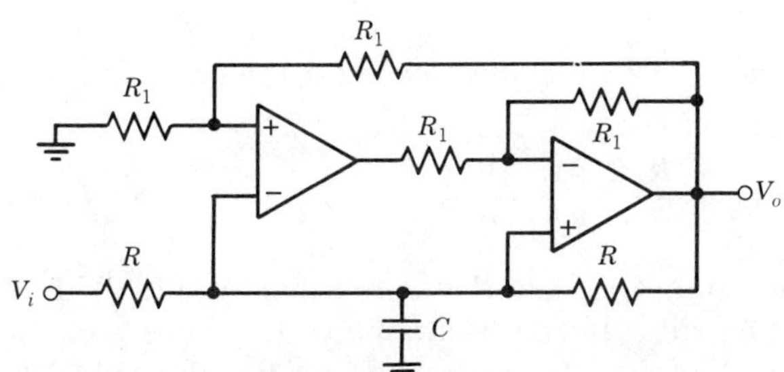

$$\frac{V_o}{V_i} = \frac{2}{RCs}$$

(*d*) Zero sensitivity.

Figure 4.15 Noninverting integrator structures.

which can be obtained directly from Eq. 4.18 by taking the Laplace transform of the time domain expression. Although this circuit is often used directly in feedback filter structures, such as in the state-variable filter that will be discussed in Chapter 8, it must be modified if used as a stand-alone circuit to prevent latch-up caused by either the offset voltages of the op amp or a small dc offset in V_i. This latch-up is inherently unavoidable in an integrator since the output voltage as expressed in Eq. 4.18 contains an integral that will eventually drive the output to saturation if any net dc voltage is present. A modification that eliminates the saturation problem and is applicable to all integrators of Figs. 4.14 and 4.15 involves making the integrator slightly lossy by placing a large resistor in shunt with the feedback capacitor.

The major limitations of the integrator of Fig. 4.14*a* are the excess phase shift and decrease in integrator gain caused by the roll-off in the gain of the op amp (GB effects) at high frequencies. As a rule of thumb, in most low-order (degree ≤ 4) standard filter applications where the passband lies below 5 kHz these effects can be ignored, provided that the gain-bandwidth product of the op amp exceeds 1 MHz.

Basic Inverting Summing Integrator This circuit, shown in Fig. 4.14*b*, has the same basic properties as the previous circuit but allows for multiple inputs.

High-Performance Inverting Integrator The circuit of Fig. 4.14*c* is less affected by the GB effects of the op amp than are the previous circuits. These improvements are obtained at the expense of an additional op amp. At higher frequencies or for higher-order integrator-based filters, the additional complexity may be justified. Details about the performance and applications of this design can be found in [2].

Basic Noninverting Integrator The noninverting integrator of Fig. 4.15*a* is obtained by inverting the output of the integrator of Fig. 4.14*a* with the inverting amplifier of Fig. 4.11*a*. This technique can be used to invert the output of all circuits in Fig. 4.14. A significant increase in circuit complexity for the noninverting circuits compared to the inverting counterparts is apparent. The performance limitations of this circuit are similar to those experienced by the inverting integrator of Fig. 4.14*a*.

Balanced-Time Constant Integrator The integrator of Fig. 4.15*b* has gain given by

$$\frac{V_o}{V_i} = \left(\frac{1 + R_1C_1s}{1 + R_2C_2s}\right)\frac{1}{R_1C_1s} \tag{4.20}$$

It can be seen that if the time constants R_1C_1 and R_2C_2 are equal (balanced), then the pole and zero cancel and the circuit performs as a noninverting integrator. In many applications a small mismatch of these time constants can be tolerated. This circuit uses only a single op amp but requires matching of passive components for proper operation.

The input impedance for this circuit is frequency-dependent and equal to $R + 1/sC$. If driven by an ideal voltage source, this will cause no problems but it must be included as a design parameter if driven from a source with a high output impedance.

Matched Resistor Noninverting Integrator The gain of the circuit of Fig. 4.15*c* is given by

$$\frac{V_o}{V_1} = \frac{1}{[R_3/(R_3 + R_4)]R_1Cs + [R_4/(R_3 + R_4)](R_3/R_4 - R_1/R_2)} \tag{4.21}$$

It can be seen that when $R_3/R_4 = R_1/R_2$, the circuit behaves as a noninverting integrator. If perfect matching does not occur, it can be seen from Eq. 4.21 that the pole of the filter will be located at

$$p = \left(\frac{R_4}{R_3}\right)\left(\frac{1}{R_1C}\right)\left(\frac{R_1}{R_2} - \frac{R_3}{R_4}\right) \tag{4.22}$$

When tuning this circuit, it is necessary to assure that $R_3/R_4 > R_1/R_2$ to maintain stability. Unfortunately, if one intentionally attempts to make R_3/R_4 significantly greater than R_1/R_2 to avoid the stability problem, significant deviations in integrator performance from ideal will be observed. If the integrator is used in a feedback filter structure, the tuning of this circuit is generally not as critical, since the filter poles are likely to stay near the desired value even if R_1/R_2 is slightly larger than R_3/R_4.

The input impedance of this circuit is frequency-dependent and is given by

$$Z_{\text{in}} = R\left(\frac{RCs}{RCs - 1}\right) \tag{4.23}$$

At low frequencies this circuit may cause significant loading on preceding stages. Since this circuit uses one op amp and one capacitor, it may offer some economical advantages over the other noninverting configurations.

EXAMPLE 4.5 Design a noninverting integrator using the matched resistor technique that satisfies the expression

$$V_o = 10^4 \int V_i(t)\, dt$$

PROCEDURE

1 Using the circuit of Fig. 4.15*c*, we must determined R_1, R_2, R_3, R_4, and C. Taking the Laplace transform of the above equation, it follows that the transfer function must be of the form

$$\frac{V_o}{V_i} = \frac{10^4}{s}$$

2 From the gain expression given in Fig. 4.15*c* it thus follows that

$$\frac{2}{RC} = 10^4$$

3 The choice of C is arbitrary, although it must be nonelectrolytic. A reasonable value for a capacitor is $C = 0.01\ \mu F$. Solving for R, we obtain

$$R = R_1 = R_2 = 20\ k\Omega$$

4 The value of R_A is arbitrary. It is often economically advantageous to minimize the number of different components in a design. With this in mind, a reasonable choice for R_A is 20 kΩ. Thus $R_3 = R_4 = 20$ kΩ.

High-Performance Noninverting Integrator The circuit of Fig. 4.15*d* is less affected by the GB effects of the op amp than are the previous noninverting circuits. This can be considered the noninverting counterpart to the circuit of Fig. 4.14*c* and will prove practical at higher frequencies or for higher-order integrator-based filter structures. Reference [2] provides technical details about this structure.

Generalized Immittance Converters

The building blocks considered thus far have all been designed to have a prescribed relationship between the input and output voltage variables. The passive components in these circuits have been restricted to resistors and capacitors. The active filter designer would have considerably more flexibility if inductors were available. Unfortunately, economic and performance considerations preclude the use of inductors.

In this section two-terminal building blocks that have useful input impedance characteristics are discussed. Resistors, capacitors, and op amps are configured in such a manner as to obtain these characteristics. In addition to circuits that behave as inductors (synthetic inductors), other more exotic two-terminal networks, which are used later in this book, are introduced. A more detailed discussion of their applications is presented in Chapter 10.

The basic generalized impedance converter (GIC) is shown in block diagram form in Fig. 4.16, where it is *assumed* that input impedance may be expressed as

$$Z_{in}(s) = H(s)Z_L^{\pm 1}(s) \tag{4.24}$$

where $H(s)$ is a function of s that is independent of the external load, $Z_L(s)$. The requirement that the input impedance is either directly or inversely proportional to the termination impedance, Z_L, places restrictions on the nature of the network designated by $H(s)$. Many authors attach different names to converters that satisfy Eq. 4.24 depending upon the nature of $H(s)$ and the sign of the exponent. These

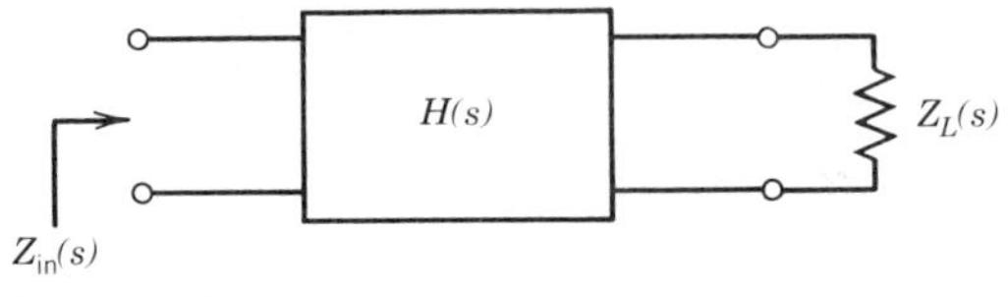

Figure 4.16 GIC structure.

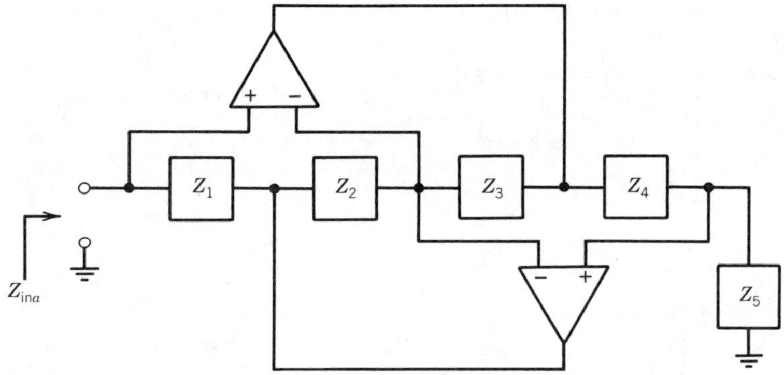

(a) Positive.

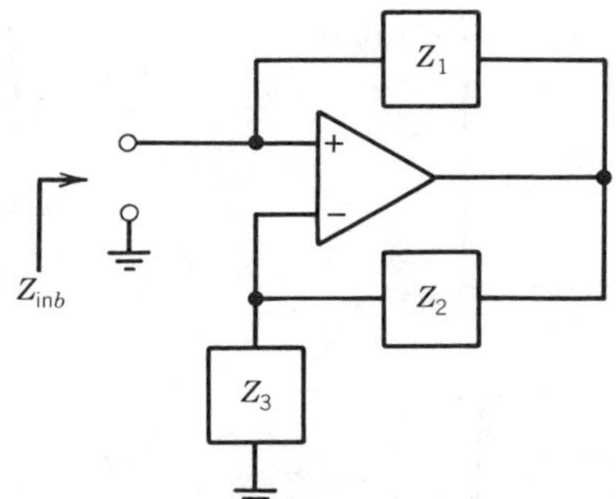

(b) Negative.

Figure 4.17 GIC circuits.

include NIC (negative immittance converter or negative impedance converter), PIC (positive immittance converter or positive impedance converter), Gyrator, GIV (generalized immittance inverter), and so on. These are all special cases of the GIC defined above.

Two popular GICs are shown in Fig. 4.17. A more extensive discussion of these and other GICs can be found in [3] to [7]. The input impedance of these circuits are given, respectively, by

$$Z_{\text{in}a} = \frac{Z_1 Z_3 Z_5}{Z_2 Z_4} \tag{4.25}$$

$$Z_{\text{in}b} = \frac{-Z_1 Z_3}{Z_2} \tag{4.26}$$

Impedance multiplication and inversion possibilities are apparent with these structures as well as the negative impedances attainable with the circuit of Fig. 4.17*b*. Several specific and useful applications of these circuits are presented in Fig. 4.18. A brief discussion of each of these circuits follows.

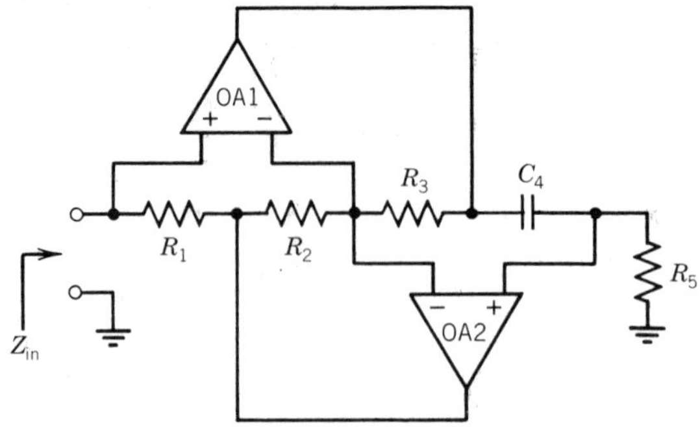

(*a*) Synthetic inductor.

$$Z_{in} = sL_{eq}, \text{ where } L_{eq} = \frac{C_4 R_1 R_3 R_5}{R_2}$$

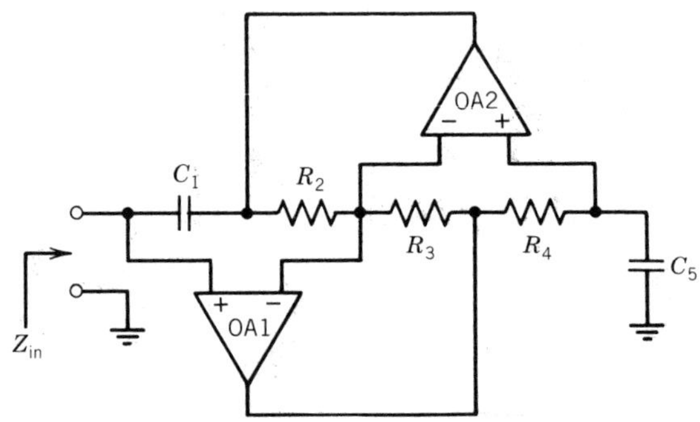

(*b*) FDNR.

$$Z_{in} = \frac{R_3}{s^2 C_1 C_5 R_2 R_4}$$

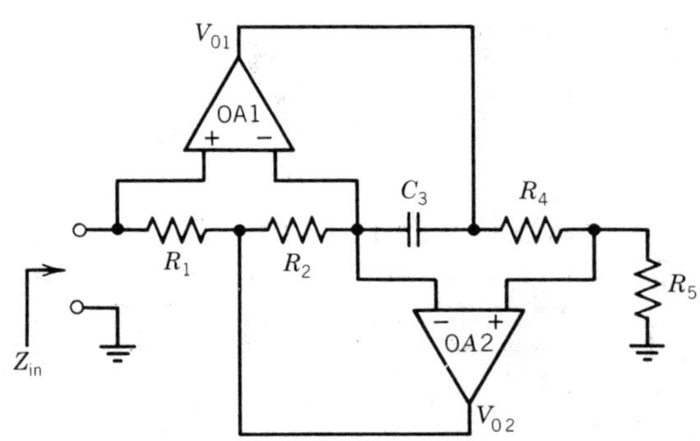

(*c*) Capacitance multiplication.

$$Z_{in} - \frac{1}{sC_{eq}}, \text{ where } C_{eq} = \frac{R_2 R_4 C_3}{R_1 R_5}$$

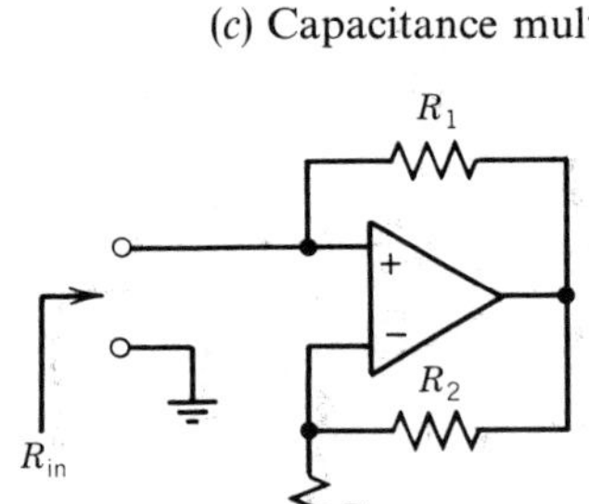

$$R_{in} = -\frac{R_1 R_3}{R_2}$$

(*d*) Negative resistor.

Figure 4.18 Impedance simulation circuits.

Synthetic Inductors The synthetic inductor shown was obtained by picking $Z_4 = 1/sC_4$, $Z_1 = R_1$, $Z_2 = R_2$, $Z_3 = R_3$, and $Z_5 = R_5$ in the GIC of Fig. 4.17*a*. It follows from Eq. 4.25 that the input impedance of this circuit is equivalent to that of an inductor of value

$$L_{\text{eq}} = \frac{C_4 R_1 R_3 R_5}{R_2} \tag{4.27}$$

This circuit is recognized as one of the better inductance simulation schemes [3]. Reasonable performance can also be obtained by interchanging R_2 and C_4.

The two major factors that ultimately limit the use of this synthetic inductor are the finite GB of the op amp and the fact that one terminal is grounded. The former limits the frequency range and performance of active filters employing the device, whereas the latter restricts the topologies in which the circuit can be used. Floating inductors that involve additional circuit complexity have been proposed, but performance and applications are generally less impressive.

EXAMPLE 4.6 (a) Convert the passive second-order bandpass filter shown in Fig. 4.19 to an active RC filter with a center frequency (natural frequency) of 2 kHz and a 3 dB bandwidth of 200 Hz using 741 type op amps by replacing the inductor with the synthetic inductor of Fig. 4.18*a*. (b) What are the effects of the operational amplifier-gain bandwidth products on the inductor?

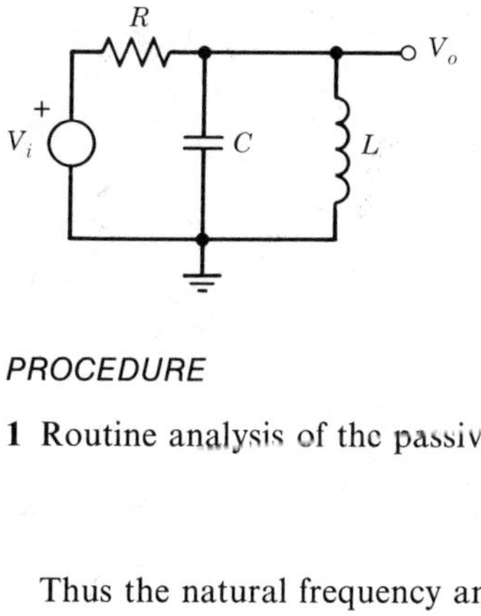

Figure 4.19 Second-order passive bandpass filter.

PROCEDURE

1 Routine analysis of the passive filter yields the gain expression

$$\frac{V_o}{V_i} = \frac{s/RC}{s^2 + s/RC + 1/LC}$$

Thus the natural frequency and 3 dB bandwidth are given by the expressions

$$f_0 = \frac{1}{2\pi\sqrt{LC}} = 2000 \text{ Hz}$$

$$\text{BW} = \frac{f_0}{Q} = \frac{1}{2\pi RC} = 200 \text{ Hz}$$

2 If we select $C = 0.01\ \mu\text{F}$ (a reasonable capacitor value), these last two expressions can be solved for R and L to yield

$$R = 79.6\ \text{k}\Omega$$

$$L = 0.633\ \text{H}$$

3 For the conversion to the active RC structure, it remains to equate L and L_{eq} of the synthetic inductor of Fig. 4.18a and determine R_1, R_2, R_3, R_5, and C_4. Selecting $C_4 = 0.01\ \mu F$ and assuming that $R_1 = R_2 = R_3 = R_5$, we obtain from the expression for L_{eq}

$$0.633\ \text{H} = (10^{-8}\ \text{F})(R_1^2)$$

which can be solved for R_1 to yield

$$R_1 = 7.96\ \text{k}\Omega$$

The completed circuit is shown in Fig. 4.20.

4 To determine the effects of the gain bandwidth products of the op amps on the synthetic inductance, it will be assumed that $R_1 = R_2 = R_3 = R_5$ and that the op amps are matched and modeled with the single-pole model with gain bandwidth products $\omega_{t1} = \omega_{t2} = \omega_t$.

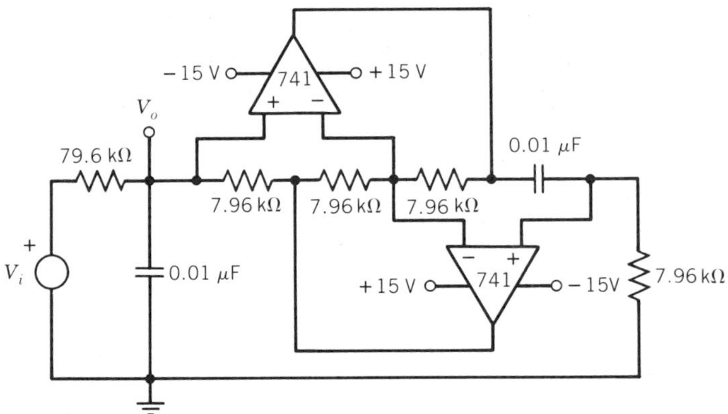

Figure 4.20 Second-order active bandpass filter using synthetic inductor.

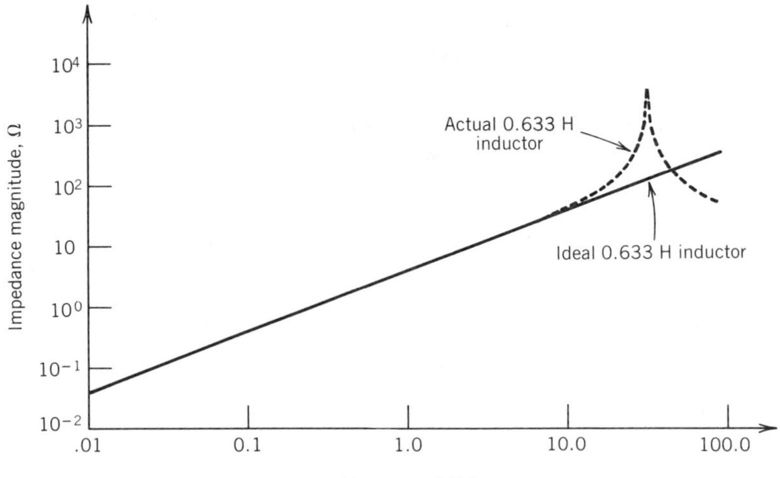

Figure 4.21 High-frequency limitations of synthetic inductor of Fig. 4.20.

5 From a somewhat tedious circuit analysis, we obtain the expression for the input impedance of the synthetic inductor

$$Z_{in}(s) = R_1 \left[\frac{(s^2/\omega_t^2)(1 + R_1C_4s) + (s/\omega_t)(1 + R_1C_4s) + R_1C_4s/2}{(s^2/\omega_t^2)(1 + R_1C_4s) + (s/\omega_t)(1 + R_1C_4s) + \frac{1}{2}} \right]$$

6 From Table 4.1, ω_t of the 741 is 6.28×10^6 rad/s. $Z_{in}(j\omega)$ for $R_1 = 7.96$ kΩ and $C_4 = 0.01$ μF is plotted in Fig. 4.21 and compared to the ideal response of the 0.633 H inductor. Note that at 2 kHz the synthetic inductor performance deviates very little from the desired response, but that at higher frequencies the deviations caused by the finite-gain–bandwidth product are very significant, thus limiting the use of the synthetic inductor to low-frequency applications.

Frequency-Dependent Negative Resistors (FDNR) A FDNR is defined to be a two-terminal device with input impedance given by

$$Z_i(s) = \frac{K}{s^2} \tag{4.28}$$

where K is some positive constant. Evaluation at $s = j\omega$ shows that

$$Z_i(j\omega) = \frac{K}{-\omega^2} \tag{4.29}$$

The $-\omega^2$ expression in the denominator, which is real, makes the ohm the fundamental unit of the FDNR and gives rise to the designation "frequency-dependent negative resistance." Three FDNR circuits can be obtained directly from the circuit of Fig. 4.17*a* by selecting exactly two of (Z_1, Z_3, and Z_5) to be capacitors and the remainder of the impedances to be resistors. The circuit of Fig. 4.18*b* is recognized as one of the better FDNRs [3]. Applications of this element are described in Chapter 10.

As in the inductance simulation scheme, the two major factors that ultimately limit the FDNR performance are the finite GB of the op amp and the fact that one terminal of the FDNR is grounded.

Capacitance Multiplication In some filter applications there are requirements for large-valued capacitors. These requirements are most common in low-frequency circuits. Large-valued capacitors are typically physically large and expensive. Furthermore, if electrolytic capacitors are used, they are typically too lossy and can be charged only in one direction; this precludes their use in many filter applications.

Capacitance multiplication schemes can be used to multiply the value of a small capacitor by a large positive constant to obtain a large effective capacitance. One such capacitance multiplication scheme is shown in Fig. 4.18*c*. Large multiplicative constants are possible with the ratio R_2R_4/R_1R_5. Capacitance multiplication circuits are not without limitations, however. The multiplication scheme shown in Fig. 4.18 is restricted to applications with one capacitor terminal grounded.

When large capacitors are rapidly discharged, a significant current flows. For proper operation of the capacitance multiplication circuit, the discharge rate must be low enough so that the maximum output current drive of the op amp is not exceeded. Often the voltage at internal nodes in capacitance multiplication schemes is much higher than the voltage across the equivalent capacitance.

To prevent saturation of the op amps, it is necessary to limit the signal applied to the circuit. The peak op amp output voltages for a sinusoidal voltage of $V_m \sin 2\pi ft$ across the input terminals of the equivalent capacitance are

$$\text{OA1:} \quad V_{o,\text{peak}} = V_m\left(1 + \frac{R_4}{R_5}\right) \tag{4.30}$$

$$\text{OA2:} \quad V_{o,\text{peak}} = V_m\sqrt{1 + (R_1 C_{\text{eq}} 2\pi f)^2} \tag{4.31}$$

As with other active filter structures, the maximum rate of change of the output voltage for each op amp must not exceed the slew rate of the op amp. It follows from Eqs. 4.13, 4.30, and 4.31 that for a sinusoidal signal of amplitude V_m and frequency f that exists across the input terminals of C_{eq}, the amplitude and frequency must be constrained by

$$V_m < \frac{SR}{2\pi f}\frac{R_5}{R_4 + R_5} \tag{4.32}$$

and

$$V_m < \frac{SR}{2\pi f\sqrt{1 + (R_1 C_{\text{eq}} 2\pi f)^2}} \tag{4.33}$$

to circumvent SR distortion.

EXAMPLE 4.7 The capacitance multiplication scheme of Fig. 4.18*c* is used to realize a 100 μF capacitor by selecting $C_3 = 0.01\ \mu$F, $R_4 = 100$ kΩ, $R_2 = 100$ kΩ, $R_1 = 1$ kΩ and $R_5 = 1$ kΩ and using 741 type op amps biased at ± 15 V as shown in Fig. 4.22. The maximum

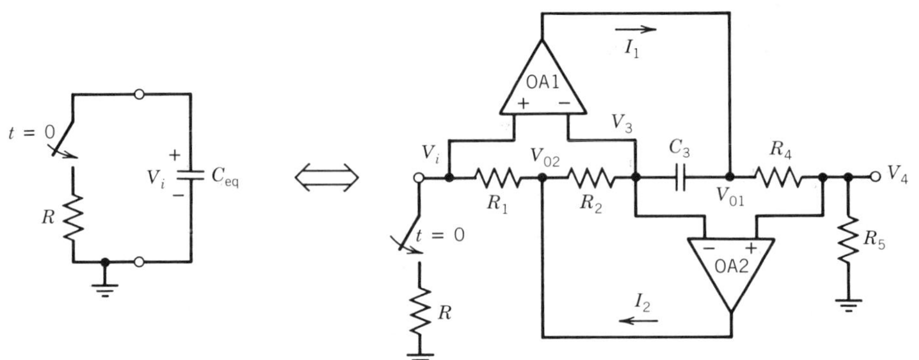

Figure 4.22 Capacitance multiplication circuit.

applied voltage and discharge rate of capacitance multiplication circuits are affected, to a large extent, by the parameters $I_{o,\max}$ $V_{o,\max}$ and the SR of the op amps. Assuming that the gain-bandwidth product of both op amps is infinite:

a. Determine the maximum voltage that can be applied to the capacitance multiplication circuit.

b. The 100μF simulated capacitor is to be discharged by closing the switch in the circuit of Fig. 4.22. Determine the minimum value of R permissible for normal operation (i.e., all op amps operating linearly) of the circuit shown if the initial capacitor voltage is 100 mV and the switch is closed at $t = 0$.

c. Determine the time it takes to reduce the capacitor voltage to 1% of its initial value with the minimum size resistor determined in (b).

PROCEDURE

1 In general, an investigation of $V_{o,\max}$, $I_{o,\max}$ and SR effects for multiple op amp active filters is somewhat tedious. This circuit is no exception. Since GB is assumed infinite, $V_i(t)$ must be sufficiently constrained to circumvent $I_{o,\max}$, $V_{o,\max}$, and SR limitations for *each* op amp. These conditions are

$$I_{1,2} < I_{o,\max} \tag{4.34}$$

$$V_{o,\min} < V_{01,02} < V_{o,\max} \tag{4.35}$$

$$\frac{dV_{01,02}}{dt} < SR \tag{4.36}$$

2 From Table 4.1 the conditions in this case become

$$I_{1,2} < 20 \text{ mA} \tag{4.37}$$

$$V_{01,02} < 13 \text{ V} \tag{4.38}$$

$$\left|\frac{dV_{01,02}}{dt}\right| < 0.5 \text{ V}/\mu\text{s} \tag{4.39}$$

3 Routine analysis yields

$$I_1 = \frac{V_i}{R_5} + \frac{C_3 R_4}{R_5}\frac{dV_i}{dt} \tag{4.40}$$

$$I_2 = -C_{eq}\left(1 + \frac{R_1}{R_2}\right)\frac{dV_i}{dt} \tag{4.41}$$

$$V_{01} = \left(1 + \frac{R_4}{R_5}\right)V_i \tag{4.42}$$

$$V_{02} = V_i - R_1 C_{eq}\frac{dV_i}{dt} \tag{4.43}$$

$$\frac{dV_{01}}{dt} = \left(1 + \frac{R_4}{R_5}\right)\frac{dV_i}{dt} \tag{4.44}$$

and

$$\frac{dV_{02}}{dt} = \frac{dV_i}{dt} - R_1 C_{eq}\frac{d^2 V_i}{dt^2} \tag{4.45}$$

4 With a constant input voltage, $dV_i/dt = 0$, so by substituting Eqs. 4.40 to 4.45 into the constraints (4.37 to 4.39), we must choose V_i to satisfy

$$\frac{V_i}{R_5} < 20 \text{ mA} \tag{4.46}$$

$$0 < 20 \text{ mA} \tag{4.47}$$

$$\left(1 + \frac{R_4}{R_5}\right) V_i < 13 \text{ V} \tag{4.48}$$

$$V_i < 13 \text{ V} \tag{4.49}$$

$$0 < 0.5 \text{ V}/\mu\text{s} \tag{4.50}$$

$$0 < 0.5 \text{ V}/\mu\text{s} \tag{4.51}$$

5 With the component values listed, the $V_{o,\max}$ limitations dominate and restrict V_i to

$$V_i < 129 \text{ mv} \tag{4.52}$$

Note that this large capacitance multiplication has been obtained at the expense of serious limitations on signal swing.

6 If the switch in Fig. 4.22 closes at $t = 0$, it follows by an analysis of the assumed linear first-order network that

$$V_i(t) = V_{i0} e^{-t/RC_{eq}} \tag{4.53}$$

where V_{i0} is the input voltage at $t = 0$.

7 Again, for all $t \geq 0$, $V_i(t)$ must be constrained to satisfy Eqs. 4.37 to 4.39. In this case

$$I_1 = \frac{V_{i0}}{R_5} e^{-t/RC_{eq}} - \frac{R_1}{RR_2} V_{i0}\, e^{-t/RC_{eq}} \tag{4.54}$$

$$I_2 = \left(1 + \frac{R_1}{R_2}\right)\left(\frac{V_{i0}}{R}\right) e^{-t/RC_{eq}} \tag{4.55}$$

$$V_{01} = \left(1 + \frac{R_4}{R_5}\right) V_{i0}\, e^{-t/RC_{eq}} \tag{4.56}$$

$$V_{02} = V_{i0}\, e^{-t/RC_{eq}}\left(1 + \frac{R_1}{R}\right) \tag{4.57}$$

$$\frac{dV_{01}}{dt} = \left(1 + \frac{R_4}{R_5}\right)\left(\frac{-1}{RC_{eq}}\right) V_{i0}\, e^{-t/RC_{eq}} \tag{4.58}$$

$$\frac{dV_{02}}{dt} = V_{i0}\, e^{-t/RC_{eq}}\left(\frac{-1}{RC_{eq}}\right)\left(1 + \frac{R_1}{R}\right) \tag{4.59}$$

8 For this circuit the left-hand side of Eqs. 4.37 to 4.39 all attain their maximum value at $t = 0$, so the constraints become

$$0.1\left(\frac{1}{1K} - \frac{1K}{R100K}\right) < 20 \text{ mA} \tag{4.60}$$

$$\frac{(1.01)(0.1)}{R} < 20 \text{ mA} \tag{4.61}$$

$$(101)(0.1) < 13 \text{ V} \tag{4.62}$$

$$\left(1 + \frac{1000}{R}\right)(0.1) < 13 \text{ V} \tag{4.63}$$

$$\frac{(101)(0.1)}{\mathrm{R}(100\ \mu\mathrm{F})} < 0.5\ \mathrm{V}/\mu\mathrm{s} \tag{4.64}$$

$$\frac{(1 + 1000/R)(0.1)}{R(100\ \mu\mathrm{F})} < 0.5\ \mathrm{V}/\mu\mathrm{s} \tag{4.65}$$

9 The limitations on R are dominated by the $V_{o,\max}$ constraint of op amp 2 as given by Eq. 4.61.

$$R > 7.75\ \Omega$$

10 The time it takes the voltage to decay to 1% of its initial value, t_d, can be obtained from Eq. 4.53 by observing

$$0.01 V_{io} = V_{io}\, e^{-t_d/RC_{eq}}$$

Thus $t_d = -RC_{eq} \ln(0.01) = 3.6$ ms. This relatively long minimum discharge time may be unacceptable in some applications.

Negative Resistors A negative resistor obtained from the circuit of Fig. 4.17*b* is shown in Fig. 4.18*d*. Occasionally, negative resistors are placed in series with a passive impedance with a known parasitic series resistance, such as inductors, to effectively cancel the parasitics and thus improve the characteristics of the given device. Negative capacitors and inductors can also be obtained from the circuit of Fig. 4.17*b* by the proper choice of Z_2 and Z_3, as can be seen from Eq. 4.26.

4.4 SPECIALIZED DEVICES AND APPLICATIONS

Many op amp applications involve small low-frequency signals; require little output current; are powered by the output of a dual power supply of ± 15 V, which is driven by a 60 Hz power line; and have characteristics that are to remain fixed after design is complete. For these applications a large number of high-quality op amps that are reasonably inexpensive are available. The 741 and 356 are two of the more popular devices used.

Needs exist, however, for active filters in environments that violate some or all of the above situations. A brief discussion of applications of active filters in more demanding and/or specialized applications is presented in this section. The emphasis here is on methods of addressing these problems rather than on extensive details about specific applications.

Single-Power-Supply Operation

Most op amps are designed to operate from dual power supplies. Often only a single dc power supply is available to bias the op amps. Such situations are typical in battery-operated equipment.

There are several methods of addressing the single-power-supply problem. One method, shown in Fig. 4.23, involves using a resistor divider or a resistor and zener diode in series to generate a voltage that is intermediate between the supply terminals. This can be thought of as a method of generating a dual power supply from a single supply. The intermediate voltage serves as the system ground and the devices thus are biased with "dual supplies" relative to this ground. The

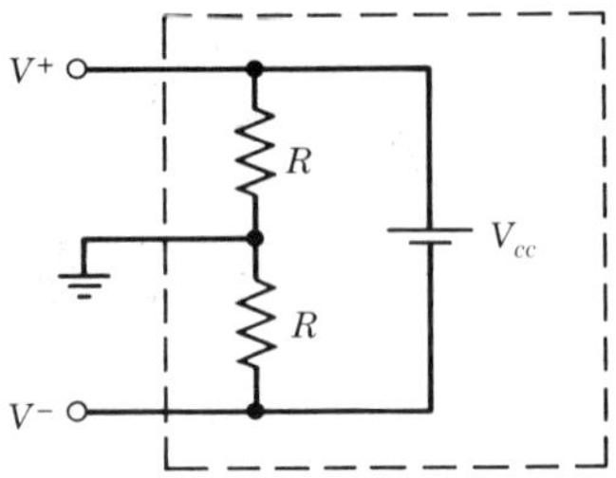

(a) Resistive divider.

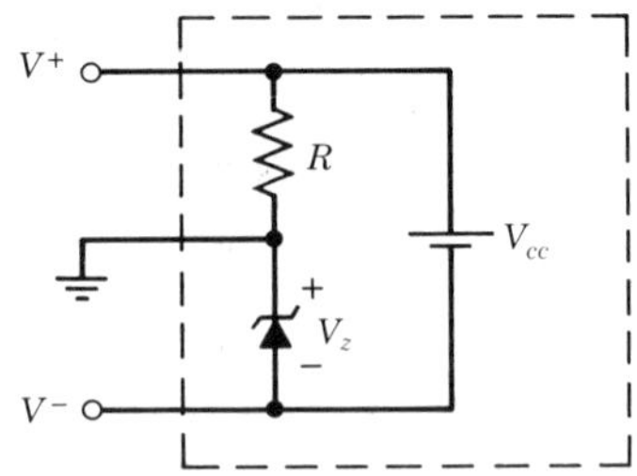

(b) Zener reference.

Figure 4.23 Generation of a "dual" power supply from a single power supply.

dc supply must be floating with respect to the system ground to use this technique. With this approach most dual supply op amps can be used.

The following points must be considered before adopting this approach.

1 Considerable current flow and consequently power dissipation may be required in the divider network to maintain the ground voltage when input signals cause varying amounts of current to flow into the ground node. This may be unacceptable in battery-operated equipment.

2 The ground established with the divider will often not be the same as the ground on the signal input. This may require capacitor coupling of the input or a subsequent level shift of the output.

3 The output voltages of the op amps are referenced with respect to the intermediate voltages established by the divider rather than to one of the terminals of the dc source.

4 The effective supply voltages, V^+ and V^-, will be much smaller than the supply voltage, V_{cc}. Typically, $V^+ = -V^- = V_{cc}/2$. This may seriously limit the output voltage swing of the op amps.

5 For proper operation with most dual supply op amps the input voltage to the op amp may not extend to the lower rail. For example, if a 741 op amp were used, the minimum voltage that could be applied to either input terminal would typically be $V^- + 2$ V. The inability to take the op amp input voltages down to V^- often causes designs to be more complicated in single-supply operations.

EXAMPLE 4.8 The amplifier shown in Fig. 4.24 employs the single to dual supply converter of Fig. 4.23 along with a 741-type operational amplifier. Determine the maximum peak-to-peak output voltage swing for this amplifier and compare it to that possible if dual ± 10 V supplies were available.

PROCEDURE

1 If dual ± 10 V supplies were available, the output voltage could swing from approximately $V^+ - 2$ V to $V^- + 2$ V $= 16$ V·p-p. With the "dual-supply" scheme shown in the figure, the voltages V^+ and V^- will be affected by the input signal itself, thus limiting the output voltage swing.

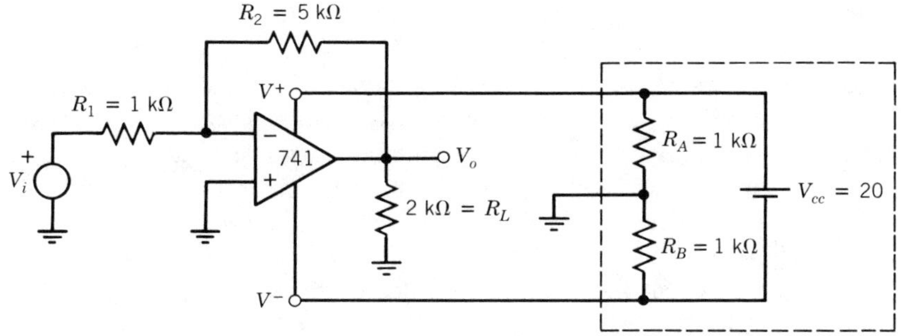

Figure 4.24 Single-supply inverting amplifier using 741 op amps.

To obtain these limits, we shall find an expression for V^+ and V^- in terms of the output voltage. From this expression and the observation that the output of the op amp must satisfy the relationship $V^+ - 2 > V_o > V^- + 2$, we can obtain the maximum output.

2 By summing *all* currents flowing into the ground lead, we obtain (the current out of the "+" terminal of the op amp has been assumed to be arbitrarily small),

$$\frac{-V_i}{R_1} + \frac{V_o}{R_L} + \frac{V^+}{R_A} + \frac{V^-}{R_B} = 0 \tag{4.66}$$

However, it is easily verified that

$$V^+ - V^- = 20 \text{ V} \tag{4.67}$$

and

$$V_o = \frac{-R_2}{R_1} V_i \tag{4.68}$$

3 Substituting these expressions into Eq. (4.66) and assuming $R_A = R_B$, we obtain

$$\frac{V_o}{R_2} + \frac{V_o}{R_L} + \frac{2V^+}{R_A} + \frac{-20}{R_A} = 0 \tag{4.69}$$

But recall that the maximum output voltage is given by $V_o = V^+ - 2$ V. Thus substituting for V^+ into Eq. 4.69 we obtain

$$V_{o,\max} = \frac{16 \text{ V}}{2 + R_A/R_2 + R_A/R_L}$$

Substituting the values for R_A, R_2, and R_L, we obtain

$$V_{o,\max} = 5.93 \text{ V}$$

4 By symmetry, it follows that

$$V_{o,\min} = -5.93 \text{ V}$$

Thus the peak-to-peak output voltage swing is 11.86 V, which is considerably less than the 16 V obtainable with dual supplies. Reducing R_A improves the signal swing at the expense of increased power dissipation by the divider network.

Another method involves using op amps that are specifically designated for single-supply operation. One such device is the LM 324 (which is actually a quad device). Both the op amp input voltages and output voltage can generally be taken to V^- with single supply op amps, although the inputs and output are still typically restricted to be from 1 V to 2 V below V^+. Single-supply op amps can also be used with dual supplies but may not be performance and/or cost competitive with standard dual-supply devices. Because of restrictions placed upon the op amp designers which result from the physical characteristics of both bipolar and MOS transistors, the user is advised to carefully read manufacturer's data sheets and application notes when using single-supply op amps in applications where either the inputs or outputs are expected to approach the rails. For example, the front page of the data sheet of the National Semiconductor LM 324 op amp [8] lists as a feature the ability of the op amp output voltage to go to ground in single-supply operation, but the data sheet later indicates that the op amp output can sink *very little* current when the output voltage is less than 1 diode drop above V^-. This limited current-sinking capability for low output voltages may be unacceptable even in some standard applications.

A third method involves generating a negative voltage directly from the single dc power supply. If the system is large, commercial voltage inverters that generate negative voltages with significant current drive capabilities may be justifiable. The better voltage inverters have a high-energy-conversion efficiency but would significantly increase the overall cost for small systems. For small systems that require modest current drive from the negative supply, the small voltage inverter ICs, such as the Intersil 7660, may be practical for generation of a negative supply voltage. Such ICs typically come in 8-pin mini-dip packages and compare in cost to a single op amp. One such device can often be used to provide the negative supply voltage for several op amps.

Power Boosting

Most general-purpose op amps do not have the current drive capability for driving large loads. One method of increasing the current drive of an op amp is to follow a standard op amp with a bipolar junction transistor (BJT). An example of this technique is shown in Fig. 4.25 where the output current drive of the basic inverting amplifier has been significantly increased. Note that the voltage gain

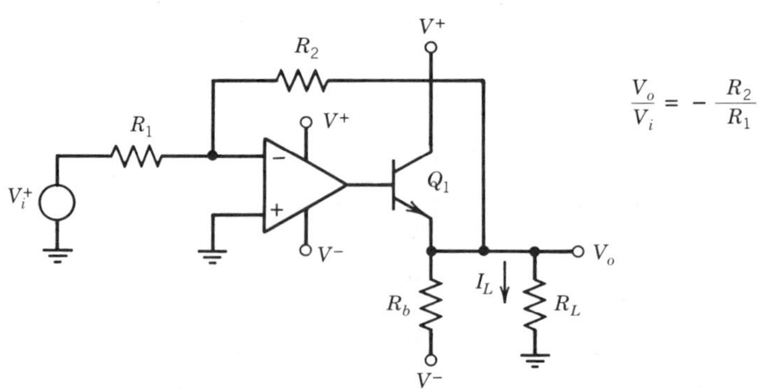

Figure 4.25 Power boosting.

expression is not affected by the BJT. One disadvantage of this circuit is the additional resistor R_b. This must be selected small enough so that whenever I_L is negative, a net positive emitter current flows to keep the BJT biased in the active region.

Special-purpose op amps that have current drive capabilities in the 1 amp range or higher are also available. Compared to the previous designs, reductions in circuit complexity are possible with these devices.

EXAMPLE 4.9 The power-boosting circuit shown in Fig. 4.26 is used to drive the 100 Ω load.

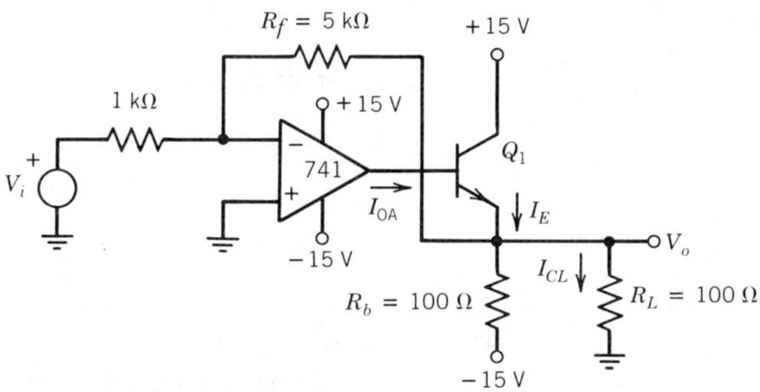

Figure 4.26 Example 4.9.

a. What is the maximum sinusoidal current that this power-boosting circuit can supply to the load without distortion? Assume that the BJT, Q_1, has a β of 150 and a sufficiently high power dissipation capability to prevent overheating.

b. How does this compare with the current that can be supplied to a 100 Ω load without the power-boosting stage?

PROCEDURE

1 The maximum and minimum output voltages will first be obtained. The smaller (in magnitude) of these will determine the maximum sinusoidal output voltage. From Ohm's law, the output current can then be readily obtained.

2 To keep the driver transistor from cutting off, I_E must be greater than zero. Summing currents at the emitter node (neglecting the current through the 5 kΩ feedback resistor), we obtain

$$I_E = \frac{V_o}{100} + \frac{V_o + 15}{100} > 0$$

or

$$V_o > -15/2 = -7.5 \text{ V}$$

3 The maximum output voltage will either be limited by $I_{o,\max}$ or $V_{o,\max}$. It can be seen that $V_{o,\max}$ restricts the output voltage to be one diode drop less than $V_{o,\max}$ due to the base-emitter voltage drop of the BJT. With ± 15 V supplies, $V_{o,\max} = 13$ V. Thus

$$V_o \leq 12.4 \text{ V}$$

4 The $I_{o,\max}$ limitation places an upper bound on the base current of Q_1. Since the base current is approximately the emitter current divided by β, it follows that to avoid $I_{o,\max}$ limitations (again neglecting current through R_f) one must require

$$\frac{V_o/100\ \Omega + (V_o + 15)/100\ \Omega}{150} < I_{o,\max}$$

which, for the 741 with $I_{o,\max} = 20$ mA, becomes

$$V_o < 142.5\ \text{V}$$

5 It thus follows that the -7.5 V restriction dominates, limiting the rms output load current to

$$I_{\text{OL,rms}} < \frac{7.5\ \text{V}}{(\sqrt{2})100\ \Omega} = 53\ \text{mA}$$

6 With no power boosting stage, $I_{o,\max}$ will limit the output current to

$$I_{\text{OL,rms}} = \frac{20\ \text{mA}}{\sqrt{2}} = 14.14\ \text{mA}$$

Low-Voltage Operational Amplifiers

Low supply voltages and/or very low power dissipation requirements place restrictions on some designs. These are particularly common on lightweight battery-operated systems. Op amps are available that can be biased at ± 1 V and dissipate very little power. These op amps typically have inputs and outputs that must allow for at least one diode drop for both input and output signals from V^+ and/or V^-. These limitations restrict the signal amplitudes quite severely, although applications still exist. Amplifiers such as the LM 358 dissipate very little power and can be operated at low supply voltages.

High-Frequency Operational Amplifiers

The GB of most general purpose op amps, which is in the 300 kHz to 5 MHz range, limits most active filters to the audio frequency range. Op amps with GBs beyond 50 MHz are available and can be used to extend the operating range of amplifiers and/or filters. These devices are considerably more expensive and challenging to apply. Special precautions relating to component placement, foil patterns, and lead lengths must be considered when using these devices. Some are stable without external compensation only for large feedback gains which makes filter design quite complicated.

Externally Compensated Amplifiers

Some operational amplifiers require an external compensation scheme, typically with a single capacitor, to maintain stability. The component values for the compensation circuit are a function of the feedback circuitry associated with the op amp. The LM 301 and LM 748 are popular general-purpose, externally compensated devices. Sufficient details are generally presented in the manufacturer's data sheets to readily compensate simple amplifiers such as those shown in

Fig. 4.11*a* and *b*. If the desired gain is large, significant improvements in amplifier bandwidth can be obtained by using externally compensated op amps rather than internally compensated devices. The use of uncompensated op amps in filter structures is very limited as a result of the complexity associated with the compensation problem.

Current-Controlled Voltage Source

The LM 3900 contains four amplifiers and is often called either a "quad op amp" or a "quad current mirror." It has an output voltage that is proportional to a difference in input currents rather than to a difference in input voltages, as is the case with a conventional op amp. As a consequence the device is a current-controlled voltage source (CCVS) and the op amp model discussed previously is no longer valid. Furthermore, it cannot be used directly in any of the building blocks considered in this chapter or as a direct replacement for an op amp in any of the filters discussed in later chapters. The device has merit, however, and can be used in filter design. Details about applications of such devices can be found in the manufacturer's application notes [9].

Operational Transconductance Amplifier

The Operational Transconductance Amplifier (OTA) is actually a voltage-controlled current source. The CA 3080 is one of the more popular OTAs. Since the output is a current rather than a voltage, it cannot be used directly in the filters presented in this book. The 3080, however, has a gain that is controllable over several decades by an external current. By using the dc current as a control variable, the OTA can be used to build voltage-controlled filters, voltage-controlled oscillators, and voltage-controlled amplifiers. Applications of the OTA can be found in the manufacturer's literature [10, 11].

4.5 TROUBLE-SHOOTING OPERATIONAL AMPLIFIER CIRCUITS

The operational amplifier is a durable device with a long expected lifetime. As such, if a piece of equipment that is in for repair has not been abused, the problems are unlikely to be with the op amps. If, however, the maximum input voltage specifications have been exceeded or a short circuit has been applied to the output of a nonprotected op amp, the amplifier itself may be defective. The following comments about trouble-shooting op amp circuits are applicable to the VCVS type (i.e., not CCVS or OTAs) op amps.

In-Circuit Testing

All op amps require that the appropriate biasing voltages at V^+ and V^- are present. As a first step in trouble-shooting it is a good idea to check to see that these dc voltages are at the specified value. A dc voltmeter or multimeter would be adequate for this check. A $\pm 10\%$ accuracy should be sufficient for this test. It is a good idea to make these measurements directly on the terminals of the op amp in case the problem is in a solder joint or socket. If these voltages are not correct, check the power supplies, sockets, and foil patterns. It is possible that excessive current flow has caused the deviation in voltage. If this or some other op amp in the circuit is causing the problem, the bad device will probably be hot because of the excessive power dissipation.

If the dc bias voltages are correct, a check of the voltage at the "+" and "−" pins of the op amp is in order. Again these can be dc measurements. The difference in these two voltages should be nearly zero (within a few mV) for linear applications. If this is not the case, the op amp is defective or some excitation is forcing these voltages to deviate from zero. Removal and external testing of the op amp are probably in order.

A third in-circuit dc test of the op amp, assuming that it passed the previous two tests, involves measuring the dc value of the output voltage. It should typically be between $V^- + 2$ V and $V^+ - 2$ V for standard dual-supply op amps.

A summary of these three steps, which can all be undertaken with an inexpensive multimeter or digital voltmeter, follows.

1 Measure the voltage at the V^+ and V^- terminals of the op amps and compare it with prescribed values.
2 Measure the voltage at "+" and "−" terminals of the op amps. If operating linearly, these voltages should be nearly equal.
3 Measure the op amp output voltage. It should be between $V^- + 2$ V and $V^+ - 2$ V in most circuits.

Operational Amplifier Test Circuits

An easy way to test an op amp is to put the device in a simple circuit where the desired relationship between the input and output is easy to verify. One such test circuit is shown in Fig. 4.27. The V^+ and V^- voltages should be chosen so as not to violate the absolute maximum values listed in the manufacturer's data sheets. For most op amps a choice of $V^+ = +15$ and $V^- = -15$ is reasonable.

If the circuit is excited with a 1 V sinusoid at 1 kHz, a 4.5 V sinusoid should appear at the output and the dc component of the output should be well under

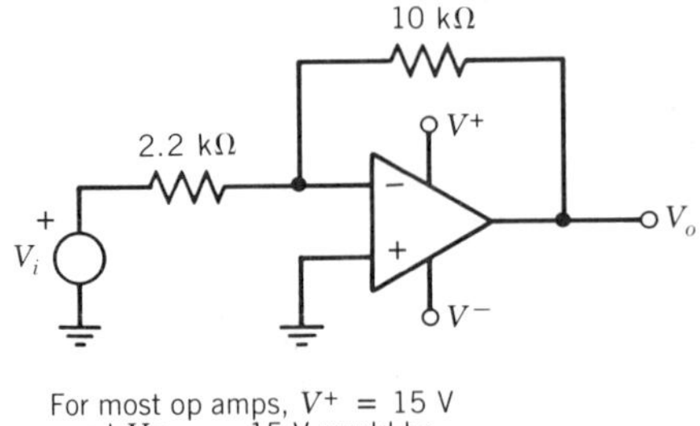

Figure 4.27 Op amp test circuit.

1 V if the source has no dc offset. An accuracy of 5 to 10% on the resistor values and voltage measurements should be adequate for these tests.

Although the gain, frequency of excitation, magnitude of excitation, and component values are not critical, it is important that they be selected so that the test circuit is not affected by the following

1 GB.

2 $I_{o,\max}$ limitations.

3 $V_{o,\max}$ limitations.

4 Slew-rate.

5 Offset voltage.

The component values listed in Fig. 4.27 will result in a test circuit that is not significantly affected by these parameters for most op amps.

Although an ac excitation is preferred, if such a source is not available a small dc excitation can be used. If a 1 V dc excitation is applied, the output should be -4.5 V.

If the op amp passes the above test, the problems are probably elsewhere in the circuit. If it fails the above test, it is defective and should be replaced.

4.6 SUMMARY

This chapter has presented essential information on the operational amplifier which forms the basic building block for almost all active filter structures.

Following a description of the basic amplifier parameters, a variety of active building blocks are introduced. These structures—namely, controlled sources, integrators, and generalized impedance converters—are used in later chapters to form active filters for realizing the various approximating functions of Chapter 2. The chapter concludes with a study of specialized devices and procedures for trouble shooting op amp circuits.

4.7 REFERENCES

[1] J. Millman and C. C. Halkias, *Integrated Electronics, Analog and Digital Circuits and Systems*, New York: McGraw-Hill, p. 386, 1972.

[2] R. L. Geiger and G. R. Bailey, "Integrator Design for High-Frequency Active Filter Applications," *IEEE Trans. on Circuits and Systems*, vol. CAS-29, pp. 595–603, September 1982.

[3] L. T. Bruton, *RC-Active Circuits, Theory and Design*, Englewood Cliffs, N.J.: Prentice-Hall, 1980.

[4] M. S. Ghausi and K. R. Laker, *Modern Filter Design*, Englewood Cliffs, N.J.: Prentice-Hall, 1981.

[5] A. S. Sedra and P. O. Brackett, *Filter Theory and Design: Active and Passive*, Portland, Ore.: Matrix, 1978.

[6] G. C. Temes and S. K. Mitra, eds., *Modern Filter Theory and Design*, New York: Wiley, 1973.

[7] P. Bowron and F. W. Stephenson, *Active Filters for Communications and Instrumentation*, London: McGraw-Hill, 1979.

[8] "LM124/LM224/LM324, LM124/LM224A/LM324A, LM2902 Low Power Quad Operational Amplifiers," *1982 Linear Databook*, Santa Clara, Calif.: National Semiconductor, 1982.

[9] T. M. Frederiksen, W. M. Howard, and R. S. Sleeth, Application Note AN. 72, "The LM3900—A New Current Differencing Quad of ± Input Amplifiers," Santa Clara, Calif.: National Semiconductor, September 1972.

[10] H. A. Wittlinger, "Application Note ICAN-6668, Applications of the CA3080 and CA3080A High-Performance Operational Transconductance Amplifiers," Somerville, N.J.: RCA, 1972.

[11] "LM13600/LM13600A Dual Operational Transconductance Amplifiers with Linearizing Diodes and Buffers," *1982 Linear Databook*, Santa Clara, Calif.: National Semiconductor, 1982.

[12] M. Yamatake, Application Note AN-24, "A Simplified Test Set for Op-Amp Characterization," *1980 Linear Applications Handbook*, Santa Clara, Calif.: National Semiconductor, 1980.

[13] J. E. Solomon, "The Monolithic Operational Amplifier: A Tutorial Study," *IEEE Journal of Solid State Circuits*, vol. SC-9, pp. 314–332, December 1974.

CHAPTER 5

Analysis of Active Filters

M. R. Lee
Sheffield City Polytechnic
England

5.1 INTRODUCTION

It is important that users of active filters assess proposed circuits for their applications. Analysis may be conducted manually or with the aid of computer programs.

Manual analysis is most appropriate at the beginning of the process. At this stage the user is faced with a wide choice of possible circuits. One circuit will eventually be chosen as being superior to the rest. It is not sufficient that the circuit simply realize a particular frequency response, but that it be capable of maintaining that response to the desired accuracy under the constraints of the construction process and the conditions of its operation. For example, as a result of tolerances, temperature, and other effects, practical circuits have component values that differ slightly from those required. However, the frequency response must remain within a specified error band around the desired response. Thus the sensitivity of the frequency response to errors in component values is an important parameter in assessing the suitability of each circuit. Similarly, certain circuit manufacturing processes impose constraints on component tolerance, so the performance of the contending circuits under these constraints must be analyzed.

A convenient starting point for these analyses is the transfer function of the circuit expressed as a function of symbols representing each component rather than numerical coefficients. This full symbolic representation is also invaluable when considering tuning procedures. The process by which these functions are derived is known as symbolic analysis.

A number of computer programs will perform symbolic analyses, but these are not widely available and tend to be expensive in computer time. The alternative approach is manual analysis and this chapter presents techniques that, for simple networks, are almost as fast as using a computer. However, the efficiency of manual analysis decreases rapidly as the complexity of the circuit increases. Thus manual analysis is recommended only for the initial appraisal of a number of possible circuits, where simple models or simplifying assumptions may be used.

Once this process has generated one or two probable circuits, further analysis is required to more accurately assess their performance under practical conditions. At this stage realistic models of the active devices must be used and the circuit extended to incorporate stray components. It is at this stage that computer-aided analysis is most valuable. Most circuit analysis programs produce their results as numerical values and will rapidly yield plots of frequency response amplitudes and phase shifts for complicated circuits. With careful modeling it is possible to produce frequency response plots that accurately predict the performance of the physical circuits.

There is no substitute for realistic modeling in the detailed assessment of a final design but, in the initial appraisal of a number of alternative designs, the symbolic transfer function is paramount. The remainder of this chapter is devoted to a number of techniques by which the symbolic transfer function may be obtained. The first technique, which uses the Nodal Admittance Matrix (NAM), is particularly suited to second-order active filters based on finite-gain voltage amplifiers (VCVS). A straightforward step-by-step formulation of this technique is presented; if followed rigorously, it produces very reliable results. Two-port matrix analysis techniques are then described for use in the analysis of the ladder structures common in LC filters. These structures are used as models when designing active filters based on the principle of inductance simulation.

Signal-flow graph analysis is then described. This technique is most useful in the analysis of filters based on analog integration and summation. These filters may be analyzed using the NAM technique, but signal-flow graph analysis is often more convenient.

Finally, sensitivity analysis is briefly covered. This will allow the reader to assess the effects of component value errors on the response of a filter circuit.

5.2 NODAL ANALYSIS

The method described in this section is ideal for the analysis of second-order filters incorporating either finite-gain voltage amplifiers or operational amplifiers. It allows rapid symbolic analysis of the transfer functions of all the circuits described in Chapters 6 through 9. Such active filters may be regarded as a passive RC network, into which one or more ideal VCVS are connected.

The analysis is performed by creating a nodal admittance matrix for the active network from that of the passive part alone. The active nodal admittance matrix is then solved to give the desired transfer function.

The nodal admittance matrix (NAM) describes the relationships between the voltages and currents at the ports of a multiple-port network. For example,

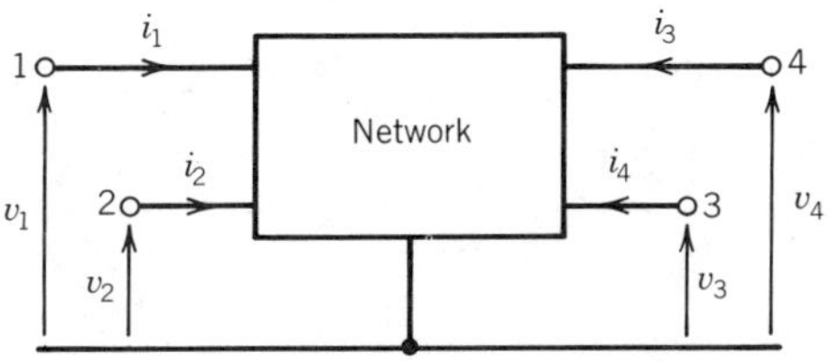

Figure 5.1 *Four-port network.*

consider the network shown in Fig. 5.1, which is a four-port network in which all ports share a common return connection or node. If port voltages and currents are defined as in Fig. 5.1, then the relationships between them may be written as in Eqs. 5.1. All the parameters y_{ij} have dimensions of admittance and are known as the admittance parameters.

$$\begin{aligned} i_1 &= y_{11}v_1 + y_{12}v_2 + y_{13}v_3 + y_{14}v_4 \\ i_2 &= y_{21}v_1 + y_{22}v_2 + y_{23}v_3 + y_{24}v_4 \\ i_3 &= y_{31}v_1 + y_{32}v_2 + y_{33}v_3 + y_{34}v_4 \\ i_4 &= y_{41}v_1 + v_{42}v_2 + y_{43}v_3 + y_{44}v_4 \end{aligned} \tag{5.1}$$

These equations may be expressed in matrix form as

$$\begin{bmatrix} i_1 \\ i_2 \\ i_3 \\ i_4 \end{bmatrix} = \begin{bmatrix} y_{11} & y_{12} & y_{13} & y_{14} \\ y_{21} & y_{22} & y_{23} & y_{24} \\ y_{31} & y_{32} & y_{33} & y_{34} \\ y_{41} & y_{42} & y_{43} & y_{44} \end{bmatrix} \begin{bmatrix} v_1 \\ v_2 \\ v_3 \\ v_4 \end{bmatrix} \tag{5.2}$$

The use of one node as the reference for all voltages yields a special case of the more general indefinite nodal admittance (INAM) equations and consequently simplifies the analysis. Furthermore, since almost all practical active filters have a common reference node, it does not restrict the application of the analysis technique.

If the ports shown in Fig. 5.1 are solely the input and output ports of a network that contains further internal nodes, then the admittance parameters are usually complicated functions of the network component values. This is the approach adopted in the two-port matrix analysis techniques described in Section 5.3. However, in forming the NAM, one must consider all interconnection nodes to be ports, including the internal nodes with no real outside connection. Figure 5.2*a* shows an example of this approach: Node 0 is the common return node; nodes 1, 3, and 4 represent true input or output ports; and node 2 is an internal node. In this case node 2 is regarded as a port, but the current i_2 flowing into the port from outside the network must be zero. It will be shown below that the zero input currents associated with internal nodes help to simplify the subsequent analysis.

The approach of regarding all nodes as ports has increased the number of equations, but it greatly simplifies the admittance parameters y_{ij}. Indeed, the parameters of the nodal admittance matrix may be written by inspection of the

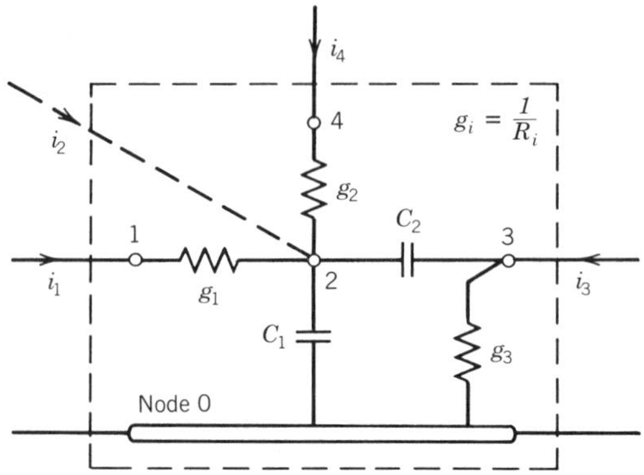

(a) Passive circuit.

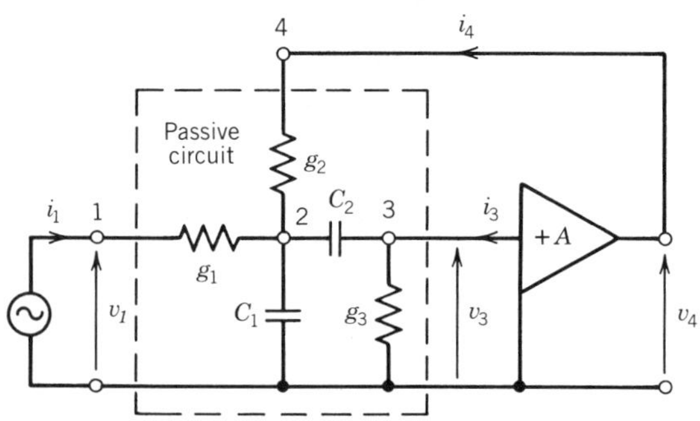

(b) Active circuit.

Figure 5.2 Second-order bandpass active filter.

passive circuit, using the following rules:

Main Diagonal Parameters

$$y_{ii} = \text{sum of all admittances connected to node } i.$$

Other Parameters

$$y_{ij} = y_{ji} = (-1)\ \{\text{sum of all admittances connected directly between nodes } i \text{ and } j\}$$

EXAMPLE 5.1 Obtain the NAM for the circuit shown in Fig. 5.2*a*.

PROCEDURE

Direct application of the above rules to the example circuit yields:

$$y = \begin{bmatrix} g_1 & -g_1 & 0 & 0 \\ -g_1 & (g_1 + g_2 + sC_1 + sC_2) & -sC_2 & -g_2 \\ 0 & -sC_2 & (g_3 + sC_2) & 0 \\ 0 & -g_2 & 0 & g_2 \end{bmatrix} \tag{5.3}$$

The parameters with value zero occur when no component is connected directly between the nodes considered.

Now consider the effect of connecting an ideal voltage amplifier into the passive circuit. The voltage at the node connected to the amplifier output is no longer independent, but is now determined by the node voltages connected to the amplifier input. The relationships between the amplifier input and output voltage are shown in Fig. 5.3. At this stage we also assume that the amplifier has zero output impedance and infinite input impedance as indicated in Fig. 5.3. Thus the presence of an ideal amplifier in the passive circuit introduces the following constraints:

$$v_o = Av_{\text{in}} \text{ for a single-input amplifier}$$

or

$$v_o = A(v_b - v_a) \text{ for a differential amplifier}$$

where v_o, v_{in}, v_a, and v_b are the voltages at the nodes to which the amplifier output and inputs are connected. In the case of the example circuit shown in Fig. 5.2*b*, the constraint becomes $v_4 = Av_3$ and Eqs. 5.1 may be rewritten as in Eqs. 5.4.

$$\begin{aligned}
i_1 &= y_{11}v_1 + y_{12}v_2 + y_{13}v_3 + y_{14}Av_3 \\
i_2 &= y_{21}v_1 + y_{22}v_2 + y_{23}v_3 + y_{24}Av_3 \\
i_3 &= y_{31}v_1 + y_{32}v_2 + y_{33}v_3 + y_{34}Av_3 \\
i_4 &= y_{41}v_1 + y_{42}v_2 + y_{43}v_3 + y_{44}Av_3
\end{aligned} \tag{5.4}$$

Note that v_4 has disappeared and that the two right-hand columns may be combined to give the terms in v_3. Furthermore, it may be shown that the fourth equation is redundant for our purposes so that a new constrained matrix may be formed as in Eq. 5.5. Note that the redundant row is that associated with the current delivered by the amplifier output. In general, all rows representing currents from amplifier outputs can be eliminated.

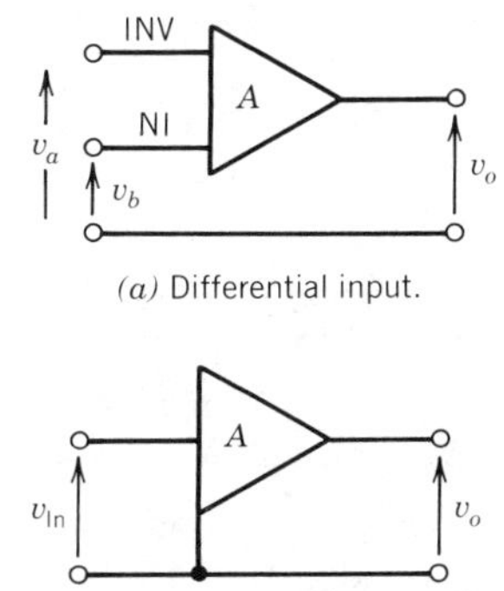

(a) Differential input.

(b) Single input.

Figure 5.3 The ideal voltage amplifier (voltage-controlled voltage source).

$$Y_{\text{constrained}} = \begin{bmatrix} y_{11} & y_{12} & (y_{13} + Ay_{14}) \\ y_{21} & y_{22} & (y_{23} + Ay_{24}) \\ y_{31} & y_{32} & (y_{33} + Ay_{34}) \end{bmatrix}$$

$$= \begin{bmatrix} y_{11} & y_{12} & y'_{13} \\ y_{21} & y_{22} & y'_{23} \\ y_{31} & y_{32} & y'_{33} \end{bmatrix} \tag{5.5}$$

EXAMPLE 5.2 Obtain the constrained NAM for the single-amplifier filter shown in Fig. 5.2*b*.

PROCEDURE

1 Note that the circuit consists of the passive part surrounded by the broken line, with an amplifier input and output terminal connected to nodes 3 and 4, respectively.

2 Obtain the NAM of the passive circuit. This was derived in Example 5.1 and the result is given in Eq. 5.3.

3 Obtain the constrained nodal admittance matrix by substituting the admittance parameters from Eq. 5.3 into Eq. 5.5. This gives Eq. 5.6, which is the constrained NAM for the active network.

$$Y_{\text{constrained}} = \begin{bmatrix} g_1 & -g_1 & 0 \\ -g_1 & (g_1 + g_2 + sC_1 + sC_2) & -(sC_2 + Ag_2) \\ 0 & -sC_2 & (g_3 + sC_2) \end{bmatrix} \tag{5.6}$$

Before attempting to use the matrix of Eq. 5.6 in the analysis of the network, we can impose further general restrictions on the form of node currents flowing into the passive circuit. If node 1 is taken to be the input node, then $i_1 \neq 0$, but the problem may be formulated such that all other nonredundant currents are zero. The redundant currents are always those connected to amplifier outputs and are generally nonzero, but take no part in the following analysis. For example, in the circuit of Fig. 5.2*b*:

1 i_1 is the nonzero input current.

2 i_2 is zero because no direct connection exists from outside the circuit to node 2.

3 i_3 is zero because node 3 is connected only to the infinite input impedance of the amplifier.

4 i_4 is nonzero but redundant.

In general, most node currents will be zero because, like node 2 in the example, no direct external connection exists. Only nodes used as inputs or outputs may have nonzero node current. In the case of outputs, this node current can be made zero (for analysis purposes) by including the terminating impedance in the passive network or by obtaining the output via an ideal voltage amplifier as in this case. Therefore, it is always possible to produce a set of equations in the following form:

$$\begin{bmatrix} i_i \\ 0 \\ 0 \end{bmatrix} = \begin{bmatrix} y_{11} & y_{12} & y'_{13} \\ y_{21} & y_{22} & y'_{23} \\ y_{31} & y_{32} & y'_{33} \end{bmatrix} \begin{bmatrix} v_1 \\ v_2 \\ v_3 \end{bmatrix} \tag{5.7}$$

Equation 5.7 may be solved to give the following relationships,* where Δ is the determinant of $y_{\text{constrained}}$:

$$\frac{v_1}{i_1} = \frac{\begin{vmatrix} y_{22} & y'_{23} \\ y_{32} & y'_{33} \end{vmatrix}}{\Delta} \tag{5.8}$$

$$\frac{v_3}{i_1} = \frac{\begin{vmatrix} y_{21} & y_{22} \\ y_{31} & y_{32} \end{vmatrix}}{\Delta} \tag{5.9}$$

$$\frac{v_3}{v_1} = \frac{\begin{vmatrix} y_{21} & y_{22} \\ y_{31} & y_{32} \end{vmatrix}}{\begin{vmatrix} y_{22} & y'_{23} \\ y_{32} & y'_{33} \end{vmatrix}} \tag{5.10}$$

Note that $v_4/v_1 = Av_3/v_1$.

Note that the expression for the voltage transfer function of a five-node network requires only the solution of determinants of order two. Solving the determinants results in the following expression for v_4/v_1

$$\frac{v_4}{v_1} = \frac{Av_3}{v_1} = \frac{A(y_{21}y_{32} - y_{31}y_{22})}{y_{22}y'_{33} - y_{32}y'_{23}} \tag{5.11}$$

EXAMPLE 5.3 Obtain the symbolic form of the transfer function v_4/v_1 for the active filter shown in Fig. 5.2*b*.

PROCEDURE

1 Obtain the constrained nodal admittance matrix for the circuit. This was obtained in Example 5.2 and is shown in Eq. 5.6.

2 Substitute the admittance parameters from Eq. 5.6 into Eq. 5.11. The required transfer function is evaluated as follows:

$$\frac{v_4(s)}{v_1(s)} = \frac{AsC_2g_1}{s^2C_1C_2 + s[C_1g_3 + C_2g_1 + C_2g_2(1-A) + C_2g_3] + g_3(g_1+g_2)} \tag{5.12}$$

When forming the constrained matrix for a circuit incorporating a differential amplifier, the columns associated with the voltages at both input terminals must be modified. As before, the row associated with the amplifier output is eliminated. The resulting constrained matrix may then be solved to give the ratio

* This analysis technique is summarized at the end of this section, where simple rules for the evaluation of 2 × 2 and 3 × 3 determinants are also presented.

of any node voltage, except the amplifier output node, to that at the input node. However, the amplifier output node voltage is usually the circuit output voltage and must be analyzed. This problem is easily overcome for the single-ended amplifier by analyzing for the amplifier input voltage and then simply multiplying the input voltage by the amplification. This procedure can be used with a differential amplifier, but it requires analysis of both amplifier input voltages.

An alternative device is to connect a dummy component between the amplifier output and the circuit output. This circuit can then be analyzed for the output voltage. The output voltage of the original circuit is then obtained by letting the dummy component approach infinite admittance.

EXAMPLE 5.4 Obtain the symbolic form of the voltage transfer function v_3/v_1 for the circuit of Fig. 5.4*a*.

PROCEDURE

1 To facilitate analysis, the dummy conductance g_3 is added and v_3 becomes the new output voltage as shown in Fig. 5.4*b*. Note that the highest node number is assigned to the amplifier output.

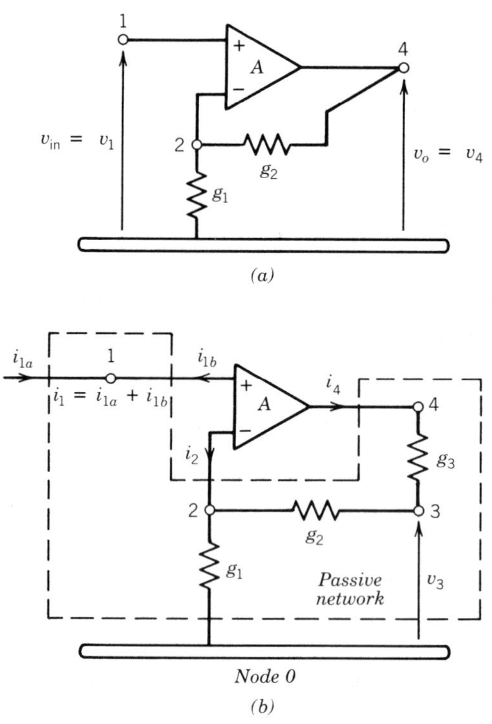

Figure 5.4 Noninverting amplifier: (a) Original, (b) with dummy conductance, g_3.

2 Obtain the NAM for the passive part of Fig 5.4*b* shown inside the broken line. Note that the passive network must include all the nodes to which the amplifier is connected, including node 1. Using the procedure of Example 5.1 gives:

$$Y = \begin{bmatrix} 0 & 0 & 0 & 0 \\ 0 & (g_1 + g_2) & -g_2 & 0 \\ 0 & -g_2 & (g_2 + g_3) & -g_3 \\ 0 & 0 & -g_3 & g_3 \end{bmatrix} \tag{5.13}$$

3 Write the constraint imposed by the amplifier on the passive circuit node voltages. By inspection and comparison with Fig. 5.3 the constraint is

$$v_4 = A(v_1 - v_2) \tag{5.14}$$

4 Substitute Eq. 5.14 into Eq. 5.13 and eliminate row 4 to yield

$$Y_{\text{constrained}} = \begin{bmatrix} 0 & 0 & 0 \\ 0 & g_1 + g_2 & -g_2 \\ -Ag_3 & Ag_3 - g_2 & g_2 + g_3 \end{bmatrix} \tag{5.15}$$

5 Solve for v_3/v_1 by substituting into Eq. 5.11:

$$\frac{v_3}{v_1} = \frac{Ag_3(g_1 + g_2)}{Ag_3g_2 + g_1g_2 + (g_1 + g_2)g_3} \tag{5.16}$$

6 Let the admittance g_3 in Eq. 5.16 approach infinity to yield the transfer function of the original circuit:

$$\frac{v_4}{v_1} = \lim_{g_3 \to \infty} \frac{v_3}{v_1} = \frac{A(g_1 + g_2)}{Ag_2 + g_2 + g_1} \tag{5.17}$$

The circuit in Fig. 5.4*a* is the usual noninverting amplifier connection for an operational amplifier. The expression in Eq. 5.17 may be further simplified by assuming that $A \to \infty$, as shown in Eq. 5.18.

$$\lim_{A \to \infty} \frac{v_4}{v_1} = \frac{g_1 + g_2}{g_2} \tag{5.18}$$

Although this expression is usually quoted as the circuit amplification, it must be realized that the simplifying assumption $A \to \infty$ may not be valid under some conditions. This is often the case in active filters. It is therefore recommended that the reader analyze for the full expression, rather than rely on techniques such as the virtual earth principle.

Summary of the NAM Analysis Technique

1 Form a definite nodal admittance matrix for the passive components using the rules

y_{ii} = the sum of all admittances connected to node i

$y_{ij} = y_{ji}$ = the negative of the sum of all the admittances connected directly between nodes i and j

2 Form the constrained nodal admittance matrix by following thcsc rules for every voltage amplifier:
  a. If node p is connected to the amplifier output, multiply column p by the signed amplification to form a new column p.
  b. If node q is connected to the same amplifier input, add the elements of the new column p to the corresponding elements of column q to form a new column q.
  c. If the amplifier has differential inputs connected to nodes a and b, form new columns a and b as for column p above but reverse the sign of that corresponding to the inverting input.
  d. Eliminate row p.

3 The voltage transfer function may then be evaluated as

$$\frac{v_o}{v_{\text{in}}} = \frac{\Delta_{\text{out}}}{\Delta_{\text{in}}}$$

where Δ_{out} and Δ_{in} are the cofactors obtained by eliminating the row and column corresponding to the output and input nodes, respectively.

4 If the output of the circuit is also the output of an amplifier, the transfer function is obtained by solving for the voltages at the amplifier inputs and thus calculating the amplifier output voltage.

Values of Determinants

$$\begin{vmatrix} a_{11} & a_{12} \\ a_{21} & a_{22} \end{vmatrix} = a_{11}a_{22} - a_{21}a_{12}$$

$$\begin{vmatrix} a_{11} & a_{12} & a_{13} \\ a_{21} & a_{22} & a_{23} \\ a_{31} & a_{32} & a_{33} \end{vmatrix} = a_{11}a_{22}a_{33} + a_{12}a_{23}a_{31} + a_{13}a_{21}a_{32} - a_{13}a_{22}a_{31} - a_{12}a_{21}a_{33} - a_{23}a_{32}a_{11}$$

5.3 TWO-PORT MATRIX ANALYSIS

The active components described in Chapter 4 have one feature in common: They all have one pair of input terminals and one pair of output terminals. In some cases the input and output may share one common terminal, but this may be regarded as a special case of the more general statement. Traditionally, such circuit structures were termed “four terminal networks,” but this led to confusion in cases where two of those terminals were connected and the circuit could be drawn with only three terminals. The modern approach is to describe each pair of terminals as a “port” and to allow ports to have common terminals; these structures thus become “two-port networks.” For purposes of analysis it is often convenient to ignore the detailed circuitry inside the device and concentrate on the relationship between the input and output voltages and currents. Thus the device is regarded as a “black box” with only a pair of input terminals (the input port) and a pair of output terminals (the output port) protruding from it. If the relationships between all of the input and output voltages and currents can be written down, then the device is fully characterized for purposes of analysis.

However, it is not only individual devices that may be characterized in this way, but also complete filters and parts of filters. Thus, if we can obtain the two-port characteristics of a complete filter from those of its constituent parts, we have a means of analyzing the filter. The most convenient method of performing this analysis is to make use of a variety of matrices that describe the relationships between the port voltages and currents. For this reason the technique is given the generic title of "two-port matrix analysis."

The four most commonly used matrix descriptions are shown in Eqs. 5.19 through 5.22. All these matrices contain the same essential description of the network, but each was developed to facilitate analysis under particular connection conditions. For example, as described below, the transmission matrix $[a]$ is particularly convenient for the analysis of several two-port networks connected in cascade. Similarly, it is shown here that the short-circuit admittance matrix $[y]$ allows ready analysis of two-port networks connected in parallel.

$$\begin{bmatrix} v_1 \\ i_2 \end{bmatrix} = \begin{bmatrix} h_{11} & h_{12} \\ h_{21} & h_{22} \end{bmatrix} \begin{bmatrix} i_1 \\ v_2 \end{bmatrix} \tag{5.19}$$

$$\begin{bmatrix} v_1 \\ v_2 \end{bmatrix} = \begin{bmatrix} z_{11} & z_{12} \\ z_{21} & z_{22} \end{bmatrix} \begin{bmatrix} i_1 \\ i_2 \end{bmatrix} \tag{5.20}$$

$$\begin{bmatrix} i_1 \\ i_2 \end{bmatrix} = \begin{bmatrix} y_{11} & y_{12} \\ y_{21} & y_{22} \end{bmatrix} \begin{bmatrix} v_1 \\ v_2 \end{bmatrix} \tag{5.21}$$

$$\begin{bmatrix} v_1 \\ i_1 \end{bmatrix} = \begin{bmatrix} a_{11} & a_{12} \\ a_{21} & a_{22} \end{bmatrix} \begin{bmatrix} v_2 \\ -i_2 \end{bmatrix} \tag{5.22}$$

Since these various matrices contain the same essential information, they are closely related, and there are simple rules by which one matrix may be obtained from a knowledge of another. These conversion rules are shown in Table 5.1. During this conversion process, and generally in the use of these matrices, it is essential that a universal convention be adopted for the directions of the voltages and currents. Figure 5.5 shows that convention. It must be used even if it conflicts with the directions known to exist in practice.

A full treatment of two-port matrix analysis is beyond the scope of this book. Instead, this section describes some of the uses of two of the matrices in applications relevant to the analysis of high-order inductance simulation filters. The first is the transmission matrix, $[a]$, which can be used to analyze LC ladder networks and is also useful in investigating the properties of the GIC. The second is the

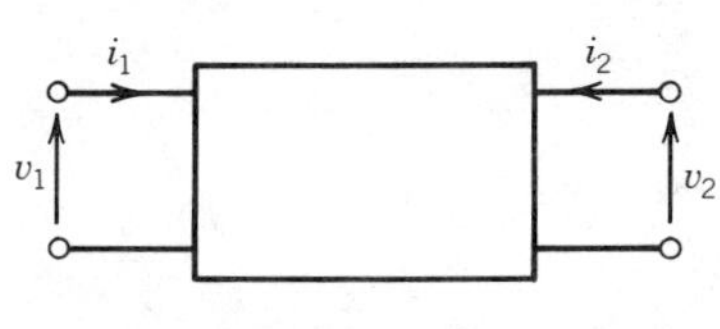

Figure 5.5 General two-port.

Table 5.1 **Matrix conversion chart**

	[z]		[y]		[a]		[h]	
[z]	z_{11}	z_{12}	$\frac{y_{22}}{\Delta_y}$	$\frac{-y_{12}}{\Delta_y}$	$\frac{a_{11}}{a_{21}}$	$\frac{\Delta_a}{a_{21}}$	$\frac{\Delta_h}{h_{22}}$	$\frac{h_{12}}{h_{22}}$
	z_{21}	z_{22}	$\frac{-y_{21}}{\Delta_y}$	$\frac{y_{11}}{\Delta_y}$	$\frac{1}{a_{21}}$	$\frac{a_{22}}{a_{21}}$	$\frac{-h_{21}}{h_{22}}$	$\frac{1}{h_{22}}$
[y]	$\frac{z_{22}}{\Delta_z}$	$\frac{-z_{12}}{\Delta_z}$	y_{11}	y_{12}	$\frac{a_{22}}{a_{12}}$	$\frac{-\Delta_a}{a_{12}}$	$\frac{1}{h_{11}}$	$\frac{-h_{12}}{h_{11}}$
	$\frac{-z_{21}}{\Delta_z}$	$\frac{z_{11}}{\Delta_z}$	y_{21}	y_{22}	$\frac{-1}{a_{12}}$	$\frac{a_{11}}{a_{12}}$	$\frac{h_{21}}{h_{11}}$	$\frac{\Delta_h}{h_{11}}$
[a]	$\frac{z_{11}}{z_{21}}$	$\frac{\Delta_z}{z_{21}}$	$\frac{-y_{22}}{y_{21}}$	$\frac{-1}{y_{21}}$	a_{11}	a_{12}	$\frac{-\Delta_h}{h_{21}}$	$\frac{-h_{11}}{h_{21}}$
	$\frac{1}{z_{21}}$	$\frac{z_{22}}{z_{21}}$	$\frac{-\Delta_y}{y_{21}}$	$\frac{-y_{11}}{y_{21}}$	a_{21}	a_{22}	$\frac{-h_{22}}{h_{21}}$	$\frac{-1}{h_{21}}$
[h]	$\frac{\Delta_z}{z_{22}}$	$\frac{z_{12}}{z_{22}}$	$\frac{1}{y_{11}}$	$\frac{-y_{12}}{y_{11}}$	$\frac{a_{12}}{a_{22}}$	$\frac{\Delta_a}{a_{22}}$	h_{11}	h_{12}
	$\frac{-z_{21}}{z_{22}}$	$\frac{1}{z_{22}}$	$\frac{y_{21}}{y_{11}}$	$\frac{\Delta_y}{y_{11}}$	$\frac{-1}{a_{22}}$	$\frac{a_{21}}{a_{22}}$	h_{21}	h_{22}

$\Delta_z = z_{11}z_{22} - z_{12}z_{21}$; $\Delta y = y_{11}y_{22} - y_{12}y_{21}$; $\Delta a = a_{11}a_{22} - a_{12}a_{21}$; $\Delta h = h_{11}h_{22} - h_{12}h_{21}$.

short-circuit admittance matrix, $[y]$, which has a number of uses in filter analysis but is particularly helpful in analyzing the responses of networks connected in parallel.

The reader desiring a full treatment of this topic will find it covered in most modern texts on network analysis, of which [2] is an example.

Transmission Matrix The transmission matrix relates the input voltage and current to the output voltage and current as shown in Eq. 5.22. Its most useful property is that if two or more networks are connected in cascade, then the transmission matrix of the cascade is the product of the individual matrices, multiplied in the same order as the cascade connection. This is illustrated for the cascade of Fig. 5.6 by the matrix expression of Eq. 5.23. The proof of this property is self-evident when it is realized that the output voltage of network (a) becomes the input voltage of

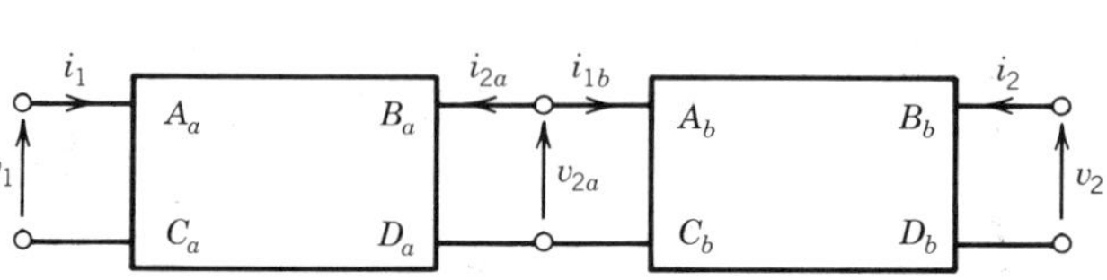

Figure 5.6 Two-port cascade.

Table 5.2 **Matrix descriptions of simple passive two-port networks**

Network	$[a]$		$[y]$	
1. 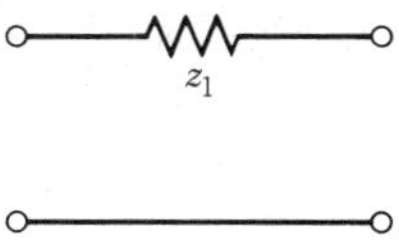	1	z_1	$\frac{1}{z_1}$	$-\frac{1}{z_1}$
	0	1	$-\frac{1}{z_1}$	$+\frac{1}{z_1}$
2. 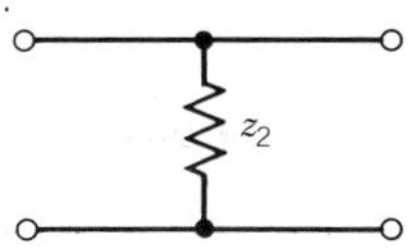	1	0	∞	∞
	$\frac{1}{z_2}$	1	∞	∞
3. 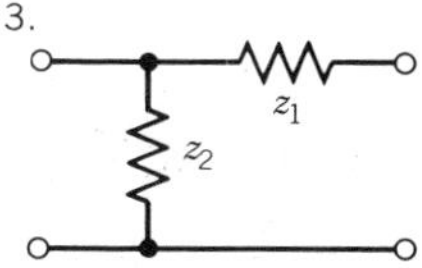	1	z_1	$\frac{z_1 + z_2}{z_1 z_2}$	$-\frac{1}{z_1}$
	$\frac{1}{z_2}$	$\left(1 + \frac{z_1}{z_2}\right)$	$-\frac{1}{z_1}$	$+\frac{1}{z_1}$
4. 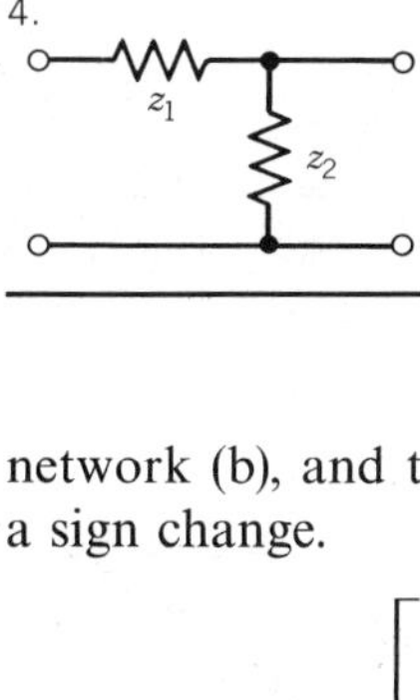	$1 + \frac{z_1}{z_2}$	z_1	$\frac{1}{z_1}$	$-\frac{1}{z_1}$
	$\frac{1}{z_2}$	1	$-\frac{1}{z_1}$	$+\left(\frac{z_1 + z_2}{z_1 z_2}\right)$

network (b), and the output current of (a) becomes the input current of (b) with a sign change.

$$\begin{bmatrix} v_1 \\ i_1 \end{bmatrix} = \begin{bmatrix} a_{11a} & a_{12a} \\ a_{21a} & a_{22a} \end{bmatrix} \cdot \begin{bmatrix} a_{11b} & a_{12b} \\ a_{21b} & a_{22b} \end{bmatrix} \begin{bmatrix} v_2 \\ -i_2 \end{bmatrix} \tag{5.23}$$

Table 5.2 shows the $[a]$ and $[y]$ matrices for some simple two-port networks.

EXAMPLE 5.5 Show that a gyrator in cascade with a shunt capacitor can simulate a shunt inductor.

PROCEDURE

1 As described in Chapter 4, the gyrator is a two-port device used in active filters to simulate inductance. The transmission matrix for a gyrator is shown in Eq. 5.24, where G is the gyration conductance:

$$\begin{bmatrix} v_1 \\ i_1 \end{bmatrix} = \begin{bmatrix} 0 & \frac{1}{G} \\ G & 0 \end{bmatrix} \begin{bmatrix} v_{2a} \\ -i_{2a} \end{bmatrix} \tag{5.24}$$

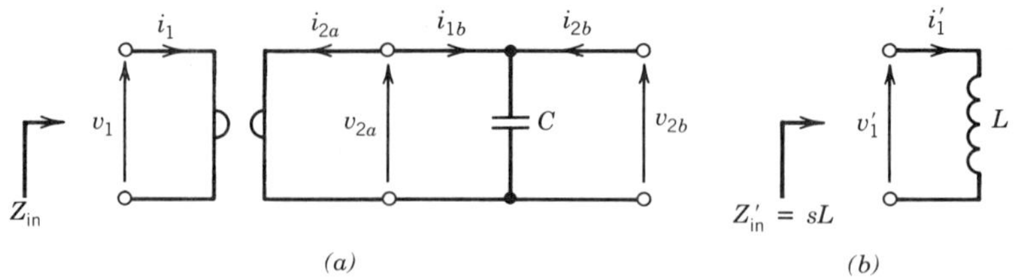

Figure 5.7 Gyrator simulation of grounded inductance: (a) Cascade, (b) simulated inductor.

2 Figure 5.7*a* shows a gyrator connected in cascade with a simple two-port network containing a shunt capacitor. The $[a]$ matrix for the capacitor network may be obtained from Table 5.2 by substituting $z_2 = 1/sC_L$ for network 2. Performing this substitution gives

$$\begin{bmatrix} v_{1b} \\ i_{1b} \end{bmatrix} = \begin{bmatrix} 1 & 0 \\ sC_L & 1 \end{bmatrix} \begin{bmatrix} v_{2b} \\ -i_{2b} \end{bmatrix} \tag{5.25}$$

3 The transmission matrix for the cascade is then given by

$$\begin{bmatrix} v_1 \\ i_1 \end{bmatrix} = \begin{bmatrix} 0 & \dfrac{1}{G} \\ G & 0 \end{bmatrix} \begin{bmatrix} 1 & 0 \\ sC_L & 1 \end{bmatrix} \begin{bmatrix} v_{2b} \\ -i_{2b} \end{bmatrix} \tag{5.26}$$

$$= \begin{bmatrix} \dfrac{sC_L}{G} & \dfrac{1}{G} \\ G & 0 \end{bmatrix} \begin{bmatrix} v_{2b} \\ -i_{2b} \end{bmatrix} \tag{5.27}$$

4 Figure 5.7*b* shows the shunt inductance that we are attempting to simulate. The input impedance is

$$z_{\text{in}}' = \frac{v_1'}{i_1'} = sL \tag{5.28}$$

5 If the circuit of Fig. 5.7*a* can be made to have its input impedance z_{in} of the same form as z_{in}', then the simulation is complete.

6 Writing out in full the equations represented by the matrix equation (5.27) gives

$$v_1 = \frac{sC_L}{G} v_{2b} - \frac{1}{G} i_{2b} \tag{5.29}$$

$$i_1 = G v_{2b} \tag{5.30}$$

7 Setting $i_{2b} = 0$ in Eq. 5.29 yields

$$v_1 \big|_{i_{2b}=0} = \frac{sC_L}{G} v_{2b} \tag{5.31}$$

8 Under this condition the input impedance is obtained by dividing Eq. 5.31 by Eq. 5.30 so that

$$z_1 \big|_{i_{2b}=0} = \frac{v_1}{i_1}\bigg|_{i_{2b}=0} = \frac{sC_L}{G^2} \tag{5.32}$$

9 Comparison of Eqs. 5.29 and 5.32 shows that when $i_{2b} = 0$, the circuit of Fig. 5.7*a* has the same input impedance as an inductance of value:

$$L = \frac{C_L}{G^2} \tag{5.33}$$

The condition $i_{2b} = 0$ is readily achieved in practice by leaving the right-hand port of Fig. 5.7*a* unconnected.

This is the expected form for a simulated inductance. In this special case the simulated inductance is a one-port network, and port 2 is redundant; hence $i_{2b} = 0$. However, by the addition of a further gyrator in the cascade as shown in Fig. 5.8*a*, an inductive two-port can be simulated.

EXAMPLE 5.6 Show that the gyrator-capacitor-gyrator cascade of Fig. 5.8*a* can simulate the series inductor of Fig. 5.8*b*.

PROCEDURE

1 Obtain the $[a]$ matrix for the cascade. From Eqs. 5.23, 5.24, and 5.27:

$$\begin{bmatrix} v_1 \\ i_1 \end{bmatrix} = \begin{bmatrix} \dfrac{sC_L}{G} & \dfrac{1}{G} \\ G & 0 \end{bmatrix} \cdot \begin{bmatrix} 0 & \dfrac{1}{G} \\ G & 0 \end{bmatrix} \begin{bmatrix} v_2 \\ -i_2 \end{bmatrix} \tag{5.34}$$

$$= \begin{bmatrix} 1 & \dfrac{sC_L}{G^2} \\ 0 & 1 \end{bmatrix} \begin{bmatrix} v_2 \\ -i_2 \end{bmatrix} \tag{5.35}$$

2 Obtain the $[a]$ matrix for the series inductor in Fig. 5.8*a*. Making the substitution $z_1 = sL$ in the $[a]$ matrix of network 1 of Table 5.2 gives the matrix of the inductor:

$$\begin{bmatrix} v_1 \\ i_1 \end{bmatrix} = \begin{bmatrix} 1 & sL \\ 0 & 1 \end{bmatrix} \begin{bmatrix} v_2 \\ -i_2 \end{bmatrix} \tag{5.36}$$

3 Comparison of Eqs. 5.35 and 5.36 shows that the arrangement in Fig. 5.8*a* simulates a series inductance of value $L = sC_L/G^2$, and that such networks are readily analyzed using the transmission matrix.

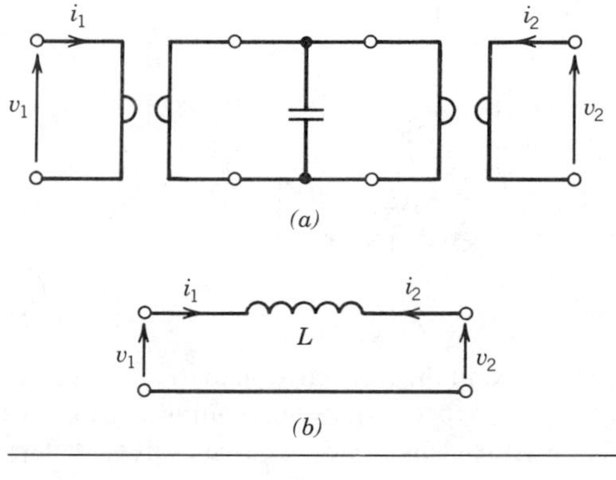

Figure 5.8 Floating inductance simulation: (a) Cascade, (b) simulated inductor.

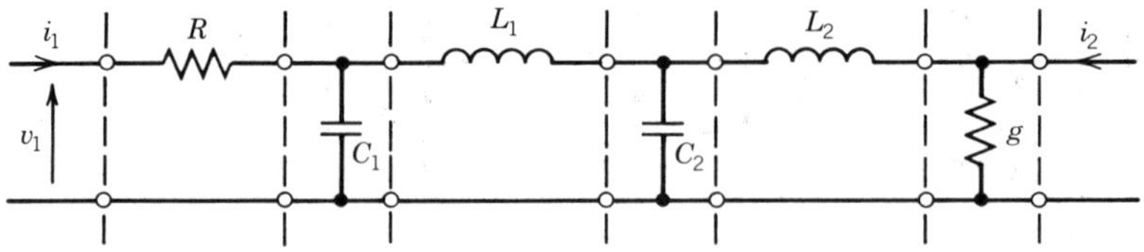

Figure 5.9 LC ladder network.

The transmission matrix may also be used to analyze the ladder networks common in LC filters. Figure 5.9 shows such a ladder structure, which has been drawn as a cascade of series and shunt elements.

The matrices for series and shunt elements are those for networks 1 and 2 in Table 5.2. They are as follows:

Shunt Admittance $1/z_2$

$$\begin{bmatrix} v_1 \\ i_1 \end{bmatrix} = \begin{bmatrix} 1 & 0 \\ \dfrac{1}{z_2} & 1 \end{bmatrix} \begin{bmatrix} v_2 \\ -i_2 \end{bmatrix} \tag{5.37}$$

Series Impedance z_1

$$\begin{bmatrix} v_1 \\ i_1 \end{bmatrix} = \begin{bmatrix} 1 & z_1 \\ 0 & 1 \end{bmatrix} \begin{bmatrix} v_2 \\ -i_2 \end{bmatrix} \tag{5.38}$$

EXAMPLE 5.7 Obtain a symbolic expression for the voltage transfer function v_2/v_1 of the circuit of Fig. 5.9.

PROCEDURE

1 Obtain the $[a]$ matrix for each element by substituting $z_2 = 1/sC_1$, $z_2 = 1/sC_2$, and $z_2 = 1/g$ into Eq. 5.37; and $z_1 = R$, $z_1 = sL_1$ and $z_1 = sL_2$ into Eq. 5.38. Writing these as a matrix product with the same order as the cascade gives Eq. 5.39.

2 Multiplying out the matrix Eq. 5.39 results in the $[a]$ matrix for the complete ladder.

$$\begin{bmatrix} v_1 \\ i_1 \end{bmatrix} = \begin{bmatrix} 1 & R \\ 0 & 1 \end{bmatrix} \begin{bmatrix} 1 & 0 \\ sC_1 & 1 \end{bmatrix} \begin{bmatrix} 1 & sL_1 \\ 0 & 1 \end{bmatrix} \begin{bmatrix} 1 & 0 \\ sC_2 & 1 \end{bmatrix} \begin{bmatrix} 1 & sL_2 \\ 0 & 1 \end{bmatrix} \begin{bmatrix} 1 & 0 \\ g & 1 \end{bmatrix} \begin{bmatrix} v_2 \\ -i_2 \end{bmatrix} \tag{5.39}$$

$$= \begin{bmatrix} 1 + sC_1R & R \\ sC_1 & 1 \end{bmatrix} \begin{bmatrix} 1 + s^2L_1C_2 & sL_1 \\ sC_2 & 1 \end{bmatrix} \begin{bmatrix} 1 + sL_2g & sL_2 \\ g & 1 \end{bmatrix} \begin{bmatrix} v_2 \\ -i_2 \end{bmatrix} \tag{5.40}$$

$$= \begin{bmatrix} \{(1 + sC_1R)(1 + s^2L_1C_2) + sC_2R\} & \{(1 + sC_1R)sL_1 + R\} \\ \{sC_1(1 + s^2L_1C_2) + sC_2\} & \{s^2L_1C_1 + 1\} \end{bmatrix} \cdot \begin{bmatrix} 1 + sL_2g & sL_2 \\ g & 1 \end{bmatrix} \begin{bmatrix} v_2 \\ -i_2 \end{bmatrix} \tag{5.41}$$

3 One further matrix multiplication is required to obtain the transmission matrix. However, if only the voltage transfer function is required, then, because the load admittance g was incorporated in the network to give $i_2 = 0$, only the element a_{11} need be evaluated. The voltage transfer function is then given by

$$\frac{v_2}{v_1} = \frac{1}{a_{11}}$$

$$= \frac{1}{(1 + sL_2g)[(1 + sC_1R)(1 + s^2L_1C_2) + sC_2R] + g[1 + sC_1R)sL_1 + R]}$$

$$= \frac{1}{\begin{array}{l} s^4(L_1L_2C_1C_2gR) + s^3(L_1RC_1C_2 + L_1L_2C_2g) \\ \quad + s^2(C_1L_2Rg + C_2L_2Rg + L_1C_2 + L_1C_1Rg) \\ \quad + s(L_2g + C_1R + C_2R + L_1g) + (1 + Rg) \end{array}} \tag{5.42}$$

Equation 5.42 shows that the ladder network in Fig. 5.9 is a fourth-order filter. The extension of this technique to longer-ladder structures is straightforward. However, even the analysis of this fourth-order ladder is tedious and the manual analysis of long-ladder networks is a daunting task, offering much room for mistakes. Further, the symbolic expressions resulting from such analyses are lengthy and are functions of many component values. As a result, the subsequent analysis procedures—for example, sensitivity analysis—become lengthy and complex. For this reason it is usual to resort to numerical analysis by computer when assessing high-order networks.

Short-Circuit Admittance Matrix The short-circuit admittance matrix has many uses in active filter analysis. Indeed, the NAM discussed in Section 5.2 may be regarded as the general N-port version of this two-port matrix. The two-port matrix is particularly suited for analyzing networks connected in parallel as shown in Fig. 5.10. Such connections are common in filter networks.

The short-circuit admittance matrix of such a parallel connection is simply the sum of the matrices of the individual networks as shown in Eq. 5.43. This is self-evident when it is considered that the input and output currents of the parallel connection are simply the sums of those of the individual networks, and the connection imposes the constraint that the input and output voltages are the same for both networks:

$$\begin{bmatrix} i_1 \\ i_2 \end{bmatrix} = \begin{bmatrix} y_{11a} + y_{11b} & y_{12a} + y_{12b} \\ y_{21a} + y_{21b} & y_{22a} + y_{22b} \end{bmatrix} \begin{bmatrix} v_1 \\ v_2 \end{bmatrix} \tag{5.43}$$

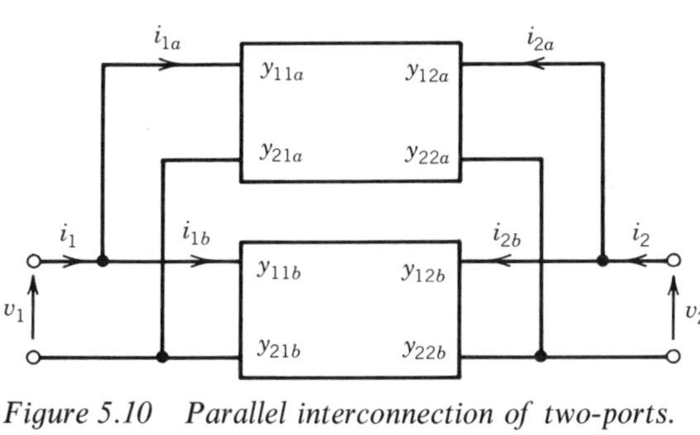

Figure 5.10 Parallel interconnection of two-ports.

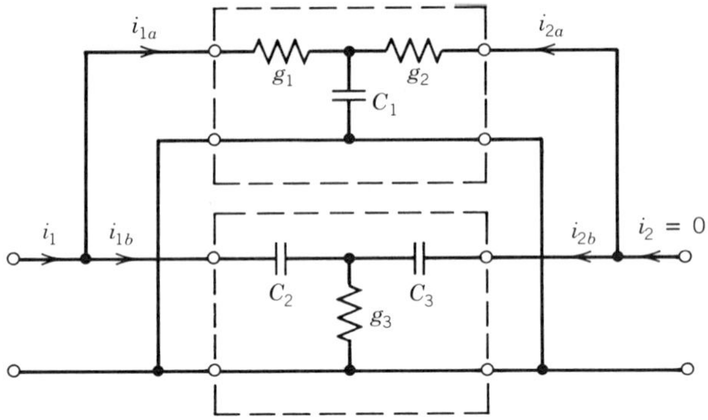

Figure 5.11 *Twin-T as a parallel connection of two-ports.*

The utility of this expression can be illustrated by analyzing the twin-T circuit shown in Fig. 5.11, which has been drawn to emphasize the parallel connection of two networks.

EXAMPLE 5.8 Obtain a symbolic expression for the voltage transfer function v_2/v_1 of the twin-T network shown in Fig. 5.11.

PROCEDURE

1 Obtain the short-circuit admittance matrices for networks *a* and *b* separately.

The individual matrices may be derived using the following rules:

$$y_{11} = \left.\frac{i_1}{v_1}\right|_{v_2=0} \tag{5.44}$$

$$y_{12} = \left.\frac{i_1}{v_2}\right|_{v_1=0} \tag{5.45}$$

$$y_{21} = \left.\frac{i_2}{v_1}\right|_{v_2=0} \tag{5.46}$$

$$y_{22} = \left.\frac{i_2}{v_2}\right|_{v_1=0} \tag{5.47}$$

The physical representation of the conditions (5.44) through (5.47) applied to network (a) are shown in Fig. 5.12. Performing each analysis indicated in Fig. 5.12 gives the following expressions for the parameters of network (a).

$$y_{11a} = \frac{g_1(g_2 + sC_1)}{g_1 + g_2 + sC_1} \tag{5.48}$$

$$y_{12a} = y_{21a} = \frac{-g_1 g_2}{g_1 + g_2 + sC_1} \tag{5.49}$$

$$y_{22a} = \frac{g_2(g_1 + sC_1)}{g_1 + g_2 + sC_1} \tag{5.50}$$

Similarly, the matrix parameters for network (b) are given by

$$y_{11b} = \frac{sC_2(sC_3 + g_3)}{s(C_2 + C_3) + g_3} \tag{5.51}$$

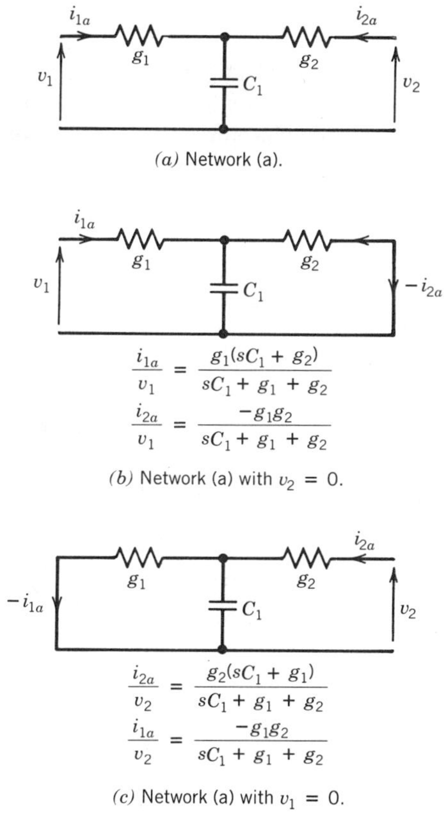

(a) Network (a).

$$\frac{i_{1a}}{v_1} = \frac{g_1(sC_1 + g_2)}{sC_1 + g_1 + g_2}$$

$$\frac{i_{2a}}{v_1} = \frac{-g_1 g_2}{sC_1 + g_1 + g_2}$$

(b) Network (a) with $v_2 = 0$.

$$\frac{i_{2a}}{v_2} = \frac{g_2(sC_1 + g_1)}{sC_1 + g_1 + g_2}$$

$$\frac{i_{1a}}{v_2} = \frac{-g_1 g_2}{sC_1 + g_1 + g_2}$$

(c) Network (a) with $v_1 = 0$.

Figure 5.12 Twin-T analysis: (a) Network a; (b) network a with $V_2 = 0$; (c) network a with $V_1 = 0$.

$$y_{12b} = y_{21b} = \frac{-s^2 C_2 C_3}{s(C_2 + C_3) + g_3} \tag{5.52}$$

$$y_{22} = \frac{sC_3(sC_2 + g_3)}{s(C_2 + C_3) + g_3} \tag{5.53}$$

2 Obtain the voltage transfer function from the y-parameters.

Note that $i_2 = 0$ so that the voltage transfer function is given by

$$\left.\frac{v_2}{v_1}\right|_{i_2=0} = \frac{-y_{12}}{y_{22}} \tag{5.54}$$

For parallel-connected networks this becomes

$$\left.\frac{v_2}{v_1}\right|_{i_2=0} = \frac{-(y_{12a} + y_{12b})}{y_{22a} + y_{22b}} \tag{5.55}$$

Substituting the parameters from Eqs. 5.48 through 5.53 and rearranging gives

$$\left.\frac{v_2}{v_1}\right|_{i_2=0} = \frac{s^3 C_1 C_2 C_3 + s^2 C_2 C_3(g_1 + g_2) + s(C_2 + C_3)g_1 g_2 + g_1 g_2 g_3}{s^3 C_1 C_2 C_3 + s^2\{C_1(C_2 + C_3)g_2 + C_1 C_3 g_3 + C_2 C_3(g_1 + g_2)\} + s\{C_1 g_2 g_3 + g_1 g_2(C_2 + C_3) + C_3 g_3(g_1 + g_2)\} + g_1 g_2 g_3} \tag{5.57}$$

5.4 SIGNAL-FLOW-GRAPH ANALYSIS

A formal introduction to the topic of signal-flow graphs is presented in Section 13.2. Readers unfamiliar with the subject should first refer to Chapter 13 for a discussion of the basic concepts. The technique, as applied in this chapter, concentrates on circuit analysis in which the graph variables are voltages and currents.

As an illustration, consider the circuit shown in Fig. 5.13*a*. The voltages v_1 and v_3 are source voltages while the voltage v_2 is given by

$$v_2 = \frac{g_1}{g_1 + g_2 + g_3} v_1 + \frac{g_3}{g_1 + g_2 + g_3} v_3 \tag{5.58}$$

The flow graph representing the circuit voltages is shown in Fig. 5.13*b*.

The value of a node variable in a signal-flow graph is equal to the sum of the products of the incoming branch values and the value of the node at the other end. Outgoing branches do not affect the value of the node but indicate that it affects other node values. Note that the nodes representing v_1 and v_3 have no incoming branches. This indicates that they are independent of the other node values, which is to be expected of source voltages.

In general, node values both affect and are affected by other node values. This is illustrated by the circuit in Fig. 5.14*a* and its flow graph in Fig. 5.14*b*. The values

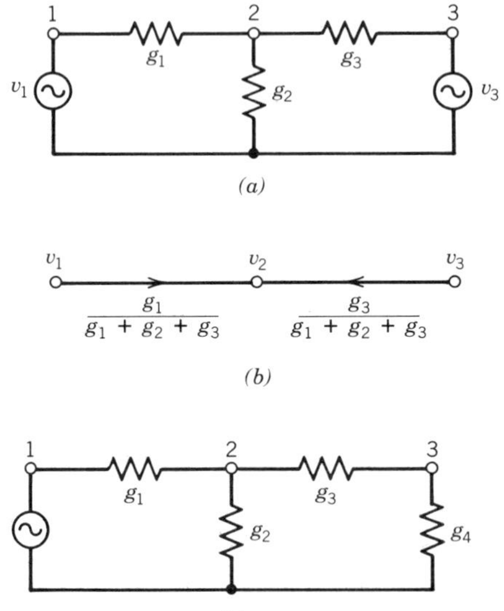

Figure 5.13 Flow graph of T-network: (a) T-network, (b) flow graph.

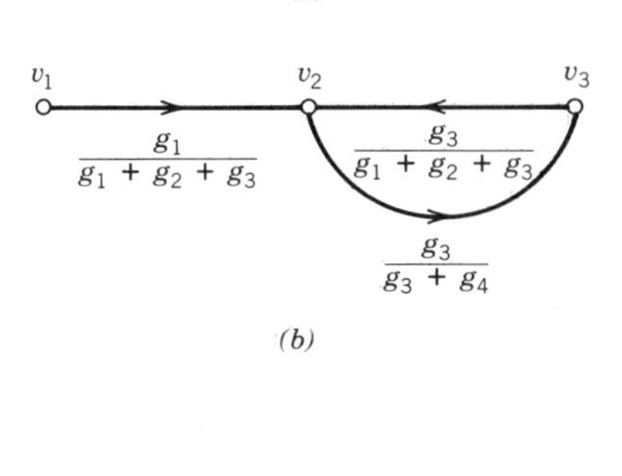

Figure 5.14 Flow graph of loaded T-network: (a) T-network, (b) flow graph.

of v_1 and v_2 remain unchanged, but v_3 is now a dependent variable. The two branches between nodes v_2 and v_3 form a "loop" of the signal flow. Such loops are common in signal-flow graphs.

In the above examples only voltages were represented by the nodes, but nodes can also represent currents. Signal-flow graphs having nodes representing both voltages and currents may be used to represent ladder networks.

EXAMPLE 5.9 Derive the signal-flow graph of the ladder network in Fig. 5.15*a* including nodes representing both circuit node voltages and branch currents.

PROCEDURE

1 Identify all circuit node voltages and branch currents. These are v_1, v_2, v_3, i_1, i_2, i_3, and i_4 as shown in Fig. 5.15*a*.

2 Write equations giving the values of all voltages and currents in terms of the other voltages and currents and the circuit element values. For the circuit of Fig. 5.15*a* these are as follows:

$$v_1 \text{ is an independent variable} \tag{5.59}$$

$$v_2 = \frac{1}{sC} i_2 \tag{5.60}$$

$$v_3 = R_L i_4 \tag{5.61}$$

$$i_1 = \frac{1}{R_S}(v_1 - v_2) = \frac{1}{R_S} v_1 - \frac{1}{R_S} v_2 \tag{5.62}$$

$$i_2 = i_1 - i_3 \tag{5.63}$$

$$i_3 = \frac{1}{sL}(v_2 - v_3) = \frac{1}{sL} v_2 - \frac{1}{sL} v_3 \tag{5.64}$$

$$i_4 = i_3 \tag{5.65}$$

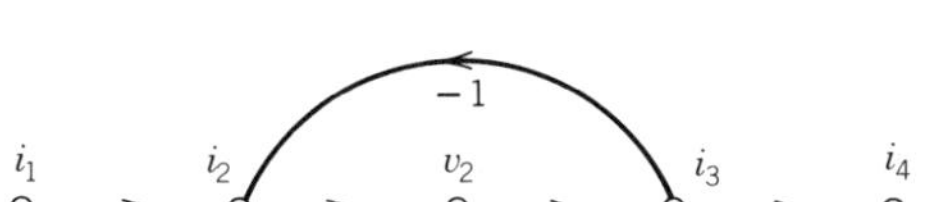

(*a*)

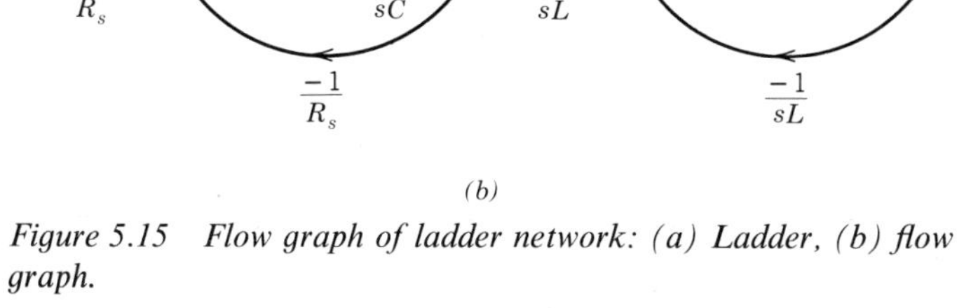

(*b*)

Figure 5.15 Flow graph of ladder network: (a) Ladder, (b) flow graph.

3 Draw the signal-flow graph describing these equations. Start by drawing the signal-flow graph nodes. Every voltage or current on the left-hand sides of Eqs. 5.59 through 5.65 must be represented. It is helpful to position the nodes in a pattern similar to the occurrences of the quantities in the circuit diagram; but expect to rearrange these positions to give a neater diagram when the graph is complete.

Now draw directed branches flowing *into* each node in turn by progressing through Eqs. 5.59 to 5.65. The resulting graph is shown in Fig. 5.15*b* and the branches flowing into each node are obtained as follows:

Equation 5.59 v_1 is an independent variable and therefore has no branches flowing into it.

Equation 5.60 v_2 has only one branch of value $1/sC$ flowing into it from i_2.

Equation 5.61 v_3 has only one branch of value R_L from i_4.

Equation 5.62 i_1 has two branches into its node. One is of value $1/R_S$ from v_1 and the other of value $-1/R_S$ from v_2.

Equation 5.63 i_2 has two branches into its node. One is of value 1 from i_1 and the other of value -1 from i_3.

Equation 5.64 i_3 has two branches into its node. One is of value $1/sL$ from v_2 and the other of value $-1/sL$ from v_3.

Equation 5.65 i_4 has one branch into its node. It is of value 1 from i_3.

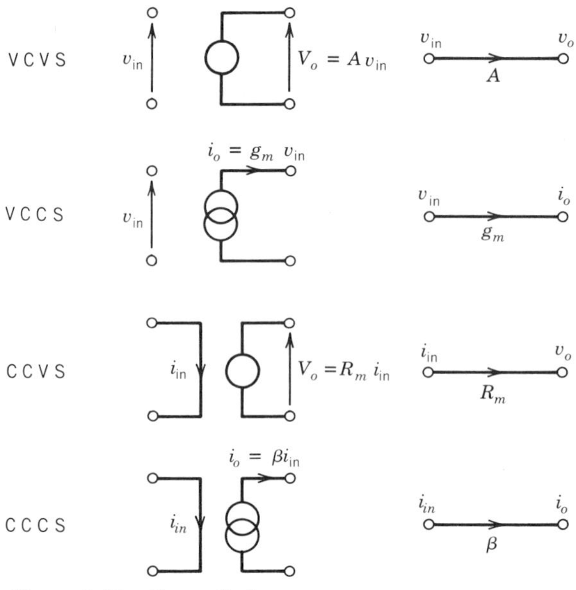

Figure 5.16 Controlled source representations.

The above examples show how signal-flow graphs may be formed to represent networks of passive components. Active components may also be represented by flow graphs as shown in Fig. 5.16. In addition to these four basic controlled sources, it is often convenient to represent some basic operational amplifier circuits by their signal-flow graphs as shown in Fig. 5.17. Note that nodes representing the input currents have been included for completeness, but that in practice they are often redundant and not shown.

Having formed a signal-flow graph for a network, it is now possible to apply standard procedures to analyze the graph. In many cases the graph can be reduced to an equivalent single path from which the transfer function may be written as

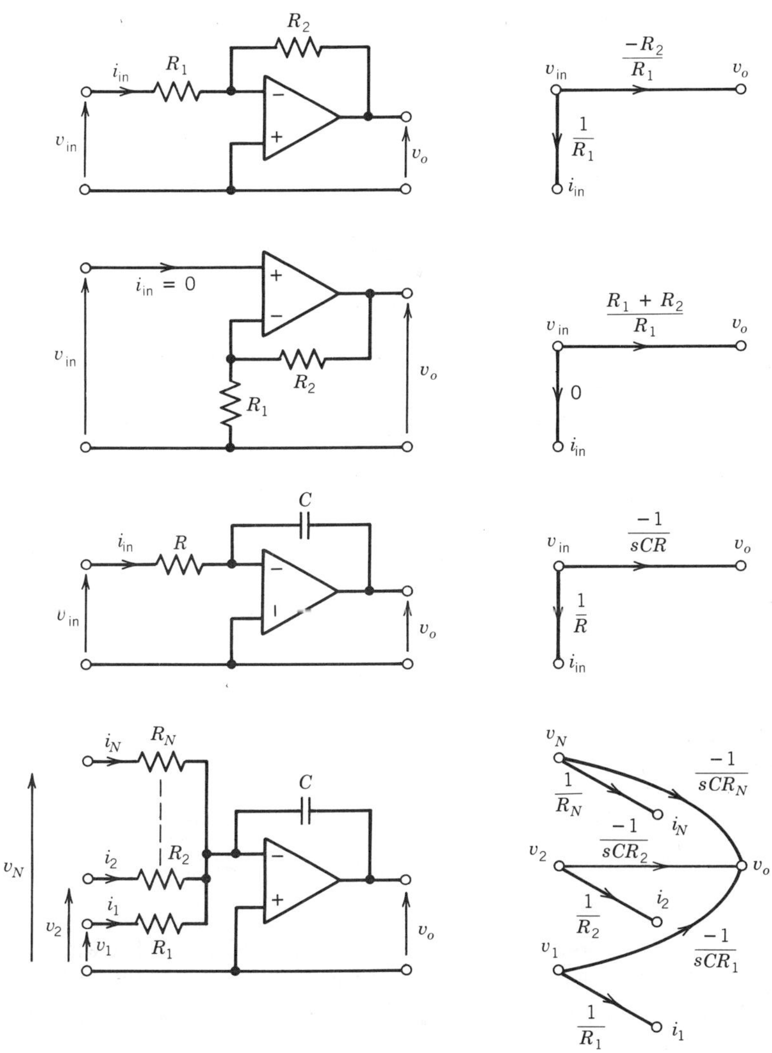

Figure 5.17 Flow graphs of simple amplifier circuits.

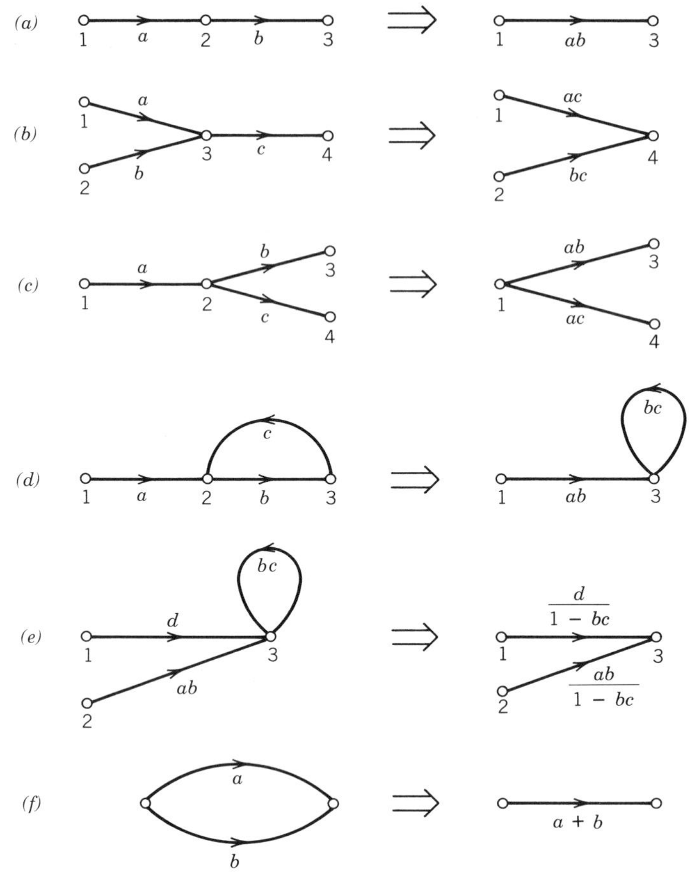

Figure 5.18 Flow-graph transformations.

the product of the branches forming the path between the input and output nodes. This is achieved by applying a series of transformations to the original graph. Some of these transformations are shown in Fig. 5.18. In applying them, the relationship between the input and output node variables is maintained, but those between other variables are lost.

EXAMPLE 5.10 Reduce the signal-flow graph shown in Fig. 5.19*a* to a single branch linking nodes 1 and 5.

PROCEDURE

1 Reduce loops to self-loops by applying transformation (*d*) from Fig. 5.18. There are two loops and self-loops may be formed on nodes 3 and 5 as shown in Fig. 5.19*b*.

2 Eliminate the self-loops by applying transformation (*e*) from Fig. 5.18. This results in the graph of Fig. 5.19*c*.

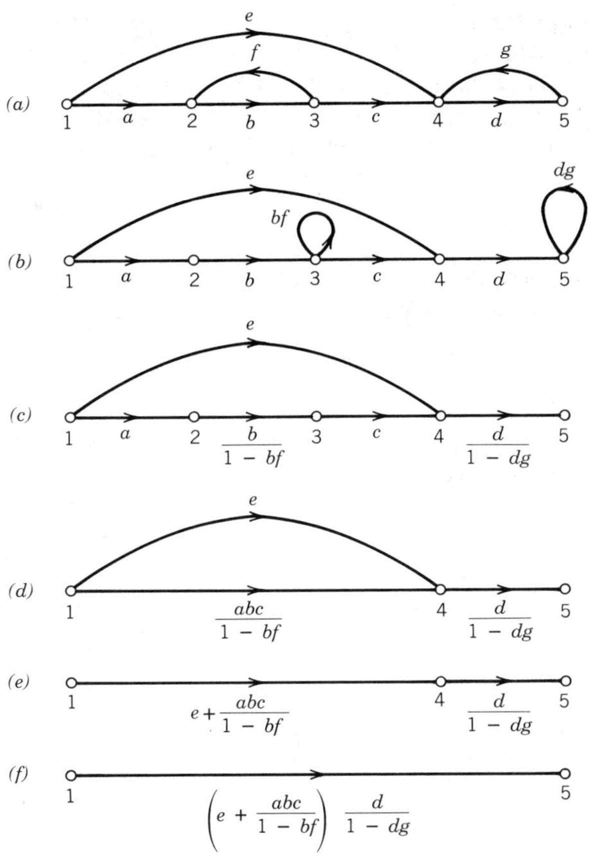

Figure 5.19 Illustration of flow-graph reduction.

3 Condense single paths by using transformation (*a*) from Fig. 5.18. This gives the graph of Fig. 5.19*d*.

4 Condense parallel paths by using transformation (*f*), resulting in the graph of Fig. 5.19*e*.

5 Condense the remaining single path into a single branch by using transformation (*a*). This single branch represents the equation

$$v_5 = \left[e + \frac{abc}{1 - bf} \right] \frac{d}{1 - dg} \cdot i_1 \tag{5.66}$$

where v_5 and i_1 are assumed to be the quantities represented by nodes 5 and 1, respectively.

For complex graphs, particularly those having a number of interacting loops, the ad hoc application of these transformations becomes tedious and prone to error. An alternative approach is to apply only those transformations which result in obvious simplifications, and then to apply a general reduction rule. Only one such rule will be presented here. This is the general multiloop, single-path reduction rule, which applies to graphs where only one forward path may be traced between the input and output nodes and where this path touches all loops. More general rules do exist, but the single-path case applies to the majority of filter graphs. A description of the more general rules can be found in Chapter 13, Section 13.2.

The general single-path, multiloop reduction rule may be written as follows:

$$T_{ij} = \frac{P_{ij}}{\Delta} \tag{5.67}$$

where T_{ij} is the transfer function relating the node variables i and j, P_{ij} is the product of the branches forming the forward path from node i to node j, and Δ is given as follows:

$$\Delta = [(1 - L_1)(1 - L_2)(1 - L_3) \cdots (1 - L_n)]' \tag{5.68}$$

where $L_1, L_2, \ldots, L_n$ represent the loop transmission of all the loops in the graph, and where the prime indicates that all terms containing products of the values of the transmission of touching loops are discarded. Two loops are said to touch if they share one or more nodes.

EXAMPLE 5.11 Obtain the transfer function T_{18} relating the variables of nodes 1 and 8 in the signal-flow graph of Fig. 5.20.

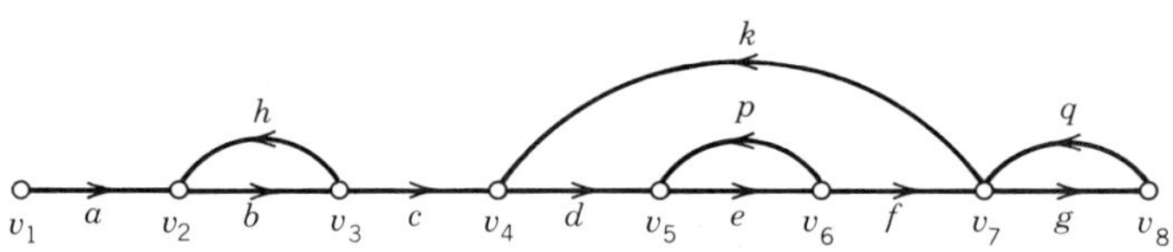

$L_1 = bh$; $L_2 = defk$; $L_3 = ep$; $L_4 = gq$

$P_{18} = abcdefg$

Figure 5.20 Illustration of Mason's rule.

PROCEDURE

1 Identify the forward path P_{18} and all the loops. These are shown in Fig. 5.20.

2 Identify all of the touching loop pairs. These are L_2L_3 and L_2L_4.

3 Identify all of the nontouching loop pairs. These are L_1L_2, L_1L_3, L_1L_4, and L_3L_4.

4 Obtain Δ as shown in Eq. 5.68. This is given by

$$\Delta = 1 - (L_1 + L_2 + L_3 + L_4) + (L_1L_2 + L_1L_3 + L_1L_4 + L_3L_4) - (L_1L_3L_4) \tag{5.69}$$

The following terms were discarded because they include the products of touching loop pairs L_2L_3 and L_2L_4:

$$L_2L_3,\ L_2L_4,\ L_1L_2L_3,\ L_1L_2L_4,\ L_2L_3L_4,\ L_1L_2L_3L_4$$

5 Substitute P_{18} and Δ into Eqs. 5.67 to give T_{18} as follows:

$$T_{18} = \frac{abcdefg}{1 - (L_1 + L_2 + L_3 + L_4) + (L_1L_2 + L_1L_3 + L_1L_4 + L_3L_4) - (L_1L_3L_4)} \tag{5.70}$$

Signal-flow graphs may be used as an effective aid in the analysis of filter circuits. However, the representation of a passive LC circuit in signal-flow graph form can also be used directly to generate an active *RC* simulation of its operation without necessarily performing a full analysis. This technique is discussed in Chapter 10.

5.5 SENSITIVITY ANALYSIS

In designing a practical active filter, it is very important to be aware of the effects of the errors that will exist between the design values of the components and their actual values. In some active filters a tiny error in the value of a component can cause a considerable error in the resulting frequency response.

The techniques by which the effects of component errors are calculated are known as sensitivity analysis. The term *sensitivity* refers to the sensitivity of a filter parameter—for example, frequency response—to errors in a component value. This sensitivity may be defined as in Eq. 5.71 where x is the filter parameter and e is the component value.

$$S_e^x = \frac{e}{x} \cdot \frac{\partial x}{\partial e} \tag{5.71}$$

The function S_e^x is referred to as the sensitivity of x to e. The utility of S_e^x stems from the relationship shown in Eq. 5.72.

$$\frac{\Delta x}{x} = S_e^x \frac{\Delta e}{e} \tag{5.72}$$

This expression allows the calculation of the per unit or percentage change in x due to a known per unit or percentage change in e. Although the expression is only approximate except for the case where Δx approaches zero, it is sufficiently accurate for most filter applications.

EXAMPLE 5.12 Calculate the percentage error in ω_n of a series-tuned LC circuit when the capacitor value is in error by 1%.

PROCEDURE

1 The ω_n of a tuned *LC* circuit is given by $\omega_n = 1/\sqrt{LC}$.

2 Obtain the sensitivity of ω_n to C.

$$S_C^{\omega_n} = \frac{C}{\omega_n} \frac{\partial \omega_n}{\partial C} = -\frac{1}{2}$$

3 Obtain the percentage error in ω_n. The error in C is given by $\Delta C/C = 0.01$. Substitute this value and $S_C^{\omega_n} = -\frac{1}{2}$ into Eq. 5.72, giving

$$\frac{\Delta\omega_n}{\omega_n} \simeq S_C^{\omega_n} \frac{\Delta C}{C} = -\frac{1}{2} \cdot 0.01$$

$$= -0.005 \qquad \text{or} \qquad -0.5\%$$

This shows that an error of $+1\%$ in C would cause an error of -0.5% in ω_n. Thus a negative-valued sensitivity implies that if the component value increases, the filter parameter decreases.

The following expressions are useful in simplifying the calculation of sensitivity.

$$S_e^x = S_e^N - S_e^D \tag{5.73}$$

where $x = N/D$

$$S_{1/e}^x = -S_e^x \tag{5.74}$$

$$S_e^{|F(j\omega)|} = \text{real } S_e^{F(j\omega)} \tag{5.75}$$

$$\phi(\omega)S_e^{\phi(\omega)} = \text{imaginary } S_e^{F(j\omega)} \tag{5.76}$$

The latter two expressions apply to the sensitivities of frequency response magnitude $|F(j\omega)|$ and phase shift $\phi(\omega)$.

EXAMPLE 5.13 Calculate the sensitivity of the frequency-response magnitude of the circuit of Fig. 5.2*b* to the amplification A at the frequency ω_n. The component values are $g_1 = g_2 = g_3 = 10^{-4}$S, $C_1 = C_2 = 10^{-7}$ F and $A = 3.8$.

PROCEDURE

1 The symbolic transfer function for this circuit is given in Eq. 5.12. Writing the numerator and denominator separately gives

$$N(s) = AsC_2g_1 \tag{5.77}$$

$$D(s) = s^2C_1C_2 + s[C_1g_3 + C_2g_1 + C_2g_2(1 - A) + C_2g_3] + g_3(g_1 + g_2) \tag{5.78}$$

2 Obtain the sensitivities of $N(s)$ and $D(s)$ to A. Applying Eq. 5.71 gives

$$S_A^{N(s)} = 1 \tag{5.79}$$

$$S_A^{D(s)} = -\frac{sC_2g_2A}{D(s)} \tag{5.80}$$

3 Obtain the sensitivity of $F(s)$ to A, where $F(s) = N(s)/D(s)$. Substituting into Eq. 5.73 gives

$$S_A^{F(s)} = 1 + \frac{sC_2g_2A}{D(s)} \tag{5.81}$$

$$= \frac{s^2C_1C_2 + s[C_1g_3 + C_2g_1 + C_2g_2 + C_2g_3] + g_3(g_1 + g_2)}{D(s)} \tag{5.82}$$

4 Now substitute numerical values for the component symbols. Note that the sensitivity analysis is not possible if numerical values are substituted before this point.

$$S_A^{F(s)} = \frac{10^{-14}s^2 + 4 \times 10^{-11}s + 2 \times 10^{-8}}{10^{-14}s^2 + 0.2 \times 10^{-11}s + 2 \times 10^{-8}} \tag{5.83}$$

5 Convert the sensitivity of the transfer function to that of the frequency response by substituting $s = j\omega$.

$$S_A^F(j\omega) = \frac{(2 \times 10^{-8} - 10^{-14}\omega^2) + j(4 \times 10^{-11}\omega)}{(2 \times 10^{-8} - 10^{-14}\omega^2) + j(0.2 \times 10^{-11}\omega)} \tag{5.84}$$

6 The natural frequency $\omega_n = \sqrt{2} \cdot 10^3$ rad/s. Evaluating $S_A^{F(j\omega)}$ at this frequency by substituting $\omega = \omega_n$ into Eq. 5.84 gives

$$S_A^{F(j\omega)}\big|_{\omega=\omega_n} = 20 + j0 \tag{5.85}$$

At this frequency the sensitivity of $F(j\omega)$ is wholly real, but at other frequencies it would include a nonzero imaginary part.

7 Obtain the sensitivity of the frequency response magnitude to A by applying Eq. 5.75

$$S_A^{|F(j\omega)|}\big|_{\omega=\omega_n} = 20 \tag{5.86}$$

Note that a sensitivity of 20 implies that if A had an error of only 1%, then $|F(j\omega)|$ would be in error by approximately 20%. Sensitivities of this order of magnitude can easily arise with some types of active filter, and the designer must always make an assessment of the sensitivities to ensure that unacceptable response errors will not result from the tolerances allowed on component values.

5.6 SUMMARY

This chapter has introduced several techniques for analyzing active and passive networks. The NAM is extremely powerful and can incorporate nonideal models of the active elements in the most general case. It has the advantage of an ordered, step-by-step approach that simplifies the analysis of complex circuits.

The two-port matrix method is particularly useful when networks are interconnected in parallel or cascade fashion. Applications of this method will be seen in later chapters, often as alternatives to the analysis described.

The signal-flow graph is invaluable when analyzing multiloop structures such as those outlined in Chapter 10.

Finally, the concept of network sensitivity has been introduced. This is of vital importance in active filters, since it is essential that practical designs be capable of operating satisfactorily in the presence of component variation.

5.7 REFERENCES

[1] A. Budak, *Circuit Theory Fundamentals and Applications*, Englewood Cliffs, N.J.: Prentice-Hall, 1978, Section 4.2 and Appendix 1.

[2] M. E. Van Valkenburg, *Network Analysis*, 3rd ed., Englewood Cliffs, N.J.: Prentice-Hall, 1974, Chapter 11.

[3] S. J. Mason and H. J. Zimmermann, *Electronic Circuits Signals and Systems*, New York: Wiley, 1960, Sections 4.8 through 4.15.

CHAPTER 6

Sallen–Key Filters

Artice M. Davis
San José State University

6.1 INTRODUCTION

This chapter presents several active filter structures, known as Sallen–Key networks [1], which have proved themselves in practice. These circuits incorporate a single amplifier imbedded in a passive RC network to generate a second-order transfer function of any of the common types (see Chapter 3). These second-order transfer functions are useful in their own right; they can also be incorporated with others, as shown in Chapter 10, to yield higher-order filters.

The basic Sallen–Key structures hinge upon a single finite-gain amplifier (a voltage-controlled voltage source). The design of such finite-gain amplifiers, of both inverting and noninverting varieties, has already been considered in Chapter 4. In what follows, a general justification of the Sallen–Key approach will be given, a method of classifying such filters will be outlined, and a number of examples presented.

6.2 SALLEN–KEY TOPOLOGY

The Transfer Function

The basic Sallen–Key structure is shown in Fig. 6.1. The box labeled "Second-order passive RC network" contains resistors and two (or sometimes three) capacitors. It is assumed to be a grounded network having three terminals, in addition to an understood ground connection, as shown. V_i is assumed to be generated by

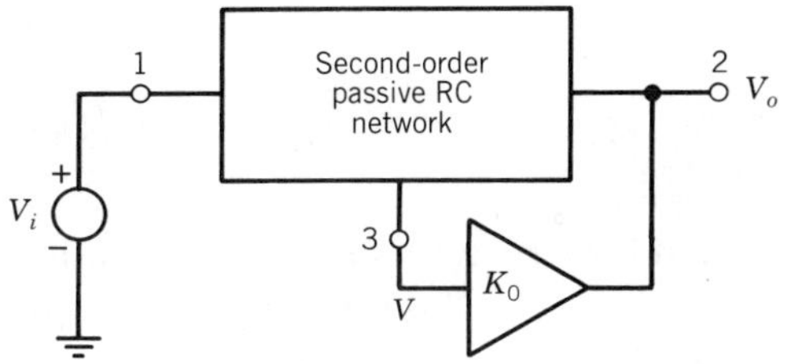

Figure 6.1 The basic Sallen–Key topology.

an ideal voltage source: V_o is the output voltage of an amplifier having gain K_0, and V is the amplifier input voltage. The amplifier is assumed to have zero output impedance. For this reason, the output of the Sallen–Key section will drive a succeeding section without experiencing loading effects. The input of the amplifier is assumed to possess infinite impedance, so the amplifier does not load the passive network.

If we apply superposition (recalling that the amplifier output is an ideal voltage source), we obtain

$$V = T_{FF}V_i + T_{FB}V_o \tag{6.1}$$

where

$$T_{FF} = \left.\frac{V}{V_i}\right|_{V_0=0} \tag{6.2}$$

and

$$T_{FB} = \left.\frac{V}{V_o}\right|_{V_i=0} \tag{6.3}$$

T_{FF} is called the feedforward gain of the passive network; T_{FB} is called the feedback gain. The interpretation of these two network functions is shown in Fig. 6.2, which may also be thought of as experimental test configurations for measuring T_{FF} and T_{FB}. Note that the amplifier is completely removed and an independent voltage source is applied to an appropriate terminal, while another is grounded.

Applying the amplifier relation

$$V_o = K_0V \tag{6.4}$$

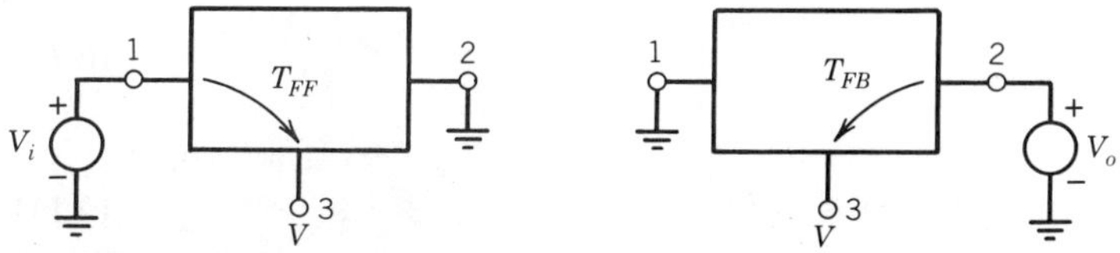

(*a*) Determination of T_{FF}. (*b*) Determination of T_{FB}.

Figure 6.2 Circuit interpretation of the feedforward and feedback transfer functions.

we obtain

$$V_o = \frac{K_0 T_{\text{FF}}}{1 - K_0 T_{\text{FB}}} V_i \tag{6.5}$$

Now, unless the RC network is degenerate (and the structures under consideration will be designed so as not to be), all voltage transfer functions of the passive network will have the same denominator polynomial:

$$D(s) = s^2 + b_1 s + b_0 \tag{6.6}$$

Therefore, we can write Eqs. 6.2 and 6.3 as

$$T_{\text{FF}} = \frac{N_{\text{FF}}(s)}{D(s)} \tag{6.7}$$

and

$$T_{\text{FB}} = \frac{N_{\text{FB}}(s)}{D(s)} \tag{6.8}$$

where $N_{\text{FF}}(s)$ and $N_{\text{FB}}(s)$ are polynomials with degree of at most two.

By using Eq. 6.5 and the two preceding relations, we can derive the overall voltage gain of the Sallen–Key filter block:

$$T(s) = \frac{V_o(s)}{V_i(s)} = \frac{K_0 N_{\text{FF}}(s)}{D(s) - K_0 N_{\text{FB}}(s)} \tag{6.9}$$

It should be clear that the zeros of $D(s)$ are the poles of the passive structure—and, as such, have values of Q that are always less than, or equal to, 0.5. Since $K_0 N_{\text{FB}}(s)$ has the effect of modifying the coefficients of the characteristic polynomial of $T(s)$, it is evident that the closed-loop poles can be placed anywhere in the complex plane by appropriately choosing K_0. Such choices are discussed below.

Classification

It is possible to identify specific configurations of the circuits shown in Fig. 6.1 with various types of transfer function by considering the form of $N_{\text{FF}}(s)$ and $N_{\text{FB}}(s)$. In order to do this, we must be more specific about notation. For this purpose, we write

$$N_{\text{FF}}(s) = a_2 s^2 + a_1 s + a_0 \tag{6.10}$$

and

$$N_{\text{FB}}(s) = c_2 s^2 + c_1 s + c_0 \tag{6.11}$$

It should be noted parenthetically that all the a and c coefficients, as well as the b coefficients in Eq. 6.6 (by the Fialkow–Gerst conditions [2]) are nonnegative. The characteristic polynomial (the denominator of Eq. 6.9) of the filter block then becomes

$$P(s) = As^2 + Bs + C = (1 - K_0c_2)s^2 + (b_1 - K_0c_1)s + (b_0 - K_0c_0) \quad (6.12)$$

The first classification scheme that we consider depends upon the poles of $T(s)$, that is, the zeros of $P(s)$.

KRC Suppose that the passive circuit configuration of Fig. 6.1 is selected such that $c_2 = c_0 = 0$. This corresponds to requiring that

$$T_{FB}(s) = \frac{N_{FB}(s)}{D(s)} = \frac{c_1 s}{s^2 + b_1 s + b_0} \quad (6.13)$$

or, in other words, that the feedback network is bandpass in nature. We can then rewrite Eq. 6.12 as

$$P(s) = s^2 + (b_1 - K_0c_1)s + b_0 \quad (6.14)$$

Thus the pole parameters of the second-order filter are (see Chapter 3):

$$\omega_0 = \sqrt{b_0} \quad (6.15)$$

and

$$Q = \frac{\sqrt{b_0}}{b_1 - K_0c_1} \quad (6.16)$$

If $K_0 = 0$, the poles are identical with those of the passive RC network. Hence $Q \leq \frac{1}{2}$. Thus, if one makes $K_0 = K > 0$, then Q can be adjusted to any desired value. This defines the KRC, or positive gain, Sallen–Key active filter structure.

A few practical comments are appropriate at this juncture. The coefficients b_0, b_1, and c_1 depend only upon the passive RC network. Since discrete resistors and capacitors can be selected for precise values, the coefficients are precisely known. On the other hand, the amplifier voltage gain $K_0 = K$ depends upon the gain of an operational amplifier as well as upon a resistor ratio (as discussed in Chapter 4). Thus, if K_0C_1 is approximately equal to b_1 (high Q), this parameter will be relatively imprecise and dependent upon temperature and other factors. Study of Eqs. 6.15 and 6.16 indicates that the tuning frequency ω_0 is set by the passive components and will therefore be a relatively precise parameter; Q, on the other hand, depends upon the active element and will thus be relatively imprecise.

−KRC Now suppose that the passive network topology is selected such that $c_1 = 0$. This implies that

$$T_{FB}(s) = \frac{N_{FB}(s)}{D(s)} = \frac{c_2 s^2 + c_0}{s^2 + b_1 s + b_0} \quad (6.17)$$

Now if $c_2 = 0$, the feedback network is low pass; if $c_0 = 0$, it is high pass; and if neither c_2 nor c_0 are zero, it is bandstop or notch. Equation 6.9 now implies that

$$P(s) = (1 - K_0c_2)s^2 + b_1 s + (b_0 - K_0c_0) \quad (6.18)$$

In order to increase Q (to something larger than 0.5), one must either increase the coefficient of s^2 and/or the coefficient of s^0, or decrease the coefficient of s. The latter is impossible, for the amplifier gain does not appear in the coefficient of s. Thus we must make $K_0 = -K$, where $K > 0$. Then we can easily find that

$$\omega_0 = \sqrt{\frac{b_0 + Kc_0}{1 + Kc_2}} \tag{6.19}$$

and

$$Q = \frac{1}{b_1}\sqrt{(b_0 + Kc_0)(1 + Kc_2)} \tag{6.20}$$

If we alternately let K go to zero and infinity, the following bounds are obtained:

$$\sqrt{b_0} \le \omega_0 \le \sqrt{\frac{c_0}{c_2}} \tag{6.21}$$

and

$$\frac{1}{b_1}\sqrt{b_0} \le Q < \infty \tag{6.22}$$

From Eqs. 6.19 and 6.20 it is apparent that both ω_0 and Q depend upon the active element parameters. Hence we could perhaps anticipate that this circuit might not be as precise as the KRC network when amplifier variability is considered.

Infinite Gain If we allow K_0 to approach $-\infty$ in Eq. 6.9, we obtain the so-called "infinite gain" realization:

$$T(s) = \frac{-N_{FF}(s)}{N_{FB}(s)} \tag{6.23}$$

The important issue here is that the poles of $T(s)$ are the same as the zeros of $N_{FB}(s)$, and these zeros can be placed anywhere in the left half-plane. Further discussion of the infinite gain case will be postponed until Chapter 7, but it should be noted that $T(s)$ (hence ω_0 and Q) is independent of the amplifier gain—as long as it is so large as to be negligible.

Thus far the classification scheme has only considered the poles of $T(s)$. Looking back at Eq. 6.9, however, it becomes clear that the filter type (HP, LP, BP, etc.) is determined by $T_{FF}(s)$. If

$$T_{FF}(s) = \frac{N_{FF}(s)}{D(s)} = \frac{a_0}{s^2 + b_1 s + b_0}$$

for instance, the filter is low pass. Similar considerations hold for each of the other types of second-order filter.

In summary, then, if the amplifier gain is positive, the feedback network must be bandpass; if it is negative, the feedback must be low pass, high pass, or notch.

***Table 6.1* Summary of pole-dependent classification**

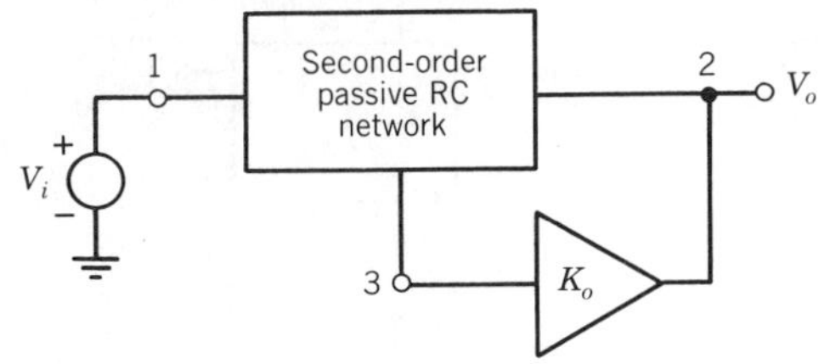

$$T(s) = \frac{K_0 N_{FF}(s)}{D(s) - K_0 N_{FB}(s)} = \frac{Q(s)}{P(s)}$$

KRC: $P(s) = s^2 + (b_1 - Kc_1)s + b_0$

$$\omega_0 = \sqrt{b_0} \qquad Q = \frac{\sqrt{b_0}}{b_1 - Kc_1}$$

−KRC: $P(s) = (1 + Kc_2)s^2 + b_1 s + (b_0 + Kc_0)$

$$\omega_0 = \sqrt{\frac{b_0 + Kc_0}{1 + Kc_2}} \qquad Q = \frac{1}{b_1}\sqrt{(b_0 + Kc_0)(1 + Kc_2)}$$

Typically, either $c_0 = 0$ or $c_2 = 0$

The feedforward-network type determines the overall filter function. Table 6.1 summarizes these results.

6.3 RAPID ANALYSIS OF ACTIVE AND PASSIVE RC NETWORKS

Before considering design examples involving Sallen–Key active filters, it is perhaps appropriate to review a quick method of analyzing active or passive RC networks [3] (see Chapter 5). Figure 6.3 shows a (perhaps active) RC network having a single capacitor. The driving source is assumed to be ideal—any source resistance having been incorporated in the nondynamic network. The same is true of any load resistance. Hence

$$T(s) = \frac{V_o(s)}{V_i(s)} = \frac{T_\infty s + T_0 P}{s + P} \tag{6.24}$$

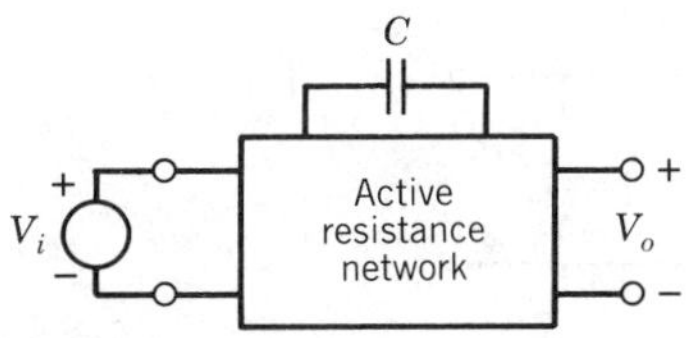

Figure 6.3 Single-capacitor network.

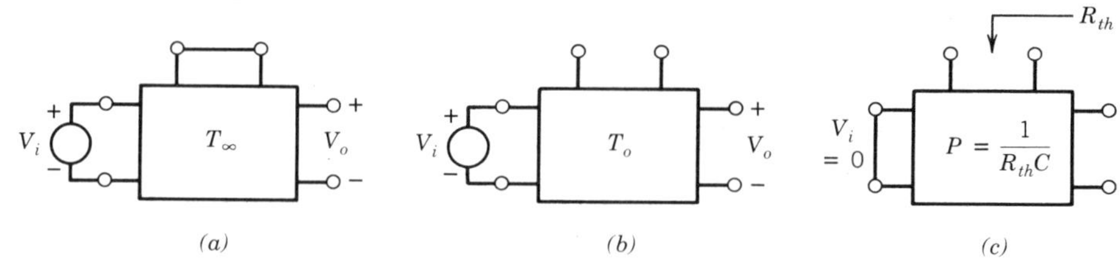

Figure 6.4 Single-capacitor network algorithm.

where T_∞ is the high-frequency gain, T_0 is the low-frequency gain, and P is the reciprocal of the circuit time constant. Thus T_∞ is computed with the capacitor shorted and T_0 with the capacitor open-circuited as shown in Fig. 6.4*a* and 6.4*b*. The "pole" (actually the negative of the network pole) is computed as indicated in Fig. 6.4*c*.

EXAMPLE 6.1 Find the voltage transfer function of the single-capacitor network of Fig. 6.5*a*.

PROCEDURE

1 Short the capacitor, as in Fig. 6.5*b*. Clearly, $T_\infty = 1$.

2 Open the capacitor, as in Fig. 6.5*c*. Again, it is clear that $T_0 = R_2/(R_1 + R_2)$.

3 Short the input voltage source, open the capacitor, and compute R_{th}. Since R_1 and R_2 are in parallel in Fig. 6.5*d*, we get

$$R_{\text{th}} = R_1 \| R_2 = \frac{R_1 R_2}{R_1 + R_2}$$

Thus

$$P = \frac{1}{R_{\text{th}} C} = \frac{R_1 + R_2}{R_1 R_2 C}$$

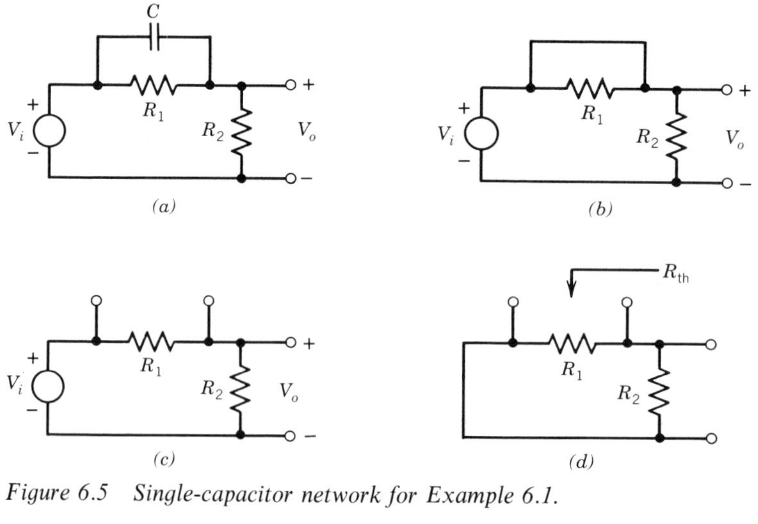

Figure 6.5 Single-capacitor network for Example 6.1.

4 Now insert these parameters into Eq. 6.24:

$$T(s) = \frac{1 \cdot s + \dfrac{R_2}{R_1 + R_2} \times \dfrac{R_1 + R_2}{R_1 R_2 C}}{s + \dfrac{R_1 + R_2}{R_1 R_2 C}} = \frac{s + \dfrac{1}{R_1 C}}{s + \dfrac{1}{(R_1 \| R_2) C}}$$

This technique also works when the network has controlled sources (amplifiers), but it will be used in this chapter merely to analyze the passive RC networks associated with Sallen–Key filters.

The preceding technique can be extended to the analysis of multiple-capacitor networks [3]. For present purposes, only the extension to two capacitor circuits will be required. In order to appreciate the following development of such a "two-capacitor method," glance at the circuit of Fig. 6.6. By the methods of Chapter 5 (and in [3]), the voltage transfer function is given by

$$T(s) = \frac{V_o(s)}{V_i(s)} = \frac{T_{12} s^2 + (T_1 P_{2s} + T_2 P_{1s}) s + T_0 \{P_{1s} \cdot P_{2o} \text{ or } P_{2s} \cdot P_{1o}\}}{s^2 + (P_{1s} + P_{2s}) s + \{P_{1s} \cdot P_{2o} \text{ or } P_{2s} \cdot P_{1o}\}} \tag{6.25}$$

where

P_{1s} = pole due to C_1 with C_2 shorted

P_{2s} = pole due to C_2 with C_1 shorted

P_{1o} = pole due to C_1 with C_2 open

P_{2o} = pole due to C_2 with C_1 open

and

T_{12} = voltage gain with C_1 shorted and C_2 shorted

T_1 = voltage gain with C_1 shorted and C_2 open

T_2 = voltage gain with C_2 shorted and C_1 open

T_0 = voltage gain with C_1 open and C_2 open

Note that the coefficient of s in the denominator is the "sum of the shorted poles," while the coefficient of s^0 is the product of an open pole and a shorted pole, due to different capacitors. The numerator parameters are the denominator poles multiplied by a voltage gain—that voltage gain being computed under the same condition (short or open) as for computation of the corresponding pole.

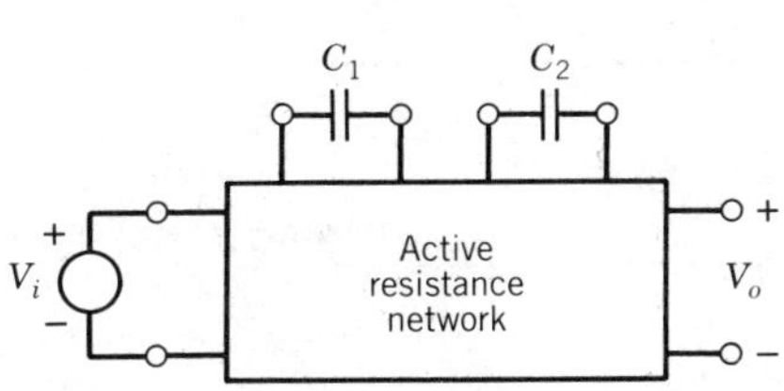

Figure 6.6 Two-capacitor network.

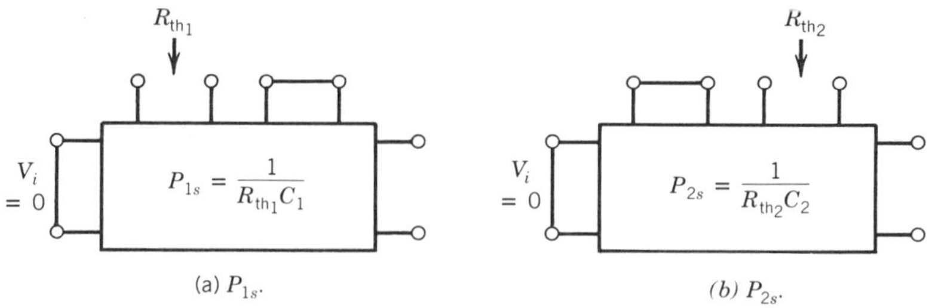

Figure 6.7 Equivalent circuits for computing the two "shorted poles."

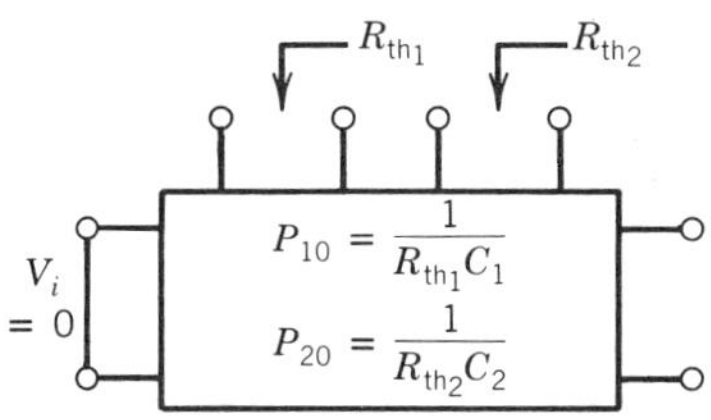

Figure 6.8 Equivalent circuit for computing the two "open poles."

Figure 6.7 shows the equivalent networks for computing the shorted poles; Fig. 6.8 shows the technique for computing the open poles. Fig. 6.9 indicates the appropriate equivalents for computing the various voltage-gain functions. The appeal of this technique lies in the fact that each calculation involves only a simple resistive network; the complex frequency never enters the calculation until all parameters are assembled into the final form of the transfer function. In fact, a little practice will allow one to calculate the various parameters without actually sketching the equivalent circuits.

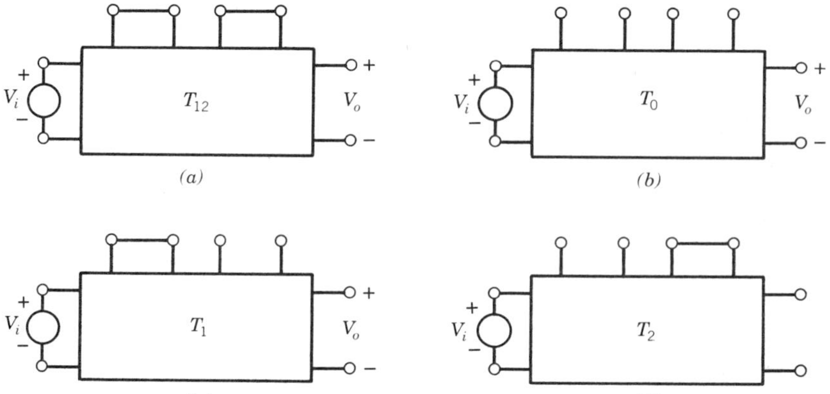

Figure 6.9 Equivalent circuits for computing the various voltage gains.

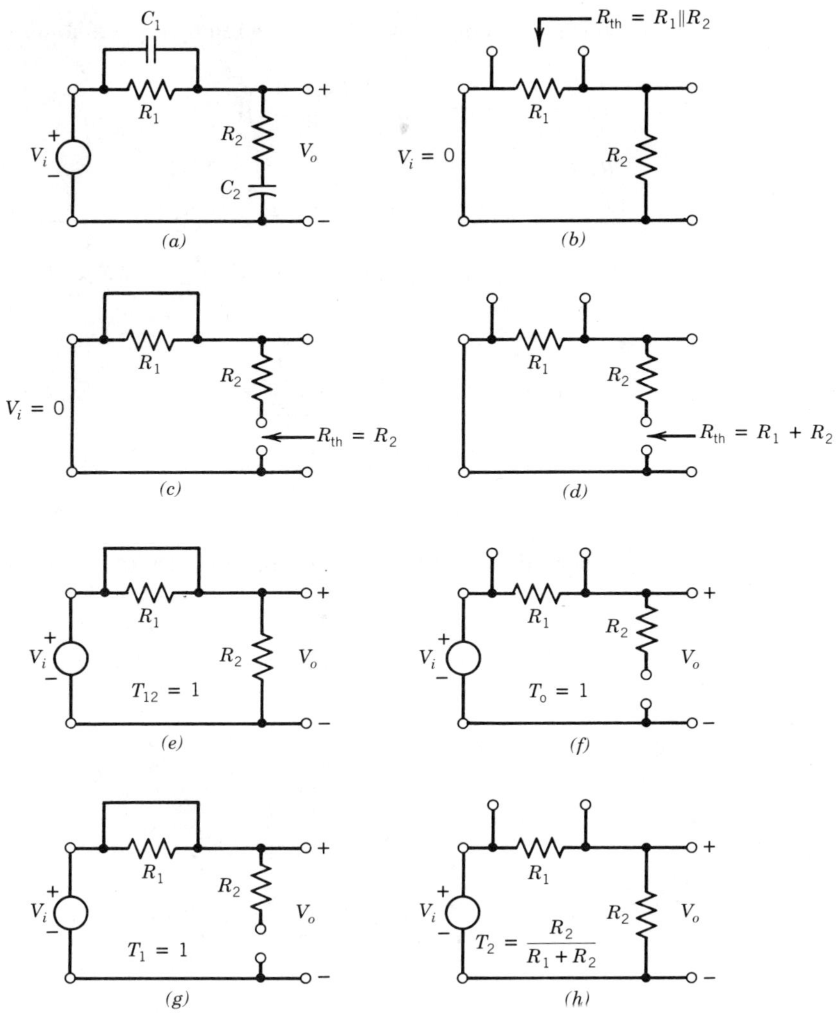

Figure 6.10 Two-capacitor network for Example 6.2.

EXAMPLE 6.2 Find the voltage transfer function of the two-capacitor RC circuit of Fig. 6.10*a*.

PROCEDURE

1 Find the two shorted poles, as in Fig. 6.10*b* and *c*. Thus $P_{1s} = 1/(R_1 \| R_2)C_1 = (R_1 + R_2)/(R_1 R_2 C_1)$ and $P_{2s} = 1/R_2 C_2$.

2 Find the two open poles, as in Fig. 6.10*d*. Here $P_{1o} = 1/R_1 C_1$ and $P_{2o} = 1/(R_1 + R_2)C_2$.

3 Find the various voltage gains, as in Figs. 6.10*e* through *h*. From these figures, it should be clear that $T_{12} = T_0 = T_1 = 1$, while $T_2 = [R_2/(R_1 + R_2)]$.

4 Assemble all of the above parameters into the transfer function given by Eq. 6.25. In this case

$$T(s) = \frac{1 \cdot s^2 + \left(1 \cdot \frac{1}{R_2C_2} + \frac{R_2}{R_1 + R_2} \times \frac{R_1 + R_2}{R_1R_2C_1}\right)s + 1 \cdot \left\{\frac{R_1 + R_2}{R_1R_2C_1} \times \frac{1}{(R_1 + R_2)C_2} \text{ or } \frac{1}{R_2C_2} \times \frac{1}{R_1C_1}\right\}}{s^2 + \left(\frac{1}{R_2C_2} + \frac{R_1 + R_2}{R_1R_2C_1}\right)s + \left\{\frac{R_1 + R_2}{R_1R_2C_1} \times \frac{1}{(R_1 + R_2)C_2} \text{ or } \frac{1}{R_2C_2} \times \frac{1}{R_1C_1}\right\}}$$

or

$$T(s) = \frac{s^2 + \left(\frac{1}{R_1C_1} + \frac{1}{R_2C_2}\right)s + \frac{1}{R_1R_2C_1C_2}}{s^2 + \left(\frac{1}{R_2C_2} + \frac{R_1 + R_2}{R_1R_2C_1}\right)s + \frac{1}{R_1R_2C_1C_2}}$$

KRC Low-Pass Filters

Several examples are now presented to illustrate the design process of low-pass filters.

EXAMPLE 6.3 Classify the Sallen–Key active filter of Fig. 6.11 and derive its transfer function $T(s) = V_o(s)/V_i(s)$.

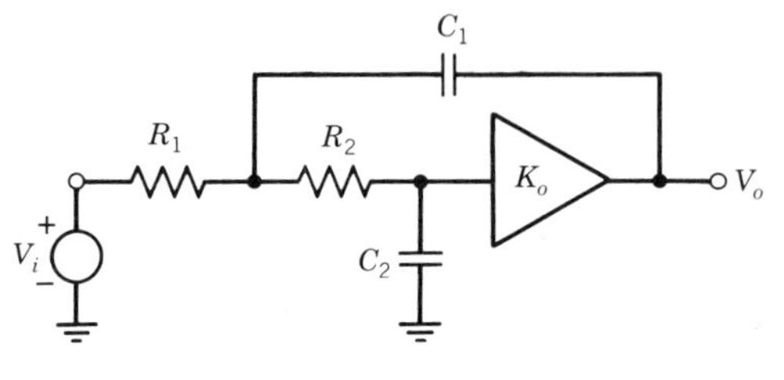

Figure 6.11 Active filter circuit for Example 6.3.

PROCEDURE

1 Determine the type of feedback network by forcing V_i to zero and replacing the amplifier output with an independent voltage source. Identify the amplifier input voltage as V. The result is shown in Fig. 6.12.

Clearly, the feedback network is of the bandpass variety, for the series capacitor C_1 forces the gain to zero at zero frequency and the shunt capacitor C_2 forces it to zero at infinite frequency. Hence the filter type is KRC, and we must make $K_0 = +K$.

2 Compute $T_{FB}(s)$. By applying the two-capacitor algorithm, we can easily verify that $T_{12} = T_2 = T_0 = 0$ and $T_1 = 1$. Furthermore, $P_{1s} = 1/(R_1\|R_2)C_1 = (R_1 + R_2)/R_1R_2C_1$, $P_{2s} = 1/R_2C_2$, $P_{1o} = 1/R_1C_1$, and $P_{2o} = 1/(R_1 + R_2)C_2$. Hence

$$T_{FB}(s) = \frac{\frac{1}{R_2C_2}s}{s^2 + \left[\frac{1}{R_2C_2} + \frac{R_1 + R_2}{R_1R_2C_1}\right]s + \frac{1}{R_1R_2C_1C_2}} = \frac{N_{FB}(s)}{D(s)} \tag{6.26}$$

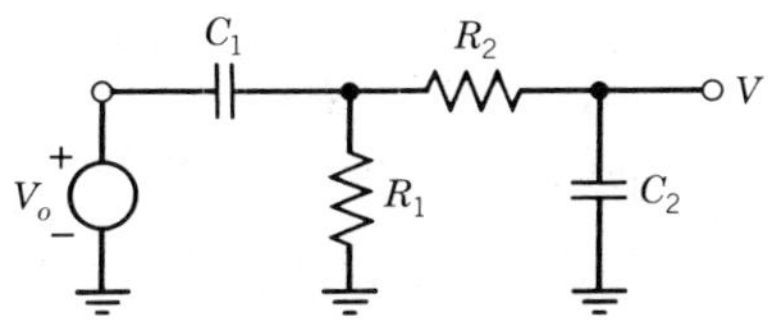

Figure 6.12 Feedback network for the circuit of Fig. 6.11.

3 Determine the type of feedforward network by forcing V_o to zero. The result is given in Fig. 6.13. Since both C_1 and C_2 are connected in shunt, the feedforward network is low pass. Thus the circuit in Fig. 6.11 is a KRC low-pass filter.

4 Compute $T_{FF}(s)$. Since the poles have already been computed, we need only calculate the voltage gains. By inspection of Fig. 6.13, $T_{12} = T_1 = T_2 = 0$ and $T_0 = 1$. Thus

$$T_{FF}(s) = \frac{\dfrac{1}{R_1R_2C_1C_2}}{s^2 + \left[\dfrac{1}{R_2C_2} + \dfrac{R_1 + R_2}{R_1R_2C_1}\right]s + \dfrac{1}{R_1R_2C_1C_2}} = \frac{N_{FF}(s)}{D(s)} \tag{6.27}$$

5 Compute $T(s)$ by using Eq. 6.9 with $K_0 = +K$. Thus

$$T(s) = \frac{\dfrac{K}{R_1R_2C_1C_2}}{s^2 + \left[\dfrac{1-K}{R_2C_2} + \dfrac{R_1 + R_2}{R_1R_2C_1}\right]s + \dfrac{1}{R_1R_2C_1C_2}} = \frac{K\omega_0^2}{s^2 + \dfrac{\omega_0}{Q}s + \omega_0^2} \tag{6.28}$$

Now by direct computation, or by use of Eqs. 6.15 and 6.16,

$$\omega_0 = \frac{1}{\sqrt{R_1R_2C_1C_2}} \tag{6.29}$$

and

$$Q = \frac{\dfrac{1}{\sqrt{R_1R_2C_1C_2}}}{\dfrac{1-K}{R_2C_2} + \dfrac{R_1 + R_2}{R_1R_2C_1}} = \frac{1}{(1-K)\sqrt{\dfrac{R_1C_1}{R_2C_2}} + \left(1 + \dfrac{R_2}{R_1}\right)\sqrt{\dfrac{R_1C_2}{R_2C_1}}} \tag{6.30}$$

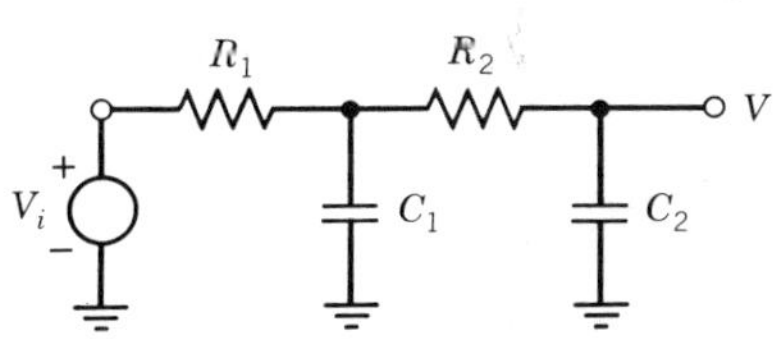

Figure 6.13 Feed-forward network for the circuit of Fig. 6.11.

EXAMPLE 6.4 Select the component values for the low-pass *KRC* filter of Fig. 6.11 such that the tuning frequency is $f_0 = 1000$ Hz and $Q = 5$.

PROCEDURE

1 First note that there are five adjustable parameters in the circuit of Fig. 6.11; the passive component values (R_1, R_2, C_1, and C_2) and the active parameter (the amplifier gain $K_0 = K$). Thus we can arbitrarily choose the values of three. Each such choice leads to a filter with different properties. The first choice that will be considered is the one for which $R_1 = R_2 = R$ and $C_1 = C_2 = C$. The rationale for this choice relies on two principal

aims: mathematical convenience and low element spread. The first needs no elaboration. The second means that it is desirable that resistance and capacitance values should not be spread too widely. For instance, we can appreciate the practical difficulty of implementing a circuit in which $C_1 = 0.01$ pF, while $C_2 = 10\mu$F.

2 Rewrite the expressions for ω_0 and Q (Eqs. 6.29 and 6.30) as

$$\omega_0 = \frac{1}{\mathrm{RC}} \tag{6.31}$$

and

$$Q = \frac{1}{3 - K} \tag{6.32}$$

Hence, using $f_0 = 1000$ Hz and $Q = 5$, we have

$$\mathrm{RC} = 159.2 \text{ microseconds } (\mu\text{s})$$

and

$$K = 3 - \frac{1}{Q} = 2.8$$

Thus, if C is arbitrarily picked to be 0.1 μF, then

$$R = 1.59 \text{ k}\Omega$$

3 Note that we must automatically accept the fact that the dc gain is equal to 2.8. If we wish the dc gain to be unity, for instance, a voltage divider can be used to replace R_1 in Fig. 6.11.

A complete design of the above active filter (with unity dc gain) appears in Fig. 6.14. Note that we must require that (Thévenin equivalent) $R' \| R'' = 1.59$ kΩ and that $R''/(R' + R'') = 1/2.8$. Hence $R' = 4.45$ kΩ and $R'' = 2.47$ kΩ. Also, noting that this circuit incorporates a noninverting op amp circuit, we must make $R_b = 1.8R_a$. In Fig. 6.14, it has been assumed that $R_a = 1$ kΩ so that $R_b = 1.8$ kΩ.

4 Although the element spread for the equal R–equal C filter is excellent, it turns out that the filter Q is quite strongly sensitive (see Chapter 11) to variations in component values [4]. Thus it is of some interest to investigate an alternative design.

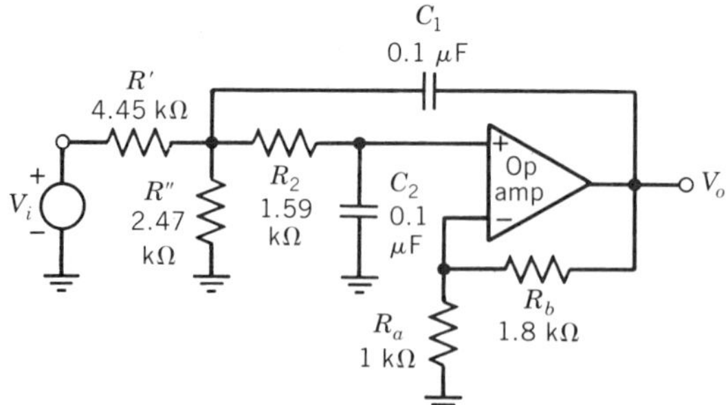

Figure 6.14 Equal R–equal C KRC low-pass filter design for $f_0 = 1000$ Hz and $Q = 5$.

EXAMPLE 6.5 Select the component values for the low-pass KRC filter of Example 6.4 if the amplifier gain is given by $K = 1$.

PROCEDURE

1 The equal R–equal C assumption will be abandoned in favor of the supposition of unity gain. Since the bandwidth of a unity gain amplifier is larger than one of higher gain, we might expect that a filter with better high-frequency capabilities would result.

2 Although the gain has been specified, we still have some latitude in choosing component values. Suppose that some of this freedom is exercised by choosing equal resistance values, $R_1 = R_2 = R$. Thus this design is usually referred to as the equal R–unity gain KRC low-pass filter. Equations 6.29 and 6.30 then yield

$$\omega_0 = \frac{1}{R\sqrt{C_1 C_2}} \tag{6.33}$$

and

$$Q = \frac{1}{2}\sqrt{\frac{C_1}{C_2}} \tag{6.34}$$

3 If we arbitrarily choose $R = 1\ \text{k}\Omega$, then these equations can be solved to give $C_1 = 1.59\ \mu\text{F}$ and $C_2 = 0.016\ \mu\text{F}$. The circuit is shown in Fig. 6.15.

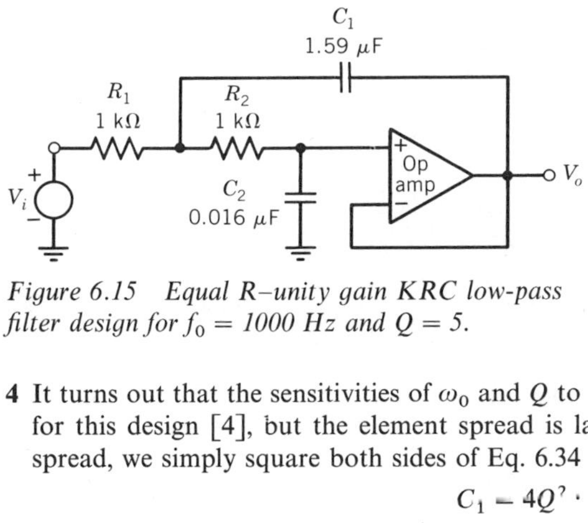

Figure 6.15 Equal R–unity gain KRC low-pass filter design for $f_0 = 1000$ Hz and $Q = 5$.

4 It turns out that the sensitivities of ω_0 and Q to the various element values are very low for this design [4], but the element spread is large. To see this problem with element spread, we simply square both sides of Eq. 6.34 and solve for C_1:

$$C_1 = 4Q^2 \cdot C_2 \tag{6.35}$$

This equation clearly shows that for quality factors of even modest size, the capacitor value spread is large.

There is an optimum compromise between element spread and sensitivity known as the Saraga circuit. The reader is referred to [4] and Saraga's paper [5] for a complete discussion.

KRC High-Pass Filters

As shown in Chapter 3, the general second-order low-pass filter function may be written as

$$T(s) = \frac{H\omega_0^2}{s^2 + (\omega_0/Q)s + \omega_0^2} \tag{6.36}$$

where H is the dc gain, and ω_0 and Q are the complex conjugate pole-pair parameters. Similarly,

$$T(s) = \frac{H(\omega_0/Q)s}{s^2 + (\omega_0/Q)s + \omega_0^2} \tag{6.37}$$

is the general second-order bandpass filter function. In this case, H is the peak gain that occurs at the frequency ω_0. Finally,

$$T(s) = \frac{Hs^2}{s^2 + (\omega_0/Q)s + \omega_0^2} \tag{6.38}$$

is the general second-order high-pass filter function. Here H is the high-frequency gain of the filter.

This section develops the design of high-pass KRC filters; the following one will concentrate on bandpass networks. The method used here is known as the RC:CR transformation [4, 6]. In order to follow the development of this result, consider once more the KRC low-pass filter of Fig. 6.11 and its associated voltage transfer function, Eq. 6.28. The tuning frequency and quality factor were given by Eqs. 6.29 and 6.30, which are repeated below for ease of reference:

$$\omega_0 = \frac{1}{\sqrt{R_1R_2C_1C_2}} \tag{6.29}$$

and

$$Q = \frac{1}{(1 - K)\sqrt{R_1C_1/R_2C_2} + [1 + (R_2/R_1)]\sqrt{R_1C_2/R_2C_1}} \tag{6.30}$$

Now suppose that all resistors and capacitors are multiplied by a factor, that is,

$$R_i \rightarrow aR_i \tag{6.39}$$

and

$$C_i \rightarrow aC_i \tag{6.40}$$

where the arrow means "is replaced by." Since Q is a function of resistor and capacitor ratios only (see Eq. 6.30), it is unaffected by the above procedure. However, a quick computation involving Eq. 6.29 yields the following replacement:

$$\omega_0 \rightarrow \frac{\omega_0}{a^2} \tag{6.41}$$

At this stage, suppose that the factor a is given by

$$a = \frac{\omega_0}{s} \tag{6.42}$$

Then

$$R_i \rightarrow \frac{\omega_0 R_i}{s} = \frac{1}{(1/\omega_0 R_i)s} \tag{6.43}$$

$$\frac{1}{C_i s} \rightarrow \frac{1}{(\omega_0 C_i/s)s} = \frac{1}{\omega_0 C_i} \tag{6.44}$$

and

$$\omega_0 \rightarrow \frac{s^2}{\omega_0} \tag{6.45}$$

Thus capacitors are replaced by resistors and resistors by capacitors (of appropriate values). Furthermore, the parameter ω_0 is replaced by s^2/ω_0. This constitutes the RC:CR transformation.

If we now investigate the effect of this transformation on the low-pass filter function given in Eq. 6.36, we obtain

$$T(s) = \frac{K(s^4/\omega_0^2)}{s^2 + (s^3/\omega_0 Q) + (s^4/\omega_0^2)} = \frac{Ks^2}{s^2 + (\omega_0/Q)s + \omega_0^2} \tag{6.46}$$

The low-pass function becomes a high-pass function under the RC:CR transformation with the same ω_0 and Q values but with a high-frequency gain that is the same as the dc gain of the original low-pass function. These results can be shown to hold in general for any RC network [6].

EXAMPLE 6.6 Design a second-order equal R–equal C KRC high-pass filter having $f_0 = 1000$ Hz, $Q = 5$, and unity high-frequency gain.

PROCEDURE

1 Because the specifications for this high-pass filter are analogous to those for the low-pass filter of Example 6.4, we simply resketch the circuit of Fig. 6.14 with capacitors replaced by resistors and vice versa. The resulting network is shown in Fig. 6.16. Note

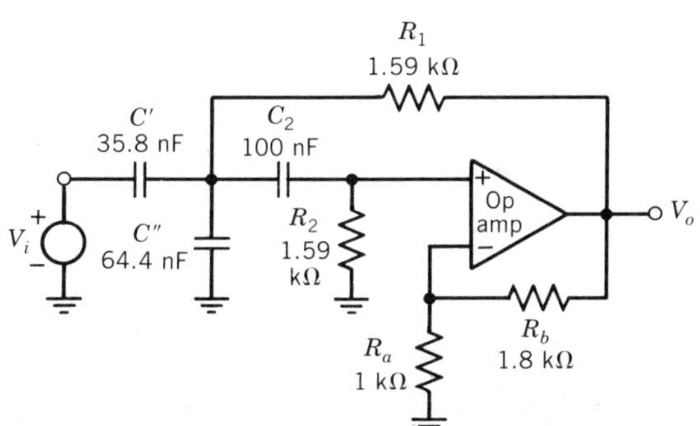

Figure 6.16 Equal R–equal C high-pass filter design for $f_0 = 1000$ Hz and $Q = 5$.

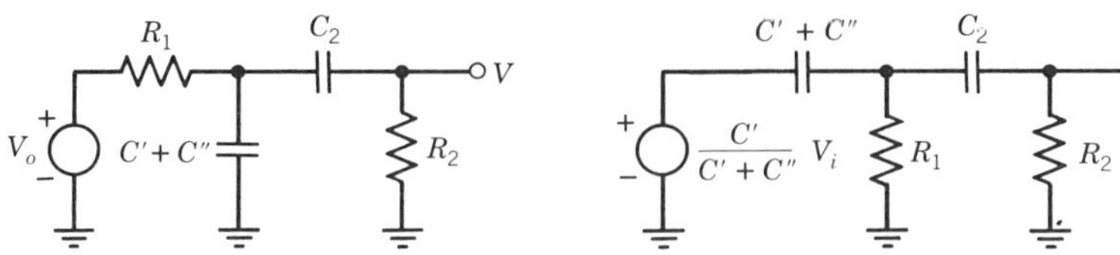

(a) Feedback network (b) Feedforward network

Figure 6.17 Constituent networks for the circuit of Fig. 6.16.

that resistors are still incorporated to set the op amp gain. (Capacitors could *theoretically* be used, but they present dc biasing problems for the op amp.)

2 Compute the element values, noting that $\omega_0 = 2\pi \times 1000$ rad/s. Thus

$$C' = \frac{1}{\omega_0 R'} = 35.8 \text{ nF},$$

$$C'' = \frac{1}{\omega_0 R''} = 64.4 \text{ nF},$$

$$C_2 = \frac{1}{\omega_0 R_2} = 100 \text{ nF},$$

$$R_1 = \frac{1}{\omega_0 C_1} = 1.59 \text{ k}\Omega,$$

and

$$R_2 = \frac{1}{\omega_0 C_2} = 1.59 \text{ k}\Omega,$$

as shown in Fig. 6.16.

3 In order to doublecheck the classification scheme previously presented, the feedback and feedforward networks are shown in Fig. 6.17. Note that the input source and the capacitors, C' and C'', have been replaced by their Thévenin equivalents in part (*b*) of that figure.

KRC Bandpass Filters

The design of KRC high-pass filters presented in Example 6.6 relied upon the RC:CR transformation. This transformation in turn relies upon the symmetry that exists between the second-order low-pass and high-pass filter transfer functions. Unfortunately, such symmetry does not exist for the bandpass function. Thus the approach must be more direct.

It is straightforward, though somewhat algebraically involved, to show that the network illustrated in Fig. 6.18 is a KRC bandpass filter and that its transfer function is given by

$$T(s) = \frac{K \cdot \dfrac{1}{R_1 C_1} s}{s^2 + \left[\dfrac{1}{R_2 C_2} + \left(1 - K + \dfrac{R_3}{R_1} + \dfrac{R_3}{R_2}\right)\dfrac{1}{R_3 C_1}\right] s + \dfrac{R_1 + R_3}{R_1 R_2 R_3 C_1 C_2}} \tag{6.47}$$

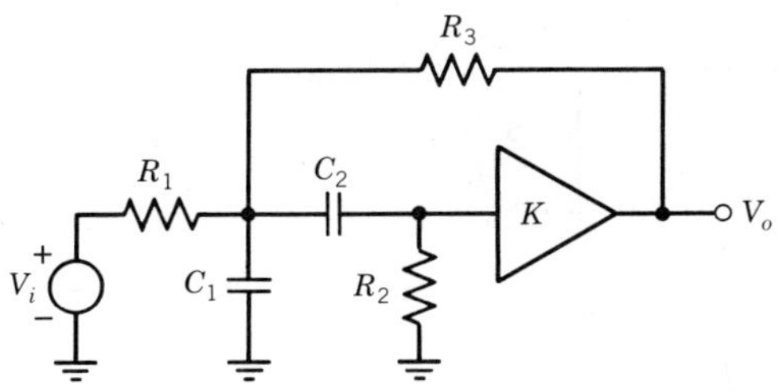

Figure 6.18 KRC bandpass filter.

from which

$$\omega_0 = \sqrt{\frac{R_1 + R_3}{R_1 R_2 R_3 C_1 C_2}} \tag{6.48}$$

and

$$Q = \frac{\sqrt{\dfrac{R_2}{R_1} + \dfrac{R_2}{R_3}}}{\sqrt{\dfrac{C_1}{C_2}} + \sqrt{\dfrac{C_2}{C_1}} \cdot \left[1 + \dfrac{R_2}{R_1} + (1 - K)\dfrac{R_2}{R_3}\right]} \tag{6.49}$$

EXAMPLE 6.7 Determine the component values for the KRC bandpass filter of Fig. 6.18 if the center frequency is to be 1000 Hz, the bandwidth 100 Hz, and the peak gain unity.

PROCEDURE

1 Since there are six parameters in the original circuit and only two parameters to be specified (ω_0 and Q), some freedom of choice is possible. Suppose that $R_1 = R_2 = R_3 = R$ and $C_1 = C_2 = C$. Then Eqs. 6.48 and 6.49 become

$$\omega_0 = \frac{\sqrt{2}}{RC} \tag{6.50}$$

and

$$Q = \frac{\sqrt{2}}{4 - K} \tag{6.51}$$

2 Compute Q from

$$Q = \frac{\omega_0}{\Delta\omega} = \frac{f_0}{\Delta f} = \frac{1000 \text{ Hz}}{100 \text{ Hz}} = 10$$

3 Solve Eqs. 6.50 and 6.51 to yield

$$RC = \frac{\sqrt{2}}{\omega_0} = 225.1 \ \mu\text{s}$$

and

$$K = 4 - \frac{\sqrt{2}}{Q} = 3.86$$

4 If one arbitrarily chooses $C = 0.1 \ \mu\text{F}$, then $R = 2251 \ \Omega$.

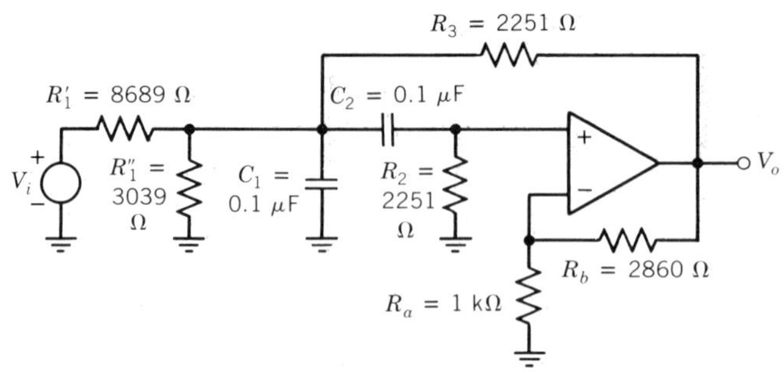

Figure 6.19 KRC bandpass filter with $f_0 = 1000$ Hz, BW = 100 Hz, and unity peak gain.

5 Since the peak gain is $K = 3.86$, and unity is desired, one must once more split R_1 into a voltage divider as shown in the complete circuit diagram of Fig. 6.19. Again, one must make $R_1''/(R_1' + R_1'') = 1/3.86$ and $R_1' \| R_1'' = 2251\ \Omega$ so $R_1' = 8689\ \Omega$ and $R_1'' = 3038\ \Omega$. Similarly, $R_b/R_a = 2.86$, so if $R_a = 1\ \text{k}\Omega$, then $R_b = 2.86\ \text{k}\Omega$.

KRC Notch Filters

Although Sallen and Key did not present a notch (or bandstop topology) in their original paper, such a network fits the general classification used here and thus will be explored in this section. This type of filter requires bandpass feedback and bandstop feedforward. The bandstop feedforward network must produce purely imaginary complex conjugate zeros. Such zeros can be produced by the well-known twin-T circuit (see Chapter 3).

Such a notch filter is shown in Fig. 6.20. By applying the previously outlined techniques, we can show that

$$T(s) = \frac{K\left(s^2 + \dfrac{1}{(\mathrm{RC})^2}\right)}{s^2 + \dfrac{4-K}{\mathrm{RC}}s + \dfrac{1}{(\mathrm{RC})^2}} \tag{6.52}$$

$$\omega_0 = \frac{1}{\mathrm{RC}} \tag{6.53}$$

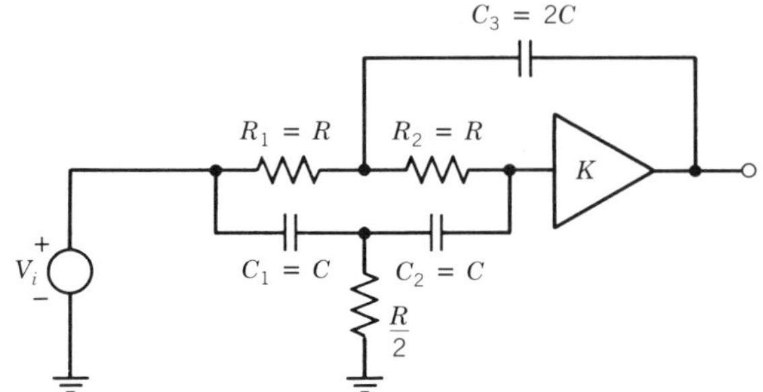

Figure 6.20 KRC notch (or bandstop) filter.

and

$$Q = \frac{1}{4 - K} \tag{6.54}$$

EXAMPLE 6.8 Design a KRC notch filter having a notch center frequency of $f_0 =$ 60 Hz and a bandwidth of 5 Hz. Assume that the high- and low-frequency gains are not important.

PROCEDURE

1 Compute the required Q from

$$Q = \frac{f_0}{\Delta f} = \frac{60 \text{ Hz}}{5 \text{ Hz}} = 12$$

2 Select K by the use of Eq. 6.54:

$$K = 4 - \frac{1}{Q} = 3.92$$

Note that the high- and low-frequency gains will be 3.92.

3 Compute the RC product from

$$\text{RC} = \frac{1}{\omega_0} = 2.65 \text{ ms}$$

4 Pick the value of capacitance arbitrarily. For convenience, suppose that $C = 0.1\ \mu\text{F}$. Then the resistance value is given by $R = 26.5\ \text{k}\Omega$.

The final design is shown in Fig. 6.21.

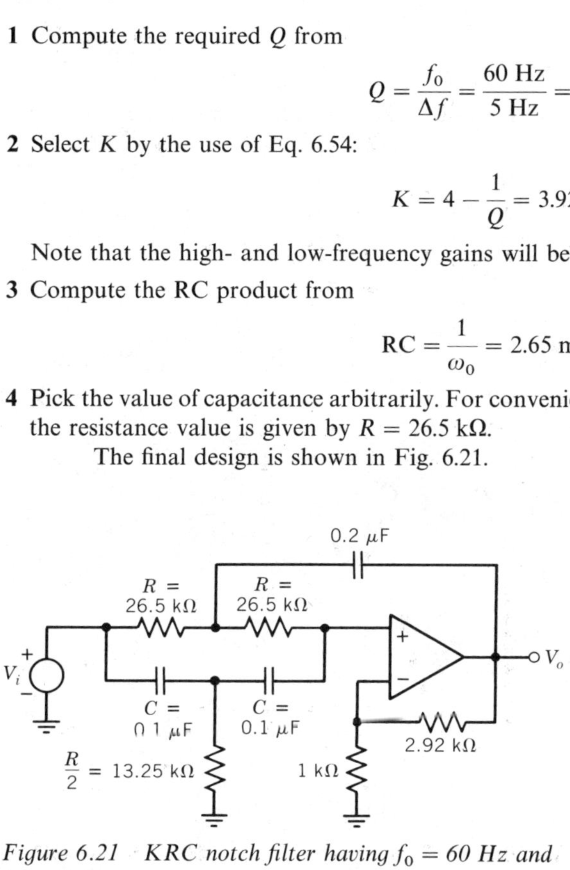

Figure 6.21 KRC notch filter having $f_0 = 60$ Hz and $\Delta f = 5$ Hz.

6.4 – KRC FILTERS

This section will present the design of −KRC realizations of the four common filter types covered in the last section (KRC Filters). The discussion relative to each will be somewhat abbreviated, for most of the comments previously given for the KRC realizations apply to the −KRC structures as well. One item, however,

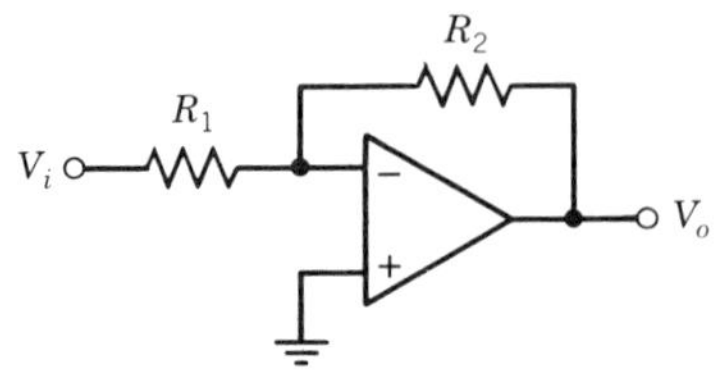

(a) Inverting amplifier.

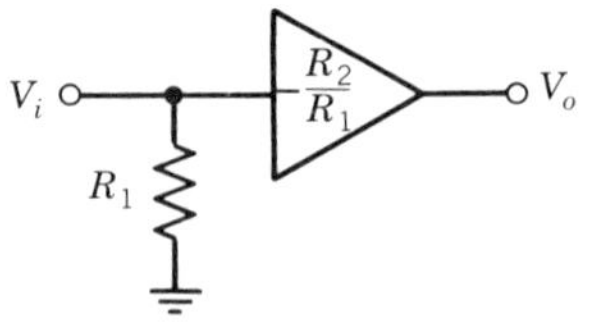

(b) Model incorporating an ideal amplifier.

Figure 6.22 Inverting amplifier and its ideal amplifier model.

is worthy of note. Figure 6.22 contains the single op amp realization of an inverting amplifier. Note that the input resistor R_1 must be an actual element in the passive RC network. Thus one must always sample (terminal 3 in Fig. 6.1) the voltage across a *resistor* in the passive circuit. (An alternative, approximate approach that removes this constraint is outlined in [7].)

−KRC Low-Pass Filters

We can readily show that the circuit of Fig. 6.23 is a −KRC low-pass filter and compute its transfer function to be

$$T(s) = \frac{\dfrac{-K}{R_1R_2C_1C_2}}{s^2 + \left[\dfrac{1}{R_3C_1} + \dfrac{1}{R_1C_1} + \dfrac{1}{R_2C_1} + \dfrac{1}{R_4C_2} + \dfrac{1}{R_2C_2}\right]s + \left[\dfrac{1+K}{R_2R_3} + \dfrac{1}{R_1R_2} + \dfrac{1}{R_3R_4} + \dfrac{1}{R_1R_4} + \dfrac{1}{R_2R_4}\right] \cdot \dfrac{1}{C_1C_2}} \tag{6.55}$$

so

$$\omega_o = \sqrt{\left(\frac{1+K}{R_2R_3} + \frac{1}{R_1R_2} + \frac{1}{R_3R_4} + \frac{1}{R_1R_4} + \frac{1}{R_2R_4}\right) \cdot \frac{1}{C_1C_2}} \tag{6.56}$$

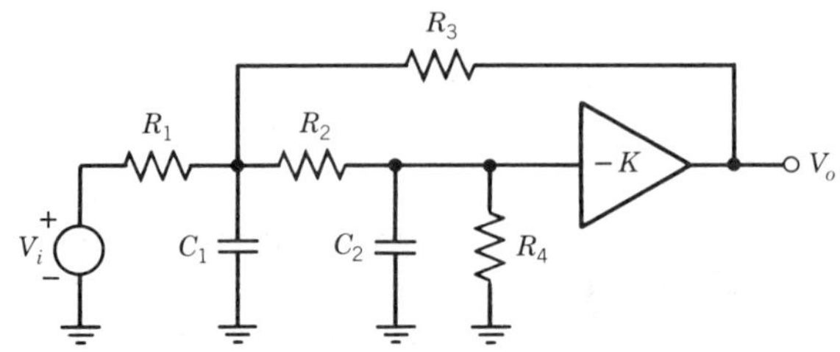

Figure 6.23 −KRC low-pass filter.

and

$$Q = \frac{\sqrt{\dfrac{1+K}{R_2R_3} + \dfrac{1}{R_1R_2} + \dfrac{1}{R_3R_4} + \dfrac{1}{R_1R_4} + \dfrac{1}{R_2R_4}}}{\left(\dfrac{1}{R_1} + \dfrac{1}{R_2} + \dfrac{1}{R_3}\right)\sqrt{\dfrac{C_2}{C_1}} + \left(\dfrac{1}{R_2} + \dfrac{1}{R_4}\right)\cdot\sqrt{\dfrac{C_1}{C_2}}} \tag{6.57}$$

EXAMPLE 6.9 Design the $-$KRC low-pass filter of Fig. 6.23 to have $f_0 = 2000$ Hz and $Q = 5$. Assume that the magnitude of the dc gain is to be unity.

PROCEDURE

1 For analytical convenience, we assume that $R_1 = R_2 = R_3 = R_4 = R$ and $C_1 = C_2 = C$. Then Eqs. 6.56 and 6.57 become

$$\omega_0 = \frac{\sqrt{K+5}}{RC} \tag{6.58}$$

and

$$Q = \frac{\sqrt{K+5}}{5} \tag{6.59}$$

2 Solve Eq. 6.59 for K:

$$K = 25Q^2 - 5 = 620.$$

3 Solve Eq. 6.58 for RC:

$$\mathrm{RC} = \frac{\sqrt{K+5}}{\omega_0} = 1.989 \text{ ms}$$

4 If we assume that $C = 0.1\ \mu$F, then

$$R = 19.89 \text{ k}\Omega$$

5 Adjust the dc gain by splitting R_1 as shown in Fig. 6.24. Then

$$\frac{R_1''}{R_1' + R_1''} = \frac{1}{K} = \frac{1}{620} \quad \text{and} \quad R_1''\|R_1' = 19.89 \text{ k}\Omega$$

which yields $R_1' = 12.33$ MΩ and $R_1'' = 19.92$ kΩ.

6 Note that $R_0 = KR_4 = 620R_4 = 12.33$ MΩ.

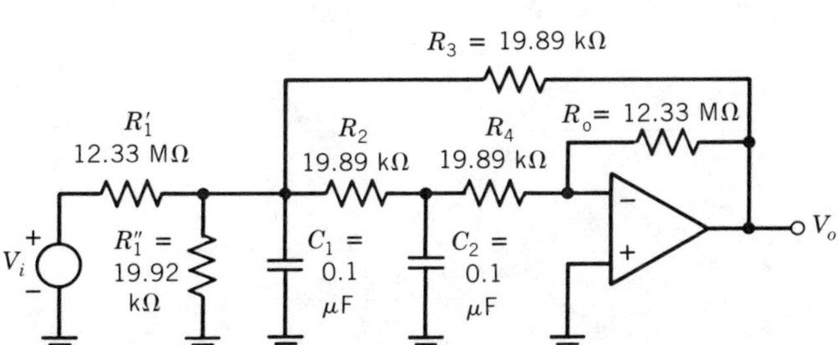

Figure 6.24 $-$*KRC low-pass filter with* $f_0 = 2000$ *Hz and* $Q = 5$.

−KRC High-Pass Filters

EXAMPLE 6.10 Design a −KRC high-pass filter such that $f_o = 2000$ Hz and $Q = 5$, with unity high-frequency gain.

PROCEDURE

1 Redraw the circuit of Fig. 6.23 using the RC:CR transformation. Note that this replaces all resistors with capacitors, and vice versa, as shown in Fig. 6.25.

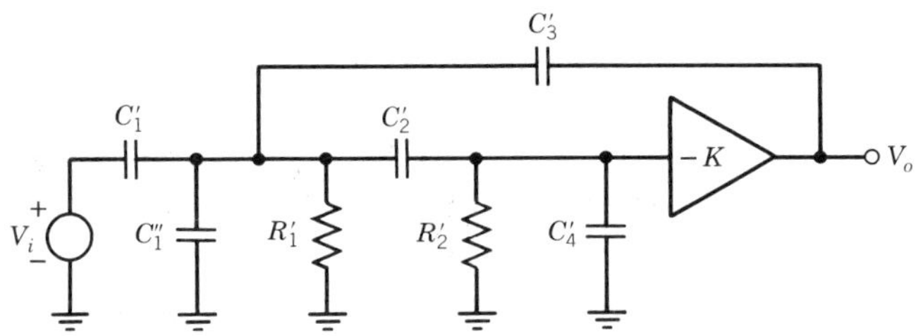

Figure 6.25 RC:CR transformed version of Fig. 6.23.

2 Compute the new component values (using the values of Fig. 6.24).

$$R_1' = \frac{1}{\omega_0 C_1} = (2\pi \times 2000 \times 0.1 \times 10^{-6})^{-1} = 0.796 \text{ k}\Omega$$

$$R_2' = \frac{1}{\omega_0 C_2} = 0.796 \text{ k}\Omega$$

$$C_1' = \frac{1}{\omega_0 R_1'} = (2\pi \times 2000 \times 12.33 \times 10^{6})^{-1} = 6.45 \text{ pF}$$

$$C_1'' = \frac{1}{\omega_0 R_1''} = (2\pi \times 2000 \times 19.92 \times 10^{3})^{-1} = 4 \text{ nF}$$

$$C_2' = C_3' = C_4' = (2\pi \times 2000 \times 19.89 \times 10^{3})^{-1} = 4 \text{ nF}$$

3 Noting that an ideal amplifier symbol is used in Fig. 6.25, we must convert to an inverting op amp circuit by using R_2' as the input resistance. This leads to $R_0 = KR_2' = 620 \times 0.796 \text{ k}\Omega = 494 \text{ k}\Omega$, as shown in the complete circuit diagram of Fig. 6.26.

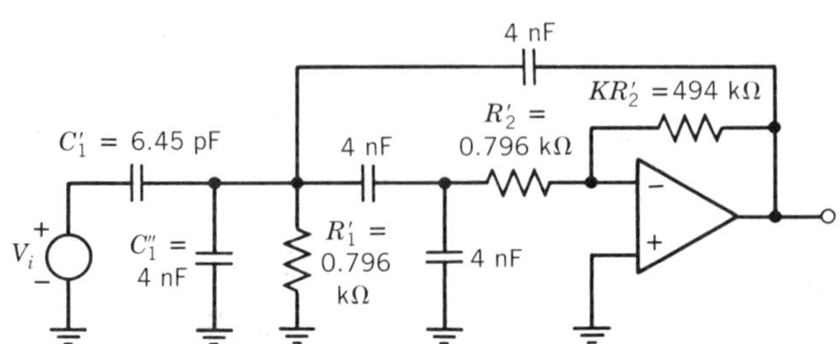

Figure 6.26 −KRC high-pass filter having $f_0 = 2000$ Hz and $Q = 5$.

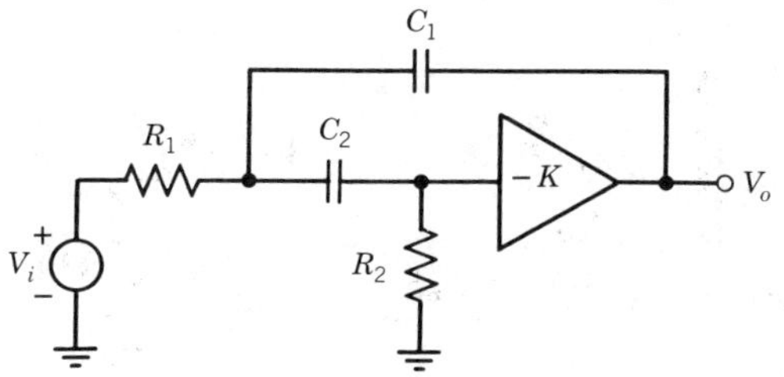

Figure 6.27 −KRC bandpass filter.

−KRC Bandpass Filters

It is straightforward to show that the network depicted in Fig. 6.27 is a −KRC bandpass filter and solve for the transfer function. The result is

$$T(s) = \frac{-\dfrac{K}{K+1} \cdot \dfrac{1}{R_1 C_1} s}{s^2 + \dfrac{1}{K+1}\left[\dfrac{R_1 + R_2}{R_1 R_2 C_1} + \dfrac{1}{R_2 C_2}\right] s + \dfrac{1}{(K+1) R_1 R_2 C_1 C_2}} \tag{6.60}$$

Also

$$\omega_0 = \frac{1}{\sqrt{(K+1) R_1 R_2 C_1 C_2}} \tag{6.61}$$

and

$$Q = \frac{\sqrt{K+1}}{\sqrt{\dfrac{R_1 C_2}{R_2 C_1}} + \sqrt{\dfrac{R_2 C_2}{R_1 C_1}} + \sqrt{\dfrac{R_1 C_1}{R_2 C_2}}} \tag{6.62}$$

EXAMPLE 6.11 Design a −KRC bandpass filter to have a peak frequency of $f_0 = 200$ Hz and a 3dB bandwidth of 20 Hz. Assume that the peak gain is not constrained.

PROCEDURE

1 Assume, for analytical convenience, that $R_1 = R_2 = R$ and $C_1 = C_2 = C$. This transforms Eqs. 6.61 and 6.62 into

$$\omega_0 = \frac{1}{\sqrt{K+1}\, RC} \tag{6.63}$$

and

$$Q = \tfrac{1}{3}\sqrt{K+1} \tag{6.64}$$

2 Use the fact that $Q = f_0/\Delta f$ to compute $Q = 200$ Hz/20 Hz $= 10$. Then use Eq. 6.64 to arrive at $K = 9Q^2 - 1 = 899$.

3 Compute the circuit time constant to be

$$RC = \frac{1}{\sqrt{K+1}\,\omega_0} = 26.5\ \mu\text{s}$$

Arbitrarily let $C = 0.1\ \mu$F. This implies that $R = 265\ \Omega$.

The complete circuit is shown in Fig. 6.28.

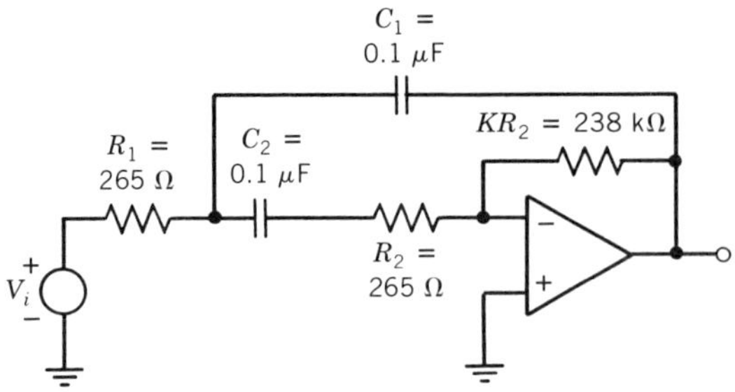

Figure 6.28 —*KRC bandpass filter with* $f_0 = 200$ *Hz and* $f = 20$ *Hz.*

—KRC Notch Filters

A slightly different approach will be taken with the design of the —KRC bandstop filter. The design will hinge on the following fact:

$$T_{BR}(s) = a - T_{BP}(s) \tag{6.65}$$

where $T_{BR}(s)$ is the bandstop filter response, $T_{BP}(s)$ is the bandpass response, and a is the magnitude of the peak gain of the bandpass response function. Thus one is led to the notch filter circuit of Fig. 6.29.

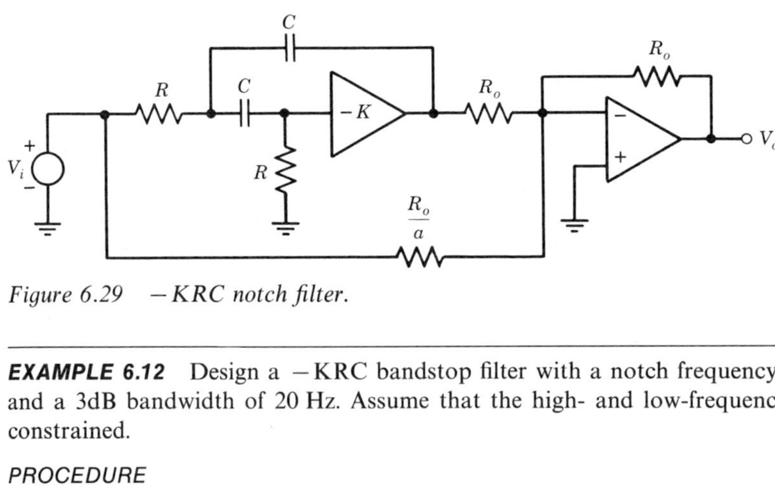

Figure 6.29 —*KRC notch filter.*

EXAMPLE 6.12 Design a —KRC bandstop filter with a notch frequency of $f_0 = 200$ Hz and a 3dB bandwidth of 20 Hz. Assume that the high- and low-frequency gains are not constrained.

PROCEDURE

1 From Eq. 6.60, we note that

$$T(s) = \frac{-(K/\sqrt{K+1})\omega_0 s}{s^2 + (\omega_0/Q)s + \omega_0^2} \tag{6.65}$$

or

$$T(s) = \frac{-\frac{1}{3}(9Q^2 - 1)(\omega_0/Q)s}{s^2 + (\omega_0/Q)s + \omega_0^2} \tag{6.66}$$

Thus the peak gain (which occurs at $\omega = \omega_0$) is $a = -\frac{1}{3}(9Q^2 - 1) = 299.7$.

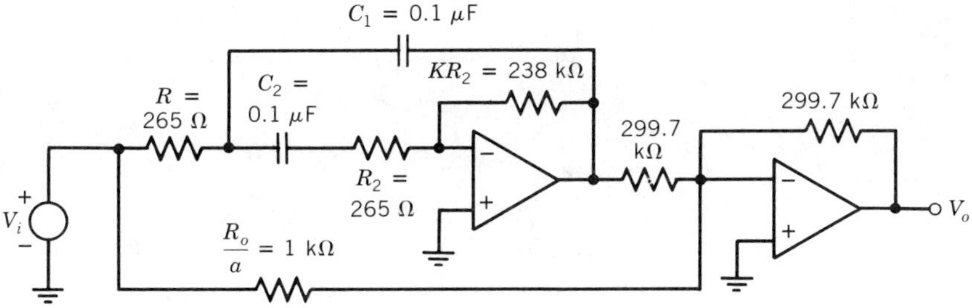

Figure 6.30 *−KRC notch filter with $f_0 = 200$ Hz and $\Delta f = 20$ Hz.*

2 Use the preceding bandpass filter design along with $R_0 = 299.7$ kΩ to yield the notch filter circuit of Figure 6.30.

A concise summary of all of the filter topologies and their design equations is given in Tables 6.2 through 6.8. In some cases, these circuits will only realize Q values of less than 0.5.

Table 6.2 **Sallen-Key KRC low-pass filter**

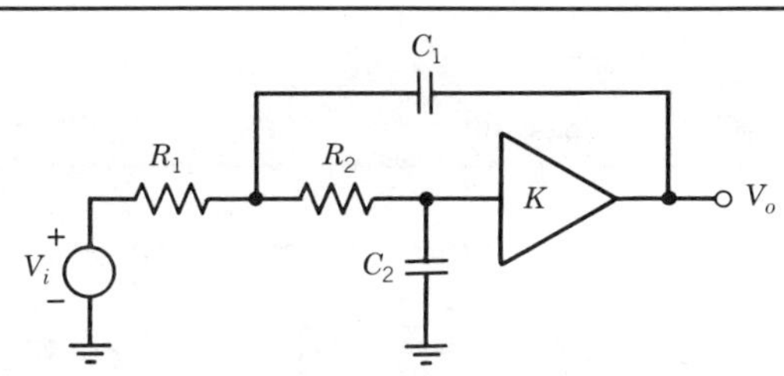

$$\frac{V_o}{V_i} = \frac{K\omega_o^2}{s^2 + \frac{\omega_o}{Q}s + \omega_o^2}$$

$$\omega_0 = \frac{1}{\sqrt{R_1R_2C_1C_2}} \qquad Q = \frac{1}{(1-K)\sqrt{\frac{R_1C_1}{R_2C_2}} + \left(1 + \frac{R_2}{R_1}\right)\sqrt{\frac{R_1C_2}{R_2C_1}}}$$

Design (*a*): Equal R, equal C ($R_1 = R_2 = R$, $C_1 = C_2 = C$)

$$\omega_0 = \frac{1}{RC} \qquad Q = \frac{1}{3-K}$$

Design (*b*): Equal R, unity gain ($R_1 = R_2 = R$, $K = 1$)

$$\omega_0 = \frac{1}{R\sqrt{C_1C_2}} \qquad Q = \frac{1}{2}\sqrt{\frac{C_1}{C_2}}$$

Table 6.3 **Sallen–Key KRC high-pass filter**

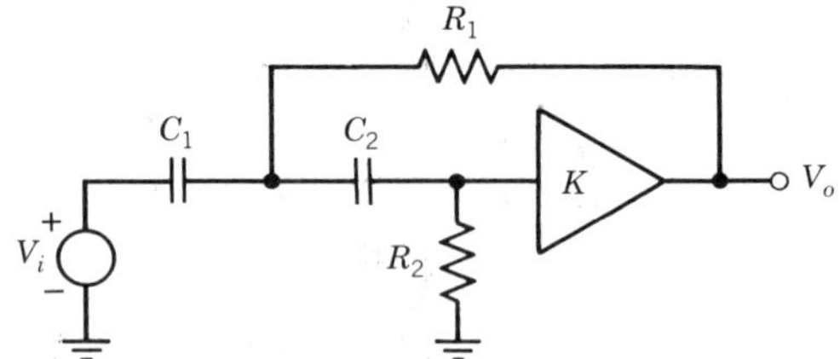

$$\frac{V_o}{V_i} = \frac{Ks^2}{s^2 + \dfrac{\omega_o}{Q}s + \omega_o^2}$$

$$\omega_0 = \frac{1}{\sqrt{R_1R_2C_1C_2}} \qquad Q = \frac{1}{(1-K)\sqrt{\dfrac{R_2C_2}{R_1C_1}} + \sqrt{\dfrac{R_1C_2}{R_2C_1}} + \sqrt{\dfrac{R_1C_1}{R_2C_2}}}$$

Design (*a*): Equal R, equal C ($R_1 = R_2 = R$, $C_1 = C_2 = C$)

$$\omega_0 = \frac{1}{RC} \qquad Q = \frac{1}{3-K}$$

Design (*b*)*: Equal R, unity gain ($R_1 = R_2 = R$, $K = 1$)

$$\omega_0 = \frac{1}{R\sqrt{C_1C_2}} \qquad Q = \frac{1}{\sqrt{\dfrac{C_2}{C_1}} + \sqrt{\dfrac{C_1}{C_2}}}$$

* Maximum value of $Q = 0.5$.

Table 6.4 **Sallen–Key KRC bandpass filter**

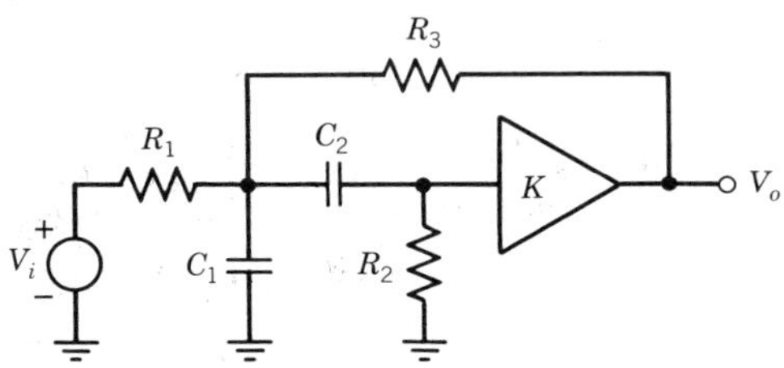

$$\frac{V_o}{V_i} = \frac{\dfrac{K}{R_1C_1}s}{s^2 + \dfrac{\omega_o}{Q}s + \omega_o^2}$$

$$\omega_0 = \sqrt{\frac{R_1 + R_3}{R_1R_2R_3C_1C_2}} \qquad Q = \frac{\sqrt{\dfrac{R_2}{R_1} + \dfrac{R_2}{R_3}}}{\sqrt{\dfrac{C_1}{C_2}} + \sqrt{\dfrac{C_2}{C_1}} \cdot \left[1 + \dfrac{R_2}{R_1} + (1-K)\dfrac{R_2}{R_3}\right]}$$

Special case: Equal R, equal C ($R_1 = R_2 = R_3 = R$, $C_1 = C_2 = C$)

$$\omega_0 = \frac{\sqrt{2}}{RC} \qquad Q = \frac{\sqrt{2}}{4-K}$$

Table 6.5 **Sallen–Key KRC notch filter**

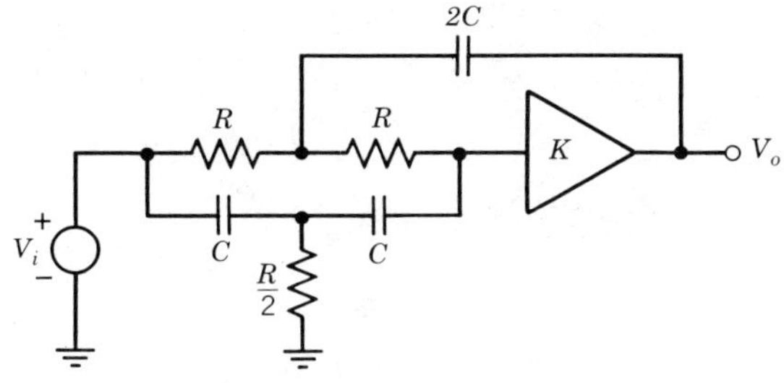

$$\frac{V_o}{V_i} = \frac{K(s^2 + \omega_o^2)}{s^2 + \dfrac{\omega_o}{Q}s + \omega_o^2}$$

$$\omega_0 = \frac{1}{RC} \qquad Q = \frac{1}{4 - K}$$

Table 6.6 **Sallen–Key −KRC low-pass filter**

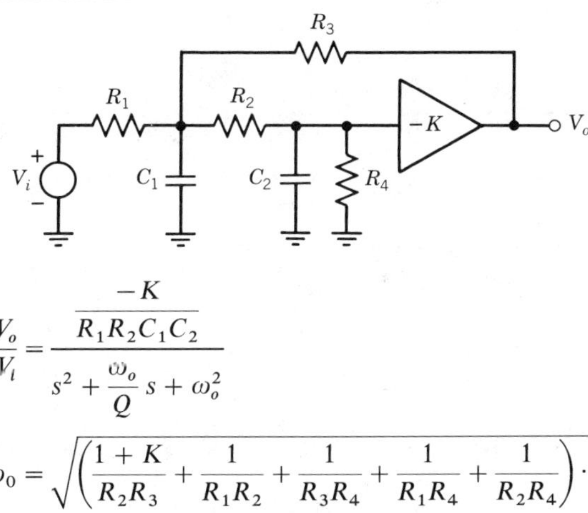

$$\frac{V_o}{V_i} = \frac{\dfrac{-K}{R_1R_2C_1C_2}}{s^2 + \dfrac{\omega_o}{Q}s + \omega_o^2}$$

$$\omega_0 = \sqrt{\left(\frac{1+K}{R_2R_3} + \frac{1}{R_1R_2} + \frac{1}{R_3R_4} + \frac{1}{R_1R_4} + \frac{1}{R_2R_4}\right) \cdot \frac{1}{C_1C_2}}$$

$$Q = \frac{\sqrt{\dfrac{1+K}{R_2R_3} + \dfrac{1}{R_1R_2} + \dfrac{1}{R_3R_4} + \dfrac{1}{R_1R_4} + \dfrac{1}{R_2R_4}}}{\left(\dfrac{1}{R_1} + \dfrac{1}{R_2} + \dfrac{1}{R_3}\right)\sqrt{\dfrac{C_2}{C_1}} + \left(\dfrac{1}{R_2} + \dfrac{1}{R_4}\right)\sqrt{\dfrac{C_1}{C_2}}}$$

Special case: Equal R, equal C ($R_1 = R_2 = R_3 = R_4 = R$, $C_1 = C_2 = C$)

$$\omega_0 = \frac{\sqrt{K+5}}{RC} \qquad Q = \frac{\sqrt{K+5}}{5}$$

Table 6.7 **Sallen–Key −KRC high-pass filter**

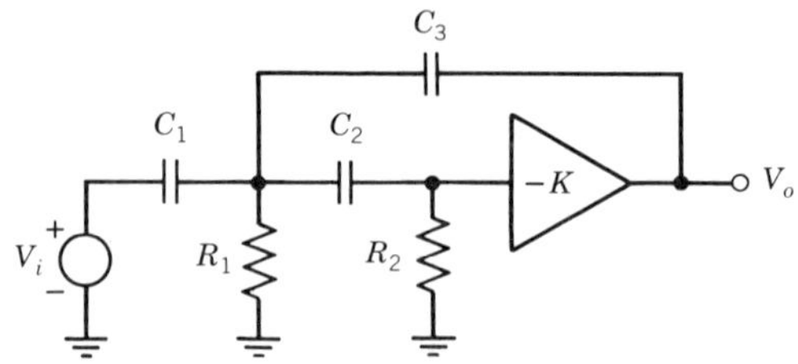

$$\frac{V_o}{V_i} = \frac{\dfrac{-K}{1 + (1+K)(C_3/C_1)}}{s^2 + \dfrac{\omega_o}{Q}s + \omega_o^2}$$

$$\omega_0 = \frac{1}{\sqrt{R_1 R_2 C_2 [C_1 + C_3(1+K)]}} \qquad Q = \frac{\sqrt{R_1 R_2 C_2 [C_1 + C_3(1+K)]}}{(R_1 + R_2)C_2 + R_1(C_1 + C_3)}$$

Special case: Equal R, equal C ($R_1 = R_2 = R$, $C_1 = C_2 = C_3 = C$)

$$\omega_0 = \frac{1}{RC\sqrt{K+2}} \qquad Q = \frac{\sqrt{K+2}}{4}$$

Table 6.8 **Sallen–Key −KRC bandpass filter**

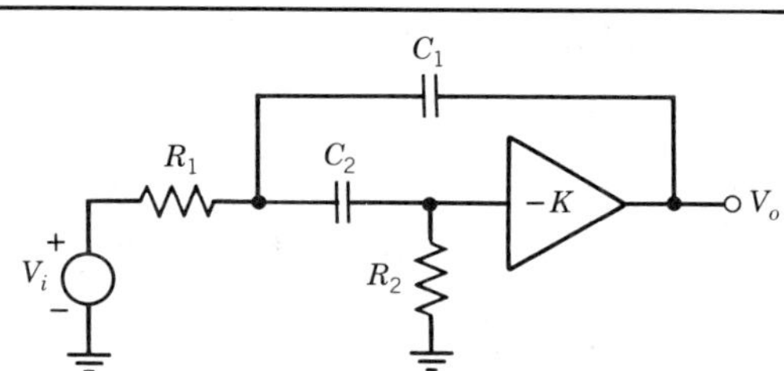

$$\frac{V_o}{V_i} = \frac{-\dfrac{K}{K+1} \cdot \dfrac{1}{R_1 C_1} s}{s^2 + \dfrac{\omega_o}{Q}s + \omega_o^2}$$

$$\omega_0 = \frac{1}{\sqrt{(K+1) R_1 R_2 C_1 C_2}} \qquad Q = \frac{\sqrt{K+1}}{\sqrt{\dfrac{R_1 C_2}{R_2 C_1}} + \sqrt{\dfrac{R_2 C_2}{R_1 C_1}} + \sqrt{\dfrac{R_1 C_1}{R_2 C_2}}}$$

Special case: Equal R, equal C ($R_1 = R_2 = R$, $C_1 = C_2 = C$)

$$\omega_0 = \frac{1}{RC\sqrt{K+1}} \qquad Q = \frac{1}{3}\sqrt{K+1}$$

6.5 MEASUREMENT TECHNIQUES

It is perhaps appropriate to mention some of the standard techniques for measuring the parameters of a second-order filter. This will be done for each different type of transfer function.

Low Pass

The basic low-pass filter transfer function is given by Eq. 6.36 (repeated here for convenience):

$$T_{LP}(s) = \frac{H\omega_0^2}{s^2 + (\omega_0/Q)s + \omega_0^2} \tag{6.67}$$

Now

$$T_{LP}(j\omega) = \frac{H\omega_0^2}{\omega_0^2 - \omega^2 + j(\omega_0/Q)\omega} = \frac{H}{1 - (\omega/\omega_0)^2 + j(1/Q)(\omega/\omega_0)} \tag{6.68}$$

Thus $T_{LP}(j\omega_0) = -jHQ$, whereas $T_{LP}(j0) = H$. Hence, if $H > 0$, ω_0 is the frequency at which the phase shift through the filter is $-90°$. Also, the ratio of the gain magnitude at ω_0 to the dc value is equal to Q. For filters of moderate Q, this represents the most viable method of measuring ω_0 and Q.

Bandpass

In this case (Eq. 6.37 repeated):

$$T_{BP}(s) = \frac{H(\omega_0/Q)s}{s^2 + (\omega_0/Q)s + \omega_0^2} \tag{6.69}$$

Clearly, $T_{BP}(j\omega_0) = H$. Thus, if $H > 0$, ω_0 is the frequency at which the phase shift is zero degrees. If one notes that

$$T_{BP}(j\omega) = \frac{H}{1 + jQ[(\omega/\omega_0) - (\omega_0/\omega)]} \tag{6.70}$$

it can be shown by a straightforward manipulation that

$$Q = \frac{\omega_0}{\Delta\omega} \tag{6.71}$$

where $\Delta\omega$ is the frequency interval between the $\pm 45°$ phase shift points.

High Pass

The high-pass function is (Eq. 6.38 repeated)

$$T_{HP}(s) = \frac{Hs^2}{s^2 + (\omega_0/Q)s + \omega_0^2} \tag{6.72}$$

Thus

$$T_{\text{HP}}(j\omega) = \frac{H}{1 - (\omega_0/\omega)^2 - j(\omega_0/Q\omega)} \tag{6.73}$$

Again, ω_0 is the frequency at which the phase shift is $+90^\circ$ (if $H > 0$). Further, Q is the ratio of the gain magnitude at $\omega = \omega_0$ to the high-frequency value.

Bandstop or Notch

For this type of filter

$$T_{\text{Br}}(s) = H \frac{s^2 + \omega_0^2}{s^2 + (\omega_0/Q)s + \omega_0^2} \tag{6.74}$$

Thus,

$$T_{\text{BR}}(j\omega) = \frac{H}{1 - j[1/Q(\omega/\omega_0 - \omega_0/\omega)]} \tag{6.75}$$

It is straightforward to show that ω_0 is the frequency at which the gain magnitude is zero (a minimum, practically speaking). Furthermore,

$$Q = \frac{\omega_0}{\Delta\omega} \tag{6.76}$$

where $\Delta\omega$ is the frequency interval between the $\pm 45^\circ$ phase points.

6.6 SUMMARY

This chapter has developed the design of Sallen–Key active filters. These networks are based upon a single amplifier configuration and can realize any of the standard second-order transfer functions.

One aspect to be considered in the selection of a topology for a filter is its sensitivity, that is, the degree to which the transfer function (or its ω_0 and Q) varies when the constituent element values are varied. This has been briefly alluded to; for a more complete discussion, see Chapter 11. We can, however, state as a rule of thumb that single amplifier structures are generally useful in the range over which $Q < 20$. For higher Q factors a multiple amplifier configuration is typically employed.

A very good general reference that covers the material presented in this chapter is the book by Budak [8].

6.7 REFERENCES

[1] R. P. Sallen and E. L. Key, "A Practical Method for Designing RC Active Filters," *IRE Transactions on Circuit Theory*, vol. CT-2, pp. 74–85, March 1955.

[2] S. Karni, *Network Theory: Analysis and Synthesis*, Boston: Allyn and Bacon, pp. 234–238, 1966.

[3] A. M. Davis and E. A. Moustakas, "Analysis of Active RC Networks by Decomposition," *IEEE Transactions on CAS*, vol. CAS-27, pp. 417–419, May 1980.

[4] G. Daryanani, *Principles of Active Network Synthesis and Design*, New York: Wiley, 1976.

[5] W. Saraga, "Sensitivity of 2nd Order Sallen–Key Type Active RC Filters," *Electronics Letters*, vol. 13, pp. 442–444, October 1967.

[6] S. K. Mitra, "A Network Transformation for Active RC Networks," *Proc. IEEE*, vol. 55, pp. 2021–2022, November 1967.

[7] P. Bowron and F. W. Stephenson, *Active Filters for Communication and Instrumentation*, New York: McGraw-Hill, 1979.

[8] A. Budak, *Passive and Active Network Analysis and Synthesis*, Boston: Houghton Mifflin, 1974.

CHAPTER 7

Multiple-Loop Feedback Filters

L. P. Huelsman
University of Arizona

7.1 INTRODUCTION

The preceding chapter showed how an RC-active filter could be developed by using a passive-RC network in combination with an amplifier (a voltage-gain block). In this chapter we investigate the particular class of filter that results when the amplifier gain is very large and for all practical purposes may be considered infinite. In practice this means that an operational amplifier is directly used as a gain element without the feedback resistors that permit the realization of lower-valued gains (see Chapter 4).

7.2 LOW-PASS MULTIPLE-LOOP FEEDBACK FILTER

The basic circuit configuration for the multiple-loop feedback filter is shown in Fig. 7.1. It is similar to the configuration analyzed in Chapter 6. There are, however, two major differences. The first is that the active element is an opera-

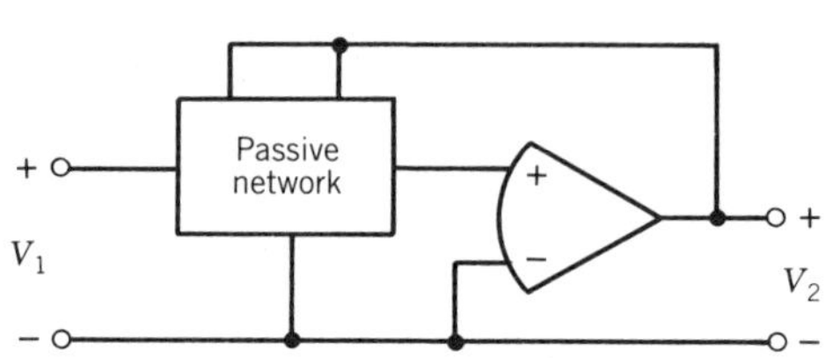

Figure 7.1 General multiple-loop feedback filter configuration.

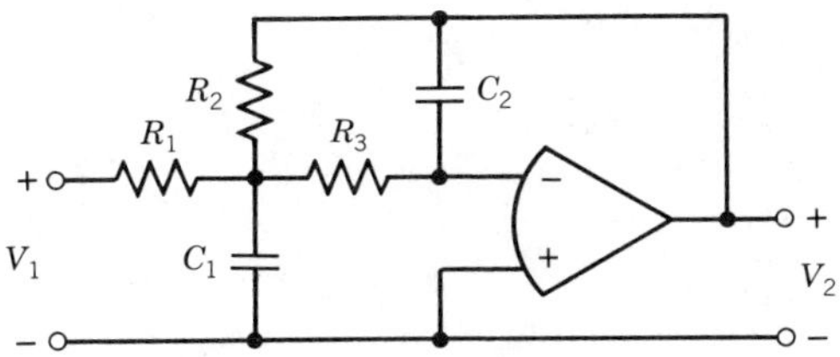

Figure 7.2 *Low-pass multiple-loop feedback filter.*

tional amplifier rather than a finite-gain one. In addition, in this circuit the operational amplifier is used in an inverting configuration, that is, the positive input terminal is grounded. The second difference is that there are two feedback paths (rather than one) from the output of the amplifier to the network. This is the reason for the name *multiple-loop feedback* filter. The basic configuration shown in Fig. 7.1 can be used to realize low-pass, bandpass, and high-pass filter functions. The circuit for the low-pass multiple-loop feedback filter is shown in Fig. 7.2. Applying the analysis techniques developed in Chapter 5, we obtain the voltage transfer function

$$\frac{V_2(s)}{V_1(s)} = \frac{-1/R_1R_3C_1C_2}{s^2 + s(1/R_1 + 1/R_2 + 1/R_3)/C_1 + 1/R_2R_3C_1C_2} \tag{7.1}$$

Note that, because of the minus sign, this circuit can be used only to realize an inverting filter function. Using the quantities Q and ω_0 defined for the general second-order low-pass function in Chapter 3, we obtain the relations

$$\omega_0 = 1/\sqrt{R_2R_3C_1C_2}$$
$$\frac{1}{Q} = \sqrt{\frac{C_2}{C_1}}\left(\frac{\sqrt{R_2R_3}}{R_1} + \sqrt{\frac{R_3}{R_2}} + \sqrt{\frac{R_2}{R_3}}\right) \tag{7.2}$$

together with $|H_0| = R_2/R_1$, the passband gain at dc. The design procedure for this filter is based on the solution of these equations and is summarized in the following example.

EXAMPLE 7.1 Design a second-order Butterworth ($Q = 0.7071$) low-pass filter having a cutoff frequency $f_0 = 100$ Hz and a dc gain of -1 ($|H_0| = 1$).

PROCEDURE

1 Choose a convenient value (call it C) for the capacitor C_1. The choice will be determined both by the frequency range for which the filter is designed and the impedance level required for the other elements of the circuit. For filters with cutoff frequencies in the audio range, values of a few tenths of a microfarad are usually appropriate. Here we let $C_1 = C = 0.1\ \mu$F.

2 Determine a constant m that affects the spread of the passive element values. This constant is bounded by the relation

$$m \le \frac{1}{4Q^2(1 + |H_0|)} = \frac{1}{4 \times 0.5 \times (1 + 1)} = 0.25$$

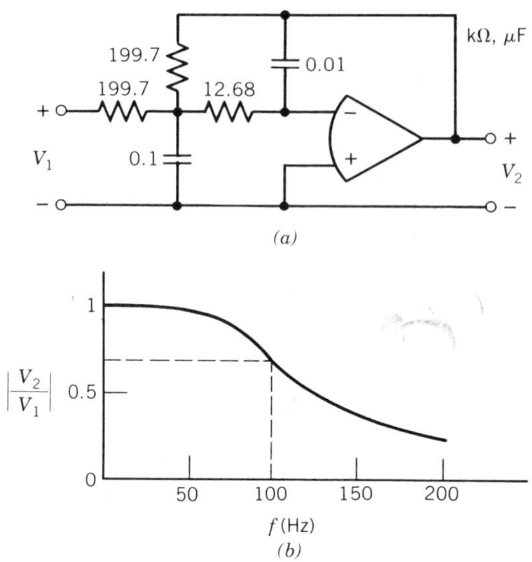

Figure 7.3 (a) Filter for Example 7.1; (b) magnitude plot.

The value of the second capacitor is now found as $C_2 = mC$. For convenience in using standard element values, let us choose $m = 0.1$. Thus $C_2 = mC = 0.01\ \mu$F.

3 Determine the value of R_2 using the relation (note that $\omega_0 = 2\pi f_0 = 2\pi \times 100$)

$$R_2 = \frac{1 + \sqrt{1 - 4mQ^2(1 + |H_0|)}}{2\omega_0 CmQ}$$

$$= \frac{1 + \sqrt{1 - 4 \times 0.1 \times 0.5 \times (1 + 1)}}{2 \times 2\pi \times 100 \times 0.1 \times 10^{-6} \times 0.1 \times 0.7071} = 199.7\ \text{k}\Omega$$

4 Determine the value of R_1 using the relation

$$R_1 = \frac{R_2}{|H_0|} = 199.7\ \text{k}\Omega$$

5 Determine the value of R_3 using the relation

$$R_3 = \frac{1}{\omega_0^2 C^2 R_2 m}$$

$$= \frac{1}{(2\pi \times 100)^2 \times (0.1 \times 10^{-6})^2 \times 199.7 \times 10^3 \times 0.1} = 12.68\ \text{k}\Omega$$

The resulting circuit and its magnitude characteristic are shown in Fig. 7.3.

EXAMPLE 7.2 Repeat the design of Example 7.1 but change the dc gain specification to -10.

PROCEDURE

1 Following the procedure outlined in Example 7.1, we let $C_1 = C = 0.1\ \mu$F.

2 Determining the element spread, we obtain $m \leq 0.04545$. Let $m = 0.02$; thus $C_2 = mC = 0.002\ \mu$F.

3 Determining R_2, we obtain 98.379 kΩ.
4 Determining R_1, we obtain 9.8379 kΩ.
5 Determining R_3, we obtain 12.874 kΩ.

This problem illustrates one disadvantage of the multiple-loop feedback filter configuration. For a given set of specifications, as the desired gain is increased, the spread of element values also increases. As an example of this, note that in Example 7.1 (with a gain of -1), the ratio of capacitor values is 10, a practical situation. In Example 7.2, however, the capacitor ratio is 50. Further increases in the gain requirement would make the capacitor ratio so large as to limit the practicality of the design.

EXAMPLE 7.3 Repeat the design of Example 7.1 but change the Q specification to 4.

PROCEDURE

1 If we follow the procedure outlined in Example 7.1 using $C = 0.1\ \mu$F, the resulting resistor values are too large to be practical. To remedy this, we must use a larger value for C. Let $C_1 = C = 1\ \mu$F.
2 Determining the element spread, we obtain $m \le 0.007812$. Let $m = 0.005$, thus $C_2 = mC = 0.005\ \mu$F.
3 Determining R_2, we obtain 63.66 kΩ.
4 Determining R_1, we obtain 63.66 kΩ.
5 Determining R_3, we obtain 7.957 kΩ.

In this example we see another disadvantage of the multiple-loop feedback filter. For a given set of specifications, as the desired Q is increased, the spread of the element values also increases. In this case, the minimum ratio of capacitor values that can be obtained is $1/m = 1/0.007812 = 128$.

Despite the disadvantages of the multiple-loop feedback filter pointed out above, it is a useful configuration for several reasons. One of these is its ability to simultaneously realize all the specifications of a low-pass filter function, namely, the constants H_0, Q, and ω_0. In general, this is not possible with the Sallen–Key filters discussed in Chapter 6. Another reason for the importance of the multiple loop feedback filter is its low sensitivities. In general, the magnitude of the percentage change in either the Q or ω_0 parameters is less than (or at worst equal to) the magnitude of the percentage change in any of the network elements. This is very different from the situation in the Sallen–Key circuits, in which the percentage change in the Q and ω_0 parameters may be many times that of the network elements.

7.3 HIGH-PASS MULTIPLE-LOOP FEEDBACK FILTER

The basic multiple-loop feedback filter configuration shown in Fig. 7.1 can also be used to realize a high-pass filter function. The circuit is shown in Fig. 7.4. The voltage transfer function is

$$\frac{V_2(s)}{V_1(s)} = \frac{-s^2 C_1/C_2}{s^2 + s(C_1/C_2C_3 + 1/C_2 + 1/C_3)/R_2 + 1/R_1R_2C_2C_3} \tag{7.3}$$

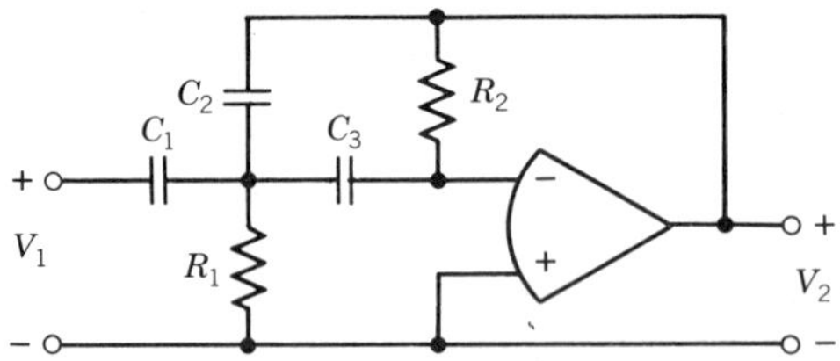

Figure 7.4 High-pass multiple-loop feedback filter.

Just as with the low-pass filter, the high-pass function that is realized has a negative sign; that is, it is an inverting one. Using the quantities Q and ω_0 defined for the second-order bandpass function in Chapter 3, we obtain the relations

$$\omega_0 = \frac{1}{\sqrt{R_1 R_2 C_2 C_3}}$$

$$\frac{1}{Q} = \sqrt{\frac{R_1}{R_2}}\left[\frac{C_1}{\sqrt{C_2 C_3}} + \sqrt{\frac{C_3}{C_2}} + \sqrt{\frac{C_2}{C_3}}\right] \tag{7.4}$$

together with $|H_0| = C_1/C_2$, the high-frequency gain.

The design procedure for the filter is based on the solution of these equations and is summarized in the following example.

EXAMPLE 7.4 Design a second-order high-pass Butterworth ($Q = 0.7071$) filter with a cutoff frequency $f_0 = 100$ Hz and a high-frequency gain of -1 ($|H_0| = 1$).

PROCEDURE

1 Choose a convenient value (call it C) for the capacitors C_1 and C_3 using the general criteria given in step 1 of the procedure for Example 7.1. Let $C_1 = C_3 = C = 0.1\ \mu$F.

2 Determine the value of R_1 using the relation

$$R_1 = \frac{|H_0|}{\omega_0 Q C(2|H_0| + 1)}$$

$$= \frac{1}{2\pi \times 100 \times 0.7071 \times 0.1 \times 10^{-6} \times (2 + 1)} = 7.502\ \text{k}\Omega$$

3 Determine the value of R_2 using the relation

$$R_2 = \frac{Q(2|H_0| + 1)}{\omega_0 C}$$

$$= \frac{0.7071 \times (2 + 1)}{2\pi \times 100 \times 0.1 \times 10^{-6}} = 33.761\ \text{k}\Omega$$

4 Determine the value of C_2 using the relation

$$C_2 = \frac{C}{|H_0|} = \frac{0.1 \times 10^{-6}}{1} = 0.1\ \mu\text{F}$$

The resulting circuit and its magnitude characteristic is shown in Fig. 7.5.

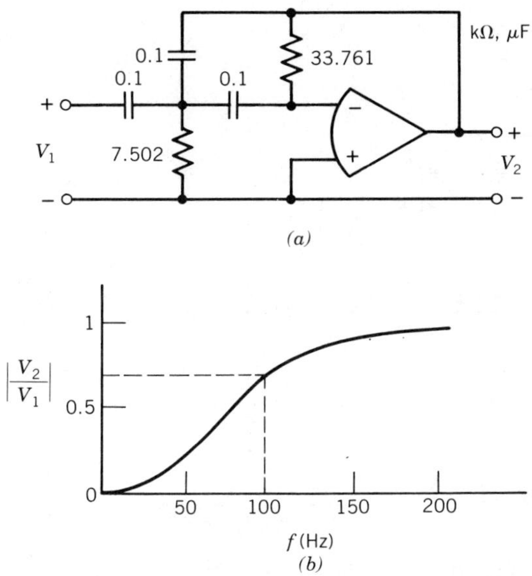

Figure 7.5 (a) *Filter for Example 7.4;* (b) *magnitude plot.*

EXAMPLE 7.5 Repeat the design of Example 7.4 using a specification of -10 for the high-frequency gain.

PROCEDURE

1 Let $C_1 = C_3 = C = 0.1\ \mu\text{F}$.

2 Determining R_1, we obtain 10.718 kΩ.

3 Determining R_2, we obtain 236.33 kΩ.

4 Determining C_2, we obtain 0.01 μF.

EXAMPLE 7.6 Repeat the design of Example 7.4 using a Q specification of 4.

PROCEDURE

1 Let $C_1 = C_3 = C = 0.1\ \mu\text{F}$.

2 Determining R_1, we obtain 1.326 kΩ.

3 Determining R_2, we obtain 190.99 kΩ.

4 Determining C_2, we obtain 0.1 μF.

In Examples 7.5 and 7.6, we see that the high-pass multiple-loop feedback filter has disadvantages similar to those of the low-pass filter; namely, large values of gain and Q lead to large spreads of element values. As with the low-pass filter, however, the high-pass one has the advantage of being able to independently realize values of H_0, Q, and ω_0, and it also has very low sensitivities.

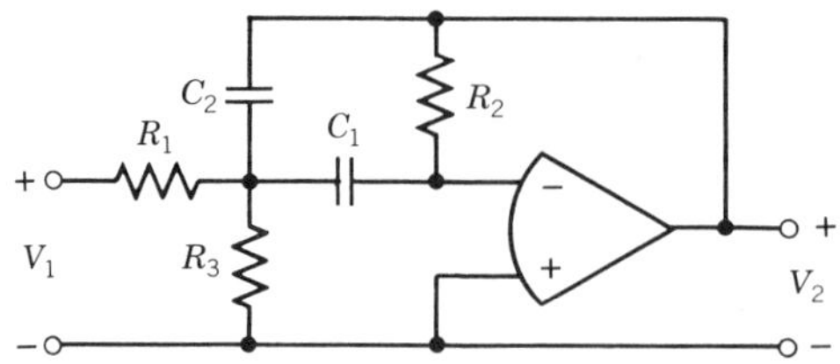

Figure 7.6 Bandpass multiple-loop feedback filter

7.4 BANDPASS MULTIPLE-LOOP FEEDBACK FILTER

The basic multiple-loop feedback filter configuration shown in Fig. 7.1 can also be used to realize a bandpass function. The circuit is shown in Fig. 7.6. Its voltage transfer function is

$$\frac{V_2(s)}{V_1(s)} = \frac{-s/R_1C_2}{s^2 + s(1/C_1 + 1/C_2)/R_2 + (1/R_1 + 1/R_3)/R_2C_1C_2} \tag{7.5}$$

As with the low-pass and high-pass functions, the bandpass function has a negative sign, that is, an inverting one. Using the quantities Q and ω_0 defined for the general second-order bandpass function in Chapter 3, we obtain the relations:

$$\omega_0 = \sqrt{\frac{1 + R_3/R_1}{R_2R_3C_1C_2}}$$

$$Q = \frac{\sqrt{1 + R_3/R_1}}{\sqrt{R_3C_1/R_2C_2} + \sqrt{R_3C_2/R_2C_1}} \tag{7.6}$$

together with

$$|H_0| = \frac{R_2/R_1}{1 + C_2/C_1}$$

the midband gain. The design procedure for the filter is based on the solution of these equations and is given in the following example.

EXAMPLE 7.7 Design a second-order bandpass filter with a resonant frequency of 1.592 kHz ($\omega_0 = 10^4$ rad/s), a Q of 2, and a gain at resonance of -2 ($|H_0| = 2$).

PROCEDURE

1 In the bandpass multiple-loop feedback filter it is convenient to use an equal-valued capacitor design in which the capacitors C_1 and C_2 both have the same value. Choose a suitable value (call it C) for the capacitors using the same general criteria given in step 1 of the procedure for Example 7.1. Let $C_1 = C_2 = C = 0.01\ \mu$F.

2 Determine the value of R_1 using the relation

$$R_1 = \frac{Q}{\omega_0 C|H_0|} = \frac{2}{10^4 \times 0.01 \times 10^{-6} \times 2} = 10\ \text{k}\Omega$$

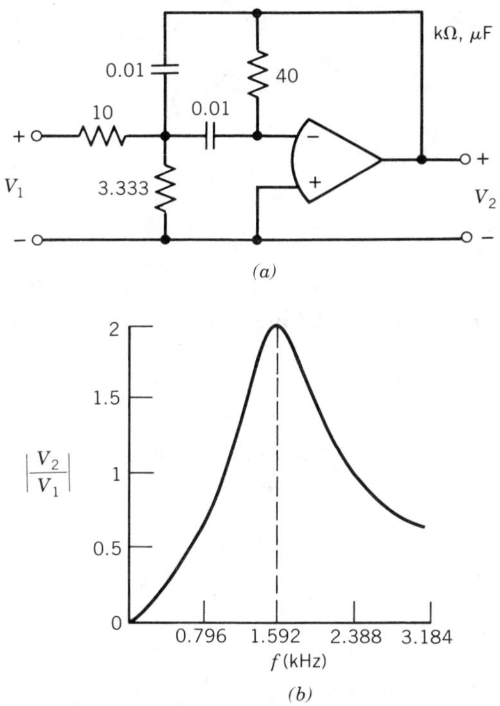

Figure 7.7 (a) Filter for Example 7.7; (b) magnitude plot.

3 Determine the value of R_2 using the relation

$$R_2 = \frac{2Q}{\omega_0 C} = \frac{4}{10^4 \times 0.01 \times 10^{-6}} = 40\ \text{k}\Omega$$

4 Determine the value of R_3 using the relation

$$R_3 = \frac{Q}{\omega_0 C(2Q^2 - |H_0|)}$$

$$= \frac{2}{10^4 \times 0.01 \times 10^{-6} \times (2 \times 4 - 2)} = 3.333\ \text{k}\Omega$$

Note that in this relation $|H_0|$ must be less than $2Q^2$ in order that R_2 be finite and positive. The resulting circuit and its magnitude characteristic are shown in Fig. 7.7.

EXAMPLE 7.8 Repeat the design of Example 7.7 using a gain specification of -7.5 for the gain at resonance.

PROCEDURE

1 Let $C_1 = C_2 = C = 0.01\ \mu$F.

2 Determining R_1, we obtain 2.667 kΩ.

3 Determining R_2, we obtain 40 kΩ.

4 Determining R_3, we obtain 40 kΩ.

EXAMPLE 7.9 Repeat the design of Example 7.7 using a Q specification of 5.

PROCEDURE

1 Let $C_1 = C_2 = C = 0.01\ \mu\text{F}$.

2 Determining R_1, we obtain 25 kΩ.

3 Determining R_2, we obtain 100 kΩ.

4 Determining R_3, we obtain 1.042 kΩ.

In Examples 7.8 and 7.9 we see that the bandpass multiple-loop feedback filter has disadvantages similar to those of the low-pass and high-pass filters, namely, large values of gain and Q lead to large spreads of element values. In the following section we shall show how such element spreads may be reduced by the use of positive feedback. Note that the bandpass filter, like the low-pass one, has the advantage of being able to independently realize H_0, Q and ω_0, and it also has extremely low sensitivities.

7.5 USE OF POSITIVE FEEDBACK

A modification of the bandpass circuit discussed in Section 7.4 can be used to reduce the spread of element values for high-Q realizations. This is accomplished by providing positive feedback to the "+" input terminal of the operational amplifier in addition to the negative (multiple-loop) feedback already present at the "−" terminal. A circuit with this provision is shown in Fig. 7.8. The positive feedback is provided by the voltage divider consisting of the two resistors R_A and R_B. To simplify the equations describing the circuit, we may represent the effect of these two resistors by defining a constant K, where

$$K = \frac{R_A}{R_B} \tag{7.7}$$

The voltage transfer function for the circuit is

$$\frac{V_2(s)}{V_1(s)} = \frac{-s(K+1)/R_1C_2}{s^2 + s(1/R_2C_2 + 1/R_2C_1 - K/R_1C_2) + 1/R_1R_2C_1C_2} \tag{7.8}$$

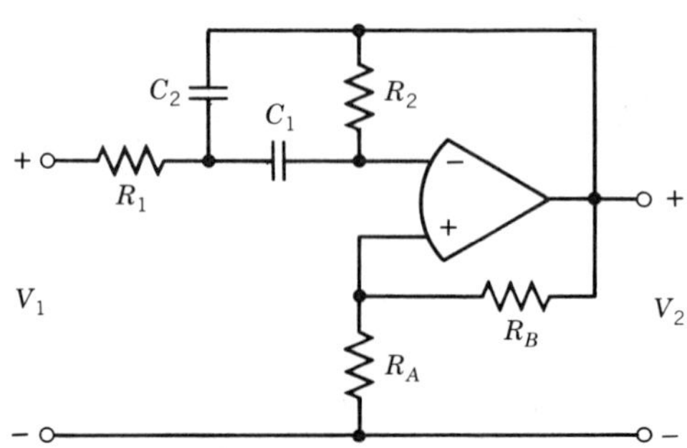

Figure 7.8 Bandpass multiple-loop positive feedback filter. Note: This is part a of Fig. 7.12.

Using the quantities Q, and ω_0 defined for the second-order bandpass function in Chapter 3, we obtain the relations

$$\omega_0 = \frac{1}{\sqrt{R_1 R_2 C_1 C_2}}$$

$$\frac{1}{Q} = \sqrt{\frac{R_1}{R_2}}\left[\sqrt{\frac{C_1}{C_2}} + \sqrt{\frac{C_2}{C_1}} - K\frac{R_2}{R_1}\sqrt{\frac{C_2}{C_1}}\right] \tag{7.9}$$

together with

$$|H_0| = \frac{(K+1)/R_1C_1}{1/R_2C_2 + 1/R_2C_1 - K/R_1C_1}$$

the midband gain. The design procedure for the filter is based on the solution of these equations and is summarized in the following example.

EXAMPLE 7.10 Design a second-order bandpass filter with a resonant frequency of 1.592 kHz ($\omega_0 = 10^4$ rad/s), a Q of 10, and a maximum resistance-value spread of 100.

PROCEDURE

1 Just as was done for the bandpass circuit described in Section 7.4, it is convenient to use an equal-valued capacitor design in which the capacitors C_1 and C_2 have the same value. Choose a suitable value (call it C) for the capacitors using the same general criteria that were given in step 1 of the procedure for Example 7.1. Let $C_1 = C_2 = C = 0.01\ \mu$F.

2 Determine the resistance ratio parameter m_0 that would be required if there were no positive feedback. This is given by the relation

$$m_0 = \frac{1}{4Q^2} = \frac{1}{4 \times 100} = 0.0025$$

3 Select the desired resistor ratio m (where m is less than 1 and greater then m_0) and use it to determine the amount of positive feedback K using the relation (and the value $m - 0.01$).

$$K = \frac{R_A}{R_B} = 2m - \frac{\sqrt{m}}{Q} = 2 \times 0.01 - \frac{0.1}{10} = 0.01$$

4 Choose a convenient value for R_B and calculate R_A. Let $R_B = 100$ kΩ.

$$R_A = KR_B = 0.01 \times 10^5 = 1\ \text{k}\Omega$$

5 Determine the value of R_2 by the relation

$$R_2 = \frac{1}{\omega_0 C\sqrt{m}} = \frac{1}{10^4 \times 0.01 \times 10^{-6} \times 0.1} = 100\ \text{k}\Omega$$

6 Determine the value of R_1 using the relation

$$R_1 = mR_2 = 0.01 \times 10^5 = 1\ \text{k}\Omega$$

7 The value of $|H_0|$, the gain at resonance, is determined by the relation

$$|H_0| = \frac{K+1}{2m - K} = \frac{0.01 + 1}{2 \times 0.01 - 0.01} = 101$$

The circuit and its magnitude characteristic are shown in Fig. 7.9.

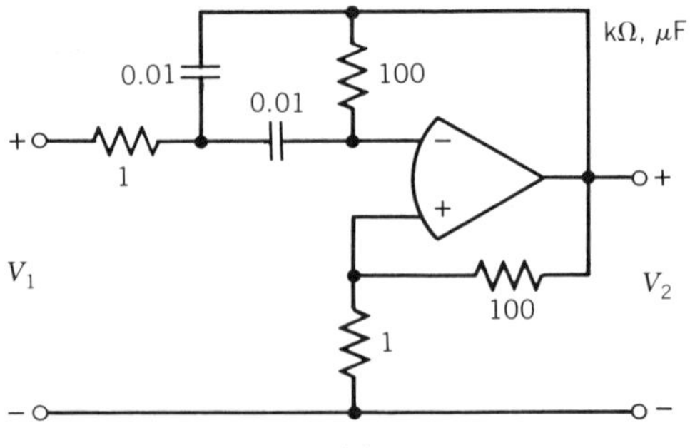

(a)

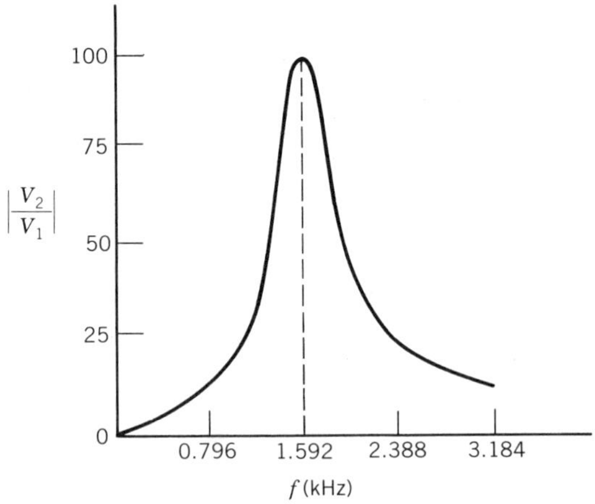

(b)

Figure 7.9 (a) Filter for Example 7.10; (b) magnitude plot.

8 In some cases the value of $|H_0|$ that results from the procedure given may be too large to be practical. If it is desired to reduce this value, it may be done by adding a resistor R_3 to the circuit as shown in Fig. 7.10. In this figure the designation for the input resistor has been changed from R_1 to R_1'. If $|H_0'|$ is the desired gain (where $|H_0'|$ is less than $|H_0|$), the values of R_1' and R_3 are found as follows. (The calculations assume that we want $|H_0'|$ to equal 50.)

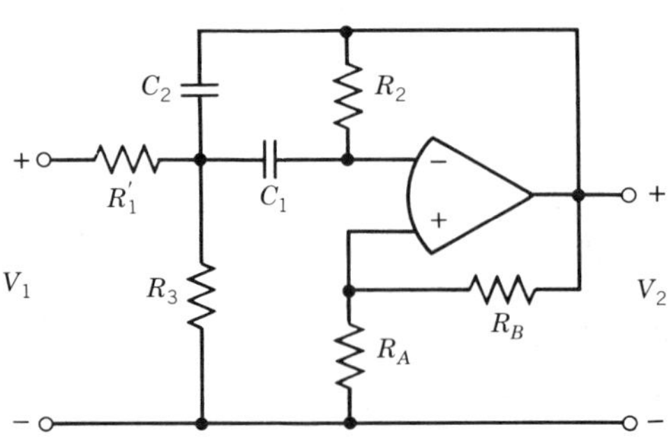

Figure 7.10 Bandpass filter with reduced gain.

$$R_1' = R_1 \frac{|H_0|}{|H_0'|} = \frac{10^3 \times 101}{50} = 2.02 \text{ k}\Omega$$

$$R_3 = \frac{R_1}{1 - |H_0'|/|H_0|} = \frac{10^3}{1 - 50/101} = 1.98 \text{ k}\Omega$$

Note that the addition of the resistor R_3 to the circuit effectively creates a voltage divider, which lowers the level of the input to the rest of the circuit and thus reduces its overall gain. The adjustment to the gain provided by this step is only approximate, since the calculations assume that the current flowing through the resistors R_1' and R_3 is much greater than the current flowing out of the junction of the resistors. For small values of m, the error produced by this assumption may usually be ignored.

The positive feedback circuit described in this section was first introduced by Deliyannis [1]. It provides a compromise between the low sensitivities (and high element-value spreads) of the multiple-loop (negative) feedback circuit described in Section 7.4, and the low element-value spread (and high sensitivity), positive-feedback Sallen–Key circuits described in Chapter 6. Since it allows the designer to specify the amount of element-value spread as part of the design process, this technique produces a versatile practical circuit that is well adapted to the realization of filters with Qs in the general range of 5 to 20.

7.6 BIQUADRATIC MULTIPLE-LOOP FEEDBACK FILTER

The preceding section indicates how positive feedback can be used to reduce the spread of element values that is otherwise inherent in multiple-loop feedback filters. In this section we extend the positive feedback concept to the realization of a *biquadratic function*, that is, one in which the numerator as well as the denominator may be a second-order polynomial. Such a function includes the low-pass, high-pass, and bandpass ones as special cases. The circuit is shown in Fig. 7.11 [2]. It realizes a voltage transfer function having the form

$$\frac{V_2(s)}{V_1(s)} = \frac{cs^2 + ds + e}{s^2 + as + b} \tag{7.10}$$

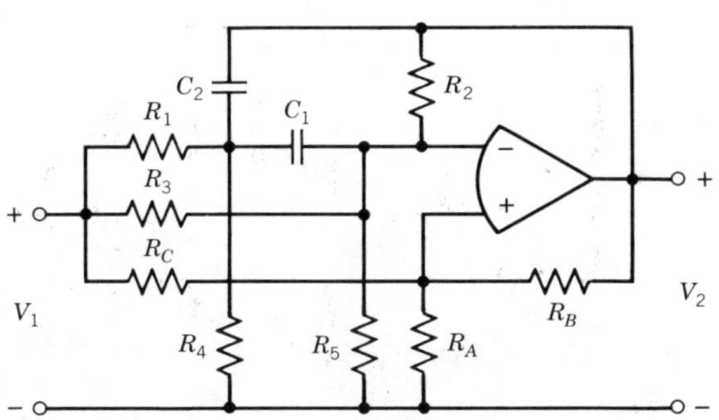

Figure 7.11 Biquadratic multiple-loop feedback filter.

The coefficients a, b, c, d, and e of this network function are related to the element values of the filter by a set of equations that are defined in the following example.

EXAMPLE 7.11 Design a second-order low-pass filter with the following specifications:

a. A passband from 0 to 1 kHz in which the maximum deviation of the magnitude is 1 dB.

b. A minimum of 17 dB attenuation (from the maximum passband gain) at all frequencies greater than 2 kHz.

c. A maximum gain in the passband of 1.0.

PROCEDURE

1 Meeting the above specifications requires the use of the Cauer (or elliptic) approximation described in Chapter 2. Tables that list the coefficients for functions having such an approximation are always given in a normalized form that assumes a passband from 0 to 1 rad/s. Thus our first step is to frequency-normalize the specifications given in (a) and (b) above to the following:

a. A passband from 0 to 1 rad/s in which the maximum deviation of the magnitude is 1 dB.

b. A minimum of 17 dB attenuation (from the maximum passband gain) at all frequencies greater than 2 rad/s.

From Table 2.6 or [3] we find that the following (normalized) function meets the specifications:*

$$\frac{V_2(s)}{V_1(s)} = \frac{0.1397s^2 + 1.0428}{s^2 + 0.9989s + 1.1701}$$

Comparing this with Eq. 7.10, we find

$$a = 0.9989 \qquad b = 1.1701 \qquad c = 0.1397 \qquad d = 0 \qquad e = 1.0428$$

2 Just as was done for the positive-feedback circuit described in Section 7.5, it is convenient for us here to use a design in which the capacitors C_1 and C_2 have the same value. Since we are dealing with frequency-normalized specifications, it is also convenient to use a (normalized) value of unity for the capacitor values. Let $C_1 = C_2 = 1$. All of the following steps assume that the capacitors have a value of unity.

3 Choose a value for the positive-feedback parameter K. This parameter is similar to the one defined in Section 7.5. In general, the larger the value of K, the smaller will be the spread of resistor values. As a first try, let us choose $K = 0.1$.

4 Select a value for the resistor R_B. To simplify future computations, it is desirable to use an impedance-normalized (see step 9) value of unity. Let $R_B = 1$.

5 Determine the values of R_C and R_A.

$$R_C = \frac{K}{c} = \frac{0.1}{0.1397} = 0.7158$$

$$R_A = \frac{K}{1-c} = \frac{0.1}{1 - 0.1397} = 0.1162$$

Note that, in the relation for R_A, we require that c be less than or equal to 1 for R_A to be nonnegative. If the original gain specifications on the network are such that c is

* Actually, this function has at least 17.095 dB attenuation at all frequencies greater than, or equal to, 2 rad/s.

greater than 1, then the overall gain of the network function must be reduced by multiplying the constants c, d, and e in Eq. 7.10 by some suitable constant. The gain that is lost by such a reduction is easily restored by following the biquadratic filter stage by an amplifier.

6 Determine the values of the resistors R_1 and R_4 by first finding the values of the constants x_1 and x_2.

$$x_1 = \frac{2K}{-a + \sqrt{a^2 + 8Kb}}$$

$$= \frac{2 \times 0.1}{-0.9989 + \sqrt{(0.9989)^2 + (8 \times 0.1 \times 1.1701)}} = 0.5105$$

$$x_2 = \frac{c + 2ex_1^2 - dx_1}{1 + K}$$

$$= \frac{0.1397 + [2 \times 1.0428 \times (0.5105)^2] - 0}{1 + 0.1} = 0.6211$$

In terms of these constants we find

$$R_1 = \frac{x_1}{x_2} = \frac{0.5105}{0.6211} = 0.8219$$

$$R_4 = \frac{R_1 x_1}{R_1 - x_1} = \frac{0.8219 \times 0.5105}{0.8219 - 0.5105} = 1.3474$$

Note that, to avoid having a negative value of R_4, if the value calculated for x_2 is greater than 1, the coefficients c, d, and e in the numerator of Eq. 7.10 must be reduced in the same manner as indicated in step 5. Also note that if these coefficients are changed (they must of course all be multiplied by the same factor) so that $x_2 = 1$, then $R_1 = x_1$ and $R_4 = \infty$—that is, an open circuit. This makes it possible to reduce the number of elements in the circuit by eliminating one resistor.

7 To determine the values of R_2, R_3, and R_5, we need first determine two constants y_1 and y_2. The constant y_1 must be selected so that it lies within the range of 0 to 1, and also so that the constant y_2 (defined below) is nonnegative. Choosing $y_1 = 0$, we find

$$y_2 = \frac{(1 + K)(c - y_1)}{x_1 b(e/b - c)}$$

$$= \frac{(1 + 0.1)(0.1397 - 0)}{0.5105 \times 1.1701 \times (1.0428/1.1701 - 0.1397)} = 0.3423$$

Using the values of y_1 and y_2, we find

$$R_2 = \frac{y_2}{x_1 y_2 b + K}$$

$$= \frac{0.3423}{(0.5105 \times 0.3423 \times 1.1701) + 0.1} = 1.124$$

$$R_3 = \frac{y_2}{y_1} = \frac{0.3423}{0} = \infty$$

$$R_5 = \frac{R_3 y_2}{R_3 - y_2} = \frac{y_2}{1 - y_2/R_3} = \frac{0.3423}{1 - 0} = 0.3423$$

Note that in this case, as a result of choosing $y_1 = 0$, we have eliminated the resistor R_3. Thus $R_5 = y_2$.

8 Frequency-denormalizing the circuit to meet the original specifications is readily accomplished by dividing the capacitor values (set to unity) by a frequency-denormalization constant Ω_n. Since the circuit has been realized with passband of 0.1592 Hz (1.0 rad/s), and the desired passband is 1 kHz, $\Omega_n = 2\pi \times 10^3$. Thus the frequency-denormalized capacitor values are $C_1 = C_2 = 1/\Omega_n = (\frac{1}{2}\pi) \times 10^{-3}$F.

9 To change the values of the resistors from their calculated values, we may perform an impedance denormalization on the circuit elements. This is done by defining an impedance-denormalization constant Z_n, which changes the values of the resistors and the capacitors as follows:

$$R_{\text{impedance denormalized}} = R_{\text{normalized}} Z_n$$

$$C_{\text{impedance denormalized}} = C_{\text{normalized}}/Z_n$$

Frequently, Z_n is chosen with a value that will permit the use of standard values for the capacitors. Letting $Z_n = 10^5/2\pi$, we obtain

$$C_1 = C_2 = 0.01\ \mu\text{F} \qquad R_1 = 13.08\ \text{k}\Omega$$

$$R_B = 15.92\ \text{k}\Omega \qquad R_4 = 21.44\ \text{k}\Omega$$

$$R_C = 11.39\ \text{k}\Omega \qquad R_2 = 17.89\ \text{k}\Omega$$

$$R_A = 1.849\ \text{k}\Omega \qquad R_5 = 5.448\ \text{k}\Omega$$

The circuit and its magnitude characteristic are shown in Fig. 7.12.

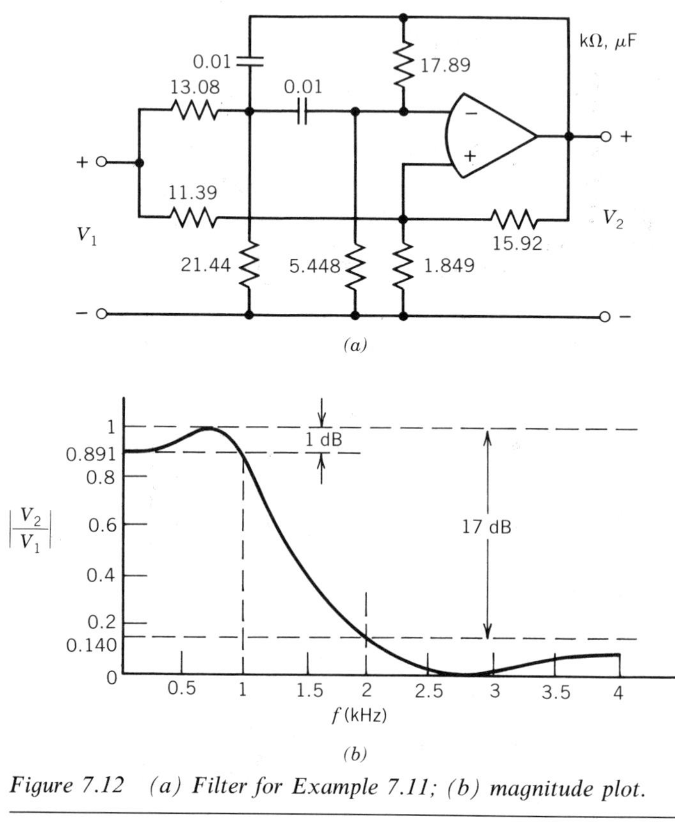

Figure 7.12 (a) Filter for Example 7.11; (b) magnitude plot.

As another example of the use of the biquadratic multiple-loop feedback filter, consider the realization of an all-pass function. Such a function is characterized by the property that, if $N(s)$ is its numerator polynomial and $D(s)$ is its denominator polynomial, then these polynomials satisfy the relation (see Chapter 3):

$$N(s) = D(-s) \tag{7.11}$$

This relation has the effect of reversing the sign of the coefficients on odd-powered denominator coefficients. The all-pass function can be shown to have a magnitude characteristic that is a constant (for all frequencies) and a phase characteristic that is two times the phase characteristic of a function that has the form $1/D(s)$. If we use the Thomson approximation (see Chapter 2) to obtain maximally flat delay (maximally linear phase) in an all-pass function, then the normalized second-order function will be

$$\frac{V_2(s)}{V_1(s)} = \frac{s^2 - 3s + 3}{s^2 + 3s + 3} \tag{7.12}$$

This function has a (normalized) delay at dc of 2 s and a constant magnitude of unity. We shall now use the biquadratic multiple-loop feedback filter in connection with this function to find a realization for the specifications given in the following example.

EXAMPLE 7.12 Design a second-order all-pass filter with a Thomson (Bessel) approximation and a delay of 2 ms (milliseconds) at dc. Base the design on the structure of Fig. 7.11.

PROCEDURE

1 In order to be able to use the function given in Eq. 7.12, it is first necessary to frequency-normalize the problem specifications by a factor of 10^{-3} (this multiplies the time domain specification by 10^{+3}) so that the required delay is 2 s at dc. Such a specification is exactly realized by the function given in Eq. 7.12. Comparing this equation with the general biquadratic one given in Eq. 7.10, we find

$a = 3 \qquad b = 3 \qquad c = 1 \qquad d = -3 \qquad e = 3$

2 Following the procedure used in Example 7.11, we let $C_1 = C_2 = 1$ F.

3 Choose $K = 0.1$.

4 Let $R_B = 1\ \Omega$.

5 Determining R_C, we obtain 0.1 Ω. Determining R_A, we obtain ∞.

6 Determining x_1, we obtain 0.53137. Determining x_2, we obtain 3.8984.

Since x_2 cannot be greater than 1, at this point we must change the specifications of Eq. 7.12 to reduce the values of the numerator coefficients. Multiplying them by 0.25652 = 1/3.8984, we obtain the (reduced gain) network function

$$\frac{V_2(s)}{V_1(s)} = \frac{0.25652s^2 - 0.76955s + 0.76955}{s^2 + 3s + 3} \tag{7.13}$$

This function has the same (normalized) delay, namely 2 s at dc, but its magnitude has been reduced to 0.25652. Comparing this function with the general biquadratic form given in Eq. 7.10, we obtain

$a = 3 \qquad b = 3 \qquad c = 0.25652 \qquad d = -0.76955 \qquad e = 0.76955$

Following the steps given in Example 7.11, we obtain

2 Let $C_1 = C_2 = 1$ F.

3 Choose $K = 0.1$.

4 Let $R_B = 1\ \Omega$.

5 Determining R_C, we obtain 0.38983 Ω. Determining R_A, we obtain 0.13450 Ω.

6 Determining x_1, we obtain 0.53137. Determining x_2, we obtain 1.0. Determining R_1, we obtain 0.53137 Ω. Determining R_4, we obtain ∞.

7 Choose $y_1 = 0$. Determining y_2, we obtain ∞. Determining R_2, we obtain 0.62731 Ω. Determining R_3, we obtain ∞. Determining R_5, we obtain ∞.

8 Frequency-denormalizing the capacitors, we obtain $C_1 = C_2 = 10^{-3}$ F.

9 Impedance-denormalizing the circuit, using $Z_n = 10^5$, we obtain

$$C_1 = C_2 = 0.01\ \mu\text{F} \qquad R_A = 13.450\ \text{k}\Omega$$
$$R_B = 100\ \text{k}\Omega \qquad R_1 = 53.137\ \text{k}\Omega$$
$$R_C = 38.983\ \text{k}\Omega \qquad R_2 = 62.731\ \text{k}\Omega$$

The resistors R_3, R_4, and R_5 are not required. The circuit and its phase characteristic are shown in Fig. 7.13. The magnitude is constant and has a value of 0.25652.

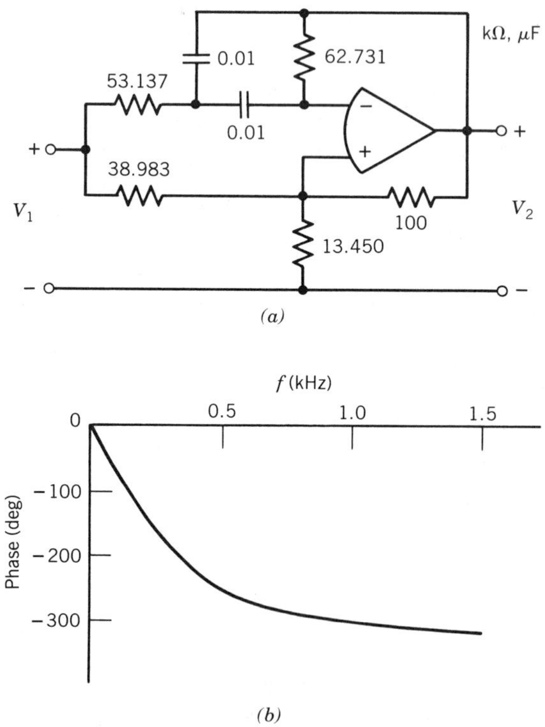

Figure 7.13 (a) Filter for Example 7.12; (b) phase plot.

The biquadratic multiple-loop feedback filter is a general and versatile filter configuration that can be used to realize a wide range of filter functions with reasonable spreads of element values and good sensitivities.

7.7 SUMMARY OF DESIGN PROCEDURES

LOW-PASS FILTER (See Section 7.2)

Network Function Realized

$$\frac{V_2}{V_1} = \frac{-|H_0|\omega_0^2}{s^2 + (\omega_0/Q)s + \omega_0^2}$$

Circuit Configuration

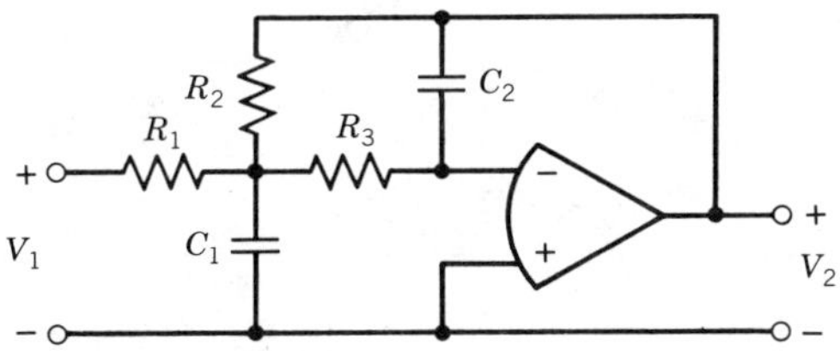

Design Steps

1 Choose a convenient capacitor value C. Let $C_1 = C$.

2 Determine the element value spread constant m such that $m \leq 1/4Q^2(1 + |H_0|)$. Determine $C_2 = mC$.

3 Determine $R_2 = [1 + \sqrt{1 - 4mQ^2(1 + |H_0|)}]/2\omega_0 CmQ$.

4 Determine $R_1 = R_2/|H_0|$.

5 Determine $R_3 = 1/\omega_0^2 C^2 R_2 m$.

HIGH-PASS FILTER (See Section 7.3)

Network Function Realized

$$\frac{V_2}{V_1} = \frac{-|H_0|s^2}{s^2 + (\omega_0/Q)s + \omega_0^2}$$

Circuit Configuration

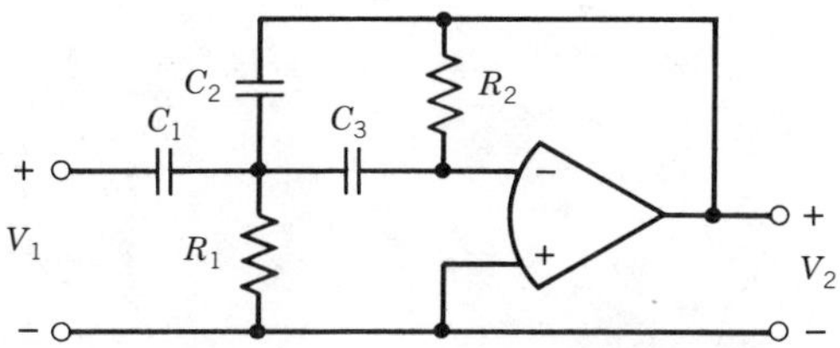

Design Steps

1 Choose a convenient capacitor value C. Let $C_1 = C_3 = C$.

2 Determine $R_1 = |H_0|/\omega_0 QC(2|H_0| + 1)$.

3 Determine $R_2 = Q(2|H_0| + 1)/\omega_0 C$.

4 Determine $C_2 = C/|H_0|$.

BANDPASS FILTER (See Section 7.4)

Network Function Realized

$$\frac{V_2}{V_1} = \frac{-|H_0|(\omega_0/Q)s}{s^2 + (\omega_0/Q)s + \omega_0^2}$$

Circuit Configuration

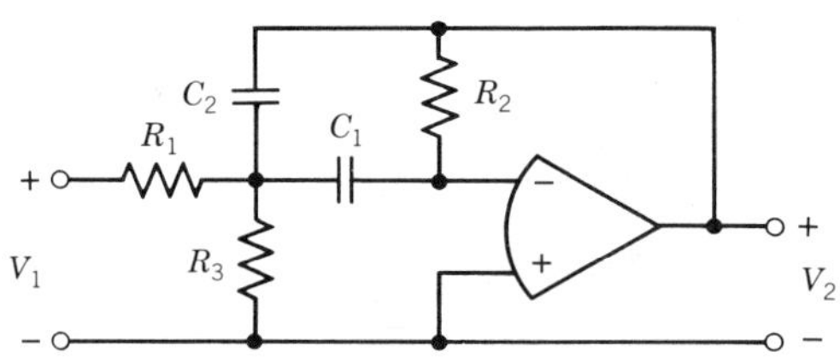

Design Steps

1 Choose a convenient capacitor value C. Let $C_1 = C_2 = C$.

2 Determine $R_1 = Q/\omega_0 C|H_0|$.

3 Determine $R_2 = 2Q/\omega_0 C$.

4 Determine $R_3 = Q/\omega_0 C(2Q^2 - |H_0|)$.

Note: $|H_0|$ must be less than $2Q^2$.

BANDPASS FILTER WITH POSITIVE FEEDBACK (See Section 7.5)

Network Function Realized

$$\frac{V_2}{V_1} = \frac{-|H_0|(\omega_0/Q)s}{s^2 + (\omega_0/Q)s + \omega_0^2}$$

Circuit Configuration

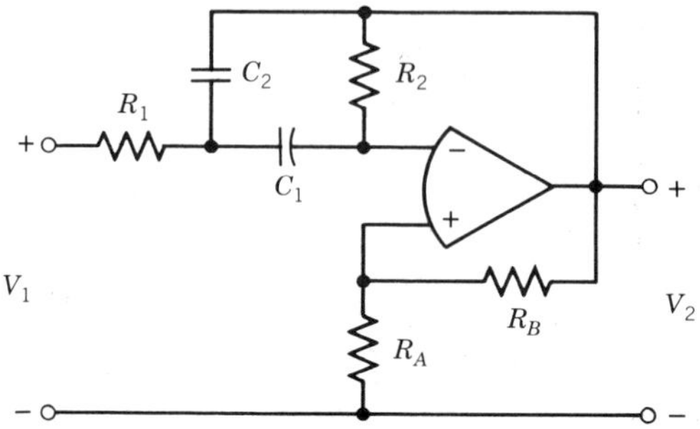

Design Steps

1. Choose a convenient capacitor value C. Let $C_1 = C_2 = C$.
2. Determine the "no positive feedback" element spread constant $m_0 = 1/4Q^2$.
3. Choose m, the element spread constant, such that $m_0 < m < 1$. Determine the positive feedback constant $K = 2m - \sqrt{m}/Q$.
4. Choose a convenient value for R_B. Let $R_A = KR_B$.
5. Determine $R_2 = 1/\omega_0 C\sqrt{m}$.
6. Determine $R_1 = mR_2$.

Note: The value of $|H_0|$ is not a free parameter. It is given by

$$|H_0| = \frac{K+1}{2m - K}$$

BIQUADRATIC FILTER (See Section 7.6)

Network Function Realized

$$\frac{V_2}{V_1} = \frac{cs^2 + ds + e}{s^2 + as + b}$$

Circuit Configuration

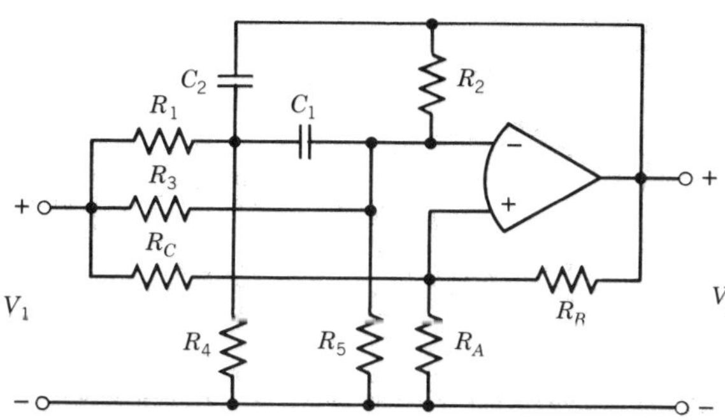

Design Steps

1. Choose normalized values for the capacitors. Let $C_1 = C_2 = 1$.
2. Choose a value for the positive feedback parameter K. As a trial choice let $K = 0.1$.
3. Choose a normalized value for the resistor R_B. Let $R_B = 1$.
4. Determine $R_C = K/c$.
5. Determine $R_A = K/(1 - c)$. *Note*: If c is greater than one, see step 5 in Example 7.11.
6. Determine the constant

$$x_1 = \frac{2K}{-a + \sqrt{a^2 + 8Kb}}$$

Determine the constant

$$x_2 = \frac{c + 2ex_1^2 - dx_1}{1 + K}$$

Determine $R_1 = x_1/x_2$. Determine $R_4 = R_1x_1/(R_1 - x_1)$. *Note*: If x_2 is greater than one, see step 6 in Example 7.11.

7 Choose a constant y_1 such that $0 \le y_1 \le 1$. *Note*: y_1 must also be chosen so that the constant y_2 is positive. Determine the constant

$$y_2 = \frac{(1 + K)(c - y_1)}{x_1 b(e/b - c)}$$

Determine $R_2 = y_2/(x_1y_2b + K)$. Determine $R_3 = y_2/y_1$. Determine $R_5 = R_3y_2/(R_3 - y_2)$.

8 Frequency and impedance denormalize the circuit as required (see steps 8 and 9 in Example 7.11).

7.8 SUMMARY

In this chapter we have presented a discussion of the multiple-loop feedback filter that uses negative feedback around an operational amplifier to realize a variety of second-order functions. This filter has the desirable property of being relatively insensitive to variations in filter components from their design values. Although specifications requiring high Q and/or large gain tend to produce large spreads of resistor values in this type of filter, such spreads can be reduced by the use of positive feedback.

7.9 REFERENCES

[1] T. Deliyannis, "High Q-Factor Circuit with Reduced Sensitivity," *Electronics Letters*, vol. 4, p. 577, December 1968.

[2] J. J. Friend, "A Single Operational-Amplifier Biquadratic Filter Section," *1970 IEEE ISCT Dig. Tech. Papers*, p. 189.

[3] L. P. Huelsman and P. E. Allen, *Introduction to the Theory and Design of Active Filters*, New York: McGraw-Hill, 1980, Chapter 2.

CHAPTER 8

Biquads I: The State-Variable Structure

E. Sánchez-Sinencio

Instituto Nacional de Astrofísica
Puebla, Mexico

8.1 INTRODUCTION

In this section a family of relatively high-Q filters is described. These filters provide flexibility, good performance, and low sensitivity; their implementation is based on the *state-variable* approach. This name is derived from the fact that state-variable methods of solving differential equations are employed in the development of the realization. In the implementation of these realizations, three basic active building blocks are generally used: the summer, the integrator, and the lossy integrator.*

Consider a second-order high-pass (HP) transfer function (see Chapter 3) characterized by

$$V_2(s) = HV_1(s) - \frac{1}{s}\frac{\omega_p}{Q_p}V_2(s) - \frac{\omega_p^2}{s^2}V_2(s) \tag{8.1}$$

where $V_2(s)$ and $V_1(s)$ are the high-pass output and input of the system, respectively. If we rearrange terms in $V_2(s)$ and $V_1(s)$ and use the inverse Laplace-transform technique, Eq. 8.1 can be written as

$$\frac{d^2v_2(t)}{dt^2} + \frac{\omega_p}{Q_p}\frac{dv_2(t)}{dt} + \omega_p^2 v_2(t) = H\frac{d^2v_1(t)}{dt^2} \tag{8.2}$$

* The description of these building blocks is given in Chapter 4.

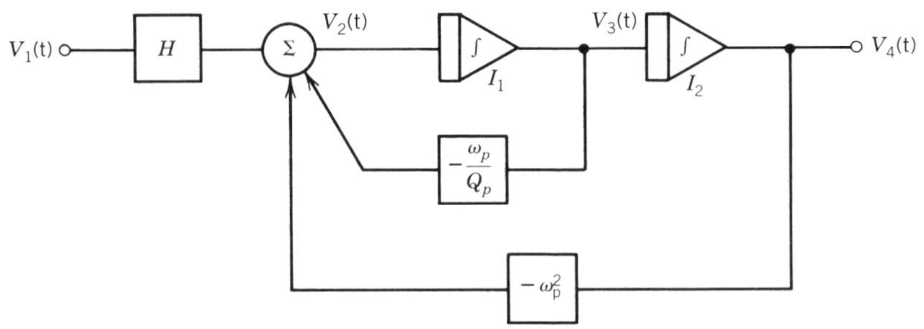

Figure 8.1 Analog computer block diagram of Eq. 8.2.

Table 8.1 **Possible transfer functions for Fig. 8.1**

	I_1	I_2	ω_p/Q_p	ω_p^2	HP	BP	LP
(a)	−	−	+	−	$\frac{Hs^2}{D(s)}$	$-\frac{Hs^2}{D(s)}$	$\frac{H}{D(s)}$
(b)	+	+	−	−	$\frac{Hs^2}{D(s)}$	$\frac{Hs}{D(s)}$	$\frac{H}{D(s)}$
(c)	+	−	−	+	$\frac{Hs^2}{D(s)}$	$\frac{Hs}{D(s)}$	$-\frac{H}{D(s)}$
(d)	−	+	+	+	$\frac{Hs^2}{D(s)}$	$-\frac{Hs}{D(s)}$	$-\frac{H}{D(s)}$

where $D(s) = s^2 + \frac{\omega_p}{Q_p} s + \omega_p^2$

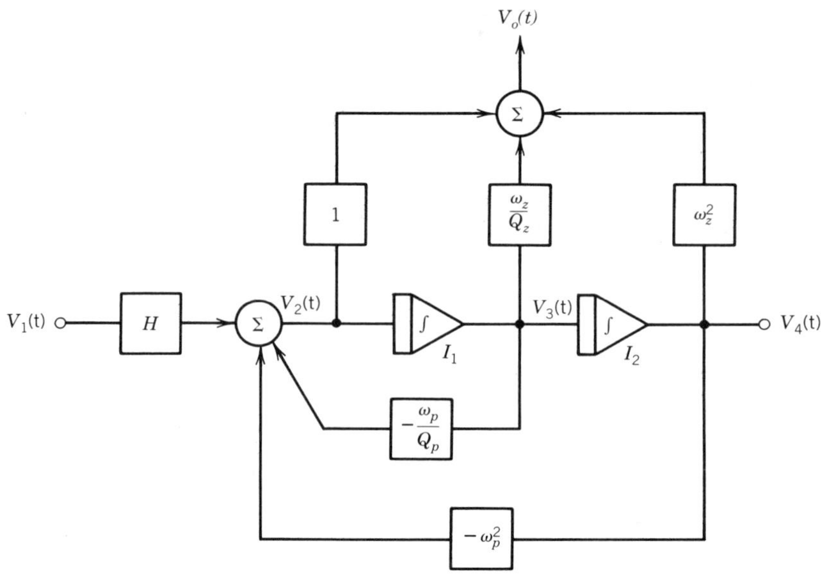

Figure 8.2 Block diagram of Eq. 8.3 using an additional summer.

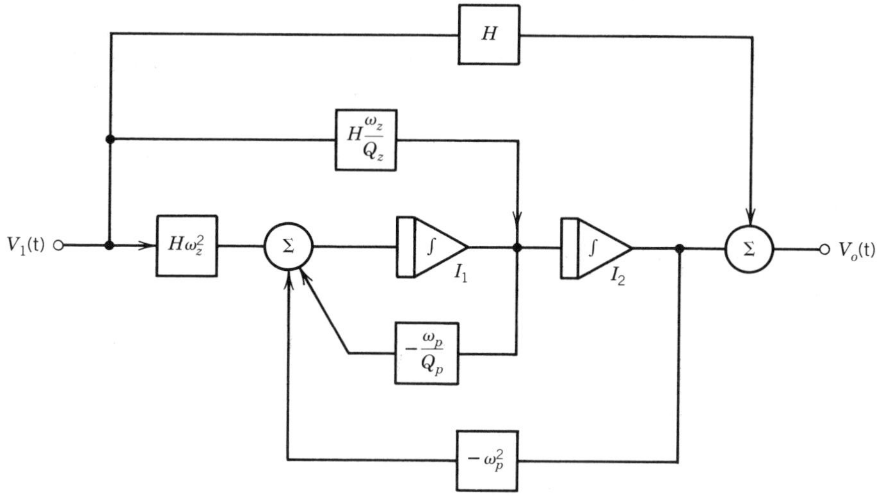

Figure 8.3 Block diagram of Eq. 8.3 using feedforward technique.

Because multiplying by $1/s$ in the complex frequency domain corresponds to integration in the time domain, Eq. 8.2 can be represented by the analog computer block diagram of Fig. 8.1. Observe that, in general, there are three possible outputs. First, $V_2(s)$ is the original high-pass output. Second, $V_3(s)$ is $V_2(s)$ after passing through the integrator I_1 and corresponds to a bandpass (BP) output, since a high-pass transfer function, when multiplied by $1/s$, becomes a bandpass transfer function. Finally, by similar reasoning, V_4 is the low-pass output (LP) because the bandpass transfer function multiplied by $1/s$ becomes a low-pass transfer function. The three functions are simultaneously available with the same poles. Furthermore, we observe that there are four possible combinations of signs for the coefficients and the integrators I_1 and I_2 if positive and negative integrators are available. The possible frequency domain transfer functions for Fig. 8.1 are described in Table 8.1. The signs of the two loops, $I_1\omega_p/Q_p$ and $I_1I_2\omega_p^2$, involve a negative feedback.

Figure 8.1 can be generalized to implement any pair of complex zeros. This is shown in Fig. 8.2., for which the transfer function is*

$$T(s) = \frac{V_o(s)}{V_1(s)} = H\frac{s^2 + (\omega_z/Q_z)s + \omega_z^2}{s^2 + (\omega_p/Q_p)s + \omega_p^2} \tag{8.3}$$

Another implementation based on the feedforward technique, in which the zeros are formed by directly introducing the input signal at the integrator I_2 and also to the summers at the input and output, is shown in Fig. 8.3, the transfer function of which is also given by Eq. 8.3.

The actual implementations of the block diagram discussed above depend on the type of the basic building blocks employed. It will be shown that by introducing the input (feedback and/or feedforward) signals at those nodes of the circuit

* Note that ω_p and Q_p are used in this chapter to distinguish between numerator and denominator parameters. When the function reduces to LP, HP, or BP forms, setting $Q_p = Q$ and $\omega_p = \omega_o$ will yield terminology consistent with that of earlier chapters.

which are at virtual ground potential, one or more summer building blocks can be saved, sometimes at the expense of losing output V_2 of Fig. 8.1. This will be shown in Section 8.3.

8.2 KERWIN–HUELSMAN–NEWCOMB (KHN) FILTER

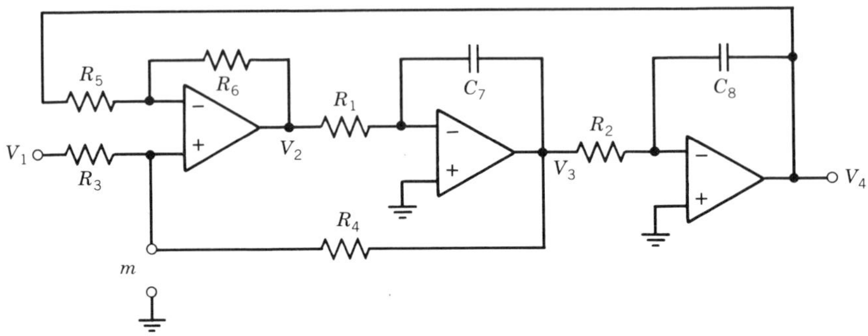

Figure 8.4 Kerwin–Huelsman–Newcomb (KHN) filter.

The implementation of this structure corresponds to Eq. (a) of Table 8.1 with two inverting integrators. The three outputs simultaneously available from the output terminals V_2, V_3, V_4 are high-pass, bandpass, and low-pass outputs, respectively, as shown in Fig. 8.4. The corresponding transfer functions are given in Table 8.2.

Table 8.2 KHN transfer functions of Fig. 8.4

$$T_{HP}(s) = \frac{V_2(s)}{V_1(s)} = \frac{H_{HP}s^2}{s^2 + (\omega_p/Q_p)s + \omega_p^2} \qquad \text{High pass}$$

$$T_{BP}(s) = \frac{V_3(s)}{V_1(s)} = \frac{-H_{BP}(\omega_p/Q_p)s}{s^2 + (\omega_p/Q_p)s + \omega_p^2} \qquad \text{Bandpass}$$

$$T_{LP}(s) = \frac{V_4(s)}{V_1(s)} = \frac{H_{LP}\omega_p^2}{s^2 + (\omega_p/Q_p) + \omega_p^2} \qquad \text{Low pass}$$

where

$$\omega_p^2 = \frac{R_6/R_5}{R_1R_2C_7C_8}$$

$$Q_p = \frac{1 + R_4/R_3}{(1 + R_6/R_5)\sqrt{R_5R_2C_8/R_6R_1C_7}}$$

and

$$H_{HP} = \frac{1 + R_6/R_5}{1 + R_3/R_4}$$

$$|H_{BP}| = R_4/R_3$$

$$H_{LP} = \frac{1 + R_5/R_6}{1 + R_3/R_4}$$

Table 8.3 **Design equations for the KHN filter of Fig. 8.4**

Design 1	Design 2
$R_5 = R_6$ $R_1 = R_2 = R$ $C_7 = a^2 C_8 = a^2 C$	$R_5 = R_6$ $C_7 = C_8 = C$ $R_1 = a^2 R_2 = a^2 R$
$\omega_p = \dfrac{1}{aRC}$	$\omega_p = \dfrac{1}{aRC}$
$Q_p = \dfrac{a}{2}\left(1 + \dfrac{R_4}{R_3}\right)$	$Q_p = \dfrac{a}{2}\left(1 + \dfrac{R_4}{R_3}\right)$
$H_{HP} = H_{LP} = 2 - \dfrac{a}{Q_p}$	$H_{HP} = H_{LP} = 2 - \dfrac{a}{Q_p}$
$\lvert H_{BP}\rvert = 2\dfrac{Q_p}{a} - 1$	$\lvert H_{BP}\rvert = 2\dfrac{Q_p}{a} - 1$

We can choose the elements depending on whether the capacitors C_7 and C_8 are equal (design case 2) or different (design case 1). These design options are illustrated in Table 8.3.

DESIGN PROCEDURE

1 Assume that ω_p, Q_p, and H are the design specifications.

2 Select convenient values for R_3, $R_5(= R_6)$, and $C(= C_7 = a^2 C_8)$ or R_3, $R_5(= R_6)$, and $R(= R_2 = R_1/a^2)$ for design 1 and design 2, respectively.

3 Calculate the following element values:

$$a = (2 - H)Q_p \qquad \text{for LP or HP} \tag{8.4}$$

$$a = \frac{2Q_p}{1 + H} \qquad \text{for BP} \tag{8.5}$$

$$R_4 = \left(2\frac{Q_p}{a} - 1\right) R_3 \tag{8.6}$$

and

$$C = \frac{1}{\omega_p a R} \qquad \text{for design 1} \tag{8.7}$$

$$R = \frac{1}{\omega_p a C} \qquad \text{for design 2} \tag{8.8}$$

EXAMPLE 8.1 Design a bandpass KHN filter having a gain $H_{BP} = 3$, $Q_p = 20$, and $\omega_p = 2\pi \times 10^3 r/s$.

PROCEDURE

1 Choose $R_3 = R_5 = R_6 = 10\ \text{k}\Omega$.

2 The voltage gain of the KHN bandpass filter is $H_{BP} = 2(Q_p/a) - 1$. Substituting the given values of H_{BP} and Q_p, we obtain $a = 10$.

3 The quality factor is $Q_p = (a/2)(1 + R_4/R_3)$, substituting the values of a, Q_p, and R_3, we obtain $R_4 = 30\ \text{k}\Omega$.

4 For the case of design 1, given $C = 0.01\ \mu\text{F}$ and the relation $\omega_p = (1/aRC)$, we obtain $R(= R_1 = R_2) = 1.59\ \text{k}\Omega$.

5 For the case of design 2, given $R = 10\ \text{k}\Omega$ and the relation $\omega_p = (1/aRC)$, we obtain $C(= C_7 = C_8) = 1.59\ \text{nF}$.

By adding a resistor R_9 between node m and the ground of Fig. 8.4, we obtain a more flexible network while maintaining the specified ω_p, Q_p, and H. If

Table 8.4 **Modified KHN transfer functions**

Definitions of $T_{\text{HP}}(s)$, $T_{\text{BP}}(s)$, and $T_{\text{LP}}(s)$ refer to Table 8.2, where

$$\omega_p^2 = \frac{R_6/R_5}{R_1 R_2 C_7 C_8}$$

$$Q_p = \frac{1 + R_4/R_3 + R_4/R_9}{(1 + R_6/R_5)\sqrt{R_2 R_5 C_8 / R_1 R_6 C_7}}$$

and

$$H_{\text{HP}} = \frac{1 + R_6/R_5}{1 + R_3/R_4 + R_3/R_9}$$

$$|H_{\text{BP}}| = \frac{R_4}{R_3}$$

$$H_{\text{LP}} = \frac{1 + R_5/R_6}{1 + R_3/R_4 + R_3/R_9}$$

Table 8.5 **Design equations for the modified KHN filter**

$$R_1 = R_2 = R'$$

$$C_7 = C_8 = C$$

$$R_4 = R_5 = b^2 R_6 = R$$

$$\omega_p = \frac{1}{bR'C}$$

$$Q_p = \frac{1 + R/R_3 + R/R_9}{(1 + 1/b^2)b}$$

$$H_{\text{HP}} = \frac{1 + 1/b^2}{1 + R_3/R + R_3/R_9}$$

$$H_{\text{BP}} = \frac{R}{R_3}$$

$$H_{\text{LP}} = \frac{1 + b^2}{1 + R_3/R + R_3/R_9}$$

we take the equations of Table 8.2 as references, the parameters become as given in Table 8.4.

We can choose some elements to obtain simpler design equations, as shown in Table 8.5 where R', R, and C are chosen.

DESIGN PROCEDURE

A simple design procedure is described as follows.

1 Assume that ω_p, Q_p, and H are the design specifications.

2 Select convenient values* of R', C, and R to determine the values of R_1, R_2, C_7, C_8, R_4, and R_6.

3 Calculate the following element values:

$$b = \frac{1}{R'C\omega_p}$$

For the HP case:

$$R_3 = \frac{R}{bQ_pH_{\mathrm{HP}}}$$

$$R_9 = \frac{R}{bQ_p[1 + (1/b^2) - H_{\mathrm{HP}}] - 1}$$

For the BP case:

$$R_3 = \frac{R}{H_{\mathrm{BP}}}$$

$$R_9 = \frac{R}{bQ_p[1 + (1/b^2) - 1 - H_{\mathrm{BP}}}$$

For the LP case:

$$R_3 = \frac{bR}{Q_pH_{\mathrm{LP}}}$$

$$R_9 = \frac{bR}{(1 + b^2)Q_p - b - Q_pH_{\mathrm{LP}}}$$

EXAMPLE 8.2 Design a bandpass modified KHN filter having a gain of $H_{\mathrm{BP}} = 3$, $Q_p = 20$, and $\omega_p = 2\pi \times 10^3 r/s$.

PROCEDURE

1 Let us choose $R = 10\ \mathrm{k\Omega}$, $C = 0.01\ \mu\mathrm{F}$, and $R' = 11.254\ \mathrm{k\Omega}$; that is, $R_1 = R_2 = 11.254\ \mathrm{k\Omega}$ and $R_4 = R_5 = 10\ \mathrm{k\Omega}$.

* A rule of thumb for choosing R' and R is to make them proportional to $10f_p$, when $Q_p > 10$, or else make R' and R proportional to f_p. Then C should be made proportional to $1/R'\omega_p$.

2 Since the values of R_1 and R_2 are 11.254 kΩ, then

$$b = \frac{1}{R'C\omega_p} = 1.414207$$

3 The expression for $R_6 = \frac{R}{b^2}$; then $R_6 = 5\ \text{k}\Omega$

4 For this case

$$R_3 = \frac{R}{H_{\text{BP}}} = 3.33\ \text{k}\Omega$$

and

$$R_9 = \frac{R}{bQ_p[1 + (1/b^2)] - 1 - H_{\text{BP}}} = 260\ \Omega$$

8.3 TOW–THOMAS (TT) FILTER

This circuit is similar to the KHN. However, notice that in Fig. 8.5 all the noninverting terminals of the op amps are grounded. The noninverting integrator is formed by an inverter $(-R_8/R_7)$ and an inverting integrator $(-1/sR_2C_{10})$. The TT implementation corresponds to Eq. (d) of Table 8.1, the summer building block being unnecessary, as the summation is carried out at the inverting terminal of the first integrator. The price of saving the summer is that the high-pass function is lost, since the node where the summation is made is a virtual ground. The components R_4, R_6, and R_5 form the feedforward path to obtain the transmission zeros. The transfer function (taking the output at V_o of Fig. 8.5) together with gain level parameters H_{HP}, H_{BP}, H_{LP}, and H_{notch} are given in Table 8.6.

A set of design equations can be generated if we fix some component values, that is, R_3, R_7, R_8, C_9, and C_{10}. Table 8.7 illustrates this design case.

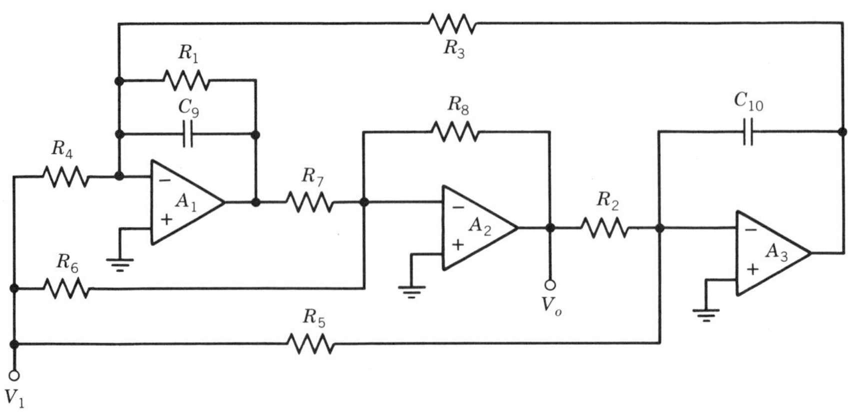

Figure 8.5 Feedforward Tow–Thomas biquad circuit.

Table 8.6 **Description of the parameters for the Tow–Thomas Filter**

General Transfer Function

$$T(s) = -\frac{R_8}{R_6}\frac{s^2 + s\left(\dfrac{1}{R_1C_9} - \dfrac{1}{R_4C_9}\dfrac{R_6}{R_7}\right) + \dfrac{R_6}{R_7}\dfrac{1}{R_3R_5C_9C_{10}}}{s^2 + s\left(\dfrac{1}{R_1C_9}\right) + \dfrac{R_8}{R_7}\dfrac{1}{R_3R_2C_9C_{10}}}$$

where

$$\omega_p^2 = \frac{R_8}{R_7R_2R_3C_9C_{10}}, \qquad \omega_z^2 = \frac{R_6}{R_3R_5R_7C_9C_{10}}$$

$$Q_p = R_1\sqrt{\frac{R_8C_9}{R_2R_3R_7C_{10}}}, \qquad Q_z = \sqrt{\frac{R_6C_9}{R_3R_5R_7C_{10}}}\bigg/\left(\frac{1}{R_1} - \frac{R_6}{R_4R_7}\right)$$

and

$$|H_{\text{HP}}| = \frac{R_8}{R_6}, \qquad \text{for} \qquad R_1 = \frac{R_4R_7}{R_6}, \quad R_5 \to \infty$$

$$|H_{\text{BP}}| = \frac{R_1R_8}{R_4R_7}, \qquad \text{for} \qquad R_5, R_6 \to \infty$$

$$|H_{\text{LP}}| = \frac{R_2}{R_5}, \qquad \text{for} \qquad R_4, R_6 \to \infty$$

For the bandstop (notch)

$$|H_{\text{notch}}| = \frac{R_8}{R_6}, \qquad \text{for} \qquad R_1 = \frac{R_4R_7}{R_6}, \quad R_5 = \frac{R_6R_2}{R_8}$$

Table 8.7 **Design equations for the Tow–Thomas filter of Fig. 8.5**

Let

$$R_3 = R$$
$$R_2 = a^2R_3$$
$$R_7 = R_8 = R'$$
$$C_9 = C_{10} = C$$

$$\omega_p = \frac{1}{aRC}, \qquad \omega_z = \frac{1}{C}\sqrt{\frac{R_6}{R_5RR'}}$$

$$Q_p = \frac{R_1}{aR}, \qquad Q_z = \sqrt{\frac{R_6}{R_5RR'}}\bigg/\left(\frac{1}{R_1} - \frac{R_6}{R_4R'}\right)$$

$$|H_{\text{HP}}| = \frac{R'}{R_6}, \qquad \text{for} \qquad R_1 = \frac{R_4R'}{R_6} = R_4|H_{\text{HP}}|, \quad R_5 \to \infty$$

$$|H_{\text{BP}}| = \frac{R_1}{R_4}, \qquad \text{for} \qquad R_5, R_6 \to \infty$$

$$|H_{\text{LP}}| = \frac{a^2R}{R_5}, \qquad \text{for} \qquad R_4, R_6 \to \infty$$

$$|H_{\text{notch}}| = \frac{R'}{R_6}, \qquad \text{for} \qquad R_1 = \frac{R_4R'}{R_6}, \quad R_5 = \frac{a^2RR_6}{R'}$$

DESIGN PROCEDURE

A design procedure can be described as follows.

1 Assume that ω_p, Q_p, and H are the design specifications.

2 Choose convenient values of R, R', and C to fix R_3, R_7, R_8, C_9, and C_{10}.

3 Compute the following component values:

$$a = \frac{1}{\omega_p RC}$$

$$R_2 = a^2 R$$

$$R_1 = aRQ_p$$

For the HP case:

$$R_5 \to \infty$$

$$R_4 = \frac{R_1}{|H_{\mathrm{HP}}|}$$

$$R_6 = \frac{R'}{|H_{\mathrm{HP}}|}$$

For the BP case:

$$R_5, \qquad R_6 \to \infty$$

$$R_4 = \frac{R_1}{|H_{\mathrm{BP}}|}$$

For the LP case:

$$R_4, \qquad R_6 \to \infty$$

$$R_5 = \frac{a^2 R}{|H_{\mathrm{LP}}|}$$

For the BS case:

$$R_1 = \frac{R' R_4}{R_6}$$

$$R_6 = \frac{R'}{|H_{\mathrm{notch}}|}$$

EXAMPLE 8.3 Design a bandpass TT filter having a gain of $H_{\mathrm{BP}} = 3$, $Q_p = 10$ and $\omega_p = 2\pi \times 10^3 r/s$.

PROCEDURE

1 Choose $R' = R = 10\ \mathrm{k\Omega}$, and $C = 0.01\ \mu\mathrm{F}$. That is, $C_9 = C_{10} = 0.01\ \mu\mathrm{F}$ and $R_3 = R_7 = R_8 = 10\ \mathrm{k\Omega}$.

2 The pole frequency is given by $\omega_p = (1/aRC)$; then $a = (1/RC\omega_p) = 1.5915$.

3 The value for R_2 is

$$R_2 = a^2 R = 25.33\ \mathrm{k\Omega}.$$

4 The quality factor is given by $Q_p = R_1/aR$; thus $R_1 = 159.15\ \text{k}\Omega$.
5 The bandpass gain is $|H_{BP}| = R_1/R_4$; then $R_4 = 53.05\ \text{k}\Omega$ and $R_5, R_6 \to \infty$.

8.4 AKERBERG–MOSSBERG (AM) FILTER

In the implementation of this topology, instead of using an inverting integrator cascade with an inverter (i.e., blocks A_2 and A_3 of Fig. 8.5), the inverter is inside the feedback loop as shown in Fig. 8.6. Hence the positive input of A_1 becomes the summing node. The transfer function for this structure is derived from the output of A_3; that is, $V_o(s)/V_{in}(s)$. The description of the transfer functions is given in Table 8.8.

By fixing R_7, R_1, R_2, C_9, and C_{10}, a set of design equations can be obtained, as shown in Table 8.9.

DESIGN PROCEDURE

A design procedure follows.

1 Assume that Q_p, ω_p, and H are the design specifications.
2 Select convenient values of R, R', and C to fix R_7, C_9, C_{10}, R_1, and R_2, as indicated in Table 8.9.
3 Calculate the following element values:

$$a = \frac{1}{RC\omega_p}$$

$$R_6 = aRQ_p$$

For the HP case:

$$C_{11} = C|H_{HP}|$$

$$R_4, R_5, R_8 \to \infty$$

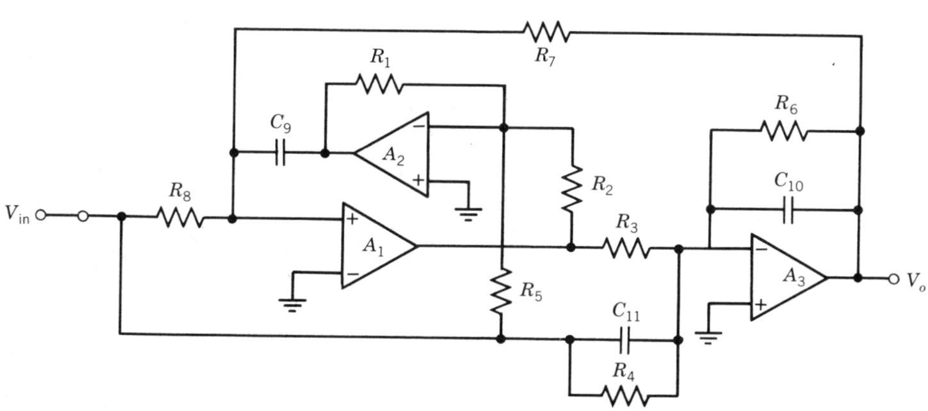

Figure 8.6 Akerberg–Mossberg (AM) filter.

Table 8.8 **Description of the parameters for the Akerberg–Mossberg filter**

General Transfer Function

$$T(s) = -\frac{C_{11}}{C_{10}} \frac{s^2 + s\left(\frac{1}{R_4} - \frac{R_2}{R_5 R_3}\right)\frac{1}{C_{11}} + \frac{R_2}{R_8 R_1 R_3 C_9 C_{11}}}{s^2 + s\frac{1}{R_6 C_{10}} + \frac{R_2}{R_7 R_1 R_3 C_9 C_{10}}}$$

where

$$\omega_p^2 = \frac{R_2}{R_7 R_1 R_3 C_9 C_{10}}, \qquad \omega_z^2 = \frac{R_2}{R_8 R_1 R_3 C_9 C_{11}}$$

$$Q_p = R_6 C_{10}\sqrt{\frac{R_2}{R_7 R_1 R_3 C_9 C_{10}}}, \qquad Q_z = \frac{\sqrt{R_2/R_8 R_1 R_3 C_9 C_{11}}}{\left(\frac{1}{R_4 C_{11}} - \frac{R_2}{R_3 R_5 C_{11}}\right)}$$

and

$$|H_{\text{HP}}| = \frac{C_{11}}{C_{10}}, \qquad \text{for} \qquad R_4, R_5, R_8 \to \infty$$

$$|H_{\text{BP}}| = \frac{R_6}{R_4}, \qquad \text{for} \qquad C_{11} = 0, \quad R_5, R_8 \to \infty$$

$$|H_{\text{LP}}| = \frac{R_7}{R_8}, \qquad \text{for} \qquad C_{11} = 0, \quad R_4, R_5 \to \infty$$

For the bandstop (notch)

$$|H_{\text{notch}}| = \frac{R_7}{R_8}, \qquad \omega_z^2 = \frac{R_2}{R_8 R_1 R_3 C_9 C_{11}} \qquad \text{for} \qquad R_4, R_5 \to \infty$$

Table 8.9 **Design equations for the Akerberg–Mossberg filter of Fig. 8.6**

Let

$$R_7 = R$$
$$R_3 = a^2 R_7$$
$$C_9 = C_{11} = C$$
$$R_1 = R_2 = R'$$

$$\omega_p = \frac{1}{aRC}, \qquad \omega_z = \frac{1}{aC}\sqrt{\frac{1}{R_8 R}}$$

$$Q_p = \frac{R_6}{aR}, \qquad Q_z = \frac{1}{a}\sqrt{\frac{1}{R_8 R}}\bigg/\left(\frac{1}{R_4} - \frac{R'}{a^2 R R_5}\right)$$

$$|H_{\text{HP}}| = \frac{C_{11}}{C}, \qquad \text{for} \qquad R_4, R_5, R_8 \to \infty$$

$$|H_{\text{BP}}| = \frac{R_6}{R_4}, \qquad \text{for} \qquad C_{11} = 0, \quad R_5, R_8 \to \infty$$

$$|H_{\text{LP}}| = \frac{R}{R_8}, \qquad \text{for} \qquad C_{11} = 0, \quad R_4, R_5 \to \infty$$

$$|H_{\text{notch}}| = \frac{R}{R_8}, \qquad \text{for} \qquad R_4, R_5 \to \infty$$

For the BP case:

$$R_4 = \frac{R_6}{|H_{\mathrm{BP}}|}$$

$$C_{11} = 0, \qquad R_5, R_8 \to \infty$$

For the LP case:

$$R_8 = \frac{R}{|H_{\mathrm{LP}}|}$$

$$C_{11} = 0, \qquad R_4, R_5 \to \infty$$

For the BS case:

$$R_8 = \frac{R}{|H_{\mathrm{notch}}|}$$

$$R_4, R_5 \to \infty$$

EXAMPLE 8.4 Design a bandpass AM filter having a gain of $H_{\mathrm{BP}} = 3$, $Q_p = 10$, and $\omega_p = 2\pi \times 10^3 r/s$.

PROCEDURE

1 Choose $R = R' = 10\ \mathrm{k}\Omega$, and $C = 0.01\ \mu\mathrm{F}$. Thus we set $R_7 = R_1 = R_2 = 10\ \mathrm{k}\Omega$, and $C_9 = C_{10} = 0.01\ \mu\mathrm{F}$.

2 The pole frequency is given by $\omega_p = (1/aRC)$; then $a = (1/RC\omega_p) = 1.5915$.

3 The value for R_6 is

$$R_6 = aRQ_p = 159.15\ \mathrm{k}\Omega$$

4 For the bandpass case:

$$R_4 = \frac{R_6}{H_{\mathrm{BP}}} = 53.05\ \mathrm{k}\Omega$$

5 The remaining component values are

$$C_{11} = 0, \qquad R_5, R_8 \to \infty \qquad \text{and} \qquad R_3 = a^2 R_7 = 25.33\ \mathrm{k}\Omega$$

8.5 PRACTICAL LIMITATIONS

The frequency response of the filters discussed in the previous sections may be designed using commercial, internally compensated operational amplifiers. However, for high Q and/or high-frequency operation, the performance will be affected by the operational amplifier characteristics (see Chapter 4). However, within certain limits of ω_p and Q_p, the filters perform close to the ideal frequency response. Table 8.10 shows the Q_p and ω_p deviations, assuming a single-pole model for the amplifiers such that

$$A_i = \frac{\omega_{t_i}}{s}$$

where ω_{t_i} is the radian gain-bandwidth product of amplifier A_i.

Table 8.10 **Frequency limitations of filters: Q_{pa} is the actual Q and Q_p is the ideal one**

Filter structure	*For different op amps*	*For identical op amps*
KHN	$\frac{Q_{pa}}{Q_p} \cong 1 + Q_p\left(\frac{\omega_p}{\omega_{t_1}}\right)\left(1 + \frac{\omega_{t_1}}{\omega_{t_2}} + 2\frac{\omega_{t_1}}{\omega_{t_3}}\right)$	$\frac{Q_{pa}}{Q_p} \cong 1 + 4Q_p\frac{\omega_p}{\omega_t}$
	$\frac{\omega_{pa}}{\omega_p} \cong 1 - \frac{\omega_p}{\omega_{t_1}}\left(1 + \frac{1}{2}\frac{\omega_{t_1}}{\omega_{t_2}}\right)$	$\frac{\omega_{pa}}{\omega_p} = 1 - \frac{3}{2}\frac{\omega_p}{\omega_t}$
Tow–Thomas	$\frac{Q_{pa}}{Q_p} \cong 1 + Q_p\left(\frac{\omega_p}{\omega_{t_1}}\right)\left(1 + \frac{\omega_{t_1}}{\omega_{t_2}} + 2\frac{\omega_{t_1}}{\omega_{t_3}}\right)$	$\frac{Q_{pa}}{Q_p} \cong 1 + 4Q_p\frac{\omega_p}{\omega_t}$
	$\frac{\omega_{pa}}{\omega_p} \cong 1 - \frac{\omega_p}{\omega_{t_1}}\left(1 + \frac{1}{2}\frac{\omega_{t_1}}{\omega_{t_2}}\right)$	$\frac{\omega_{pa}}{\omega_p} = 1 - \frac{3}{2}\frac{\omega_p}{\omega_t}$
Akerberg–Mossberg	$\frac{Q_{pa}}{Q_p} \cong \frac{1}{\frac{\omega_{pa}}{\omega_p}\left[1 + Q_p\left(\frac{2\omega_p}{\omega_{t_3}} - \frac{\omega_p}{\omega_{t_2}} - \frac{\omega_p}{\omega_{t_1}}\right) - 2\left(1 + \frac{\omega_{pa}}{\omega_p}\right)Q_p\frac{\omega_p}{\omega_{t_1}}\left(1 + \frac{\omega_{t_1}}{\omega_{t_2}}\right)\right]}$	$\frac{Q_{pa}}{Q_p} = \frac{1}{\left(1 - \frac{3}{2}\frac{\omega_p}{\omega_t}\right)\left(1 - 6Q_p\frac{\omega_p^2}{\omega_t^2}\right)}$
	$\frac{\omega_{pa}}{\omega_p} \cong 1 - \frac{\omega_p}{\omega_{t_1}}\left(1 + \frac{1}{2}\frac{\omega_{t_1}}{\omega_{t_2}}\right)$	$\frac{\omega_{pa}}{\omega_p} \cong 1 - \frac{3}{2}\frac{\omega_p}{\omega_t}$

The KHN and TT topologies present more severe limitations than the AM filter regarding Q enhancement. For example, assuming identical op amps, if $f_p = 1\,\text{kHz}$, $\omega_t = 2\pi(1\,\text{MHz})$, and $Q = 25$, then for the KHN and TT the Q changes 10% and for the AM, only 0.13%.

In fact, the AM filter of Fig 8.6, with a proper design and assuming identical op amps, can yield Q_p independent of ω_t. A more detailed expression for Q_{pa}/Q_p than the one shown in Table 8.10 is the following:

$$\frac{Q_{pa}}{Q_p} \cong \frac{1}{1 + Q_p\omega_p\left(\dfrac{1 + R_1/R_2 + R_1/R_5}{\omega_{t_2}} - \dfrac{R_2/R_1}{\omega_{t_1}} - \dfrac{1 + C_{11}/C_{10}}{\omega_{t_3}}\right)}$$

Note that the signs of the individual terms in the factor in parentheses (which multiplies $Q_p\omega_p$ in the denominator of the above equation) are both negative and positive. Therefore, with an appropriate design procedure, it is possible to cancel the effects of Q_p on the gain bandwidths of the op amps. To illustrate this, let us assume that $\omega_{t_1} = \omega_{t_2} = \omega_{t_3} = \omega_t$ and define

$$k = \frac{R_2}{R_1} + \frac{C_{11}}{C_{10}} - \frac{R_1}{R_2} - \frac{R_1}{R_5}$$

Compensation is obtained for $k = 0$. With reference to Table 8.9, it is seen that $C_{11} = 0$ and $R_5 \to \infty$ are required for the low-pass and bandpass cases. In these cases,

$$R_1 = R_2$$

For the cases of $j\omega$-axis zeros and the high-pass, where R_4, $R_5 \to \infty$:

$$k = 0 \qquad \text{for} \qquad \frac{C_{11}}{C_{10}} = \frac{R_1}{R_2} - \frac{R_2}{R_1}$$

A suitable value of R_1/R_2 to make $k = 0$ is

$$\frac{R_1}{R_2} = \frac{C_{11}}{2C_{10}} + \sqrt{1 + \left(\frac{C_{11}}{2C_{10}}\right)^2}$$

However, R_1 and R_2 might differ in magnitude significantly and thus make it impossible to obtain a signal level as uniform as possible in the filter. An alternative procedure is to detach the terminal of R_5 from V_{in} and ground it. Thus, although k is still the same, R_5 performs as if it had the value of infinity in Table 8.9, and

$$k = 0 \qquad \text{for} \qquad R_1 = R_2 \quad \text{and} \quad \frac{C_{11}}{C_{10}} = \frac{R_1}{R_5}$$

Further consideration of these effects is provided in Chapter 11.

8.6 SUMMARY

Low sensitivities, relatively high Q_p, flexibility, and easy tuning can be obtained with any one of the state-variable structures discussed in this chapter. These filters perform well for certain limited values of Q_p and ω_p, as shown in Table 8.10. Integrators that offer reduced-gain-bandwidth dependence can be employed to increase the frequency of operation and/or the quality factor. For this, the reader is referred to Chapters 4 and 11 and the references in this chapter.

8.7 REFERENCES

[1] L. P. Huelsman and P. E. Allen, *Introduction to the Theory and Design of Active Filters*, New York: McGraw-Hill, 1980.

[2] A. S. Sedra and P. O. Brackett, *Filter Theory and Design: Active and Passive*, Champaign Ill., Matrix Publishers, 1978.

[3] R. L. Geiger and A. Budak, "Design of Active Filters Independent of First- and Second-Order Operational Amplifier Time Constant Effects," *IEEE Trans. Circuits and Systems*, vol. CAS-28, pp. 749–757, August 1981.

[4] R. W. Moon and F. W. Stephenson, "Second-Order Q and ω_O Shifts in RC Active Filters Having Nonideal Amplifiers," *Electronics Letters*, vol. 18, pp. 265–266, April 1982.

[5] P. B. Aronhime, "Effects of Finite Gain-Bandwidth Product on Three Recently Proposed Quadratic Networks," *IEEE Trans. Circuits and Systems*, vol. CAS-24, pp. 657–660, November 1977.

CHAPTER 9

Biquads II: The Current Generalized Immittance (CGIC) Structure

W. B. Mikhael
West Virginia University

9.1 INTRODUCTION

In this chapter a two-amplifier biquad is presented that is based on the Antoniou CGIC (see Chapter 4). Although several two-amplifier biquads have been reported, the one presented here has proved to be clearly superior to single-amplifier biquads. This is because the design carried out [1] was constrained by stringent performance criteria satisfying important properties and features such as stability, versatility, insensitivity to component tolerances and drift, low dependence on the op amp (OA) frequency limitations, finite gain, tunability, small spread in component values, and minimum total capacitance.

In addition, the performance of the 2-OA CGIC-based biquads presented here is comparable to multiple-OA biquads. On the other hand, the 2-OA CGIC biquad does not have as many OA outputs and consequently as many simultaneous filtering functions. Also the 2-OA CGIC biquad uses the OAs in the differential mode.

In the following discussion, the generalized structure of the 2-OA CGIC biquad is presented. Illustrative examples of the element identification to realize the most commonly used biquads are tabulated. Stability and sensitivity properties are discussed. A design procedure is described that minimizes the active sensitivities while maintaining the filter's stability. Several second-order design examples are given. In addition, a sixth-order Chebyshev LPF and a sixth-order elliptic BPF are presented as illustrations of higher-order structures utilizing the suggested design values and tuning procedure. The excellent performance of the resulting realizations is experimentally verified. A universal 2-OA GIC hybrid implementation using thick film technology is also described.

9.2 BIQUADRATIC STRUCTURE USING THE ANTONIOU CGIC

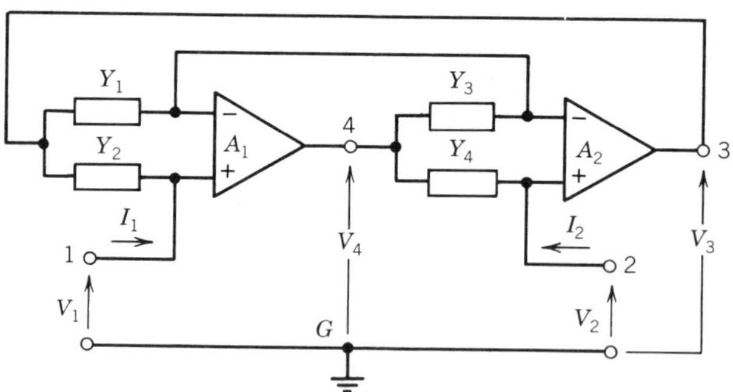

Figure 9.1 Antoniou's CGIC with additional ports 3G and 4G.

Consider the network of Fig. 9.1, which is simply Antoniou's CGIC [2], with two new ports created across 3-G and 4-G. This is represented symbolically in Fig. 9.2.

A new configuration is obtained, as shown in Fig. 9.3. A synthesis procedure is now described that uses this configuration. The transfer functions between the input and output terminals 2, 3, and 4, assuming ideal OAs are readily obtained using the methods of Chapter 5 as:

$$\frac{V_3}{V_i} = T_1 = \frac{Y_5 + h(s)[Y_7(1 + Y_6/Y_2) - Y_5 Y_8/Y_2]}{D(s)} \tag{9.1a}$$

$$\frac{V_4}{V_i} = T_2 = \frac{Y_5(1 + Y_8/Y_4) - Y_6 Y_7/Y_4 + h(s) Y_7}{D(s)} \tag{9.1b}$$

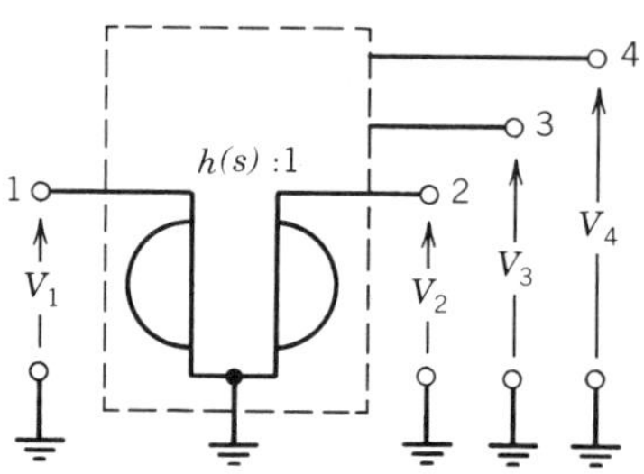

Figure 9.2 Symbolic representation of the CGIC in Fig. 9.1.

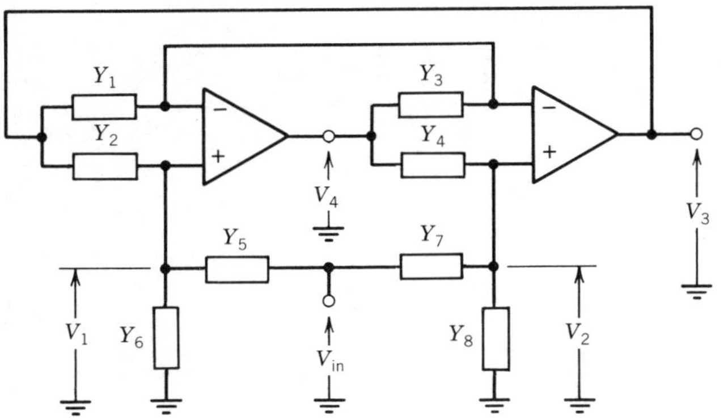

Figure 9.3 *The basic configuration.*

$$\frac{V_2}{V_i} = T_3 = \frac{Y_5 + h(s)Y_7}{D(s)} \tag{9.1c}$$

where

$$h(s) = \frac{Y_2 Y_3}{Y_1 Y_4}$$

$$D(s) = (Y_5 + Y_6) + h(s)(Y_7 + Y_8) \tag{9.1d}$$

The conversion function $h(s)$ and the admittances Y_5 to Y_8 can be selected in many different ways, and it is clear that any stable second-order transfer function with any desired zero and pole locations can be realized.

Letting

$$Y_i = sC_i + G_i \tag{9.2}$$

where $i = 1$ to 4, we have from Eq. 9.1

$$h(s) = \frac{(sC_2 + G_2)(sC_3 + G_3)}{(sC_1 + G_1)(sC_4 + G_4)} \tag{9.3}$$

Clearly, by omitting one or more conductances and/or capacitances, a number of specific conversion functions can be generated.

9.3 REALIZATION OF THE MOST COMMONLY USED BIQUADS

The most frequently used second-order transfer functions have already been described in Chapter 3. Realizations of these functions are given in Table 9.1. The LP, HP, BP, and BS sections are produced by choosing $h(s)$ in a simple manner such as $K_1 s$, $K_2 s^2$, $K_3 s + K_4 s^2$, or their reciprocals. Circuits 3, 4, and 7 can be regarded as realizations of simple RLC networks [2, 3]. Second-order AP sections can be obtained from circuits 11 and 12 of Table 9.1.

Table 9.1 **Element identification for realizing the most commonly used transfer functions**

Circuit number	$h(s)$	Y_1	Y_2	Y_3	Y_4	Y_5	Y_6	Y_7	Y_8	Transfer function	Remarks
1	$\frac{sC_2(sC_3+G_3)}{G_1G_4}$	G_1	sC_2	sC_3+G_3	G_4	G_5	0	0	G_8	$T_2=\frac{G_1G_5(G_4+G_8)}{G_1G_5G_4+sC_2G_3G_8+s^2C_2C_3G_8}$	LP
2	$\frac{G_2G_3}{sC_1(sC_4+G_4)}$	sC_1	G_2	G_3	$sC_4+G_4^*$	0	G_6	G_7	$sC_8^*+G_8^*$	$T_1=\frac{G_2G_3G_7\left(1+\frac{G_6}{G_2}\right)}{G_2G_3(G_7+G_8)+s(C_2G_4G_6+C_8G_2G_3)+s^2C_1C_4G_6}$	LP
3	$\frac{sC_3G_2}{G_1G_4}$	G_1	G_2	sC_3	G_4	0	G_6	sC_7	$sC_8^*+G_8$	$T_1=\frac{s^2C_3C_7(G_2+G_6)}{G_6G_1G_4+sC_3G_2G_8+s^2(C_7+C_8)G_2C_3}$	HP
4	$\frac{sC_3G_2}{G_1G_4}$	G_1	G_2	sC_3	G_4	0	G_6	$\frac{sC_7}{1+sC_7R_7}$	0	$T_1=\frac{s^2C_3C_7(G_2+G_6)}{G_6G_1G_4+sC_7G_1G_4G_6R_7+s^2C_3C_7G_2}$	HP
5	$\frac{s^2C_2C_3}{G_1G_4}$	G_1	sC_2	sC_3	G_4	sC_5	G_6	0	G_8	$T_2=\frac{sC_5\left(1+\frac{G_8}{G_4}\right)G_1G_4}{G_1G_4G_6+sC_5G_1G_4+s^2C_2C_3G_8}$	BP
6	$\frac{G_2G_3}{sC_1(sC_4+G_4)}$	sC_1	G_2	G_3	$sC_4+G_4^*$	0	G_6	sC_7	G_8	$T_1=\frac{sC_7G_2G_3\left(1+\frac{G_6}{G_2}\right)}{G_2G_3G_8+s(C_7G_2G_3+C_1G_4G_8)+s^2C_1C_4G_6}$	BP

7	$\frac{sC_3G_2}{G_1G_4}$	G_1	G_2	sC_3	G_4	0	G_6	G_7	$sC_8 + G_8^*$	$T_1 = \frac{sC_3G_7(G_2 + G_6)}{G_1G_4G_6 + sC_3G_2(G_7 + G_8) + s^2C_3C_8G_2}$	BP
8	$\frac{G_2G_3}{s^2C_1C_4}$	sC_1	G_2	G_3	sC_4	G_5	0	G_7	sC_8	$T_2 = \frac{s^2C_1G_5(C_4 + C_8) + G_2G_3G_7}{s^2C_1G_5C_4 + sC_8G_2G_3 + G_2G_3G_7}$	N
9	$\frac{sG_2C_3}{G_1G_4}$	G_1	G_2	sC_3	G_4	G_5	0	sC_7	G_8	$T_2 = \frac{s^2C_3C_7G_2 + G_1G_5(G_4 + G_8)}{s^2C_3C_7G_2 + sC_3G_2G_8 + G_1G_4G_5}$	N
10	$\frac{sG_2C_3}{G_1G_4}$	G_1	G_2	sC_3	G_4	G_5	G_6^*	sC_7	$sC_8^* + G_8$	$T_3 = \frac{\left(s^2 + \frac{G_1G_4G_5}{C_3C_7G_2}\right)\left(\frac{C_7}{C_7 + C_8}\right)}{s^2 + s\frac{G_8}{C_7 + C_8} + \frac{G_1G_4(G_5 + G_6)}{C_3G_2(C_7 + C_8)}}$	N
11	$\frac{G_2G_3}{s^2C_1C_4}$	sC_1	G_2	G_3	sC_4	G_5	G_6^*	G_7	sC_8	$T_1 = \frac{s^2C_1C_4G_5 - sG_5C_8G_3 + G_3G_7(G_2 + G_6)}{s^2C_1C_4(G_5 + G_6) + sC_8G_2G_3 + G_2G_3G_7}$	Nonminimum phase For all pass: $G_6 = 0$ $G_5 = G_2$
12[†]	$\frac{sG_2C_3}{G_1G_4}$	G_1	G_2	sC_3	G_4	G_5	G_6	sC_7	G_8	$T_1 = \frac{s^2C_3C_7(G_2 + G_6) - sC_3G_5G_8 + G_1G_4G_5}{s^2C_3C_7G_2 + sC_3G_2G_8 + (G_5 + G_6)G_1G_4}$	Nonminimum phase For all pass: $G_6 = 0$ $G_2 = G_5$

* These elements can be set equal to zero.

[†] A special case of Circuit 12—namely, the all-pass case—has also been independently proposed by J.T. Lim, Bell Northern Research, Canada, as communicated privately by him to Dr. A. Antoniou.

Figures 9.1 and 9.2 and Table 9.1 show that, with the exception of circuit 10, the response is obtained from the output of an OA. Owing to the low output resistance of the amplifier, any number of sections can thus be cascaded without using isolation amplifiers.

9.4 STABILITY DURING ACTIVATION AND SENSITIVITY

Stability Properties

It has been shown [4] that some networks using CGICs can be conditionally stable; that is, a circuit can lock in an unstable mode during activation (just after switching on the power supply).

For amplifiers with a finite open loop gain A, the circuit of Fig. 9.3 gives

$$\frac{V_k}{V_i} = \frac{N_k(s)}{D(s)}$$

where $k = 2, 3, 4$ and

$$D(s) = M_1 Y_1 + M_2 Y_3 + (1 + M_1)(1 + M_2)\left[\frac{Y_1}{A_1} + \frac{Y_3}{A_2} + \frac{Y_1}{A_1 A_2} + \frac{Y_3}{A_1 A_2}\right] \qquad (9.4)$$

$$M_1 = \frac{Y_5 + Y_6}{Y_2}$$

$$M_2 = \frac{Y_7 + Y_8}{Y_4}$$

It can easily be shown that for the circuit in Fig. 9.3, low-frequency unstable modes cannot arise during activation. This is due to the absence of differences and changes of sign in the denominator coefficients. Thus any combination of OA gains that occur during transients and power supply switching does not result in saturation or low-frequency instability before reaching the steady state [5].

Sensitivity Analysis

The pole Q factor Q_p, the undamped frequency of oscillation ω_p, the notch frequency ω, and the multiplier constants H_{BS}, H_{LP}, H_{HP}, and H_{BP}, as well as Q_Z, ω_Z, and H_{AP}, have been defined in Chapter 3. The sensitivity of a quantity x with respect to variations in an element e is denoted by S_e^x. For ideal amplifiers, the use of Table 9.1 leads to

$$0 \leq S_e^x \leq 1 \qquad (9.5)$$

where x represents any one of these H, ω, or Q quantities and e represents any capacitance or conductance.

9.5 DESIGN AND TUNING PROCEDURE

A design procedure is now described. Table 9.1 shows that there are several degrees of freedom in the choice of element values. These may be used to minimize $S_A^{Q_p}$ or the spread of element values. By using minimum sensitivity (to the OA parameters) constraints in circuits 1, 3, 7, 10, and 12, possible sets of element values for LP, HP, BP, BS, and AP sections have been obtained, as shown in Table 9.2. In addition, this choice of elements guarantees stable operation with real 0As. It is seen that the notch frequency ω_n, the undamped frequency of oscillation ω_p and the Q-factor Q_p can easily be adjusted by sequentially trimming three distinct resistors. A tuning sequence is also given in Table 9.2.

It is to be noted from Table 9.2, for the design of BS sections, that

$$\omega_n^2 = \frac{\omega_p G_5}{C_7} \tag{9.6}$$

where

$$G_5 + G_6 = G_1, \qquad C_7 + C_8 = C \qquad \text{and} \qquad \omega_p = \frac{G}{C}$$

Hence

$$\frac{\omega_n^2}{\omega_p^2} = \frac{G_5}{G} \cdot \frac{C}{C_7} \tag{9.7}$$

where

$$\frac{G_5}{G} \text{ is always } \leq 1 \qquad \text{and} \qquad \frac{C}{C_7} \text{ is always } \geq 1$$

It is seen from Eq. 9.7 that for $\omega_n \leq \omega_p$, C_8 can be set to zero ($C_7 = C$), and only two capacitors per section are used. For $\omega_n > \omega_p$, three capacitors are required. In all cases, regardless of the choice of C_7 and C_8, the total capacitance per section is equal to $2C$.

In most applications where notch sections are used, ω_n/ω_p is close to unity and care should be taken when choosing G_5 and C_7 in Eq. 9.7. A suitable choice of G_5, G_6, C_7, and C_8 values may be obtained by letting

$$\frac{G_5}{G} = \frac{1}{K} \qquad \text{and} \qquad \frac{C}{C_7} = 2$$

Thus K is approximately equal to 2($K < 2$ for $\omega_n > \omega_p$ and $K > 2$ for $\omega_n \leq \omega_p$). The suggested choice yields a capacitor spread of 2 and both R_5 and R_6 are approximately equal to $2R$.

Table 9.2 **Design values and tuning procedure**

Circuit number (*from Table* 9.1)	Design values	Transfer function realized	Tuning sequence ω_n	Tuning sequence ω_p	Tuning sequence Q_p
1	$G_1 = G_4 = G_5 = G_8 = G$, $G_3 = G/Q_p$, $C_2 = C_3 = C$, where $C = \dfrac{G}{\omega_p}$	$T_2 = \dfrac{2\omega_p^2}{D(s)}$	—	G_8	G_3
3	$G_1 = G_2 = G_4 = G_6 = G$, $G_8 = G/Q_p$, $C_8 = 0$, $C_3 = C_7 = C$, where $C = \dfrac{G}{\omega_p}$	$T_1 = \dfrac{2s^2}{D(s)}$	—	G_4	G_8
7	$G_1 = G_2 = G_4 = G_6 = G$, $G_7 = G/Q_p$, $G_8 = 0$, $C_3 = C_8 = C$, where $C = \dfrac{G}{\omega_p}$	$T_1 = \dfrac{(2\omega_p/Q_p)s}{D(s)}$	—	G_2	G_7
10	$G_1 = G_2 = G_4 = G_5 + G_6 = G$, $G_8 = G/Q_p$, $C_3 = C_7 + C_8 = C$, where $\omega_p = \dfrac{G}{C}$ and $\omega_n^2 = \omega_p \dfrac{G_5}{C_7}$	$T_3 = \dfrac{C_7}{C} \dfrac{(s^2 + \omega_n^2)}{D(s)}$	G_2	G_6	G_8
12	$G_1 = G_2 = G_4 = G_5 = G$, $G_6 = 0$, $C_3 = C_7 = C$, $G_8 = G/Q_p$, where $\omega_p = \dfrac{G}{C}$	$T_1 = \dfrac{D(-s)}{D(s)}$		G_4	G_8

Note: $D(s) = s^2 + (\omega_p/Q_p)s + \omega_p^2$.

9.6 DESIGN EXAMPLES

EXAMPLE 9.1 Design a second-order Butterworth ($Q_p = 0.707$) LP filter having a cutoff frequency $f_p = 20{,}000/2\pi$ Hz.

PROCEDURE

1 Circuit 1 in Table 9.2 realizes a LPF. The design equations are also given in Table 9.2.

2 First we choose an appropriate value for C, say 10 nF. Thus $C = C_2 = C_3 = 10$ nF.

3 Now,

$$R = \frac{1}{C\omega_p} = \frac{1}{20{,}000 \times 10^{-8}}$$

Therefore, $R = 5\ \text{k}\Omega$

4 Consequently, $R = R_1 = R_4 = R_5 = R_8 = 5\ \text{k}\Omega$ and

$R_3 = RQ_p = 3.535\ \text{k}\Omega$

5 The circuit is shown in Fig. 9.4*a*. It is to be noted that the low-frequency gain of the LP filter, H_{LP}, is 2.

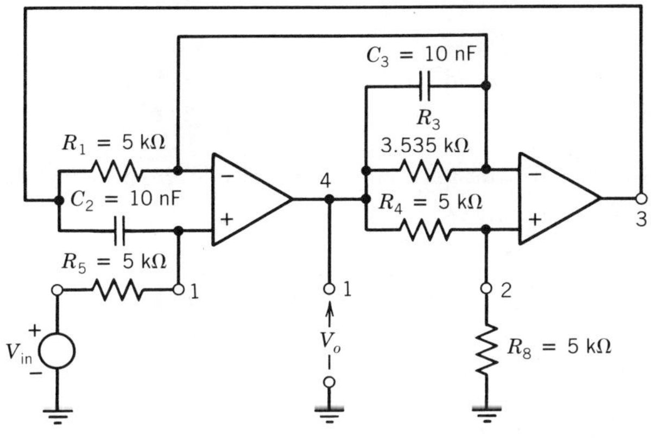

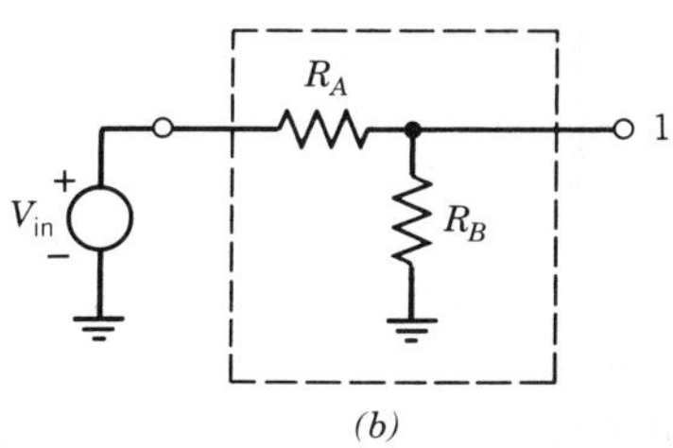

Figure 9.4 (a) Second-order Butterworth LPF design example; (b) controlling the gain factor of the LPF.

A simple procedure can be followed to scale H_{LP} by a factor x less than unity, that is, effectively multiplying the transfer function realized by x. This is done by replacing the resistance R_5 by two resistors R_A and R_B (in series with the input V_{in}) in the manner shown in Fig. 9.4*b*, where

$$\begin{aligned} R_5 &= R = 5\ \text{k}\Omega \\ &= R_A \| R_B \end{aligned}$$

The desired gain and scale factor $x = R_B/(R_A + R_B)$. Thus for $x = \frac{1}{2}$, resulting in a dc gain of the LP filter of unity, the choice of resistors R_A and R_B is $R_A = R_B = 10\ \text{k}\Omega$.

If functional tuning of the filter is desired, the tuning sequence of circuit 1 in Table 9.2 can be followed. First, ω_p is adjusted by applying a sinusoidal input at the desired ω_p frequency. Then R_8 is tuned until ω_p realized equals the desired value. This can be detected by monitoring the phase angle of the output relative to the input. When the proper ω_p is reached, the output lags the input by 90°. Next, to adjust Q_p, the filter gain H_{dc} of the LPF at a frequency much lower than ω_p is determined. Then an input at ω_p is applied. R_3 is adjusted until the gain of the LPF at ω_p is Q_p desired times H_{dc}.

EXAMPLE 9.2 Design a second-order BP filter with $Q_p = 10$ and $f_p = 10{,}000/2\pi$ Hz.

PROCEDURE

1 Circuit 7 in Table 9.2 realizes a BP filter. The design equations are also given in Table 9.2.

2 First we choose a suitable value for C, say 10 nF.

3 Thus $C_3 = C_8 = C = 10$ nF.

4 Hence $R = (1/C\omega_p) = 10\ \text{k}\Omega$. Consequently, $R = R_1 = R_2 = R_4$, $R_6 = 10\ \text{k}\Omega$ and $R_7 = RQ_p = 100\ \text{k}\Omega$.

5 The circuit is shown in Fig. 9.5. The gain at resonance, that is, at $\omega = \omega_p$, is equal to 2. To scale the gain by a factor x less than 2, the resistor R_7 is split into two resistors in a manner similar to that in Fig. 9.4*b* and explained in Example 9.1.

Again, if functional tuning is desired, the sequence in Table 9.2 can be followed.

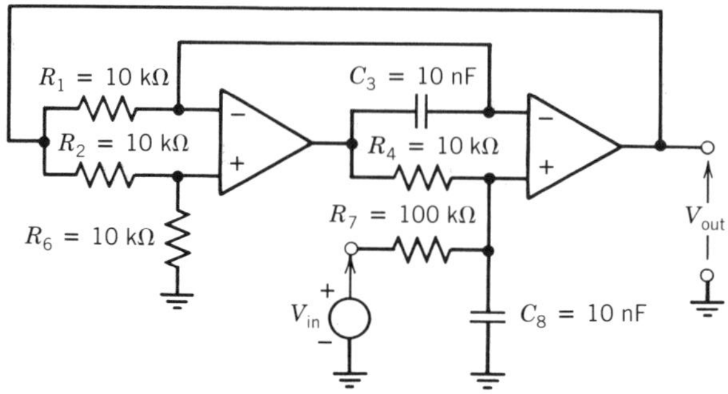

Figure 9.5 Design of a second-order BPF.

***EXAMPLE* 9.3** Design a second-order HP filter with $Q_p = 1$ and $f_p = 10{,}000/2\pi$ Hz.

PROCEDURE

1 Circuit 3 in Table 9.2 realizes an HP filter.

2 Let us choose $C_3 = C_7 = C = 5$ nF. Hence R can be computed as

$$R = \frac{1}{\omega_p C} = \frac{1}{10{,}000 \times 5 \times 10^{-9}}$$
$$= 20 \text{ k}\Omega$$

3 Consequently, $R_1 = R_2 = R_4 = R_6 = R = 20$ kΩ and $R_8 = RQ_p = 20$ kΩ.

4 The realization is shown in Fig. 9.6. The gain at high-frequency, H_{HP}, is equal to 2.

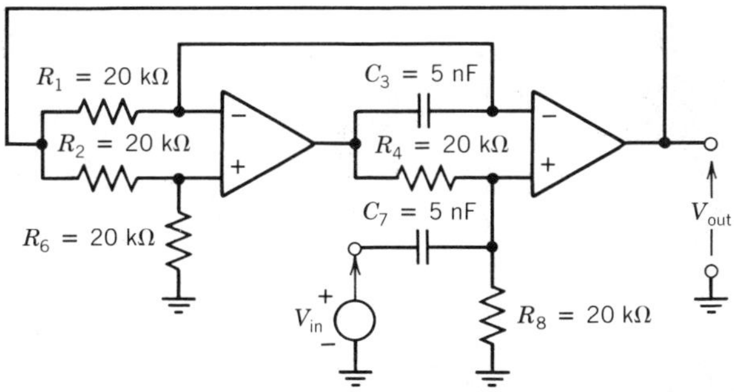

Figure 9.6 Design of a second-order HPF.

9.7 PRACTICAL HIGH-ORDER DESIGN EXAMPLES USING BIQUAD II

Using Table 9.2, ±1% metal film resistors, ±2% polystyrene capacitors, and μA741 OAs, a sixth-order Chebyshev LP filter and a sixth-order elliptic BP filter were designed and constructed. The LP filter has a maximum passband attenuation of 1.0 dB; bandwidth = 3979 Hz. The BP filter has the following specifications:

Center frequency = 1500 Hz
Passband = 60 Hz
Maximum passband attenuation = 0.3 dB
Minimum stopband attenuation outside the frequency range
    1408 → 1595 Hz = 38 dB

Low-Pass Filter

The realization uses cascaded sections of type 1, in Table 9.2, as shown in Fig. 9.7*a*. The measured frequency response (input level = 50 mV), shown in Figs. 9.7*b* and *c*, agrees with the theoretical response. The effect of dc-supply variations is illustrated in Fig. 9.7*d*. The deviation in the passband ripple is about 0.1 dB for

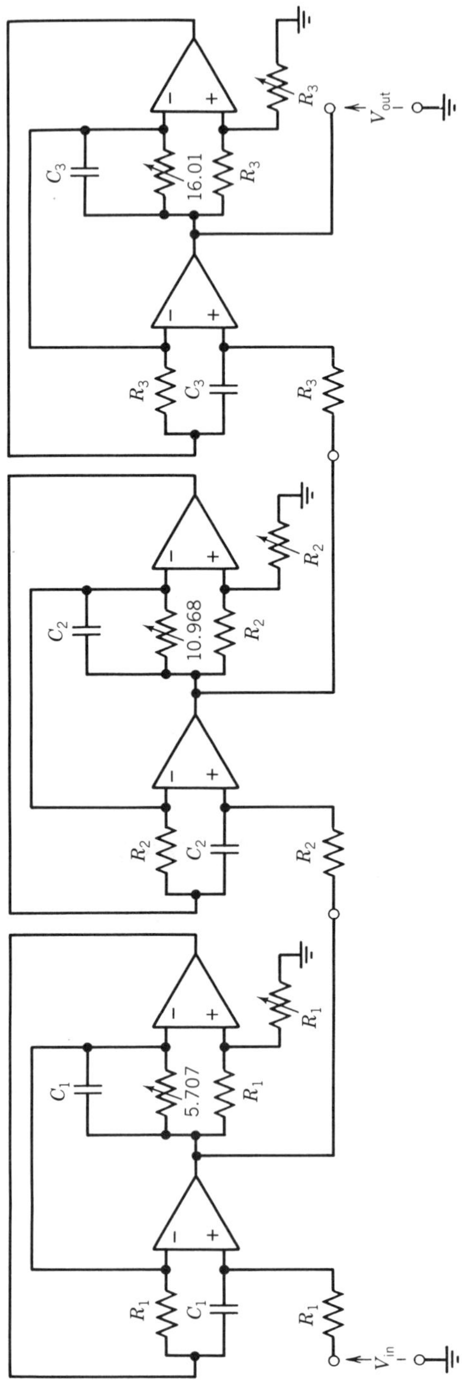

Figure 9.7(a) *Realization of the sixth-order Chebyshev low-pass filter.*

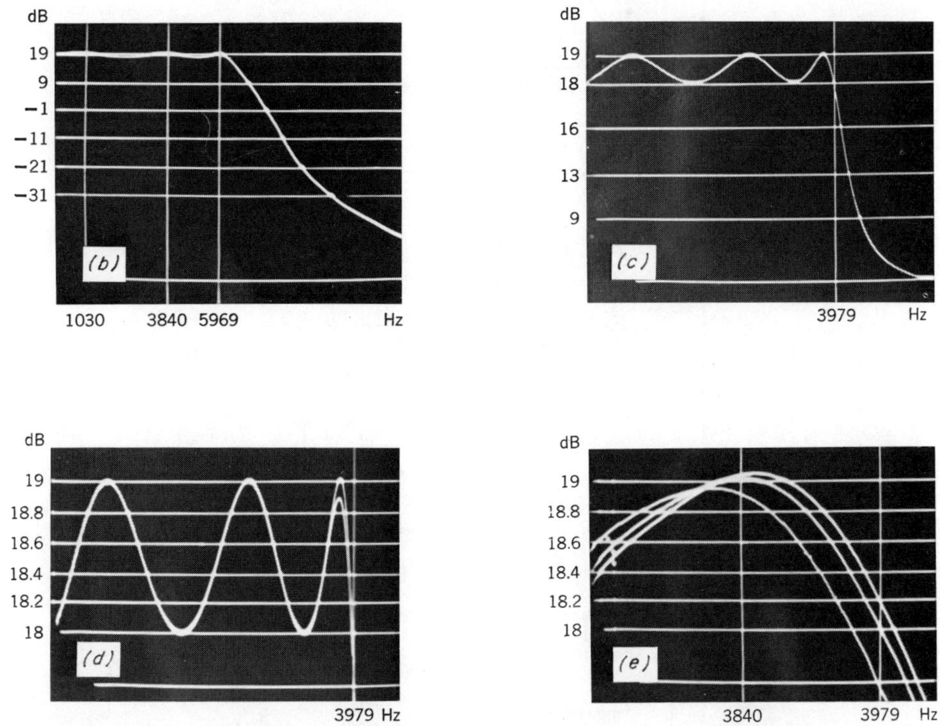

Figure 9.7b to e *Frequency responses: (b) Logarithmic gain scale and linear frequency scale; (c) linear gain and frequency scales; (d) for supply voltages ±5 V (lower curve) and ±15 V, input level = 0.05 V; (e) at temperatures −10°C (right-hand curve), 20°C and 70°C (hand-left curve).*

supply voltages in the range ±5 to ±15 V. The effect of temperature variations is illustrated in Fig. 9.7*e*, which shows the frequency response at −10°C (right-hand curve), 20°C, and 70°C (left-hand curve). The last peak has been displaced horizontally by 42 Hz, which corresponds to a change of 133 ppm/°C. The frequency displacement is due to passive element variations and is within the predicted value.

Bandpass Filter

The realization uses cascaded sections of the types 7 and 10, in Table 9.2, and is shown in Fig. 9.8*a*. The measured frequency response is shown in Figs. 9.8*b* and *c*, and it is in agreement with the theoretical response. Figure 9.8*d* shows the frequency response for supply voltages of ±7.5 V (lower curve) and ±15 V; the input is 0.3 V. The passband ripple remains less than 0.39 dB and the deviation in the

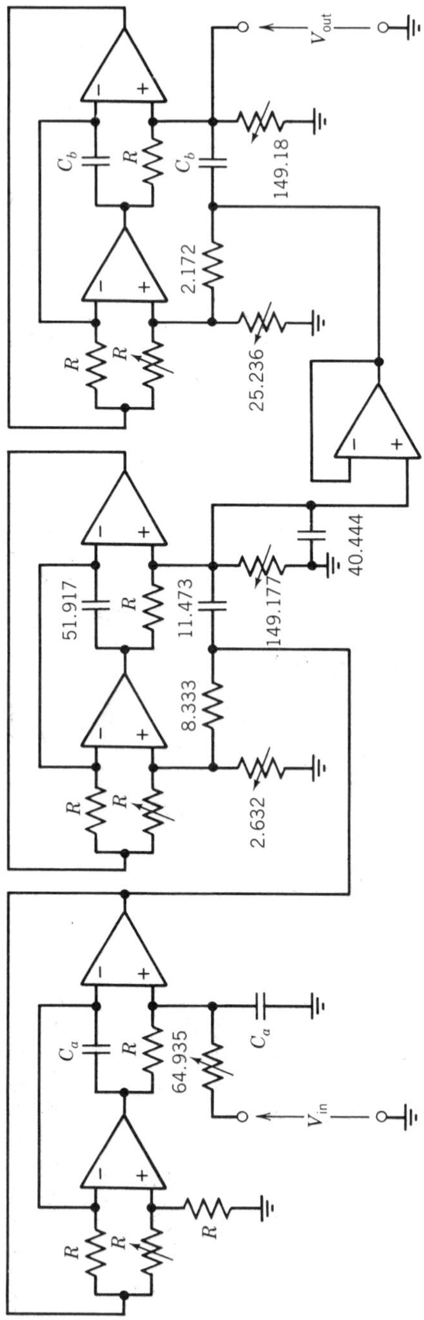

Figure 9.8(a) Realization of the sixth-order elliptic bandpass filter.

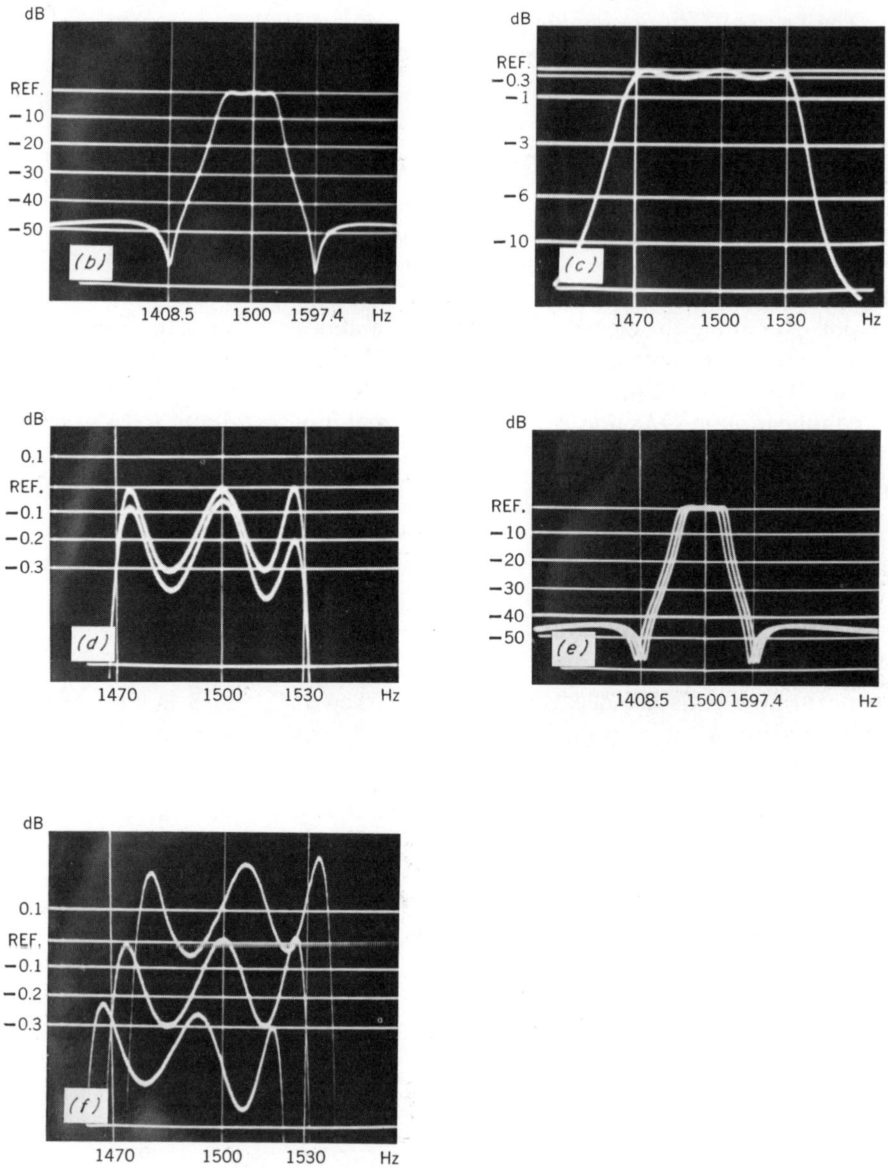

Figure 9.8b to f *Frequency responses of bandpass filter: (b) logarithmic gain and linear frequency scales; (c) linear gain and frequency scales; (d) for supply voltages of* ± 7.5 *V (lower curve) and* ± 15 *V, input level of 0.3 V; (e) at temperatures of 10°C (right-hand curve), 20°C, and 70°C (left-curve); and (f) expanded passband of Fig. 9.8(e).*

stopband is negligible. Figures 9.8*e* and *f* illustrate the effect of temperature variations. The passband ripple remains less than 0.35 dB in the temperature range −10 to 70°C. A center frequency displacement of 15 Hz has been measured, which corresponds to a change of 125 ppm/°C.

9.8 UNIVERSAL 2-OA CGIC BIQUAD

Study of Table 9.2 suggests that several circuits may be combined to form a universal biquad. This can be achieved on a single substrate using thick-film technology.

Upon examining the element identification and design values in Table 9.2, it is easy to see that one common thick-film substrate can be made to realize circuits 1, 3, 7, and 10 in Table 9.2 (other circuits from Table 9.1 can be included if desired) with no duplication in OAs and chip capacitors and minimal duplication in resistors. The superposition of circuits 1, 3, 7, and 10 is shown in Fig. 9.9. The following items should be noted.

1 Each resistor having the same subscript represents one resistor only and needs to appear once in a given biquad realization and thus once on the substrate. As an example, for $R_J = RQ_p$, only one R_J is needed with connections to several nodes. The unwanted connections may be opened during the trimming process according to the type of circuit required.

2 Three capacitor pads are needed; they are marked 1, 2, and 3 in Fig. 9.9. To obtain capacitor 4, either capacitor 2 or 3 connections are made common with capacitor 4 terminals. The capacitor pad terminals are available on the external

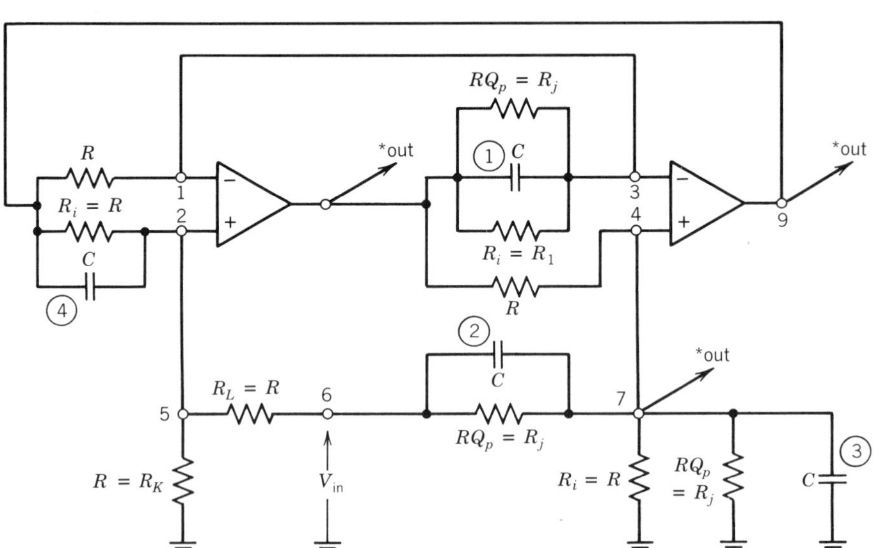

*Figure 9.9 Superposition of circuits 1, 3, 7, and 10 from Table 9.2 Note: * = output terminals.*

Figure 9.10 Dual CGIC Universal Biquad implemented using thick-film resistors, chip NPO capacitors and one quad OA.

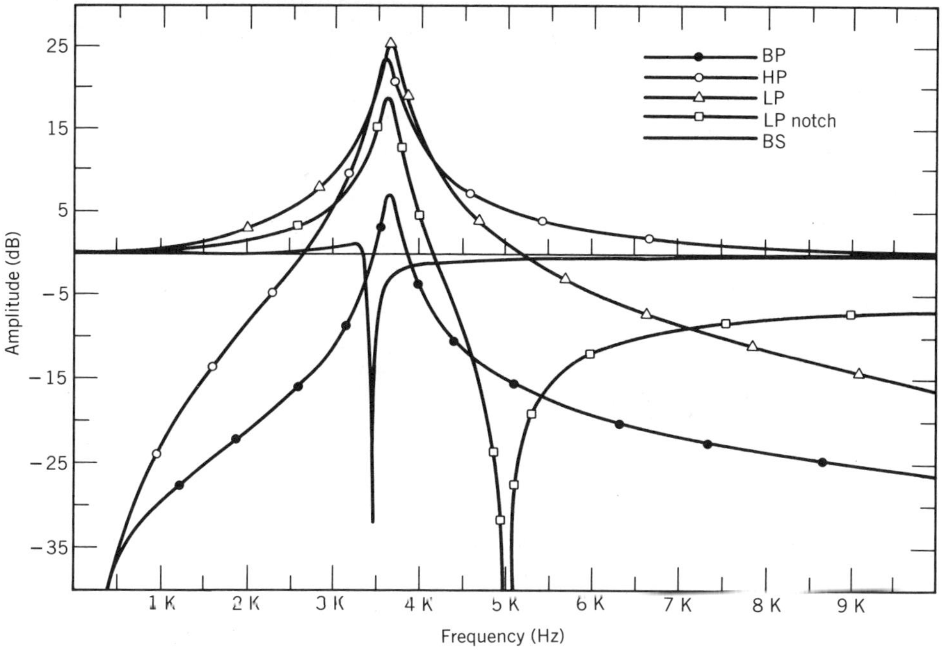

Figure 9.11 Different second-order realizations obtained from the Universal CGIC Biquad.

terminals of the substrate.* The chip capacitors are reflow-soldered in the appropriate locations based on the circuit realized.

A dual CGIC Universal Biquad implemented using thick-film resistors, chip NPO capacitors, and one quad OA is shown in Fig. 9.10. Note that this hybrid array is capable of realizing gyrators and simulating inductors and super capacitors. Sample results are given in Fig. 9.11 for realizing different biquadratic functions using this implementation.

* This can readily be understood by examining the different realizations to be obtained from this layout.

9.9 SUMMARY

A configuration has been presented for the synthesis of RC-active networks. This has been used to design a number of universal second-order sections such as low-pass, high-pass, bandpass, and bandstop sections. By using these sections, most of the practical filter specifications can be realized. Each section employs a CGIC, which can be implemented by using only 2 op amps. The sensitivities of Q_p, ω_p, Q_z, ω_z, and also the multiplier constant of the realization, have been found to be low with respect to the passive- and active-element variations. A simple functional tuning sequence in which only resistors are adjusted has been described for circuits 1, 3, 7, and 10. With the exception of one of the notch sections in the 12 circuits considered, the output can be located at an OA output. Consequently, these sections can be cascaded without the need for isolating amplifiers. Although exact matching of the gain bandwidth products of the two OAs in the CGIC biquad is not essential, optimum results are obtained when they are matched within practical ranges. This is achieved easily by using dual or quad OAs. Also, the choice of the conversion function and the element identification in a CGIC biquad are very important for obtaining realizations with high-performance properties; without these, general CGIC-based biquads are not guaranteed to have good properties.

9.10 REFERENCES

[1] W. B. Mikhael and B. B. Bhattacharyya, "A Practical Design for Insensitive RC-Active Filters," *IEEE Trans. on Circuits and Systems, CAS-21(1)*, pp. 75–78, 1974.

[2] A. Antoniou, "Realization of Gyrators Using Operational Amplifiers and Their Use in RC-Active Network Synthesis," *Proc. IEE*, vol. 116, pp. 1838–1850, November 1969.

[3] J. Valihora, "Modern Technology Applied to Network Implementation," *IEEE International Symposium on Circuit Theory*, pp. 169–173, April 1972.

[4] A. Antoniou, "Stability Properties of Some Gyrator Circuits," *Electron. Lett.*, vol. 4, pp. 510–512, 1968.

[5] B. B. Bhattacharyya, Wasfy B. Mikhael, and A. Antoniou, "Design of RC-Active Networks by Using Generalized Immittance Converters," *IEEE International Symposium on Circuit Theory*, pp. 290–294, April 1973.

CHAPTER 10

Design of High-Order Active Filters

Bert D. Nelin

Siemens Communication Systems, Inc.
Boca Raton, Florida

10.1 INTRODUCTION

The previous chapters have dealt with the design of first- and second-order filter sections. In this chapter, we present design methods for filters of higher order. The first method to be discussed is the so-called *cascade design* method outlined in Chapter 1. This method is the most commonly used because of its simplicity. It is based on coupling first- and second-order filter sections in such a manner that each section does not interact with the others. Thus the variations in the active and passive components affect only the frequency characteristic of the section that contains the particular component. Because of this, the effect of component variations on the overall frequency response of the filter is large. This may be tolerated for applications where only a few filters are constructed and manual tuning is used, or where the filter specifications are not particularly tight. However, when the filters are being produced in large quantities and according to stringent specifications, the cascade design method is not recommended.

As indicated in Chapter 1, the frequency response of a doubly terminated LC ladder filter exhibits a remarkably low sensitivity to component variations. This can be explained by the fact that the components of this type of filter are not buffered. Thus, they interact strongly with each other in such a manner that the effect of component variations on the overall frequency response is minimized.

Design methods based on the simulation of LC ladder filters will result in active filters that usually are superior to those designed using the cascaded method. (Later on in this chapter, we present active-filter design methods based on LC ladder filter structures.)

10.2 CASCADED ACTIVE FILTER DESIGN

This method begins with the pole and zero locations of the transfer function that approximates the given filter specifications. These poles and zeros can be obtained from published tables [1] or from those presented in Chapter 13. This information makes it possible to express the transfer function of the filter as the product of first- and second-order factors of the following forms.

1. First-order low-pass section:

$$T_1(s) = K_1 \frac{\omega_p}{s + \omega_p} \tag{10.1}$$

2. First-order high-pass section:

$$T_2(s) = K_2 \frac{s}{s + \omega_p} \tag{10.2}$$

3. Second-order low-pass section:

$$T_3(s) = K_3 \frac{\omega_p^2}{[s^2 + (\omega_p/Q_p)s + \omega_p^2]} \tag{10.3}$$

4. Second-order low-pass section with transmission zero (notch):

$$T_4(s) = K_4 \frac{(\omega_p^2/\omega_z^2)(s^2 + \omega_z^2)}{[s^2 + (\omega_p/Q_p)s + \omega_p^2]} \qquad \omega_z \geq \omega_p \tag{10.4}$$

5. Second-order high-pass section:

$$T_5(s) = K_5 \frac{s^2}{[s^2 + (\omega_p/Q_p)s + \omega_p^2]} \tag{10.5}$$

6. Second-order high-pass section with transmission zero (notch):

$$T_6(s) = K_6 \frac{s^2 + \omega_z^2}{[s^2 + (\omega_p/Q_p)s + \omega_p^2]} \qquad \omega_z \leq \omega_p \tag{10.6}$$

7. Bandpass section:

$$T_7(s) = K_7 \frac{(\omega_p/Q_p)s}{[s^2 + (\omega_p/Q_p)s + \omega_p^2]} \tag{10.7}$$

Here ω_p and Q_p are the radian frequency and Q value of the pole, and ω_z is the radian frequency of the transmission zero (notch), as described in Chapter 3, Section 3.3. The various Ks are gain factors. K_1, K_3, and K_4 are dc gains of low-pass sections; K_2, K_5, and K_6 are high-frequency ($\omega \gg \omega_p$) gains of high-pass sections; and K_7 is the gain at resonance ($\omega = \omega_p$) of the bandpass section.

In determining the order of the sections and the pairing of poles and zeros, it is important to consider the occurrence of "internal resonances." This refers to the phenomenon of large-voltage buildup at certain frequencies at internal nodes of the filter (usually at the output terminal of an op amp). Thus, the gain to one of these nodes from the input of the filter may be higher than the gain to the output of the filter. Consequently, clipping may occur at the output of an internal op amp before the filter output voltage shows clipping. This internal clipping will show up as a level-dependent change in the overall frequency characteristic and will restrict the range of input signal levels that the filter can handle.

In order to achieve maximum dynamic range, the filter should be designed so that clipping will first occur at the output of the last op amp of the filter. The following rules of thumb are useful in maximizing the dynamic range of cascaded design filters.

Rule 1.

Place the sections in the order of increasing Q values. (Here a first-order section is assumed to have a Q of zero.)

Rule 2.

Pair each pole with that zero (notch) which is closest in frequency, starting with the pole having the highest Q.

Rule 3.

Distribute the overall gain equally (in dB) among the sections.

In certain cases the active filter is required to reject interfering signals (in the stopband) that may have much greater amplitudes than the largest signal in the passband. In such cases it may be necessary to disobey rules 1 and 2 and begin the filter with a second-order section that has the transmission zero closest to the interfering signal and with the lowest Q.

EXAMPLE 10.1 Design a five-pole low-pass filter having dc a gain of 20 dB and the following poles and zeros.

$$f_{p1} = 3047 \text{ Hz} \qquad Q_{p1} = 6.43$$

$$f_{p2} = 2297 \text{ Hz} \qquad Q_{p2} = 1.35$$

$$f_{p3} = 1299 \text{ Hz}$$

$$f_{z1} = 4500 \text{ Hz}$$

$$f_{z2} = 6700 \text{ Hz}$$

PROCEDURE

1 The filter will consist of three cascaded sections: one first-order low-pass section and two second-order low-pass sections. The dc gain of each section is $20/3 = 6.67$ dB or $K = \text{antilog}(6.67/20) = 2.16$.

2 Arrange the sections in the order of increasing Q:

Section	*I*	*II*	*III*	*Unit*
f_p	1299	2297	3047	Hz
Q_p	—	1.35	6.43	—
K	2.16	2.16	2.16	—

3 Pair each zero with the closest pole, starting with the pole having the greatest Q:

Section	*I*	*II*	*III*	*Unit*
f_z	—	6700	4500	Hz
f_p	1299	2297	3047	Hz
Q_p	—	1.35	6.43	—
K	2.16	2.16	2.16	—

4 By using the data above, each section may be designed by the methods described in Chapters 6 to 9. Section I is a first-order low-pass section whereas sections II and III are second-order low-pass sections with transmission zeros.

EXAMPLE 10.2 Design an eight-pole high-pass filter having a gain of 0 dB in the pass-band and a notch at 60 Hz to filter out a strong interfering 60 Hz signal. The filter has the following poles and zeros.

$$fp_1 = 1669 \text{ Hz} \qquad Qp_1 = 1.612$$

$$fp_2 = 994 \text{ Hz} \qquad Qp_2 = 11.542$$

$$fp_3 = 1161 \text{ Hz} \qquad Qp_3 = 3.468$$

$$fp_4 = 3369 \text{ Hz} \qquad Qp_4 = 0.677$$

$$fz_1 = 60 \text{ Hz}$$

Six zeros at $f = 0$ Hz.

PROCEDURE

1 The filter will consist of four cascaded second-order sections. The high-frequency gain of each section is 0 dB or $K = 1$.

2 Pair the 60 Hz zero with the lowest Q pole. This will be the first section. Arrange the remaining poles in the order of increasing Q. Pair each of these with the zeros at $f = 0$ Hz.

Section	*I*	*II*	*III*	*IV*	*Unit*
f_z	60	0, 0	0, 0	0, 0	Hz
f_p	3369	1669	1161	994	Hz
Q_p	0.677	1.612	3.468	11.542	—
K	1	1	1	1	—

The first section is a second-order high-pass section with a transmission zero, and the other three sections are second-order high-pass sections.

EXAMPLE 10.3 Design a 10-pole bandpass filter having a gain of 10 dB at the center frequency. A strong interfering signal is present in the upper stopband. The filter has the following poles and zeros.

f_p	Q_p
1994	16.750
1754	6.178
1414	4.885
1140	6.178
1003	16.750

Five zeros at $f = 0$ Hz.

PROCEDURE

1 Due to the fact that a strong interference is present in the upper stopband, the filter will be designed with two second-order low-pass sections, followed by one second-order bandpass section, and lastly by two second-order high-pass sections.

2 The gain at the center frequency (1414 Hz) of the filter of each section is designed to be $\frac{10}{5} = 2$ dB or antilog $(\frac{2}{20}) = 1.26$.

3 Arrange the poles in order of increasing Q and pair the zeros at $f = 0$ Hz with the last three sections.

Section	*I*	*II*	*III*	*IV*	*V*	*Unit*
f_z	—	—	0	0, 0	0, 0	Hz
f_p	1414	1140	1754	1003	1994	Hz
Q_p	4.885	6.178	6.178	16.750	16.750	—
Gain at 1414	1.26	1.26	1.26	1.26	1.26	—

Sections I and II are second-order low-pass sections providing large attenuation in the upper stopband at the input of the filter, which will prevent the op amps of the first section from being overdriven by the large interfering signal. Section III is a second-order bandpass section, and sections IV and V are second-order high-pass sections.

4 If no large interfering signal occurs, the zeros at $f = 0$ Hz should be distributed one to each section. This will result in the filter having five second-order bandpass sections with the same ordering of the poles as before.

10.3 LC LADDER SIMULATION TECHNIQUES USING THE GENERALIZED IMPEDANCE CONVERTER

In Section 10.1 it was pointed out that active filter design methods based on the active simulation of LC ladder filters are superior to the cascaded design technique with respect to component variations. In this section we present a design method based on direct simulation of some of the elements in the LC filter by active circuits. In so doing, the topological structure of the original LC filter is preserved, and the beneficial interactions between the elements remain. To apply this method, an LC ladder filter satisfying the design requirements must first be obtained. Fortunately, extensive tables are available for certain LC ladder filters such as those with Butterworth, Chebyshev, Cauer, and Bessel frequency characteristics [2]. A selection of tables is presented in Chapter 13 for a normalized cutoff frequency and unity termination resistance. By using the appropriate frequency and impedance transformations, it is possible to obtain the LC ladder filter that satisfies the given requirements.

Active Simulation of a Grounded Inductance

Figure 10.1 shows the generalized impedance converter (GIC) used for the simulation of a grounded inductance. The input impedance is given by (see Section 4.3.3):

$$Z_{\text{in}}(\omega) = \frac{j\omega C_4 R_1 R_3 R_5}{R_2} \tag{10.8}$$

If R_2 is set equal to R_3:

$$Z_{\text{in}}(\omega) = j\omega C_4 R_1 R_5 \tag{10.9}$$

Thus the input impedance $Z_{\text{in}}(\omega)$ corresponds to the reactance of an equivalent inductance:

$$L_{\text{eq}} = C_4 R_1 R_5 \tag{10.10}$$

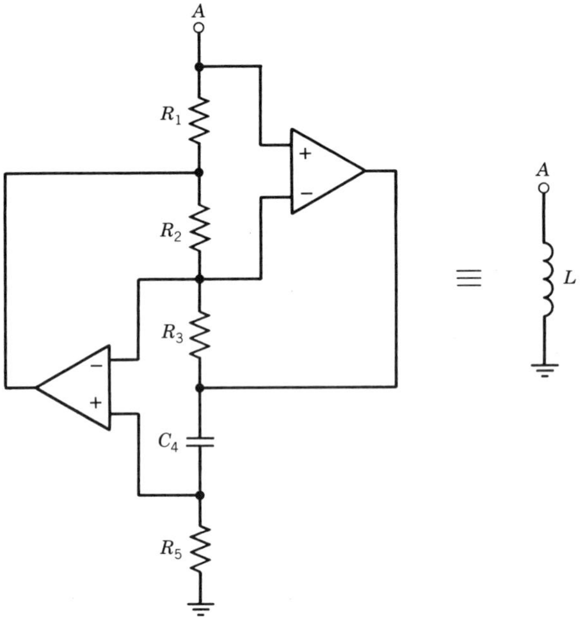

Figure 10.1 Active simulation of grounded inductor.

In the practical realization of the active inductance, the following precautions are recommended:

1 The two op amps should derive from the same integrated circuit—that is, from a dual or quad op amp chip.

2 To avoid possible oscillations due to the parasitic capacitances to ground, $R_2 = R_3$ should be selected as small as possible without presenting an appreciable load to output of the op amps: $R_2 = R_3 = 10\ \Omega$ is a good compromise.

EXAMPLE 10.4 Design a simulated inductor having an inductance of 300 mH.

PROCEDURE

1 Select standard values for R_3, R_2, and C_4; for example, $R_3 = R_2 = 10\ \text{k}\Omega$ and $C_4 = 10$ nF.

2 From Eq. 10.10

$$R_1 R_5 = \frac{L_{eq}}{C_4} = \frac{3 \cdot 10^{-1}}{10 \cdot 10^{-9}} = 3 \cdot 10^7$$

Let

$$R_1 = R_5 = \sqrt{3 \cdot 10^7} = 5.48\ \text{k}\Omega$$

3 Select the nearest 1% standard value, that is, $R_1 = R_5 = 5.49\ \text{k}\Omega$.

The resulting circuit is shown in Fig. 10.2.

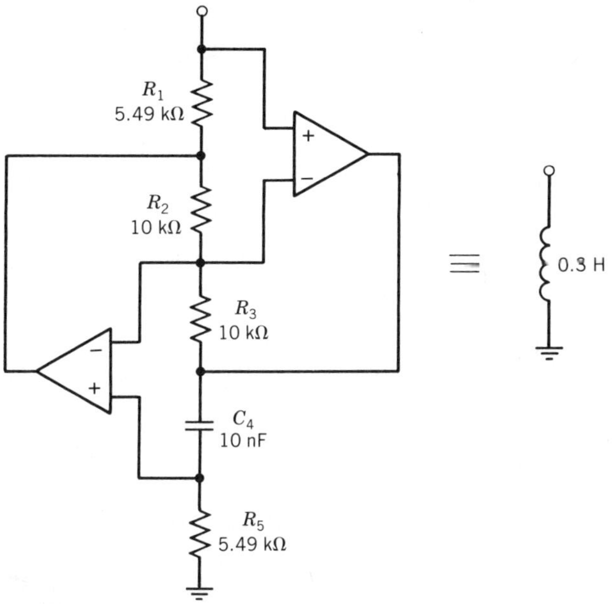

Figure 10.2 Active simulation of grounded 300 mH inductor.

Active Simulation of a Grounded Frequency-Dependent Negative Resistor

It can be shown that multiplying the impedance of each element in an LC filter by the same factor does not affect the frequency characteristic of the filter, even though the factor is frequency-dependent.* Later in this section we transform an LC ladder filter by multiplying the impedance of each element by the factor $1/j\omega$. In that case, the impedance of a capacitor of value D is transformed to

$$\frac{1}{j\omega} \cdot \frac{1}{j\omega D} = \frac{-1}{\omega^2 D} \tag{10.11}$$

This represents the impedance of a frequency-dependent negative resistor (FDNR), as introduced in Section 4.3.

The schematic symbol and the active simulation circuit of the FDNR element is shown in Fig. 10.3. The impedance of this circuit is given by

$$Z_{\text{in}}(\omega) = \frac{-R_5}{\omega^2 C_1 C_3 R_2 R_4} \tag{10.12}$$

Letting $R_4 = R_5$ results in

$$Z_{\text{in}}(\omega) = \frac{-1}{\omega^2 C_1 C_3 R_2} \tag{10.13}$$

* It may, however, change the gain factor of the filter.

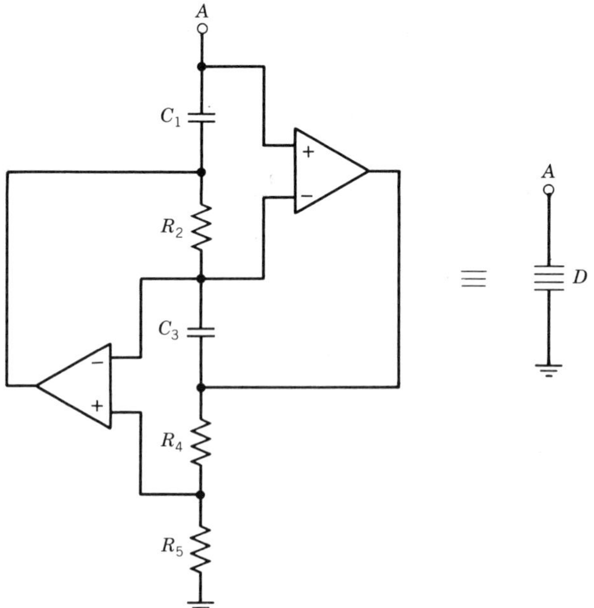

Figure 10.3 Active simulation of grounded FDNR.

Comparing this expression with Eq. 10.11 results in

$$D = C_1 C_3 R_2 \tag{10.14}$$

EXAMPLE 10.5 Design an FDNR circuit element resulting from transforming a capacitor of value 10 nF.

PROCEDURE

1 Select standard values for R_4, R_5, C_1, and C_3; for example, $R_4 = R_5 = 10\ \text{k}\Omega$ and $C_1 = C_3 = 0.33\ \mu\text{F}$.

2 From Eq. 10.14:

$$R_2 = \frac{D}{C_1 C_3} = \frac{10 \cdot 10^{-9}}{(0.33 \cdot 10^{-6})^2} = 91.83\ \text{k}\Omega$$

3 Select the nearest 1% standard value, that is, $R_2 = 92\ \text{k}\Omega$.
The resulting circuit is shown in Fig. 10.4.

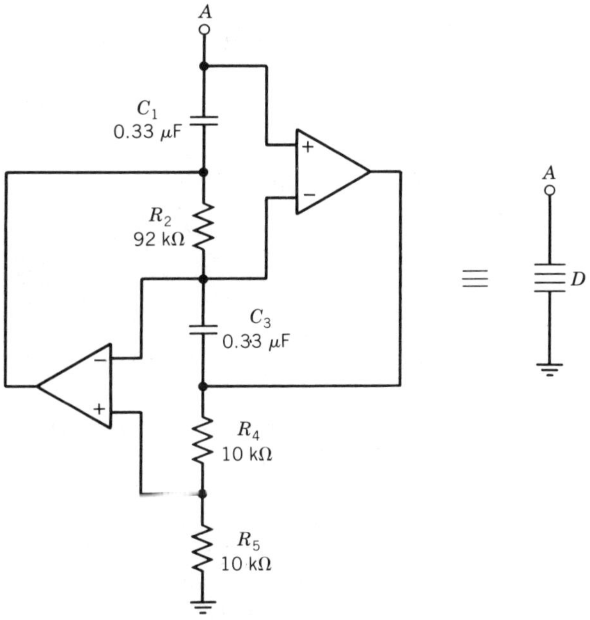

Figure 10.4 Active simulation of a grounded FDNR with $D = 10^{-8}\ F^2\Omega$.

Design of High-pass Filters Using Active Simulation of Grounded Inductors

In this section the application of the active simulation of grounded inductors to the design of high-order high-pass filters is illustrated. This type of design begins with finding the normalized, doubly terminated, LC ladder *low-pass* filter that satisfies the transformed specifications. This can be accomplished by the use of published tables, such as those in Chapter 13. The resulting low-pass filter is then transformed into a normalized high-pass filter using the low-pass to high-pass transformation. Then the elements are scaled so that the filter has the required cutoff frequency and impedance level. Finally, the grounded inductors are replaced by the active circuit of Fig. 10.1.

EXAMPLE 10.6 Design an active high-pass filter according to the following specifications:

Cutoff frequency, f_c	3 kHz
Maximum passband ripple	0.5 dB
Minimum stopband attenuation ($f \leq 2$ kHz)	45 dB
Gain in passband	0 dB
Input termination resistance	20 kΩ

PROCEDURE

1 Normalize the cutoff and stopband limit frequencies by dividing them by the cutoff frequency. This yields the normalized cutoff frequency (see Chapter 2, Section 2.7):

$$\omega_2 = 1 \text{ rad/s} \tag{10.15}$$

and the normalized stopband limit frequency:

$$\omega_1 = \tfrac{2}{3} \text{ rad/s} \tag{10.16}$$

2 Find the normalized frequencies for the equivalent low-pass filter. The normalized cutoff frequency is given by

$$\omega_p = \frac{1}{\omega_2} = 1 \text{ rad/s} \tag{10.17}$$

and the stopband limit frequency by

$$\omega_s = \frac{1}{\omega_1} = 1.5 \text{ rad/s} \tag{10.18}$$

3 Find the order of the normalized low-pass filter that satisfies the given specification. Use the nomograph of Fig. 13.11 (p. 397) to establish the need for a fifth-order ($n = 5$) Cauer-type filter.

4 Find the normalized low-pass filter that satisfies the given specification. This may be done in two ways:

a. From Table 13.10, with $n = 5$ and $\omega_s = 1.5$, the value of $A_{\min}$ satisfying the specifications is equal to 50.61 dB. The final doubly terminated LC ladder low-pass filter is obtained from Table 13.13, and shown in Fig 10.5*a*.

b. The Zverev tables [2] may be used. From [2, p. 220], a fifth-order Cauer filter with $\rho = 25\%$ ($A_{\max} = 0.28$ dB) and $\theta = 43°$ (giving $A_{\min} = 46.63$ dB) satisfies the specifications. This is the closest available approximation from the tables, since the next available value of ρ (50%) gives $A_{\min} = 1.25$ dB. Note, however, that the characteristic chosen offers a safety margin in both passband and stopband, thereby allowing for subsequent component variations. The element values and structure of the filter are given in [2, p. 221]. The resulting normalized, doubly terminated LC ladder lowpass filter is shown in Fig. 10.5*b*. For this example, both circuits will be used to develop active-filter designs.

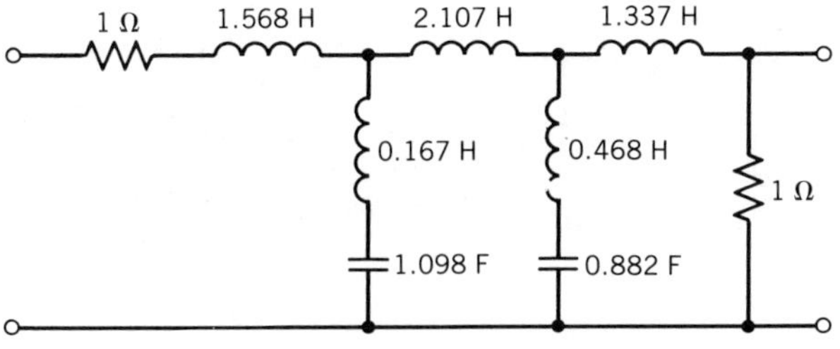

(a)

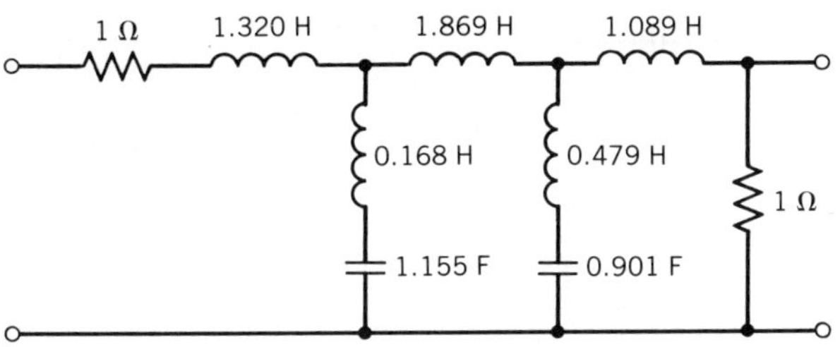

(b)

Figure 10.5 (a) Normalized low-pass filter (from Table 13.10); (b) normalized low-pass filter (from Zverev tables).

5 Transform the normalized low-pass filter into the corresponding high-pass filter by replacing each capacitor by an inductor and each inductor by a capacitor. The value of the new capacitors and new inductors are given by

$$C' = \frac{1}{L} \tag{10.19}$$

and

$$L' = \frac{1}{C} \tag{10.20}$$

where L and C are the values of the elements that were replaced. Performing this transformation on the low-pass filters of Fig. 10.5 results in the normalized high-pass filters shown in Fig. 10.6.

6 Perform frequency and impedance level scaling on the elements of the normalized high-pass filters. The equations for the denormalized element values are given by

$$R'' = R \tag{10.21}$$

$$L'' = \frac{L'R}{2\pi f_c} \tag{10.22}$$

$$C'' = \frac{C'}{R2\pi f_c} \tag{10.23}$$

Here, R is the desired value of the termination resistors. Selecting $R = 20\ \text{k}\Omega$ and applying Eqs. 10.21 to 10.23 results in the denormalized high-pass filters of Fig. 10.7.

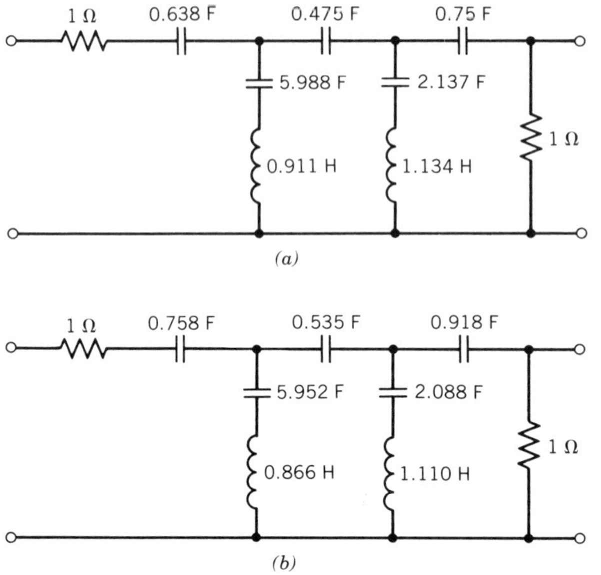

Figure 10.6 Normalized high-pass filters derived from the low-pass filters of Fig. 10.5. (a) Circuit derived from Fig. 10.5(a); (b) circuit-derived from Fig. 10.5(b).

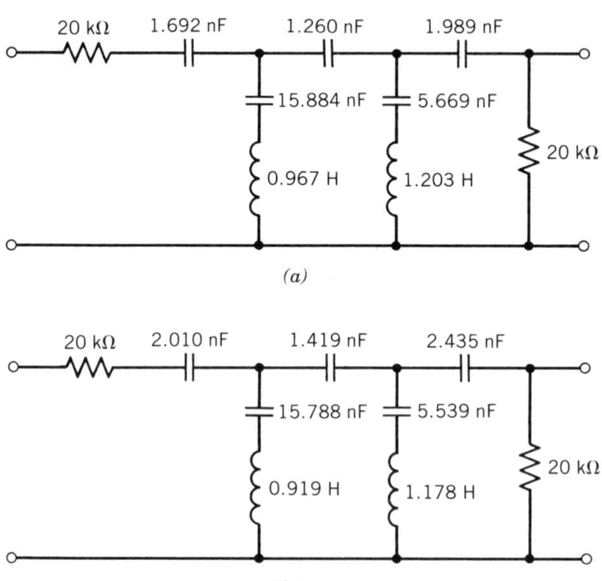

Figure 10.7 Denormalized high-pass filters from the circuits of Fig. 10.6. (a) Denormalized filter from Fig. 10.6(a); (b) denormalized filter from Fig. 10.6(b).

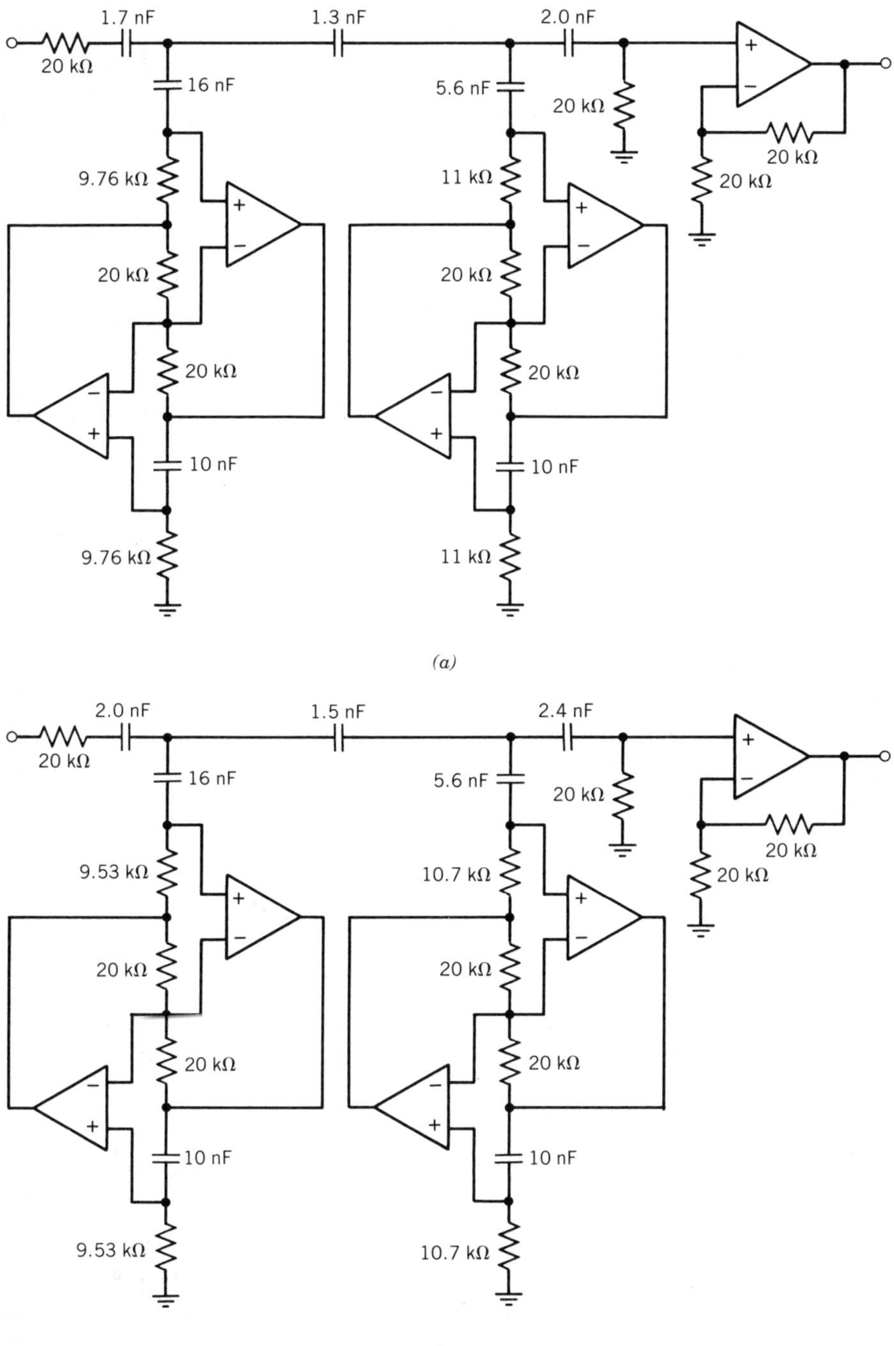

Figure 10.8 Active high-pass filters using active simulation of grounded inductors. (a) Derived from Fig. 10.7(a); (b) derived from Fig. 10.7(b).

7 Replace the grounded inductors by the active inductor simulation circuit using the procedure of Section 10.3.

8 Add an amplifier stage to the output of the filter to compensate for the loss of the LC ladder filter. A doubly terminated LC ladder filter having equal termination resistors has a loss of 6 dB in the passband. Since the specification called for no loss (0 dB gain) in the passband, the post-amplifier must have a gain of 2 (6

The schematic diagrams of the resulting active high-pass filters are shown in Fig. 10.8. Here, the closest standard values for the capacitors and resistors have been selected.

Design of Low-pass Filters Using Frequency-Dependent Negative Resistors

As mentioned in Section 10.3, it is possible to multiply the impedance of each element in an LC filter by the same factor without affecting the frequency characteristic of the filter. In this section, it will be shown that the inductors of an LC low-pass filter can be eliminated by multiplying the impedance of each element by the frequency-dependent factor $1/(j\omega)$. This transforms each element according to Table 10.1. Here it is shown that an inductor is transformed into a resistor, a resistor into a capacitor, and a capacitor into an FDNR. As an example of this transformation, the LC low-pass filter in Fig. 10.9 is transformed. Hence we have chosen the version of the filter in which the transmission zeros are implemented by series resonant circuits, since that results in the minimum number of capacitors and the smallest number of FDNRs in the transformed filter. The transformed filter is shown in Fig. 10.10. As can be seen, the objective of eliminating the inductors has been achieved.

Table 10.1 **Transformation by the factor $1/(j\omega)$**

Original element	*Original impedance*	*Transformed impedance*	*Transformed element*
Resistor R	R	$R/(j\omega) = 1/(j\omega/R)$	Capacitor $C_t = 1/R$
Inductor L	$j\omega L$	$j\omega L/(j\omega) = L$	Resistor $R_t = L$
Capacitor C	$1/(j\omega C)$	$1/[(j\omega C)\cdot(j\omega)] = 1/(-\omega^2 C)$	FDNR $D_t = C$

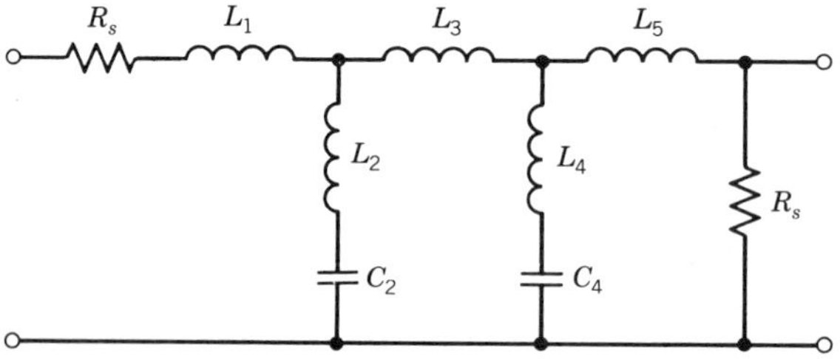

Figure 10.9 LC low-pass filter.

EXAMPLE 10.7 Design an active low-pass filter according to the following specifications:

Cutoff frequency, *fc*	4 kHz
Maximum passband ripple	0.5 dB
Minimum stopband attenuation ($f \geq 5$ kHz)	55 dB
Gain in passband	10 dB

PROCEDURE

1 Normalize the cutoff and stopband limit frequencies by dividing them by the cutoff frequency. This yields the normalized cutoff frequency

$$\omega_p = 1 \text{ rad/s} \tag{10.24}$$

and the normalized stopband limit frequency

$$\omega_s = \tfrac{5}{4} \text{ rad/s} \tag{10.25}$$

2 Find the normalized low-pass filter that satisfies the given specifications. In [2, p. 284] it is shown that a seventh-order ($N = 7$) Cauer-type filter with $\rho = 25\%$ and $\theta = 54°$ satisfies the specifications.* Using the element values given in [2, p. 286] for $K^2 = \infty$ results in the singly terminated LC ladder filter of Fig. 10.11. Choosing the singly terminated filter

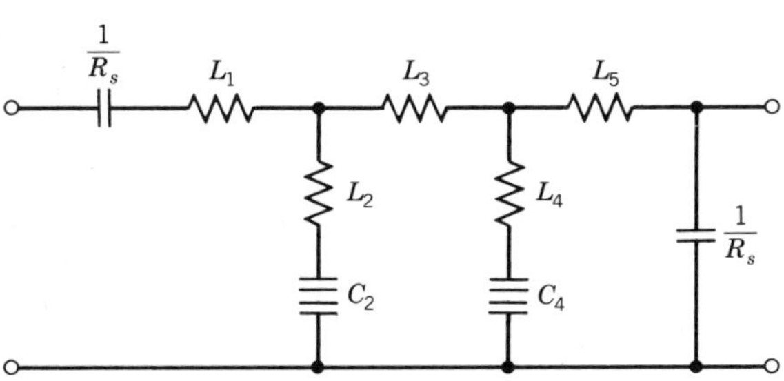

Figure 10.10 Transformed version of the filter in Fig. 10.9.

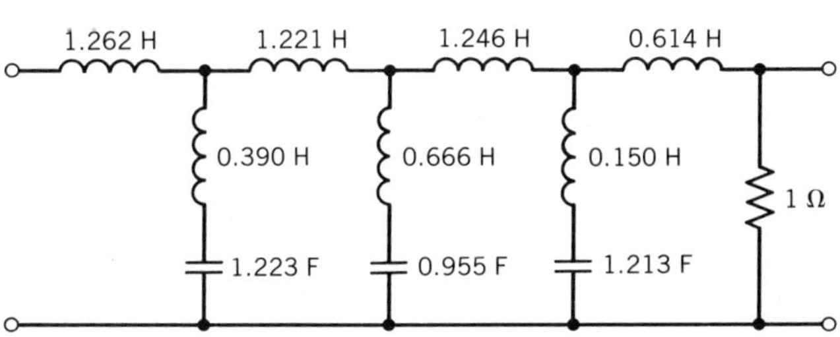

Figure 10.11 Normalized low-pass filter.

* Note that, as in Example 10.6, an alternative (doubly terminated) arrangement is available by using the tables in Chapter 13.

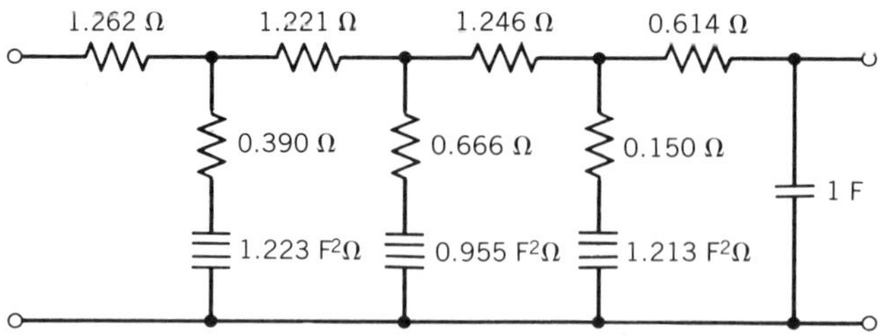

Figure 10.12 Transformed version of the low-pass filter in Fig. 10.11.

version avoids the problem associated with the dc response and saves one terminating capacitor at the expense of a somewhat greater sensitivity to element variations.

3 Perform the network transformation according to Table 10.1. This results in the transformed filter of Fig. 10.12.

4 Perform frequency and impedance level scaling on the normalized transformed low-pass filter. The equations for the denormalized element values are given by

$$C'' = C \tag{10.29}$$

$$R'' = \frac{R'}{2\pi f_c C} \tag{10.30}$$

$$D'' = \frac{D'C}{2\pi f_c} \tag{10.31}$$

Here C is the desired termination capacitor.

Selecting $C = 2$ nF and applying Eqs. 10.29 to 10.31 results in the denormalized filter of Fig. 10.13.

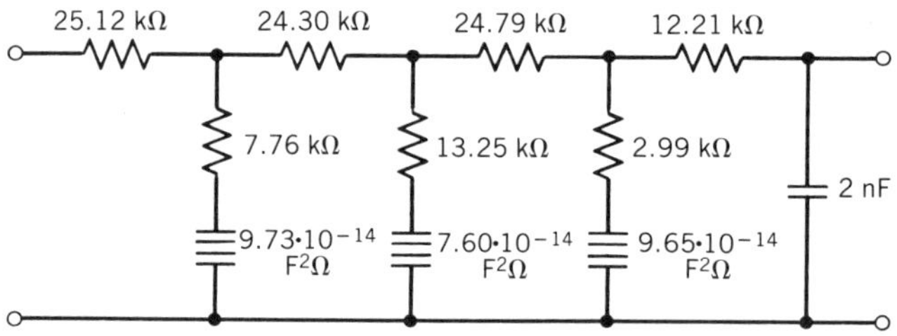

Figure 10.13 Denormalized transformed low-pass filter.

5 Implement the FNDRs using the procedure of Section 10.3.

6 Add an amplifier stage to the output of the filter to bring the gain up to 10 dB. A singly terminated filter has no loss in the passband. Thus the post-amplifier must have a gain of 10 dB.

The schematic diagram of the active low-pass filter is shown in Fig. 10.14. Here the closest standard values for the circuit components have been selected.

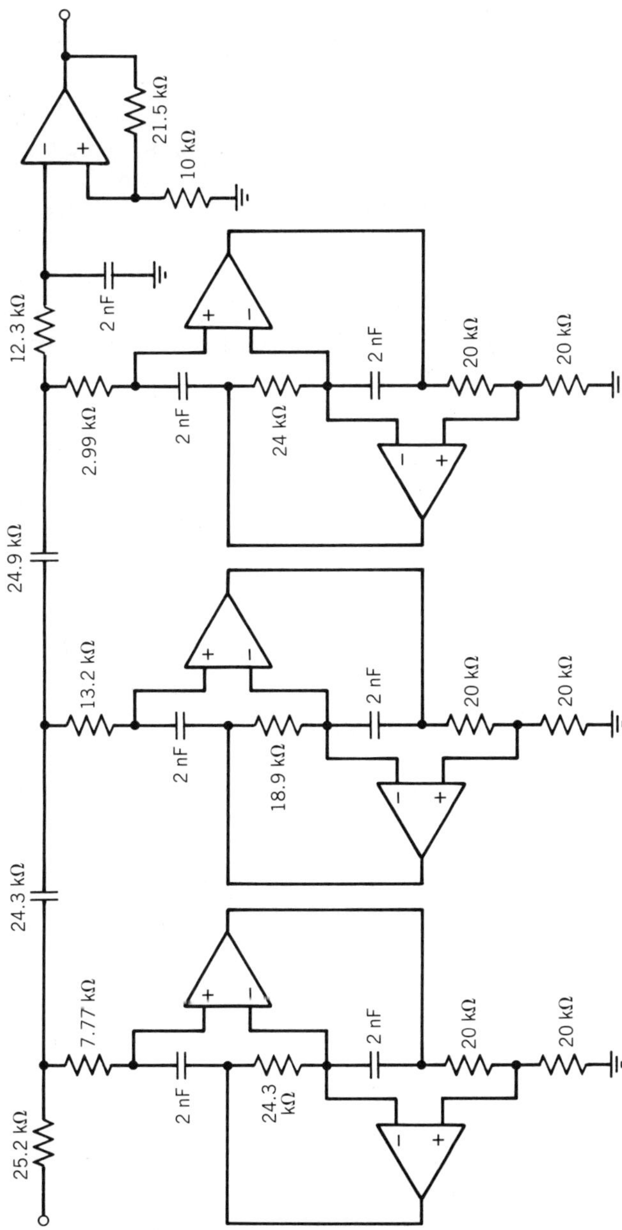

Figure 10.14 Active low-pass filter using FDNRs.

10.4 LC LADDER SIMULATION TECHNIQUES USING MULTILOOP SIMULATION DESIGN

The previous section indicated that the low-sensitivity properties of LC ladder filters can be preserved in active filters by maintaining the structure of the original LC filter. In this section an alternative design method is discussed. Although not preserving the actual structure of the original LC ladder filter, this method maintains its voltage-current relationships. Thus the interactions between the components are preserved, because the voltage-current relationships simulated in the resulting multiloop RC-active filter are identical to those of the original LC ladder network. However, in the multiloop RC-active filter, both voltages and currents in the LC filter are simulated by voltages.

Multiloop Simulation Design of Low-pass Filters

To illustrate the multiloop simulation design, the LC ladder network shown in Fig. 10.15 will be simulated. The voltage-current relationships for this filter are given by

$$I_1 = \frac{(V_{in} - V_1) \cdot (1/R_1)}{1 + sL_1/R_1} \tag{10.32}$$

$$V_1 = \frac{I_1 - I_2}{sC_1} \tag{10.33}$$

$$I_2 = \frac{V_1 - V_{out}}{sL_2} \tag{10.34}$$

$$V_{out} = \frac{I_2 \cdot R_2}{1 + sR_2 \cdot C_2} \tag{10.35}$$

These relationships can be used to develop the block diagram of the multiloop feedback simulation system resulting in Fig. 10.16. It is this structure that has given the name "leap-frog" to the design of active filters by this method.

The next step in the design process is the implementation of each block in Fig. 10.16 by active filter sections. The first and last blocks represent first-order

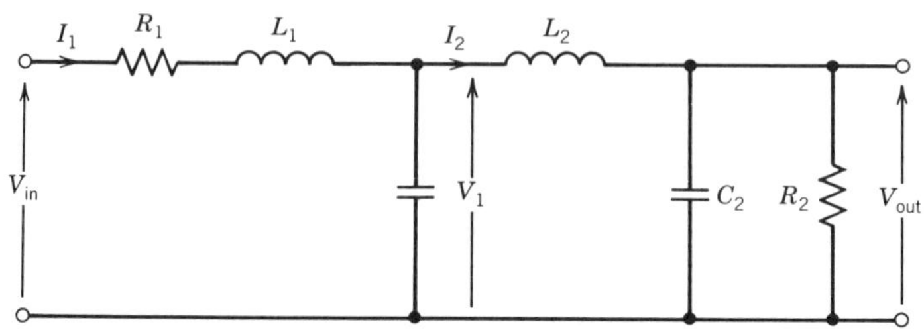

Figure 10.15 Low-pass LC ladder filter.

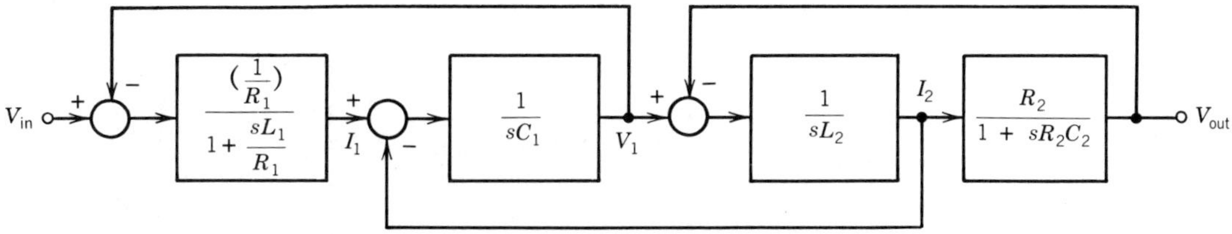

Figure 10.16 Multiloop feedback simulation of the LC ladder filter in Fig. 10.15.

sections, and all the blocks in between represent integrators. A typical first-order section is shown in Fig. 10.17; an integrator is shown in Fig. 10.18.

The transfer function of the first-order section is given by

$$\frac{V_{out}}{V_1 + V_2} = \frac{-R_2/R_1}{1 + sR_2C} \tag{10.36}$$

and that of the integrator by

$$\frac{V_{out}}{V_1 + V_2} = \frac{-1}{sRC} \tag{10.37}$$

Both of these sections are inverting and can sum two inputs. Examination of the multiloop feedback simulation circuit in Fig. 10.16 shows that each loop contains one phase inversion. It is necessary to include a unity gain inverter in every second block when utilizing the inverting sections in Figs. 10.17 and 10.18. This will be illustrated in the following design example.

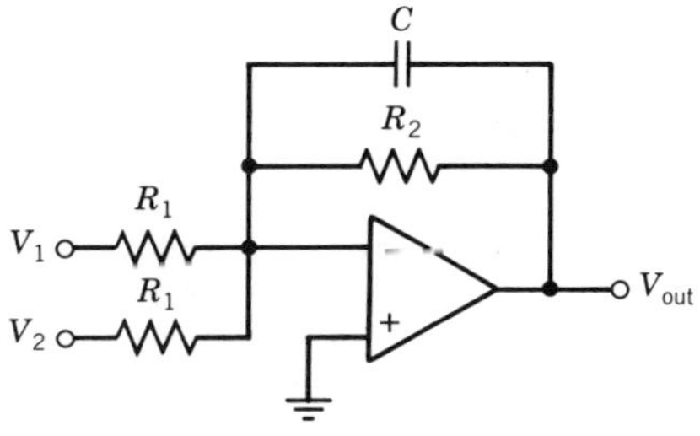

Figure 10.17 Summing inverting first-order section.

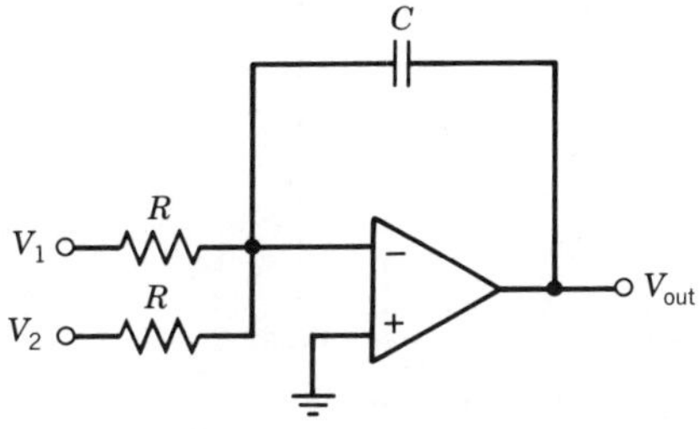

Figure 10.18 Summing inverting integrator.

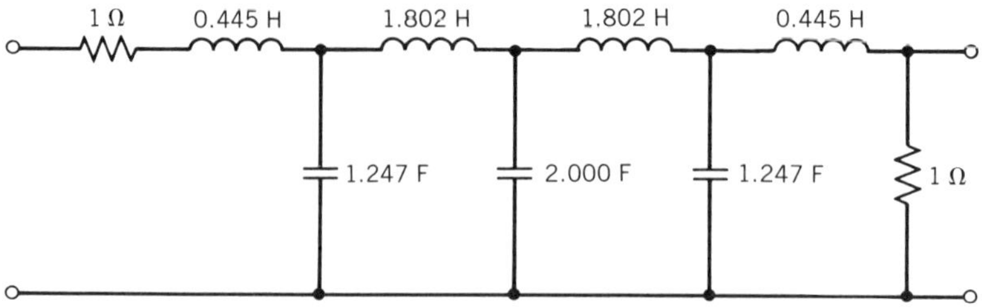

Figure 10.19 Normalized seventh-order Butterworth low-pass filter.

EXAMPLE 10.8 Design an active low-pass filter satisfying the following specifications:

Type	Butterworth
3 dB cutoff frequency	5 kHz
Stopband attenuation slope	42 dB/octave
DC gain	0 dB

PROCEDURE

1 Find the order N of the filter from

$$N = \tfrac{42}{6} = 7 \tag{10.38}$$

2 Find the structure and the element values for the normalized seventh-order Butterworth, equally terminated, low-pass LC filter from Table 13.18. The resulting filter is shown in Fig. 10.19.

3 Frequency scale the normalized LC filter from the normalized cutoff frequency of 1 rad/s to 5 kHz by dividing the values of all the reactive elements by the factor $2\pi 5000$. The resulting filter is shown in Fig. 10.20.

4 Derive the following voltage-current relationships from Fig. 10.20:

$$I_1 = \frac{V_{\text{in}} - V_2}{1 + sL_1} \tag{10.39}$$

$$V_2 = \frac{I_1 - I_3}{sC_2} \tag{10.40}$$

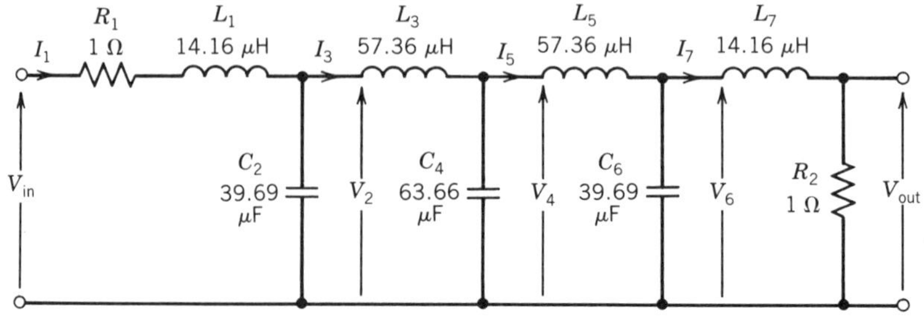

Figure 10.20 Frequency scaled seventh-order Butterworth low-pass filter.

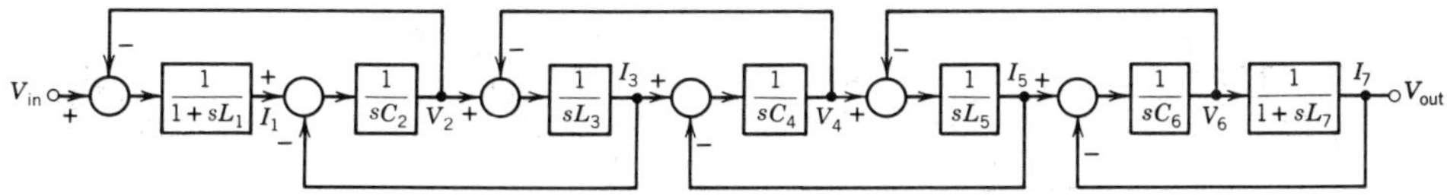

Figure 10.21 Multiloop feedback simulation block diagram of the LC network in Fig. 10.20.

$$I_3 = \frac{V_2 - V_4}{sL_3} \tag{10.41}$$

$$V_4 = \frac{I_3 - I_5}{sC_4} \tag{10.42}$$

$$I_5 = \frac{V_4 - V_6}{sL_5} \tag{10.43}$$

$$V_6 = \frac{I_5 - I_7}{sC_6} \tag{10.44}$$

$$V_{out} = \frac{V_6}{1 + sL_7} \tag{10.45}$$

Here we have taken advantage of the fact that $R_1 = R_2 = 1\ \Omega$.

5 From the equations above, form the multiloop feedback simulation block diagram as shown in Fig. 10.21.

6 Implement the units of the block diagram in Fig. 10.21 using the circuits shown in Figs. 10.18 and 10.19 and unity gain inverters. This results in the "leap-frog" filter of Fig. 10.22. Here the resistors are calculated from the following equations obtained by combining Eqs. 10.36 and 10.37 and Eqs. 10.39 to 10.45:

$$R_1 = \frac{L_1}{C} \tag{10.46}$$

$$R_2 = \frac{C_2}{C} \tag{10.47}$$

$$R_3 = \frac{L_3}{C} \tag{10.48}$$

$$R_4 = \frac{C_4}{C} \tag{10.49}$$

$$R_5 = \frac{L_5}{C} \tag{10.50}$$

$$R_6 = \frac{C_6}{C} \tag{10.51}$$

$$R_7 = \frac{L_7}{C} \tag{10.52}$$

Selecting $C = 1$ nF and applying the equations above results in the standard 1% resistance values given in Fig. 10.22. Here we have halved the value of the resistor connected to the input to increase the dc gain of the filter from the -6 dB of the original equally terminated LC filter to the required 0 dB.

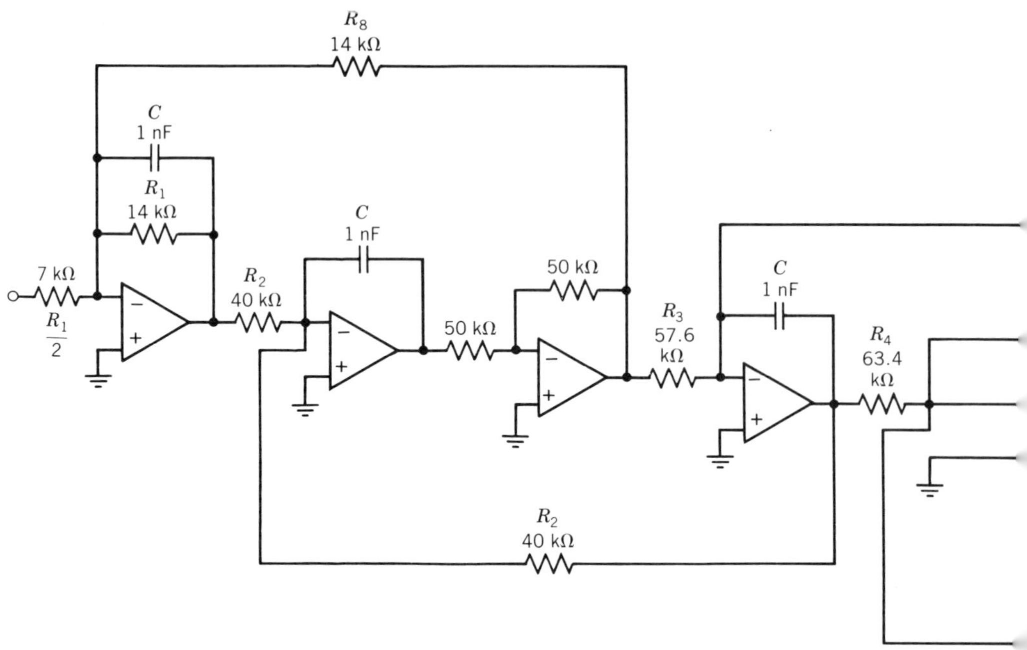

Figure 10.22 "Leap-frog" realization of the seventh-order Butterworth low-pass filter in Fig. 10.20.

Multiloop Feedback Simulation Design of Bandpass Filters

In order to arrive at a multiloop feedback simulation system of a bandpass filter, the corresponding prototype LC ladder filter must first be obtained. Since no published tables for the component values of bandpass LC ladder filters are available, the prototype filter must be derived by a low-pass to bandpass transformation of a low-pass LC ladder prototype filter. This transformation is achieved by replacing each inductor of the low-pass prototype by a series LC combination resonating at the specified center frequency of the bandpass filter. The inductor and capacitor values of the series combination is given by

$$L_s = \frac{L}{2\pi B} \tag{10.53}$$

$$C_s = \frac{1}{(2\pi f_0)^2 L_s} \tag{10.54}$$

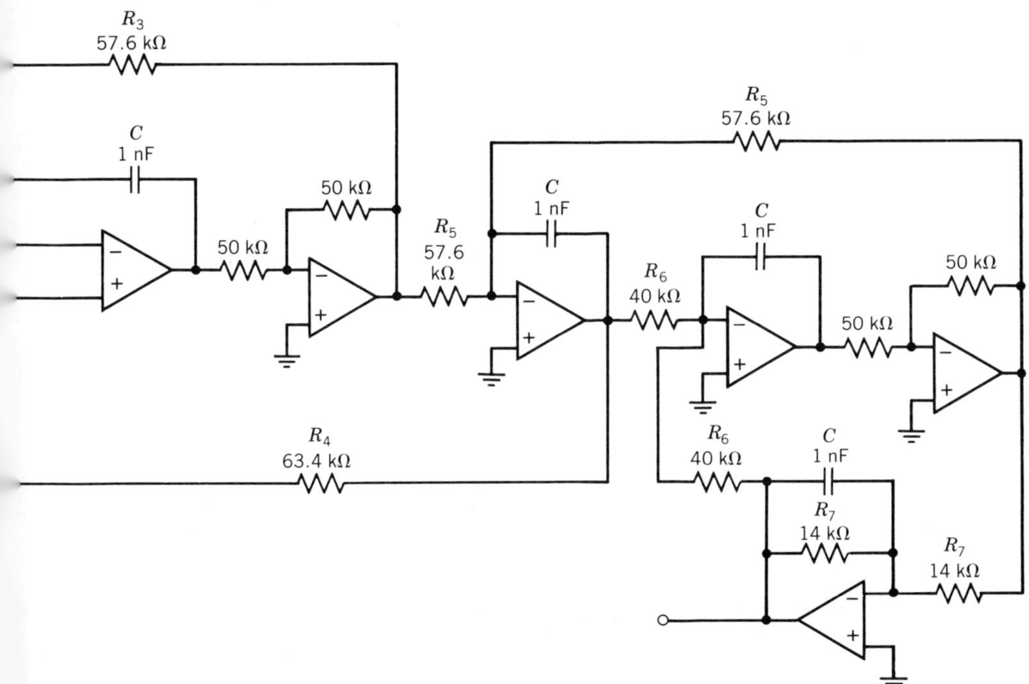

Here B and f_0 are the specified bandwidth and center frequency of the bandpass filter; and L is the inductance of the inductor in the prototype low-pass filter, which is replaced by the LC series combination. Furthermore, each capacitor in the low-pass prototype is replaced by a parallel LC combination resonating at the bandpass center frequency. The values for the components in the LC parallel combination are given by

$$C_p = \frac{C}{2\pi B} \tag{10.55}$$

$$L_p = \frac{1}{(2\pi f_0)^2 C_p} \tag{10.56}$$

where C is the value of the capacitor in the low-pass prototype that is being replaced.

Table 10.2 Lowpass to bandpass element transformation

Normalized lowpass elements	*Bandpass elements*
	 $L_s = \frac{L}{(2\pi B)}$, $C_s = \frac{1}{(2\pi f_0)^2 L_s}$
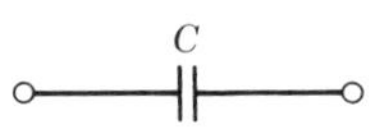	$L_p = \frac{1}{(2\pi f_0)^2 C_p}$ 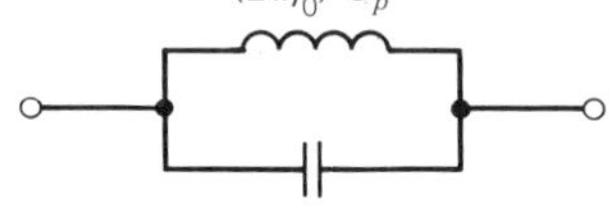 $C_p = \frac{C}{(2\pi B)}$

Table 10.2 illustrates this low-pass to bandpass transformation. The center frequency f_0 mentioned above is defined as the geometric mean of the lower and upper cutoff frequencies f_1 and f_2 of the bandpass filter and is given by

$$f_0 = \sqrt{f_1 \cdot f_2} \tag{10.57}$$

The bandwidth B of the bandpass filter is defined by

$$B = f_2 - f_1 \tag{10.58}$$

As an illustration of the low-pass to bandpass transformation, the low-pass prototype shown in Fig. 10.23 is transformed into a bandpass filter. The resulting

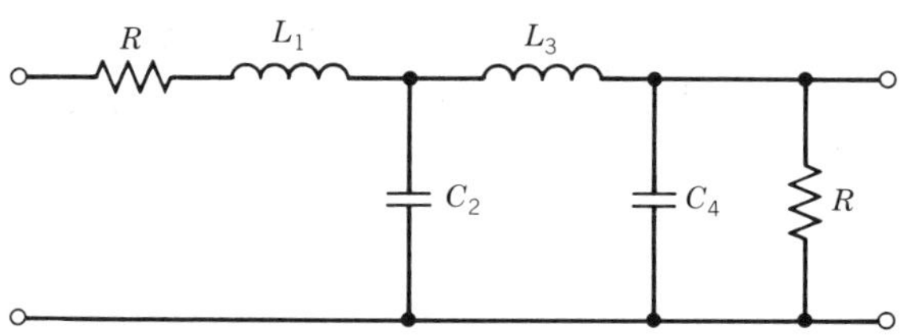

Figure 10.23 Low-pass prototype.

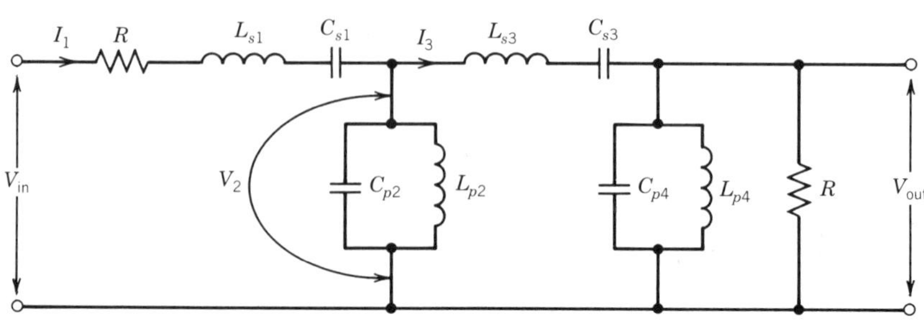

Figure 10.24 Bandpass filter obtained by transforming the low-pass prototype in Fig. 10.23.

bandpass filter is shown in Fig. 10.24. In order to derive the multiloop feedback simulation system of the bandpass filter, the following voltage-current relationships are obtained from Fig. 10.24:

$$I_1 = \frac{V_{in} - V_2}{R + sL_{s1} + 1/(sC_{s1})}$$

$$= \frac{(V_{in} - V_2)(s/L_{s1})}{s^2 + s/(L_{s1}/R) + 1/(L_{s1}C_{s1})} \tag{10.59}$$

$$V_2 = \frac{I_1 - I_3}{(sC_{p2} + 1/s)L_{p2}}$$

$$= \frac{(I_1 - I_3)(s/C_{p2})}{s^2 + 1/(L_{p2}C_{p2})} \tag{10.60}$$

$$I_3 = \frac{V_2 - V_{out}}{(sL_{s3} + 1)/(sC_{s3})}$$

$$= \frac{(V_2 - V_{out}) \cdot (s/L_{s3})}{(s^2 + 1)/(L_{s3}C_{s3})} \tag{10.61}$$

$$V_{out} = \frac{I_3}{1/R + 1/sL_{p4} + sC_{p4}}$$

$$= \frac{I_3(s/C_{p4})}{s^2 + s/(RC_{p4}) + 1/(L_{p4}C_{p4})} \tag{10.62}$$

Using these relationships, the multiloop feedback system for the bandpass filter can be developed, resulting in the block diagram in Fig. 10.25. Here it can be seen that each block corresponds to a second-order bandpass transfer function similar to the one of Eq. 10.7.

The resonant frequency of each block is the same (f_0), and the first and the last block have equal and finite Qs, while all the other blocks have infinite Qs. Normally, an infinite Q would result in an unstable second-order bandpass section capable of sustained oscillations at the resonant frequency. However, in this case, the overall system is stable because of the multiple *negative* feedback loops. These second-order blocks can be implemented with any of the bandpass biquad circuits

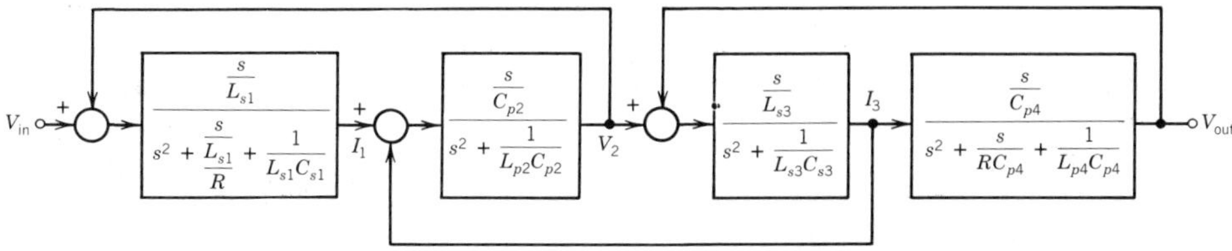

Figure 10.25 Multiple-feedback simulation block diagram of the LC bandpass filter in Fig. 10.24.

described in the previous chapters. However, in this section, we show an implementation using the Deliyannis inverting single op amp bandpass circuit described in Section 7.5. It results in the filter having a minimum number of op amps, and shows good sensitivity characteristics with respect to variations in the passive and active components. Figure 10.26 shows this biquad in a summing configuration for which $R_2 = 10R_1$ (i.e., $m = 0.1$) and $C_1 = C_2 = C$. (See Example 7.10.) R_1 is the resistance of the parallel combination of $R_1'/2$ with R_1''. In this case, the transfer function is given by

$$\frac{V_{out}(s)}{V_1(s)} = \frac{V_{out}(s)}{V_2(s)}$$

$$= \frac{-1}{2 + R_1'/R_1''} \cdot \frac{s(1 + K)/(R_1 C)}{s^2 + (s) \cdot [2/(10R_1 C) - K/(R_1 C)] + 1/(10R_1^2 C^2)} \tag{10.63}$$

where

$$K = R_A/R_B \tag{10.64}$$

In the case of one input only and in the absence of R_1'', the first term of Eq. 10.63 becomes unity. It is of interest to note that in order to obtain an infinite Q, the middle term in the denominator of Eq. 10.63 should vanish. This occurs for

$$K = \tfrac{2}{10} \tag{10.65}$$

and thus

$$R_B = 5R_A \tag{10.66}$$

The resonant frequency f_0 of each block is given by

$$f_0 = \frac{1}{2\pi\sqrt{R_1^2 C^2/m}} \tag{10.67}$$

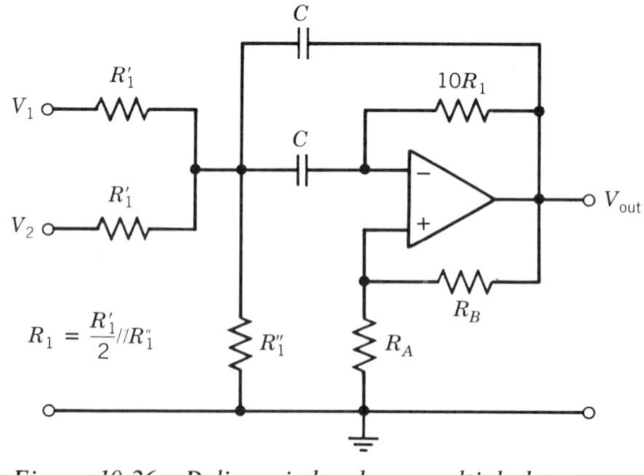

Figure 10.26 Deliyannis bandpass multiple-loop positive feedback circuit adapted from Fig. 7.8.

which yields

$$R_1 = \frac{\sqrt{m}}{2\pi f_0 C} \tag{10.68}$$

Because the circuit in Fig. 10.26 inverts the signal, it is necessary to include an inverter in each loop. This will be illustrated in the following example.

EXAMPLE 10.9 Design an eight-pole bandpass filter satisfying the following specifications:

Passband type	Butterworth
Lower 3 dB cutoff frequency	1.5 kHz
Upper 3 dB cutoff frequency	2.5 kHz
Gain at center frequency	0 dB

PROCEDURE

1 Determine the structure and element values for the normalized fourth-order (half the order of the desired bandpass filter) Butterworth low-pass prototype filter from Table 13.18. The resulting prototype low-pass filter is shown in Fig. 10.27.

2 Transform the prototype low-pass filter into a bandpass filter using Eqs. 10.53 to 10.58. Thus

$$f_0 = \sqrt{f_1 f_2} = 1937 \text{ Hz} \quad \text{and} \quad B = f_2 - f_1 = 1 \text{ kHz}$$

This results in the eighth-order bandpass filter illustrated in Fig. 10.28.

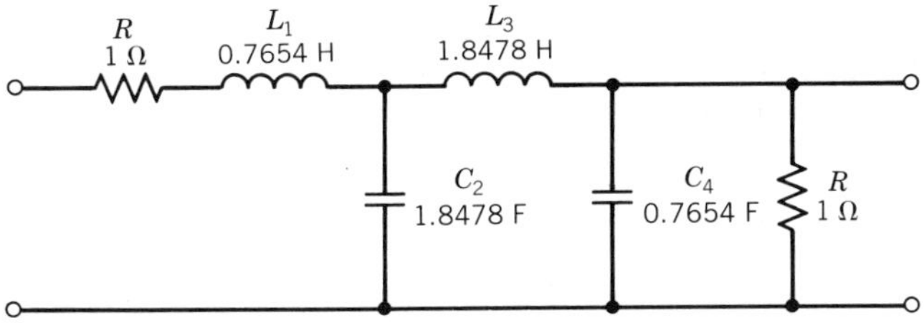

Figure 10.27 Prototype fourth-order Butterworth low-pass filter.

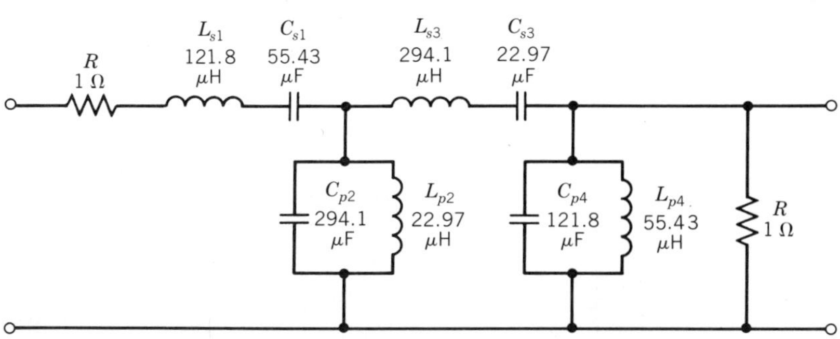

Figure 10.28 Eighth-order Butterworth bandpass filter obtained by transformation of the prototype low-pass filter in Fig. 10.27.

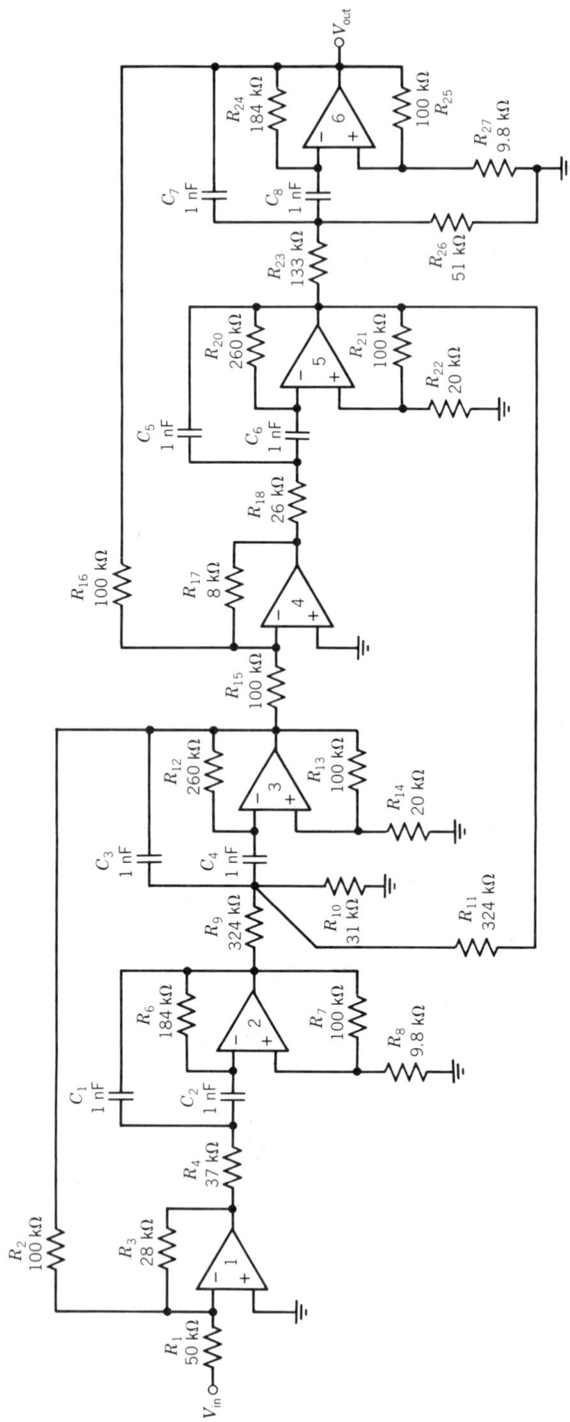

Figure 10.29 Multiple-feedback simulation bandpass filter.

3 Derive the voltage-current relationships as in Eqs. 10.59 to 10.62 and form the multiple-feedback simulation block diagram as shown in Fig. 10.25.

4 Implement each block in the diagram, utilizing the single op amp bandpass biquad circuit shown in Fig. 10.26. Include a summing voltage inverter in every other block. Select the gain of each inverter so that the required gain of the block is obtained without a voltage divider at the input of the corresponding biquad. Design all biquads for the same resonant frequency (1937 Hz). Design the Q of the first and last biquads for

$$Q = \frac{2\pi f_0 L_{s1}}{R} = 2\pi f_0 C_{p4} R = 1.482 \tag{10.69}$$

with $m = 0.2$, and the second and third biquads for $Q = \infty$ with $m = 0.1$. Design the center frequency gain of each block (from Eqs. 10.59 to 10.62) to be, respectively, R, ∞, ∞, R.

The resulting active bandpass filter is illustrated in Fig. 10.29. Here the value of the input resistor R_1 has been halved to ensure that the total gain in the passband of the filter is 0 dB as specified.

10.5 SUMMARY

This chapter has presented a practical approach to the design of higher-order filters.

Rules for the ordering of cascaded second-order sections have been considered so as to maximize dynamic range and to reject strong stopband interference signals.

With the GIC used as a basic block, realizations of grounded inductors and grounded FDNRs have been discussed. These elements are shown to be valuable in the simulation of high-order filters by direct replacement techniques. Finally, consideration has been given to the "leap-frog" method for simulating an LC ladder filter.

10.6 REFERENCES

[1] E. Christian and E. Eisenman, *Filter Design Tables and Graphs*, Knightdale, N.C.: Transmission Networks International, Inc., 1977.

[2] A. I. Zverev, *Handbook of Filter Synthesis*, New York: Wiley, 1976.

CHAPTER 11

Practical Limitations of Active Filters

R. W. Steer, Jr.
Frequency Devices, Inc.
Haverhill, Massachusetts

11.1 INTRODUCTION

The components that are used to implement an RC-active filter realization of an analytic transfer function determine the accuracy of the realization. If all resistors, capacitors, and amplifiers were "ideal," theoretical functions could be precisely realized. In practice, however, components are not ideal. They are available only in discrete incremental values; they have tolerances, temperature coefficients, voltage ratings, nonlinearities, and other properties that influence the performance of the circuits in which they are used. Active components, particularly operational amplifiers, have many properties such as phase shift, finite gain, dc offset and bias requirements, internal noise generation, slew rate limiting, and others that usually limit or degrade the filter characteristics.

In what follows, the properties of the passive and active components that limit circuit performance are examined. Some practical guidelines are established for the useful Q and frequency range of several common active filter circuits, and some practical techniques are shown to compensate for the effects of amplifier gain, bandwidth, and phase shift.

11.2 SENSITIVITY ANALYSIS

The sensitivity of a filter parameter to the value of a particular component can be determined through a classical sensitivity analysis as described in Chapter 5 (Section 5.5). The sensitivity concept can be further extended to determine the sensitivity of the gain, phase, or delay of a filter to its individual components. Such sensitivities depend upon the analytic function being realized (e.g., a multipole function such as an eight-pole Butterworth low-pass filter), the circuit configuration used (e.g., a Sallen–Key second-order low-pass section), and the components that are used in the circuit.

In addition, with the availability of high-speed computation and graphical representation, a circuit implementation can be analyzed by computation of the desired parameter variation as a function of individual component variations. In practice, computer analysis is often a more realistic approach, since some sensitivity functions are high enough, and some practical incremental values of components are coarse enough, so that there is often a substantial difference between the derivative (classical sensitivity) and the incremental (computed sensitivity) calculations.

Sensitivity of a Quadratic Function

Considerable practical insight can be gained by consideration of the gain and phase sensitivities for a quadratic transfer function for variations in frequency and Q. By examining these functions we can make a determination of the accuracy required of the frequency and Q of the sections used to implement a filter in order to realize the desired transfer function within required tolerance limits.

The magnitude and phase functions for a quadratic function* are

$$G_i(s) = s^2 + \frac{\omega_i}{Q_i} s + \omega_i^2$$

$$G_i(j\omega) = \omega_i^2 - \omega^2 + j\frac{\omega\omega_i}{Q_i}$$

$$\begin{aligned} G_i &= 20 \log_{10} |G_i(j\omega)| \\ &= 20 \log_{10} \left[(\omega_i^2 - \omega^2)^2 + \frac{\omega_i^2 \omega^2}{Q_i^2} \right]^{1/2} \\ &= 10 \log_{10} \left[(\omega_i^2 - \omega^2)^2 + \frac{\omega_i^2 \omega^2}{Q_i^2} \right] \end{aligned} \tag{11.1}$$

$$\phi_i = \tan^{-1} \frac{\omega_i \omega}{Q_i(\omega_i^2 - \omega^2)} \tag{11.2}$$

* This may be either a numerator or denominator factor.

The change in magnitude per unit change in ω_i is given by

$$\int_{\omega_i}^{G_i} = \omega_i \frac{d}{d\omega_i}\left[10 \log_{10}\left((\omega_i^2 - \omega^2)^2 + \frac{\omega_i^2\omega^2}{Q_i^2}\right)\right]$$

$$= 8.686 \frac{\left[2\omega_i^2(\omega_i^2 - \omega^2) + \dfrac{\omega_i^2\omega^2}{Q_i^2}\right]}{(\omega_i^2 - \omega^2)^2 + \dfrac{\omega_i^2\omega^2}{Q_i^2}} \tag{11.3}$$

In terms of the normalized frequency, $\Omega_i(= \omega/\omega_i)$:

$$\int_{\omega_i}^{G_i} = 8.686 \frac{\left[2(1 - \Omega_i^2) + \dfrac{\Omega_i^2}{Q_i^2}\right]}{(1 - \Omega_i^2)^2 + \dfrac{\Omega_i^2}{Q_i^2}} \tag{11.4}$$

Similarly, the sensitivity of the change in gain per unit change in Q_i is given by

$$\int_{Q_i}^{G_i} = Q_i \frac{d}{dQ_i}\left[10 \log_{10}\left([\omega_i^2 - \omega^2)^2 + \frac{\omega_i^2\omega^2}{Q_i^2}\right)\right]$$

$$= -\frac{8.686\left(\dfrac{\omega_i^2\omega^2}{Q_i^2}\right)}{(\omega_i^2 - \omega^2)^2 + \dfrac{\omega_i^2\omega^2}{Q_i^2}} \tag{11.5}$$

The expression may be arranged in terms of Ω_i as:

$$\int_{Q_i}^{G_i} = \frac{-8.686\left(\dfrac{\Omega_i^2}{Q_i^2}\right)}{(1 - \Omega_i^2)^2 + \dfrac{\Omega_i^2}{Q_i^2}} \tag{11.6}$$

In like manner we can compute the sensitivity of the phase shift for a quadratic function to unit changes in ω_i and Q_i:

$$\int_{\omega_i}^{\phi_i} = \omega_i \frac{d}{d\omega_i} \tan^{-1} \frac{\omega_i\omega}{Q_i(\omega_i^2 - \omega^2)}$$

$$= \frac{-\dfrac{\omega_i\omega}{Q_i}(\omega^2 + \omega_i^2)}{(\omega_i^2 - \omega^2)^2 + \left(\dfrac{\omega_i\omega}{Q_i}\right)^2} \tag{11.7}$$

In normalized form, Eq. 11.7 becomes

$$S^{\phi_i}_{\omega_i} = \frac{-\dfrac{\Omega_i}{Q_i}(1+\Omega_i^2)}{(1-\Omega_i^2)^2 + \left(\dfrac{\Omega_i}{Q_i}\right)^2} \tag{11.8}$$

$$S^{\phi_i}_{Q_i} = Q_i \frac{d}{dQ_i} \tan^{-1} \frac{\omega_i \omega}{Q_i(\omega_i^2 - \omega^2)}$$

$$= \frac{-\dfrac{\omega_i \omega}{Q_i}(\omega_i^2 - \omega^2)}{(\omega_i^2 - \omega^2)^2 + \left(\dfrac{\omega_i \omega}{Q_i}\right)^2} \tag{11.9}$$

In normalized form, Eq. 11.9 becomes

$$S^{\phi_i}_{Q_i} = \frac{-\dfrac{\Omega_i}{Q_i}(1-\Omega_i^2)}{(1-\Omega_i^2)^2 + \left(\dfrac{\Omega_i}{Q_i}\right)^2} \tag{11.10}$$

Similarly, sensitivities can be computed for other functions such as envelope delay, stopband floor, or any other transfer function property that can be expressed in equation form.

The normalized transfer function sensitivities for gain, phase, and delay for first- and second-order factors are summarized in Tables 11.1, 11.2, and 11.3.

Table 11.1 **Transfer function gain sensitivity**

Gain expressions (*in* dB)
Gain constant: K
$S^{G_i}_{K} = 8.686$
Single-pole or zero factor: $(s + \omega_s)$
$S^{G_i}_{\omega_s} = \dfrac{8.686\Omega_s}{1+\Omega_s^2}$
where $\Omega_s = (\omega/\omega_s)$
Quadratic factor: $\left(s^2 + \dfrac{\omega_i}{Q_i}s + \omega_i^2\right)$
$S^{G_i}_{\omega_i} = \dfrac{8.686\{2(1-\Omega_i^2) + (\Omega_i/Q_1)^2\}}{(1-\Omega_i^2)^2 + (\Omega_i/Q_i)^2}$
$S^{G_i}_{Q_i} = \dfrac{8.686(\Omega_i/Q_i)^2}{(1-\Omega_i^2)^2 + (\Omega_i/Q_i)^2}$
where $\Omega_i = \omega/\omega_i$

Table 11.2 **Transfer function phase sensitivity**

Phase expressions (in radians)
Single factor: $(s + \omega_s)$
$\int_{\omega_s}^{\phi_i} = \dfrac{\Omega_s}{1 + \Omega_s^2} \qquad \Omega_s = \dfrac{\omega}{\omega_s}$
Quadratic factor: $\left(s^2 + \dfrac{\omega_i}{Q_i} s + \omega_i^2\right)$
$\int_{\omega_i}^{\phi_i} = \dfrac{-(\Omega_i/Q_i)(1 + \Omega_i^2)}{(1 - \Omega_i^2)^2 + (\Omega_i/Q_i)^2} \qquad \Omega_i = \dfrac{\omega}{\omega_i}$
$\int_{Q_i}^{\phi_i} = \dfrac{-(\Omega_i/Q_i)(1 - \Omega_i^2)}{(1 - \Omega_i^2)^2 + (\Omega_i/Q_i)^2}$

Table 11.3 **Transfer function delay sensitivity. Note: Expressions take the form $\omega \int_x^D$.**

Delay expressions (in seconds)
Single factor: $(s + \omega_s)$
$\omega_s \int_{\omega_s}^{D} = \dfrac{(1 - \Omega_s^2)}{(1 + \Omega_s^2)^2} \qquad \Omega_s = \dfrac{\omega}{\omega_s}$
Quadratic factor: $\left(s^2 + \dfrac{\omega_i}{Q_i} s + \omega_i^2\right)$
$\omega_i \int_{\omega_i}^{D} = \dfrac{(1 - \Omega_i^2)[\Omega_i^4 + 6\Omega_i^2 + 1 - (\Omega_i/Q_i)^2]}{Q_i[(1 - \Omega_i^2)^2 + (\Omega_i/Q_i)^2]}$
$\omega_i \int_{Q_i}^{D} = \dfrac{(1 + \Omega_i^2)[(1 - \Omega_i^2)^2 - (\Omega_i/Q_i)^2]}{Q_i[(1 - \Omega_i^2)^2 + (\Omega_i/Q_i)^2]} \qquad \Omega_i = \dfrac{\omega}{\omega_i}$

Determination of Frequency and Q Tolerances

Sensitivity analysis can determine the variation in the shape of a transfer function per unit variation in the frequency and Q of a filter section and can be carried further to the components used in a particular circuit. However, in practice, what is needed is to select the circuit design and components that permit the realization of a filter transfer function within specified bounds of the theoretical function.

To accomplish this objective with analytic rigor is difficult because the equations that relate the magnitude and phase characteristics to ω and Q are logarithmic and trigonometric and the variables are nonlinearly coupled. However, design guidelines can be established by examination of the transfer function sensitivities.

Since most electronic components are relatively precise elements with tolerances in the range from 0.1 to 10%, and since intuition tells us that we cannot realize precision filters with very loose tolerance components, it is most useful to examine the transfer function sensitivities to components as a function of 1% variations in the variables ω and Q. Figures 11.1, 11.2, and 11.3 illustrate the sensitivities for first- and second-order factors, amplitude, phase, and delay as a function of a 1% variation in Ω and Q for various values of Q.

Examination of Fig. 11.1*b* reveals that, for high Q sections, the amplitude sensitivity to frequency is very high. For example, a 1% variation in the frequency of a quadratic section with a Q of 10 will produce a variation in amplitude of 0.086 dB at $\Omega = 1$ but 0.86 dB at $\Omega = 0.9$. Linear scaling would suggest that to realize the shape of a filter with a section with a pole Q of 10 within 0.086 dB of the theoretical shape, the pole frequency would have to be accurate to within 0.1%.

Figure 11.1*c* shows that the amplitude sensitivity to variations in Q has a maximum of 0.086 dB for a 1% variation in Q and that the maximum occurs at $\Omega = 1$. The maximum does not vary as a function of Q, but the frequency range over which the sensitivity has an effect does vary with Q. However, the higher the Q, the smaller the frequency range over which it is felt. The same high Q section ($Q = 10$) mentioned in the prior paragraph would only yield a 0.086 dB variation in amplitude for a 1% variation in Q; even then its influence would only be in the immediate vicinity of $\Omega = 1$.

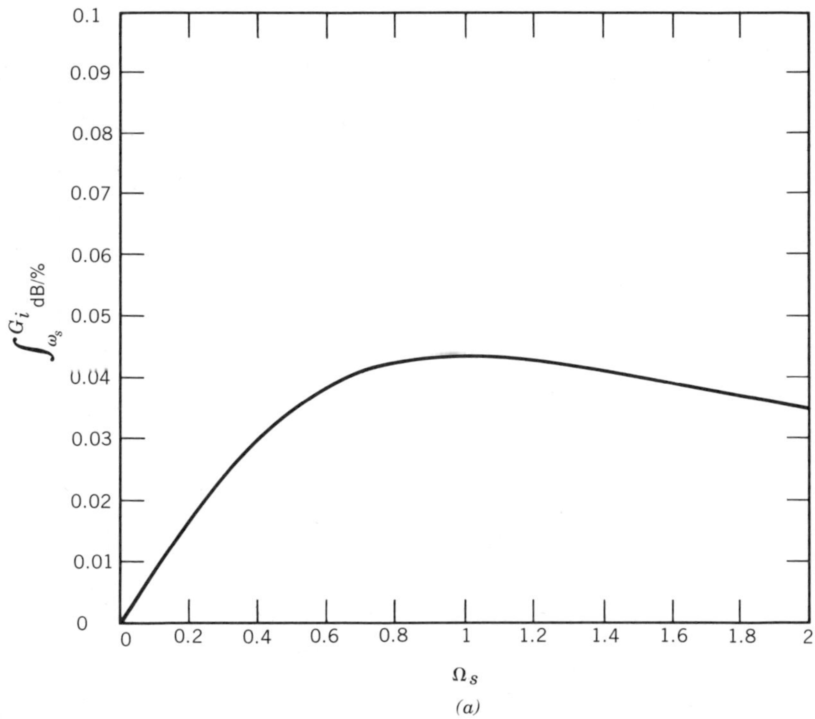

(*a*)

Figure 11.1 Normalized amplitude sensitivites: (a) First-order factor ($S_{\omega_i}^{Gi}$)*; (b) second-order factor* ($S_{\omega_i}^{Gi}$)*; (c) second-order factor* ($S_{Q_i}^{Gi}$)*.*

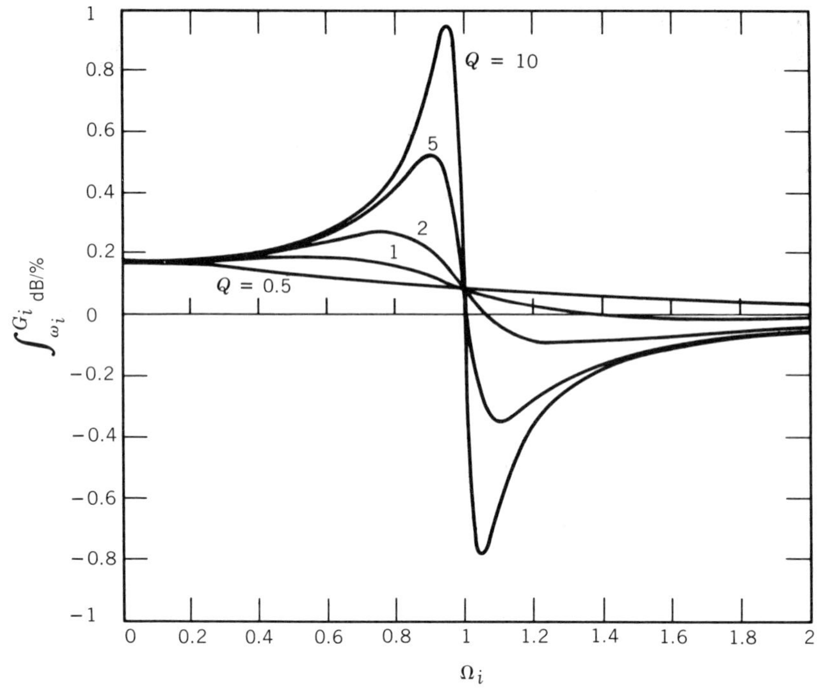

(b)

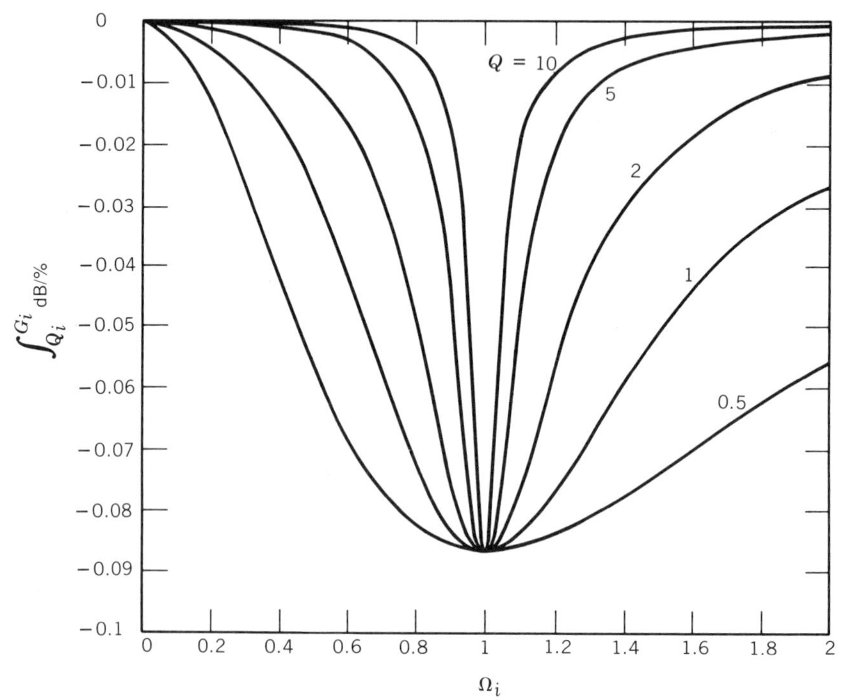

(c)

Figure 11.1 (*b*) and (*c*) (*continued*).

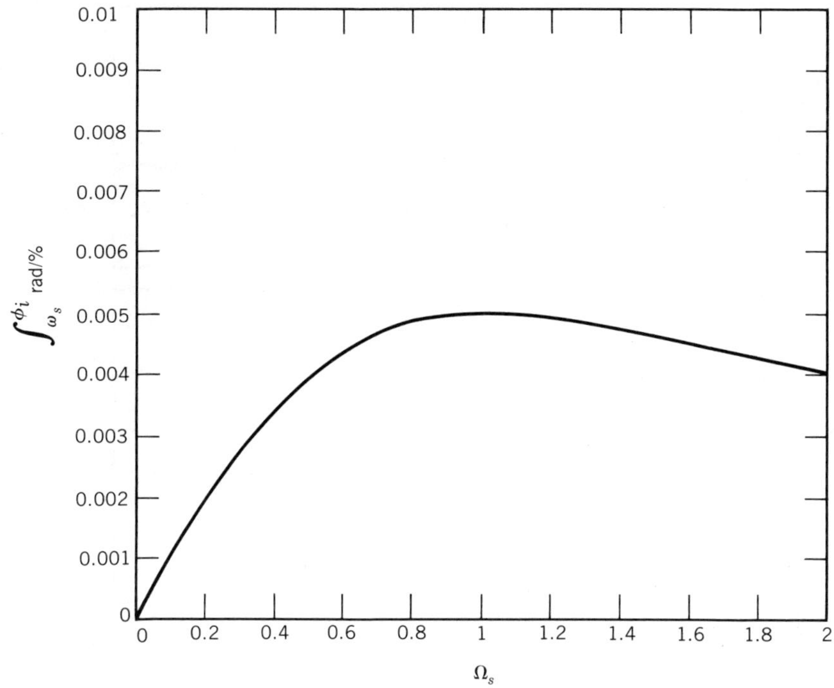

(a)

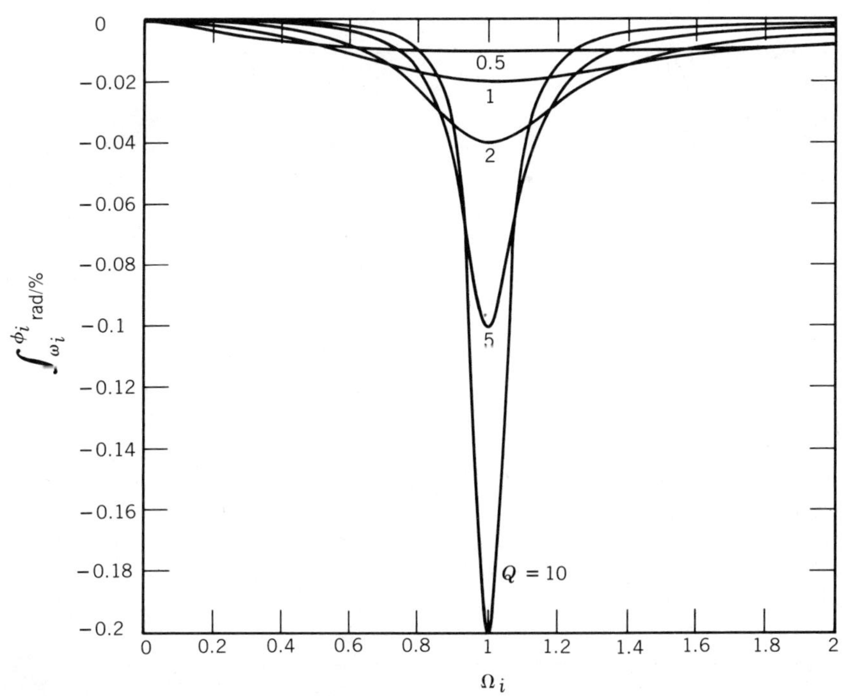

(b)

Figure 11.2 Normalized phase sensitivities: (a) First-order factor ($S_{\omega_s}^{\phi i}$); (b) second-order factor ($S_{\omega_i}^{\phi i}$); (c) second-order factor ($S_{Q_i}^{\phi i}$).

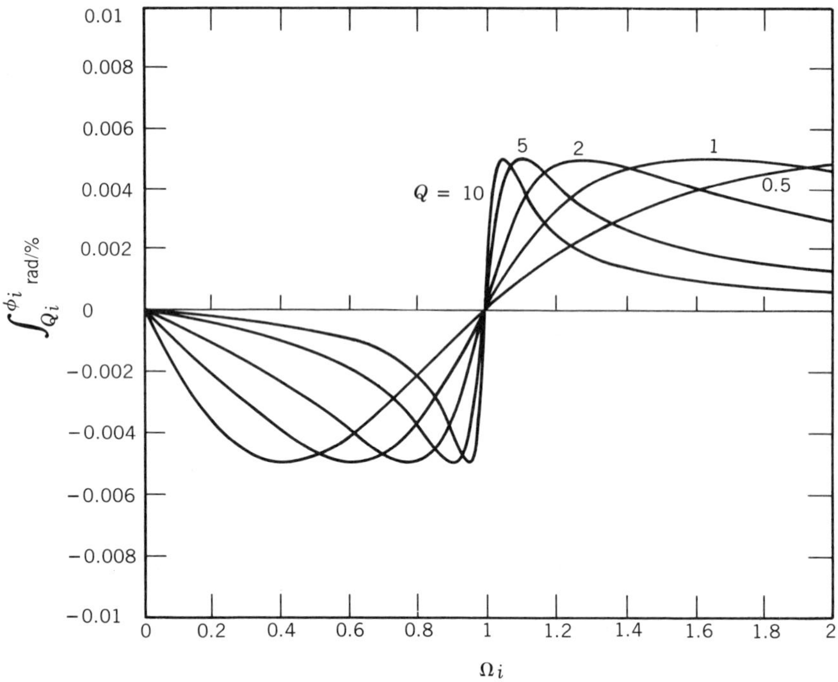

(c)

Figure 11.2 (c) (continued).

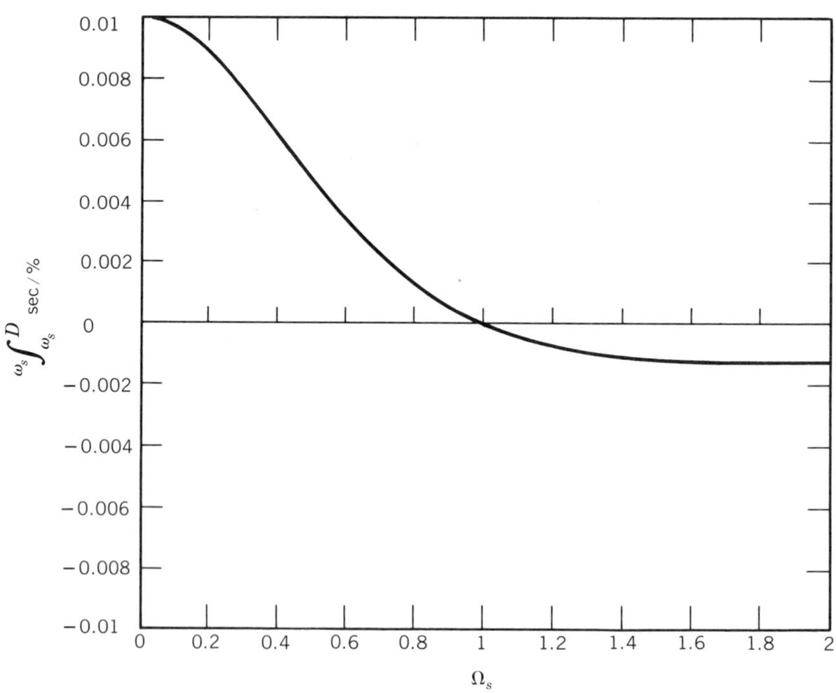

(a)

Figure 11.3 Normalized delay sensitivities: (a) First-order factor ($\omega_s \int_{\omega_s}^{D}$); (b) second-order factor ($\omega_i \int_{\omega_i}^{D}$); (c) second-order factor ($\omega_i \int_{Q_i}^{D}$).

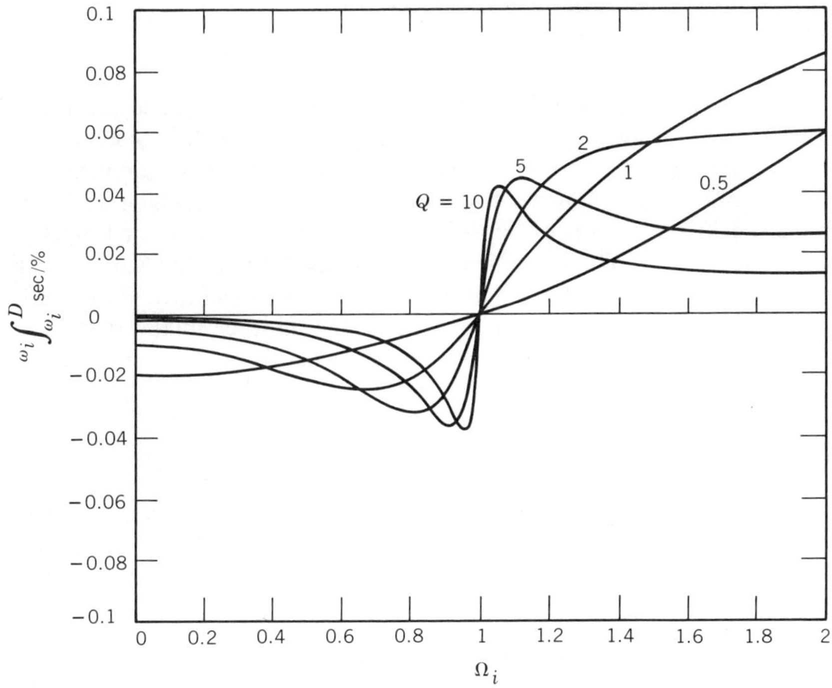

(b)

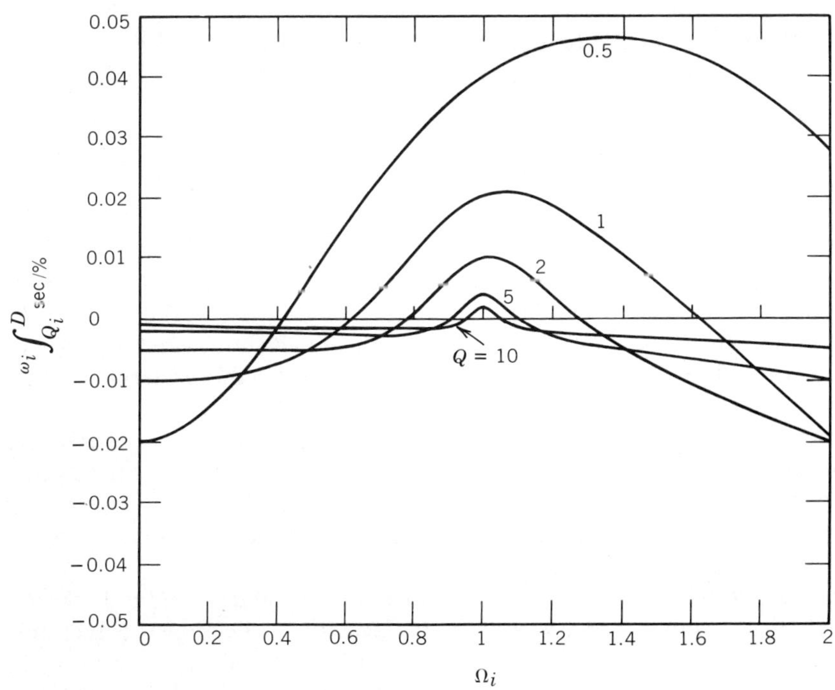

(c)

Figure 11.3 (b) and (c) (continued).

The low sensitivity to Q variation usually means that the frequency accuracy of the sections is more important than that of Q. In practice this low Q sensitivity is often offset by the high Q sensitivity to particular elements in the circuit.

One additional observation worthy of notice for design considerations is that amplitude sensitivity to ω is of opposite sign to that of Q and, for low Q, is similar in shape for $\Omega < 1$. Therefore, if a circuit is used where the variation in frequency and Q have the same sensitivity to a particular element, the amplitude variations caused by that element tend to cancel one another in the passband. This is useful in selecting which element should be varied to tune a section to the correct frequency.

An Eight-Pole Butterworth Design Example

An eight-pole Butterworth low-pass transfer function is commonly used and provides a good example of how variations in frequency and Q can affect the accuracy of the realization of the ideal function. The factored form of this function is given by

$$G_{8B}(j\omega) = K \frac{\omega_{p1}^2 \cdot \omega_{p2}^2 \cdot \omega_{p3}^2 \cdot \omega_{p4}^2}{\left(s^2 + \dfrac{\omega_{p1}}{Q_{p1}} s + \omega_{p1}^2\right)\left(s^2 + \dfrac{\omega_{p2}}{Q_{p2}} s + \omega_{p2}^2\right)\left(s^2 + \dfrac{\omega_{p3}}{Q_{p3}} s + \omega_{p3}^2\right)\left(s^2 + \dfrac{\omega_{p4}}{Q_{p4}} s + \omega_{p4}^2\right)} \tag{11.11}$$

For Butterworth transfer functions all of the section frequencies are at the -3 dB corner frequency, but the section Qs are all different

$$K = 1 \qquad \omega_{p1} = \omega_{p2} = \omega_{p3} = \omega_{p4} = 1$$

$$Q_{p1} = 0.5098 \qquad Q_{p2} = 0.6013$$

$$Q_{p3} = 0.9000 \qquad Q_{p4} = 2.5629$$

For this example the most illustrative analysis is to evaluate the shape of the magnitude of the transfer function in the passband for $\pm 1\%$ variation in each of the ω_{pi} and Q_{pi} and for the worst case, when they are all simultaneously varied $\pm 1\%$ as shown in Fig. 11.4 and 11.5. As seen in Fig. 11.4 the variation of the frequency of the highest Q pole causes a deviation of -0.140 dB from the theoretical curve at $0.82f_o$. If it were desirable to keep the magnitude curve within 0.2 dB of its theoretical value, and all other ω_{pi} and Q_{pi} were exactly correct, the frequency of the high-Q pole would have to be accurate to within 1.4%. Similarly, if they were the only errors, ω_{p1}, ω_{p2}, and ω_{p3} would each have to be accurate to 2.3%. The maximum sensitivities of ω_{p1}, ω_{p2}, and ω_{p3} all occur at f_o.

As illustrated in Fig. 11.5, all of the Q variations produce amplitude variations that are maximum at $\omega = 1$ and are only 0.086 dB/%. The lower Q sections have an effect over a wider portion of the passband than the higher Q sections. If all the sections are simultaneously 1% high in Q, the maximum deviation from

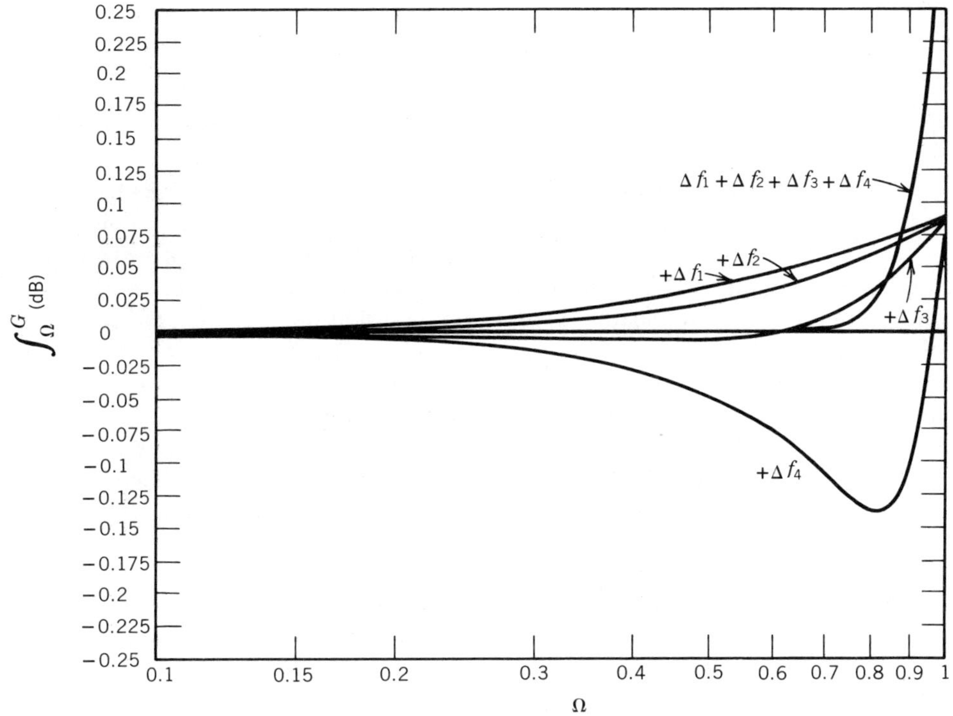

Figure 11.4 ΔG *versus* $\Delta\Omega$ *for an eight-pole Butterworth low-pass transfer function.*

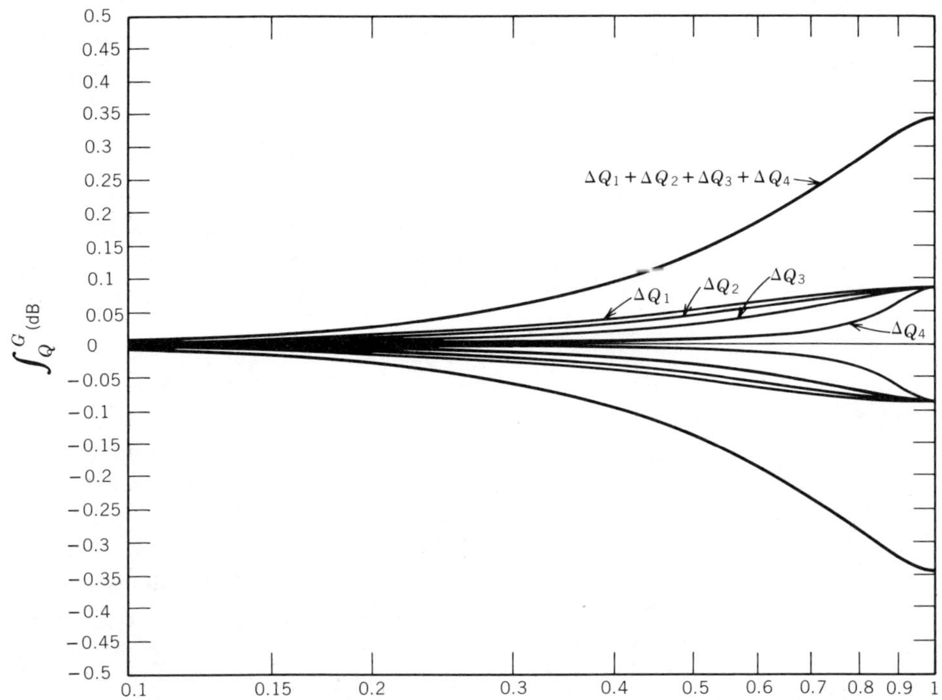

Figure 11.5 ΔG *versus* ΔQ *for an eight-pole Butterworth low-pass transfer function.*

the ideal amplitude is 0.35 dB and occurs at the corner frequency. At half the corner frequency it is approximately 0.15 dB.

For a Butterworth transfer function, all of the ω and Q variations have their maximum effect at or near the corner frequency. While this has the obvious disadvantage that errors add together, it also affords some opportunities for simplified "multistage" tuning that permits accurate realization with a minimum of testing and trimming. For example, the theoretical phase shift through the first two stages of an eight-pole Butterworth filter at the corner frequency is 180° and through all four stages, 360°. This property allows very simple tuning or adjustment of the individual stages by use of a Lissajous pattern on an oscilloscope. In addition, the similarity in shape of the transfer functions of the individual sections allows for the tuning of "two stages" at a time. If an eight-pole Butterworth filter is implement by four cascaded two-pole sections, and the first and third sections are "kitted" to a frequency accuracy of 2%, the second stage may be trimmed in frequency to give 180° of phase shift through the first two cascaded stages at f_o. Following this, the fourth stage is trimmed in frequency to give 360° of phase shift through the entire four cascaded stages at f_o. As a result, the phase shift through the filter will be within 1 to 2° and the amplitude will be within 0.1 to 0.2 dB of the theoretical value.

After completion of the sensitivity analysis and determination of the ω and Q accuracies that must be maintained to implement the function within acceptable error bounds, it is necessary to select a circuit that can realize the required ω and Q to the necessary accuracies. The ability to trim ω and Q (and in some cases the ability to trim them independently of one another) is critical to the "manufacturability" of the filter.

Design By "Error Budget" Analysis

Although the Butterworth transfer function has all of its pole frequencies located at f_c, other transfer functions have their poles distributed over the passband (Chebyshev and elliptic) and, in some cases, outside the passband (Bessel). Elliptic transfer functions have zeros in the stopband that contribute to the shape of the amplitude response.

To determine the ω and Q accuracies needed to implement a design within required tolerances, it is necessary to determine the regions of the passbands and stopbands where each filter pole or zero exerts its primary influence. At the beginning of this section it was demonstrated that the primary influence of a biquadratic function is limited to a narrow frequency range near the pole frequency for a high Q section and is spread out over a wide band of frequency for a low Q section. For small variations in ω and Q the total variation in amplitude, phase, or delay is, to a first-order approximation, linear. The overall variation of these properties to incremental variations in several ω and Q variables can, for all practical purposes, be treated by linear system analysis techniques.

A useful design tool to determine ω and Q tolerances from section sensitivities and filter specifications is to assign an "error budget" to various frequency values on the desired function curve and to use linear system mathematics (matrix algebra) to compute the permissible variations. This is best done by determining the frequency at which the specified parameter (e.g., amplitude) has its highest sensitivity for each independent variable (e.g., ω and Q) and also evaluating the sensitivity to the other variables at those frequencies. Then, using the allowed variation in the specified parameter, the following sensitivity matrix can be set up and solved for the independent variable tolerances:

$$\begin{aligned} \Delta G_{f_{1,\max}} &= \int_{f_1}^{G} \Big|_{f_{1,\max}} \Delta f_1 + \int_{f_2}^{G} \Big|_{f_{1,\max}} \Delta f_2 + \cdots + \int_{f_n}^{G} \Big|_{f_{n,\max}} \Delta f_n \\ &\vdots \\ \Delta G_{f_{n,\max}} &= \int_{f_1}^{G} \Big|_{f_{n,\max}} \Delta f_1 + \int_{f_2}^{G} \Big|_{f_{n,\max}} \Delta f_2 + \cdots + \int_{f_n}^{G} \Big|_{f_{n,\max}} \Delta f_n \end{aligned} \qquad (11.12)$$

Matrix techniques may be used to solve this array for $\Delta f_1 \cdots \Delta f_n$. A similar set of equations can be solved to determine permissible Q variations and then a judgment can be made as to how to split the error budget between $\Delta\omega$ and ΔQ.

If we use the eight-pole Butterworth example previously discussed, with a maximum allowable amplitude deviation of 0.2 dB, the equations are obtained with the aid of Fig. 11.4 as

$$G_{w=1} \qquad 0.2 = 0.086\,\Delta f_1 + 0.086\,\Delta f_2 + 0.086\,\Delta f_3 + 0.075\,\Delta f_4 \qquad (11.13a)$$

$$G_{w=1} \qquad 0.2 = 0.086\,\Delta f_1 + 0.086\,\Delta f_2 + 0.086\,\Delta f_3 + 0.075\,\Delta f_4 \qquad (11.13b)$$

$$G_{w=1} \qquad 0.2 = 0.086\,\Delta f_1 + 0.086\,\Delta f_2 + 0.086\,\Delta f_3 + 0.075\,\Delta f_4 \qquad (11.13c)$$

$$G_{w=0.82} \qquad 0.2 = 0.070\,\Delta f_1 + 0.060\,\Delta f_2 + 0.030\,\Delta f_3 + 0.145\,\Delta f_4 \qquad (11.13d)$$

Because, for this example, three of the frequency sensitivities are equal, Eqs. 11.13a, 11.13b, 11.13c are identical and the array does not have a unique solution. However, since the sensitivities are the same, we can make the assumption that $\Delta f_1 = \Delta f_2 = \Delta f_3$ so that the equations reduce to

$$0.2 \text{ dB} = 0.258\,\Delta f_1 + 0.075\,\Delta f_4 \qquad (11.13e)$$

$$0.2 \text{ dB} = 0.167\,\Delta f_1 + 0.140\,\Delta f_4 \qquad (11.13f)$$

Hence Δf_1, Δf_2, and $\Delta f_3 = 0.55$ and $\Delta f_4 = 0.77$

Thus, if Δf_1, Δf_2, and Δf_3 are kitted or tuned to within an accuracy of 0.55% and Δf_4 within 0.77% and all Qs are exactly correct, the amplitude curve will be within 0.2 dB of the theoretical value. Actually, since the accuracy was "forced" at $0.82 f_0$ and since the sensitivities below this frequency are smaller than they are above it, the deviation from ideal will be much less at lower frequencies.

Similarly, if the Qs are all varied and the frequencies held constant, the equations that relate the sensitivity to the tolerance can be evaluated. Once again, because the maximum sensitivities to Q are at the section frequencies (and all the section frequencies are at f_0), the four equations are identical:

$$0.2 = 0.086\,\Delta Q_1 + 0.086\,\Delta Q_2 + 0.086\,\Delta Q_3 + 0.086\,\Delta Q_4 \tag{11.14}$$

Assigning equal weighting to the variations in Q, yields

$$\Delta Q_1 = \Delta Q_2 = \Delta Q_3 = \Delta Q_4 = 0.58\%$$

In this case, as with the frequency variations, forcing a 0.2 dB error bound at f_0 will result in a much tighter error bound at frequencies below the corner.

Because, in this example, $\Delta\omega$ and ΔQ tolerances are all of the same order of magnitude, it is desirable to split the 0.2 dB error specification into approximately equal amounts of 0.1 dB. Therefore, to realize this 0.2 dB amplitude bound, both $\Delta\omega$ and ΔQ for all sections must be maintained to tolerances of the order of 0.28%. Accordingly, the circuit design, component tolerances, and trim procedures must be selected to achieve this accuracy.

Other transfer functions will have amplitude variations that are more spread out over the passband, and in some cases it may be desirable to constrain one or more points in the stopband to meet a specification. This can be accommodated by including an equation with the stopband tolerance and sensitivities in the set. In every case it is necessary to have as many independent equations as there are independent variables. When it is desired to constrain both amplitude and phase characteristics, the tolerances of $\Delta\omega$ and ΔQ to realize the G and ϕ specs should be calculated separately and the tighter bounds imposed on $\Delta\omega$ and ΔQ. Most often, phase constraints will impose tighter tolerances on $\Delta\omega$ and ΔQ.

Once we have determined the $\Delta\omega$ and ΔQ requirements, we must then select circuits and components to realize these specifications.

11.3 COMPONENT PROPERTIES

The ω and Q of active filters depend upon the properties of the passive and active components from which they are fabricated. In what follows, the most significant characteristics of resistors, capacitors, and amplifiers are discussed, and specific types of components are recommended. Other secondary effects such as printed circuit board layouts, stray capacitance, encapsulating materials, and so on, are also discussed.

Resistors

The resistors used in active filter circuits perform three basic functions. The most common is as a part of an RC time constant, and in that capacity it is usually a frequency-determining element. In some circuits a resistor determines the Q of the circuit. The second function is that a resistor ratio determines the gain in a circuit. Finally, resistors are used to provide amplifier bias and offset current and to adjust for offset voltage.

When used as a frequency-determining element, the absolute value, tolerance, and temperature coefficient of the resistor are critical to the performance of the circuit. When used as gain elements, the ratio tolerance and temperature coefficient are most important; and when used for bias elements, the tolerance and temperature coefficient are relatively unimportant. The following resistor types find uses as indicated.

Discrete Resistors

Metal Film Resistors. Deposited metal film resistors of the type RN55C or their commercial grade equivalents have become very popular for use in active filter circuits. They are available in $\frac{1}{4}$ and $\frac{1}{8}$ W, 1% tolerance, and 25, 50, and 100 ppm/°C temperature coefficients at very low cost (typically, $0.02 each). Tolerances as tight as 0.1% are available at a considerable cost penalty. These are frequently used as frequency- and gain-determining elements and also (for convenience of manufacturing) for the bias elements. For tight tolerance requirements (0.1%) it is common to select 1% resistors or add a frequency trim resistor to achieve the required accuracy.

Carbon Film Resistor. Deposited carbon film resistors are available in $\frac{1}{8}$ and $\frac{1}{4}$ W sizes with 2, 5, 10, and 20% tolerances with temperature coefficients from 200 to 500 ppm, and are very low in cost. They are generally not usable as frequency-determining elements but are sometimes used for gain stages because, within a lot from one manufacturer, they typically track to within 50 to 100 ppm/°C. They are commonly used for bias current and offset voltage trims.

Cermet Resistors. Ceramic-metallic composition resistors with tolerances as tight as 0.01% and temperature coefficients as low as 2 ppm/°C are available for applications where very tight accuracy or stability are required. These components are generally larger than metal film resistors and are very costly (typically, $1,00 to $5.00 each). Therefore, they are used only in applications where tight specifications override cost considerations.

Resistor Networks

Thin-Film Resistor Networks. Thin-film networks are fabricated by vapor deposition or sputtering techniques on silicon substrates. These networks have basically the same temperature coefficients (25, 50, 100 ppm/°C) as discrete metal film resistors. However, because each resistor in the network is deposited at the same time with the same material on the common substrate, their temperature coefficient tracking characteristics are excellent. Ratio temperature coefficients of the order of 2 to 5 ppm/°C are common. Laser trimming techniques yield absolute trimming accuracies as close as 0.1% and ratio trimming accuracies of better than 0.1%. These networks are usually custom-made for a particular application and are generally used in hybrid circuits that are assembled by the chip and wire technique. In very large volumes their prices per resistor approach discrete metal film resistors.

Thick-Film Resistor Networks. Thick-film resistor networks are fabricated by screening conductive ink onto ceramic substrates and firing in an oven. The resistors that result have the same general characteristics as discrete carbon film devices.

When screened and fired, the resistors are usually of very loose tolerance (15 to 20%). However, both abrasive and laser trim techniques work well, and absolute tolerances of 1% and ratio tolerances of 0.5% are readily achievable. Temperature coefficients to 100 to 200 ppm/°C are common; some manufacturers, with careful preparation of inks and processing, can achieve 50 ppm/°C. As with thin-film resistors, since all resistors on a network are screened at the same time, with the same ink, and on the same substrate, the ratio temperature coefficient is much better than the absolute temperature coefficient—typically by a factor of five.

Thick-film resistor networks are generally custom-fabricated on ceramic substrates to which chip capacitors and semiconductors are solder reflowed or chip and wire bonded to complete an active filter circuit.

Diffused Silicon Resistor Networks. Semiconductor manufacturers have developed various methods whereby diffusion techniques are used to fabricate "polysilicon" resistors in silicon substrates. These methods were devised to diffuse resistors directly into the silicon from which the semiconductors are made, thereby creating an entire circuit within a monolithic silicon block. These resistors are generally not good enough to be used as frequency-determining elements because of their loose tolerances (typically 20 to 50%) and poor temperature coefficients (typically 300 to 500 ppm/°C). However, their ratio characteristics are generally very good, and they find widespread use in applications such as multiplying D-A converters where ratio match and tracking are the important parameters. These components are not available as purchased parts at this time and are discussed here only by way of introducing the process that semiconductor manufacturers use at this time to include resistor networks in their products.

Capacitors

In active filter circuits, capacitors are used as frequency-determining elements and as compensation, bypass, or decoupling capacitors. As frequency-determining elements the tolerance, variation with temperature, and dissipation factors are critical to the operation of the filter circuits in which they are used. The tolerance and temperature coefficient directly influence the frequency stability of the filter circuit, and dissipation generally has a first-order effect on the Q of the circuit. The most commonly used capacitors for active filter circuits are now summarized.

Monolithic Ceramic Capacitors Several manufacturers produce capacitors that utilize single and multilayer ceramic dielectric with interleaved metallic conductive layers. These devices are assembled as "sandwiches" while the ceramic is in the green state and fired in high-temperature ovens to produce a solid monolithic ceramic/metallic structure that has excellent mechanical properties. End terminations to connect the parallel plates are added after the ceramic slabs are scored and broken into individual capacitors. Various lead-bonding techniques are used to produce axial- and radial-leaded devices that are protected with epoxy coatings, molded epoxy cases, and molded glass cases. The chip capacitors without leads are also available to use as components in hybrid circuits.

This type of capacitor is the most popular for use in active filters because of its small size, moderate cost, variety of lead configurations, and stability and dissipation characteristics. Several blends of ceramic with different dielectric properties are used to produce devices of this type. Of these, it is the "NPO" formulation that is the most useful in active filter circuits.

NPO Ceramic (Ultra Stable). Ceramic formulations are available that produce capacitors with very low temperature coefficients over a wide operating temperature range. These components are generally called NPO (negative-positive-about zero) and maintain a temperature coefficient of 0 ± 30 ppm/°C over the temperature range from -55 to $+125$°C. Over the commercial 0 to 70°C temperature range they are typically well within specification and 0 ± 10 ppm/°C is commonly achieved. These capacitors are available from 10 to 4700 pF in small sizes, at moderate cost, and in larger values with considerable increase in both size and cost. Tolerances from 1 to 20% are available, with 5 and 10% units being realized by normal manufacturing process control. Tighter tolerance units (1 and 2%) are usually selected and, as a result, are more costly. However, some manufacturers are experimenting with laser and abrasive trimming, and 1% capacitors at low cost may soon be available. Also, there is not much cost advantage in increasing tolerances from 10 to 20%. A common practice in the active filter industry is to purchase the lower-cost 5 or 10% components and to kit or trim the resistors that are needed to realize the tighter frequency specifications.

The dissipation factor of NPO capacitors is usually of the order of 0.1% at 1 kHz and is generally caused by the series resistance of the termination, the resistance of the plates, and the losses in the ceramic dielectric. The dissipation factor is relatively independent of temperature. The insulation resistance is typically of the order of 1000 MΩ-μF and is generally so high that it does not have any significant effect on the active filter circuit.

These components are very popular for use as the frequency-determining elements in active filters. They are available in axial leaded, radial leaded, and leadless configurations, and their mechanical integrity is such that they are virtually indestructible.

Metallized Polycarbonate Capacitors The next most popular capacitor type for use in active filters is the wound metallized polycarbonate capacitor (MPC). These devices are manufactured from a very thin metallized polycarbonate film that is wound on a bobbin; the foil ends are crimped and leads attached. An outer layer of mylar foil is applied and the ends are filled with epoxy, thus coining the term "wrap and fill" construction. They are available over the range 0.001 to 10 μF in reasonable sizes.

These components, when properly manufactured and handled during filter manufacturing, are very good for use as frequency-determining elements. They are available in 1, 2, 5, 10, and 20% tolerances, with 5 and 10% being the result of normal manufacturing control and components with tolerances of 1 and 2% being sorted from the 5 and 10% batches. As with NPO capacitors, it is common to purchase the looser tolerance components and then to kit or trim the resistors to accompany the capacitors that are used.

The dissipation of MPC capacitors is usually of the order of 0.15% at 1 kHz, and the temperature coefficient over the range from 0 to 70°C is generally 0 ± 50 ppm/°C. For temperatures below 0°C and above 70°C they vary more rapidly.

The MPC capacitors have to be handled carefully during storage and manufacture. The hermeticity of the wrap-and-fill construction technique is not very good and polycarbonate film is hygroscopic. Absorbed moisture can significantly change both the capacitance value and the dissipation. In addition, stresses on, and heating of, the ends of the capacitor during soldering can change their values and dissipation. Therefore, to maintain stable characteristics it is desirable to store MPC capacitors in closed containers with dessicant and to stress-relieve leads when bending them in the assembly process. It is also advisable to handsolder their leads and clean the solder connection by hand rather than expose them to the heat cycle and submersion cleaning that are generally used with wave soldering techniques.

Because of the large volume of their wound layers, MPC capacitors are susceptible to RF pickup; it often is desirable to use an NPO ceramic capacitor in parallel with the MPC to reduce this problem. They are also prone to change value after assembly. Therefore, it is desirable to use MPC capacitors that are encapsulated in epoxy boxes or to encapsulate the entire circuit as soon as it is trimmed in manufacturing.

Other Capacitors Several other types of capacitor are used in active filter circuits, each with their advantages and limitations. A selection of these types follows.

Metallized mylar, mylar, and polycarbonate foil are constructed by the same techniques as the MPC capacitor. However, mylar as a dielectric has somewhat higher dissipation and all three are prone to change with temperature more than MPC capacitors.

Mica capacitors, particularly dipped, silvered mica capacitors are good quality, low loss, stable components for use in filter circuits. However, they are generally larger and more expensive than their ceramic NPO counterparts.

Polystyrene capacitors have excellent temperature coefficient and dissipation characteristics and are relatively inexpensive. However, extreme care must be taken in handling them because the polystyrene melts at a very low temperature and the capacitor can permanently change value or be destroyed by ordinary soldering techniques.

Devices such as Teflon, polysulfonate foil, and others have special purpose applications for high temperature and so on, but are generally larger and more expensive than their MPC counterparts.

Amplifiers

As indicated in Chapter 4, the active elements in active filter circuits are usually integrated operational amplifiers. Although some circuits have a higher sensitivity to particular parameters than others, in general, it is the frequency response characteristics of the amplifiers that limit the performance of the circuits in which they are used.

Other properties impose constraints that bound the operational characteristics of the filter but do not have a direct influence on the degree of accuracy to which the circuit realizes the desired theoretical transfer function. The influence of various amplifier properties on active filter circuits will now be discussed.

Gain Bandwidth Product and Slew Rate Limiting The uncompensated open-loop frequency response of most amplifiers at low frequencies is primarily determined by their high impedance input circuitry. Second- and third-order break frequencies at higher frequencies are introduced by the intermediate gain stages and the output stage. To ensure stable operation in closed-loop configurations, they are internally or externally compensated to provide for adequate phase margin at their unity gain crossover frequency. For operation at frequencies below their gain bandwidth (GB), these devices can be modeled as a dc gain with a first-order rolloff that begins at a very low frequency and remains a first-order rolloff for several decades until it crosses unity gain at a frequency (f_t), called its gain bandwidth product. The GB is essentially the frequency at which the amplifier will have 3 dB of attenuation when it is used in a unity gain inverting feedback configuration.

The proximity of the GB to the corner frequency of an active filter circuit and the circuit configuration used determine by how much the amplifier modifies the performance of the filter circuit. The GB and the slew rate are determined by the geometry and fabrication method of the device, and the designer must use circuits that minimize its influence. Some amplifiers have provisions for external compensation that can increase the gain bandwidth; however, this is usually at the expense of degradation of other parameters. The two most prevalent methods of such compensation are two-pole and feed-forward compensation.

To conserve operating current, most integrated operational amplifiers utilize a class B output stage that can drive 20 to 50 times its quiescent bias current. To achieve this, a "push-pull" transistor pair is used in the output stage. When the amplifier is used in a closed-loop configuration, the intermediate drive stage must slew rapidly through a one or two diode voltage drop during the transition from push to pull and conversely. When attempting to provide the voltage swing and drive current to drive the junction capacitance of the output stage, the intermediate gain stage reaches its speed and current limits. For output voltage slew rates above its capacity, it can no longer drive the output stage fast enough to maintain linear operation and the output exhibits slew rate limiting. This specification is given in V/μs; typical values range from 0.5 to 20 V/μs for the integrated operational amplifiers available at this time. Some low current amplifiers have slower slew rates and some amplifiers use class A/B or class A output stages to achieve higher slew rates, but they do so at the expense of considerably higher quiescent currents.

Except for a few devices that have provisions for two-pole or feed-forward external compensation, the slew rate is a function of the amplifier design geometry and fabrication method and the user is required to accept this limitation. In design practice, care must be taken to ensure that the maximum slew rate of the amplifier remains within its capacity. As long as this restriction is complied with, and linear operation is maintained, the slew rate has no real effect on the operation of the amplifier in the filter circuits in which it is used.

The slew rate limits the frequency at which an amplifier can provide large voltage swings at its output and therefore determines the large signal frequency response capability of the device. The large signal response curves are often given in graphical form by the manufacturer and are usually an easier design tool to work with than the slew rate. It is incumbent upon the designer to use a device that will provide a large signal bandwidth wide enough to ensure linear operation over the frequency range required for the operation of the filter.

Direct Current Gain The low-frequency open loop gain of most integrated operational amplifiers is of the order of 100,000 to 1,000,000. When used in closed-loop unity gain configurations, the term $A/1 + A$ appears as a gain multiplier in cascade with the dynamic response and, for some circuit configurations, the ω or Q terms also are modified by this term. However, since A is such a large number, this multiplier varies from unity by only one part in 10^{-5} or 10^{-6} and therefore does not significantly affect the circuit. However, there are circuit configurations for which this term must be evaluated to determine if it makes a significant contribution.

One example of gain-limiting the performance of a circuit is in the multiple-feedback bandpass configuration, where it can be shown that A must be $\gg 4Q^2$ in order not to significantly reduce the actual Q of the circuit. In that case, a gain of 1,000,000 would only be $\gg 4Q^2$ for $Q < 50$ and for a gain of 100,000, $Q < 17$. Other circuits have similar sensitivities. In circuits where one is attempting to realize high Q or high gain, care must be taken not to "run out of gain."

Input and Output Impedance Operational amplifiers are designed with low-current, high-input impedance input stages and high-current, low-output impedance output stages. The input impedance is a function of the geometry of the device, the diffusion process, and the fabrication technology. Values from 0.5 MΩ for bipolar technology to 10^9 MΩ for FET input devices are common. In most active filter circuits the amplifier is used in a negative feedback closed-loop configuration that in effect multiplies the open-loop input impedance and divides the open-loop output impedance by the loop gain. For most active filter circuits, where the amplifier is used for frequencies two or more decades below its f_t, and at high enough frequencies so that the passive element impedance levels are two or more decades below the input impedance of the amplifier, the assumptions of infinite input impedance and zero output impedance are justifiable. However, when the frequency of the circuit approaches $0.1f_t$, there is not much gain left to reduce the output impedance, and the impedance levels of the passive circuitry are lowered to values for which the output impedance should be considered as an element in the circuit and its effect taken into account. Similarly, at very low frequencies, where the passive components have impedances in the MΩ range, the effect of the amplifier input impedance must be considered.

The input capacitance of amplifiers is generally a few picofarads. In most low-pass filter circuits it simply adds to the value of a passive capacitor. However, in high-pass filter circuits the input capacitance of the amplifier often results in a capacitor divider that degrades the performance of the filter at high frequencies. This effect is significant only when the frequency of the high-pass filter section approaches the f_t of the amplifier.

Input Offset Voltage and Bias Current The properties of the junctions of semiconductors and the requirement to provide bias current for the input stage result in input offset voltage and input bias current in operational amplifiers. These properties do not affect the ω and Q of the filter but do introduce dc voltage in the output that is not proportional to the input dc voltage. This output offset is usually on the order of a few millivolts and is generally of little importance in bandpass or high-pass filter circuits. However, for low-pass filters where the dc information is used, the offset represents an error in the output. Some amplifiers have offset adjustment capability and for those that do not the circuit in which it is used can usually be modified to provide an offset trim.

In some circuits, especially low-frequency low-pass filters where resistance values are in the MΩ range, the bias current can contribute significant offset voltage. This can often be compensated for by using comparable impedance levels in both of the inputs of the amplifier, thereby taking advantage of the fact that the input offset current, which is the difference between the + and − input bias currents, is generally a factor of 10 or more smaller than the bias current.

The input bias current of FET input amplifiers is much smaller than bipolar inputs. However, the temperature dependence of the input offset voltage of FET input devices is higher than for bipolar input devices.

Noise Voltage and Current The randomness of the flow of the electrons across the *P-N* junction of the semiconductor material and other physical phenomena manifest themselves as a randomness in the output voltage and input current of an operational amplifier, called "noise." That is, even with no input signal applied to the device, there is an output voltage that is random in nature. When spectrum-analyzed, the noise usually has very low amplitude components over a broad frequency range. This noise is generally specified in units of $\mu V/\sqrt{Hz}$. The noise generated within the device limits its usable dynamic range, since, for very small output voltages, the signal level is smaller than the noise level and is obscured by it.

Most active filter circuits modify the noise spectrum of the amplifiers used in them by enhancing the portion of the spectrum near ω_o by an amount proportional to Q. However, for multistage filters the noise of each stage is shaped by the transfer function of subsequent stages. Therefore, the output noise of an active filter is largely that of its output amplifier, enhanced by the Q of the output stage and the filtered spectrum of the prior stages.

Low noise amplifiers have been designed for instrumentation applications and are generally considerably more costly than the general-purpose devices. However, in some low-signal-level applications, where noise is the critical parameter, their cost may be justified.

Other Properties Numerous other amplifier properties constrain the specifications of their circuits. Not much can be done about them except to reflect their influence in the specifications of the filter. Some of these follow.

Input/Output Voltage Range. Most amplifiers accommodate input and output voltage swings that are within one or two diode voltage drops of the power supplies. A few have input/output swings that can go all the way to the power voltages. Most inputs can be damaged if driven beyond the power supply voltages.

Output Current Drive. The current gain of the output stage and its drive stage limit the amount of current that the amplifier can drive. Care must be exercised to ensure that the load presented to the output stage can be driven by it. In active filter circuits the feedback load must be taken into account as well as the external load that the filter must drive. Because of the influence of the amplifier as a feedback element, superposition does not necessarily apply, and it is common for the impedance of an element to be in effect divided by Q over some portion of the frequency range. Loop or nodal current analysis techniques that include the active elements must generally be used.

Many amplifiers have asymmetric drive capability because their PNP output transistor can drive more output current on the positive-going swing than their NPN output transistor can sink on the negative-going swing.

Short-Circuit Protection. Most operational amplifiers are designed with internal current limiting circuitry that prevents the destruction of the devices when the output is shorted to power ground. Therefore, accidents during use, solder shorts, or overloading of the output to ground do not usually damage the amplifier. However, an accidental connection of the output to either $+V_s$ or $-V_s$ will permanently damage most amplifiers.

Power Supply Consumption. Power supply current requirements for most devices range from 300 μA to 5 mA per amplifier. The lower-current devices also usually have lower bandwidth and output drive capability than the higher-current devices.

When selecting the power supply for a filter, consideration must be given to the fact that the current delivered to the output load might be much larger than the zero-signal-level, quiescent current requirement of the filter. Higher-frequency filters use lower internal impedance levels, and the internal current drive needs might also exceed the quiescent current.

Power Supply Line Noise Rejection. A typical specification for the amount of power supply line noise that couples into the input of an amplifier is -80 dB or 10^{-4}. That is to say, 10 mV of noise on the power line becomes 1 μV of noise on the output. This, like other input noise, is enhanced by the filter circuit so that care must be taken to minimize it. Power supply decoupling can be used to reduce high-frequency line noise.

Feedforward compensation bypasses the differential input stage at high frequencies and dramatically reduces the noise rejection capability. For some devices it can reduce it from −80 dB to 0 dB. Extreme care must be taken to high-frequency decouple the power supplies when feedforward compensation is used.

Common Mode Rejection Ratio (CMRR). Most amplifiers have a wide dynamic range for common mode input signals and a typical CMRR of 100 dB. In filter circuits, where the amplifier is used in the inverting mode, with its positive input grounded, this parameter has no effect. When used as a follower or in the non-inverting mode, both the finite open-loop gain and the CMRR reduce the closed-loop gain of the circuit. Because the frequency response characteristics of the + and − inputs vary considerably, an amplifier that has a 100 dB CMRR at 100 Hz may only have 60 dB at 10 kHz.

Amplifiers Frequently Used in Filter Circuits There is a proliferation of amplifier products available today that perform well in active filter circuits. Some long-established parts that have been used include the following.

Vendor	*Part number*
Fairchild	μA 301A
National Semiconductor	LM310, LF351
Raytheon Co.	RC4136
Texas Instruments	TL081, 82, 84

These parts are by no means all-inclusive; several other products are also available from these and other vendors. They are presented here merely as examples of products in use today.

Other Considerations

Since active filter circuits are intended to be frequency-selective and rely on capacitors to determine their transfer functions, care must be taken in their design to minimize the effects of stray capacitance. The basic design considerations are the same as those for high-frequency amplifiers and include the following precautions.

Circuit Layout The design considerations for printed circuit board layouts for discrete component assemblies and ceramic substrate layouts for hybrid assemblies are basically the same. Some practical guidelines follow.

1 Keep the length of the runs to the (−) inputs of the amplifiers as short as possible. If they have to be long, it is desirable to have a grounded run or a ground shield surrounding them.

2 Keep the runs from low impedance, high signal level outputs from other amplifiers from the same filter stage and amplifiers from other stages as far away from the (−) inputs as possible.

3 Decouple the power supply frequently. The ideal is to decouple every amplifier with capacitors located as close as possible to it; however, this is expensive and usually not necessary. For high Q circuits a pair of decoupling capacitors for every 1 to 2 inches of power run is usually adequate. For low Q circuits less decoupling is necessary. Ceramic monolithic capacitors of 0.01 μF or greater are recommended because of their low series resistance at high frequencies.

4 Ground wiring for high Q circuits is critical. A very small potential summed into a sensitive point can result in a large change in Q or create an oscillator. It is best to bring ground runs from individual amplifiers, or at least from all of the amplifiers of a filter section, back to a common ground mode on a separate run. Avoid long, strung-out ground runs, where the ground circuits from one stage sum together with those from other stages.

5 Component placement is also important. Although most capacitive coupling problems are usually associated with the printed wiring of a circuit, the relative placement of components can also be critical. The capacitive coupling between the bodies of two resistors placed in parallel and close together can exceed that of the runs on the printed wiring board. The same guidelines that apply to the layout of the wiring runs also apply to the proximity of the components on the circuit. Keep large signals from low impedance sources away from (−) inputs.

Encapsulating Materials and Protective Coatings It is generally desirable to encapsulate active filters to protect the components, to maintain a more uniform temperature distribution, and to reduce the effects of humidity on the capacitors. However, care must be taken to use low dielectric materials to minimize the added distributed capacitance. A glass microballoon-filled epoxy with a low dielectric constant, a low exotherm and low coefficient of shrinkage during the cure cycle, and a low coefficient of thermal expansion is desirable. Conformal coatings and fluidized bed application methods for epoxy coating can also be used.

Note that encapsulated polycarbonate capacitors of wrap and fill construction can change value because they are squeezed during the cure cycle.

11.4 PRACTICAL CONSIDERATIONS FOR SALLEN–KEY LOW-PASS FILTER SECTIONS

The classical Sallen–Key two-pole, low-pass filter section (see Chapter 6) shown in Fig. 11.6 is often used as a second-order building block for multipole active filters because of its simplicity of design and minimum number of circuit elements. The common simplification, where the amplifier is assumed to have no frequency dependence and an ideal gain of unity, yields a very simple solution for R_1 and R_2 in terms of C_1, C_2, ω_0, and Q.

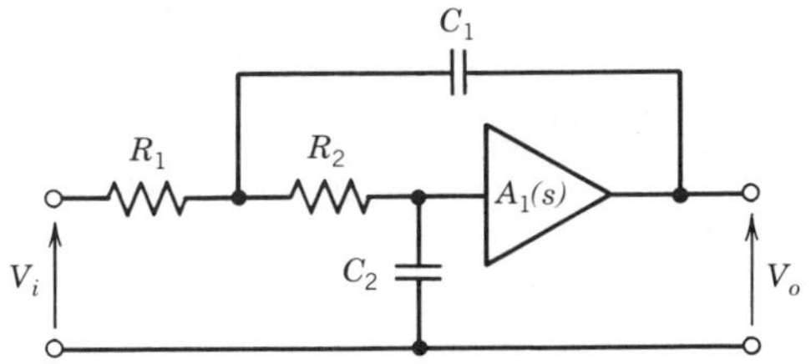

$$\omega_i = (R_1R_2C_1C_2)^{-\frac{1}{2}}$$

$$Q_i = \frac{\sqrt{R_1R_2}}{R_1 + R_2} \cdot \sqrt{\frac{C_1}{C_2}}$$

Figure 11.6 Sallen–Key low-pass single-amplifier filter.

When the amplifier is nonideal, the ω_0 and Q of the second-order section are modified and the amplifier's frequency response characteristics increase the order of the circuit to a third- (or higher-) order function. The design equations that include the amplifier frequency response characteristics, the sensitivities of ω_0 and Q to component variations, a circuit compensation method, and a design example are presented in the following sections.

Sallen–Key Transfer Function and Component Sensitivities

The transfer function for the Sallen–Key circuit given in Fig. 11.6 is

$$\frac{V_0}{V_i} = \frac{\dfrac{A_1(s)}{R_1R_2C_1C_2}}{s^2 + s\left\{\dfrac{1}{R_1C_1} + \dfrac{1}{R_2C_1} + \dfrac{[1 - A_1(s)]}{R_2C_2}\right\} + \dfrac{1}{R_1R_2C_1C_2}} \tag{11.15}$$

The amplifier's dynamic response $A_1(s)$ enters into the transfer function in the numerator as a cascaded function and as a coefficient multiplier in the denominator. If a single-pole rolloff model is used for the controlled source amplifier, then:

$$A_1(s) = \frac{K_1}{1 + \tau s} \approx K_1(1 - \tau s) \qquad \text{where } K_1 = \frac{A}{1 + A}, \tau = \frac{1}{\omega_t}$$

and A is the open-loop gain of the amplifier. Substitution into Eq. 11.15 yields

$$\frac{V_0}{V_i} = \frac{K_1(1 - s\tau)\{R_1R_2C_1C_2 + R_1C_1K_1\tau\}^{-1}}{s^2 + \left\{\dfrac{(R_1 + R_2)C_2 + (1 - K_1)R_1C_1}{R_1R_2C_1C_2 + R_1C_1K_1\tau}\right\}s + \dfrac{1}{\{R_1R_2C_1C_2 + R_1C_1K_1\tau\}}}$$

$$= \frac{\dfrac{K_1}{R_1R_2C_1C_2}}{s^2 + \left\{\dfrac{\left(1 + \dfrac{R_2}{R_1}\right)\dfrac{C_2}{C_1} + (1 - K_1)}{R_2C_2 + \tau K_1}\right\}s + \dfrac{1}{R_1C_1\{R_2C_2 + K_1\tau\}}} \tag{11.16}$$

Hence

$$Q_0 = \frac{[(R_1C_1)(R_2C_2 + K_1\tau)]^{1/2}}{(R_1 + R_2)C_2 + R_1C_1(1 - K_1)} \tag{11.17a}$$

$$\omega_0 = \{R_1C_1[R_2C_2 + K_1\tau]\}^{-1/2} \tag{11.17b}$$

and

$$\left.\frac{V_0}{V_i}\right|_{s=0} = K_1 = K = \text{dc gain} \tag{11.17c}$$

These expressions for ω_0, Q_0, and K can be used to evaluate the classical sensitivities

$$\int_x^{\omega_0}, \quad \int_x^{Q_0}, \quad \text{and} \quad \int_x^{K}$$

However, it will be shown later that the effects of the τ and K_1 of the amplifier on ω_0 and Q_0 can be easily expressed in terms of the ideal $(A(s) = 1)\,\omega_i$ and Q_i, where, if

$$A(s) = 1 \qquad \tau = 0 \qquad K_1 = 1$$

then

$$\omega_i = (R_1R_2C_1C_2)^{-1/2} \tag{11.18a}$$

$$Q_i = \frac{\sqrt{R_1R_2}}{R_1 + R_2}\sqrt{\frac{C_1}{C_2}} \tag{11.18b}$$

$$\left.\frac{V_0}{V_i}\right|_{s=0} = K_i = 1 \tag{11.18c}$$

The classical sensitivities for these ideal ω_i and Q_i are presented in Table 11.4. The effects of a nonideal $A(s)$ on ω_0 and Q are analyzed below, and a compensation method is given to minimize their influence.

Table 11.4 **Classical sensitivity expressions for the Sallen–Key single-amplifier low-pass circuit of Fig. 11.6**

Element x	$\int_x^{\omega_i}$	$\int_x^{Q_i}$	$\int_x^{K_i}$
R_1	-0.5	$0.5 - \dfrac{R_1}{R_1 + R_2}$	0
R_2	-0.5	$0.5 - \dfrac{R_2}{R_1 + R_2}$	0
C_1	-0.5	0.5	0
C_2	-0.5	-0.5	0

Component Calculations

The denominator polynomial of Eq. 11.16 can be solved in terms of the ω_0 and Q of the low-pass filter to be realized and the variables R_i, C_i, and K. Since it is generally desirable to use fixed capacitors and select or trim resistors, the most useful solution is obtained by solving for the resistors R_1 and R_2 in terms of the other variables. For the ideal case, where $A(s) = 1$:

$$R_1, R_2 = \frac{1}{4\pi f_0 Q C_2}\left[1 \pm \sqrt{1 - \frac{4Q^2 C_2}{C_1}}\right] \tag{11.19}$$

Therefore, an R_1 and an R_2 can be obtained as long as the radical is positive, which requires that $C_1/C_2 > 4Q^2$.

The most economical realization of a Sallen–Key low-pass filter section can be obtained by selecting loose tolerance capacitors with a ratio adequate to ensure the desired Q, measuring their values, and computing, selecting, or trimming the values of R_1 and R_2 to set the ω_0 and Q at their required values.

These results consider only the ideal case for $A(s) = 1$ and therefore ignore a real amplifier's effects on ω_0 and Q. These effects, and how to correct for them when they are significant, are dealt with in the following section.

Amplifier Effects on ω_0 and Q_0

If the ω_0 and Q_0 expressions of Eq. 11.17a and 11.17b are written in terms of ω_i and Q_i, then:

$$\omega_0 = \omega_i\left[1 + \frac{K_1\tau}{R_2C_2}\right]^{-1/2} \tag{11.20a}$$

$$Q_o = Q_i\frac{[1 + K_1\tau/R_2C_2]^{1/2}}{2Q_i^2(1 - K_1) + 1} \tag{11.20b}$$

As seen from 11.20a, the frequency of the two-pole section is decreased as the R_2C_2 product approaches the gain bandwidth time constant $\tau(\tau = 1/\omega_t)$. The Q_0 is increased by the same factor as the ω_0 is decreased. In addition, the Q_0 is decreased by a $2Q_i^2(1 - K_1)$ term if $K_1 \neq 1$.

Since the large signal bandwidth is usually a decade or more below ω_t, the R_2C_2 time constant is generally a factor of 10 or more larger than τ. Therefore, the frequency shift is generally less than 5% and, if the filter frequency is two decades below ω_t, then the frequency shift is less than 0.5%.

The open-loop gain of most common operational amplifiers is of the order of 10^5 to 10^6; therefore, $(1 - K_1)$ is typically $1 - A/(1 + A) = 1/1 + A$ or 10^{-5} to 10^{-6}. Thus the $2Q_i^2(1 - K_1)$ term in Q_0 is of the order of 2×10^{-3} for a Q_i of 10 and $A = 10^{-5}$ and represents a 0.2% change in Q_0.

Both the change in ω_0 and the change in Q_0 are relatively small as long as $Q_1 < 10$, $A > 10^5$, and $\omega_0 < 0.1\omega_t$. In addition, both of these parameters can be made correct by computating the R_1 and R_2 required to realize the desired ω_0

and Q. The values of K_1 and ω_t merely impose a minimum capacitor ratio C_1/C_2 above which real values of R_1 and R_2 result. In other words, these deviations from the ideal can be trimmed out.

The real culprit that interferes with the precise realization of second-order transfer functions is the cascade $A(s)$ that appears because of the amplifier. If ω_0 of the two-pole section is at $0.1\omega_t$ the real pole from the amplifier contributes 0.05 dB of attenuation and 5.7° of phase at ω_0. If several two-pole stages are used to construct a multipole filter, these errors can considerably distort the transfer function.

A Compensation Method

Fortunately, for low-pass Sallen–Key or first-order real poles, there is an extremely simple circuit modification that can virtually eliminate the influence of amplifier dynamics on the ω_0 and Q and, at the same time, remove the additional pole from the circuit. Consider separately the real pole and amplifier combination consisting of R_2C_2 and $A(s)$ from Fig. 11.6. Reconfigure it slightly as shown in Fig. 11.7 by placing a portion of the resistor R_2 in series with the capacitor C_2. The transfer function of the subcircuit is given by

$$\frac{V_o}{V_a} = \frac{R_{2a}C_2s + 1}{(R_{2a} + R_{2b})C_2s + 1} \cdot A(s) = \frac{R_{2a}C_2s + 1}{(R_{2a} + R_{2b})C_2s + 1} \cdot \frac{K}{s/\omega_t + 1}$$

If the time constant $R_{2a}C_2$ is made equal to $1/\omega_t$, the real poles cancel and the transfer function becomes

$$\frac{V_o}{V_a} = \frac{K}{(R_{2a} + R_{2b})C_2s + 1}$$

which, if $R_{2a} + R_{2b} = R_2$ and $K = 1$, is exactly the same as if $A(s) = 1$. Therefore, placing a portion of R_2 in series with C_2 and setting $K = 1$ will, to the degree that the model of the amplifier is valid and ω_t is known, compensate for the presence of the amplifier.

A substantial correction can be made by calculating the value of R_2 using a typical ω_t. However, a reduction of phase error by more than a factor of 10 can be realized by trimming R_2 and K.

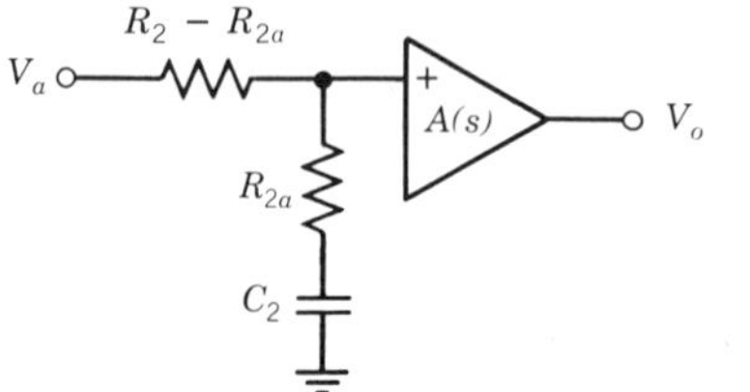

Figure 11.7 Lead-lag compensation for Sallen–Key low-pass circuit.

Trimming Methods for ω_0 and Q_0

The classical form for a second-order low-pass transfer function is

$$H(s) = \frac{V_o}{V_i} = \frac{K\omega_0^2}{s^2 + \dfrac{\omega_0}{Q_0}s + \omega_0^2} \tag{11.21}$$

For $s = j\omega$,

$$H(j\omega) = \frac{V_o}{V_i} = \frac{K\omega_0^2}{\omega_0^2 - \omega^2 + j\,\omega\omega_0/Q_0}$$

$$H(o) = K$$

$$H(j\omega_0) = \frac{KQ}{j} = KQ\underline{/-90^\circ}$$

The most convenient way to trim the ω_0 and Q_0 is to measure and record the dc gain at $\omega \ll \omega_0$ and then set the frequency by adjusting the phase to -90° at $\omega = \omega_0$ and the Q by adjusting the gain to KQ at $\omega = \omega_0$. The adjustment is made by varying R_1 and R_2. Unfortunately, the ω_0 and Q_0 are not independently adjustable because they both vary as a function of R_1 and R_2. However, an interactive trim procedure applies by which one begins with calculated design nominal values of R_1 and R_2 and then adjusts R_1 (or R_2) until a correct -90° phase is obtained at $\omega = \omega_0$. The Q is then adjusted by varying R_1 and R_2 in opposite directions until a correct value of gain (KQ) is obtained at $\omega = \omega_0$ while -90° of phase shift is maintained.

This trim procedure works well with discrete decade box trim resistors or multiturn potentiometers where the resistance value can be increased or decreased. For thick-film or thin-film resistor networks where the resistance value can only be increased by trimming, it is necessary to begin with resistor values that are too low and then sequentially increase them until the phase and gain specifications are simultaneously met.

As indicated in Section 11.4, if $\omega_0 > 0.01\omega_t$, the contribution of the pole of $A(s)$ can introduce a phase error in excess of 0.57° at $\omega = \omega_0$. This can result in inaccurate tuning of the frequency of the section if phase is used as the trim criterion. Therefore, if a highly accurate frequency trim for $\omega_0 > 0.01\omega_t$ is desired, it is necessary to trim R_{2a} to remove the phase contribution of the amplifier. This trim is difficult to achieve because it should be performed at ω_t in order to ensure that $R_{2a}C_2$ is canceling the phase shift from $A(s)$. This can best be done by driving node V_a with a small amplitude signal at ω_t and trimming R_{2a} for 90° of phase shift at V_o. Once the compensating trim is completed, R_1 and R_2 can be trimmed as before.

EXAMPLE 11.1 Design a second-order low-pass filter with a 10 kHz corner frequency and a Q of 2.5 using a LF351 amplifier. Assume that it is desirable to maintain an input impedance level of the order of 20 kΩ.

PROCEDURE

1 Determine the range and ratio of C_1 and C_2.

$$C = \frac{1}{2\pi f R} = \frac{1}{2\pi \times 10^4 \times 20 \times 10^3} = 0.0008\ \mu\text{F}$$

$$C_1 = 2QC = 2 \times 2.5 \times 800\ \text{pF} = 4000\ \text{pF}$$

$$C_2 = \frac{C}{2Q} = \frac{800}{2 \times 2.5\ \text{pF}} = 160\ \text{pF}$$

To ensure $C_1/C_2 > 4Q^2$, use $C_1 = 4700$ pF and $C_2 = 150$ pF.

2 Determine R_1 and R_2 with the aid of Eq. 11.19.

$$R_1, R_2 = \frac{1}{4\pi f_0 Q C_2}\left(1 \pm \sqrt{1 - \frac{4Q^2C_2}{C_1}}\right)$$

$$= \frac{1}{4\pi(10{,}000)(2.5)(150 \times 10^{-12})}\left[1 \pm \sqrt{1 - \frac{4(2.5)^2 150}{4900}}\right]$$

$$= \frac{10^6}{4\pi(2.5)(1.5)}\left[1 \pm \sqrt{1 \pm \frac{2.5 \times 1.5}{4.7}}\right]$$

Hence $R_1 = 30.860\ \text{k}\Omega$ and $R_2 = 11.68\ \text{k}\Omega$.

3 Check ω_0 and Q_0 to determine if amplifier K_1 and τ require that R_{2a} be used to compensate for the amplifier. For the LF351, $A = 10^5$ and $\omega_t = 4$ MHz from which we obtain

$$K_1 = \frac{A}{1 + A} \approx 1 - \frac{1}{A} = 1 - 10^{-5} = 0.99999$$

$$\tau = \frac{1}{2\pi f_c} = \frac{1}{2\pi \times 4 \times 10^6}$$

$$\omega_0 = \omega_i\left[1 + \frac{K_1\gamma}{R_2C_2}\right]^{-1/2}$$

$$= \omega_i\left[1 + \frac{(0.99999)(4 \times 10^{-8})}{(150 \times 10^{-12})(11.68 \times 10^3)}\right]^{-1/2} = 0.98888\omega_i$$

This represents approximately 1.1% reduction in ω_i.

$$Q_0 = Q_i\frac{[1 + K_1\gamma/R_2C_2]^{-1/2}}{2Q^2(1 - K_1) + 1}$$

$$= Q_i\frac{[1 + (0.99999)(4 \times 10^{-8})/(150 \times 10^{-12})(11.68 \times 10^3)]^{1/2}}{2(2.5)^2(1\text{-}0.99999) + 1} = 1.0112Q_i$$

This represents approximately 1.1% increase in Q_i.

If the frequency and Q accuracy requirements are of the order of 1% or less and the unit is not to be trimmed, it is desirable to compensate for the influence of the amplifier or recompute the R_1 and R_2 to set the ω_i 1.13% higher and the Q_i 1.13% lower.

In this case both the ω_0 and Q_0 deviations are small and, if the accuracy requirements demand that they be trimmed, the deviations should be within the normal trim range of R_1 and R_2; therefore, no additional computations are needed.

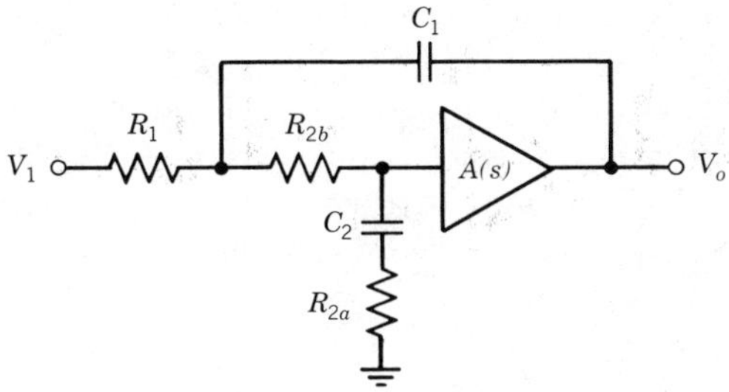

Figure 11.8 Compensated Sallen–Key low-pass filter stage.

4 If compensation for the amplifier is required because of very tight ω_0 and Q_0 accuracy requirements or to eliminate the effect of the additional pole that is introduced by the amplifier, then:

$$R_{2a}C_2 = \frac{K_1}{\omega_t}$$

$$R_{2a} = \frac{(0.99999)}{150 \times 10^{-12} \times 2\pi \times 4 \times 10^6} = 265\ \Omega$$

Note that the change in Q_0 resulting from K_1 not being exactly unity is only 0.0125%; therefore, it is not necessary to set or trim the gain K_1.

Hence the final component values to be used for the compensated Sallen–Key stage of Fig. 11.8 are

$$R_1 = 30{,}769\ \Omega \qquad C_1 = 4700\ \text{pF}$$
$$R_{2a} = 265\ \Omega \qquad C_2 = 150\ \text{pF}$$
$$R_{2b} = 11{,}415\ \Omega$$

11.5 PRACTICAL CONSIDERATIONS FOR MULTIPLE-LOOP FEEDBACK (MF) BANDPASS FILTER SECTIONS

The single-amplifier, two-pole (one-pole pair), multiple-loop feedback (MF) active filter section of Chapter 7 is shown in Fig. 11.9. It is commonly used for low to medium Q bandpass filter sections because of its simplicity of design, the minimum number of circuit elements, and the ease of tuning its center frequency. The circuit

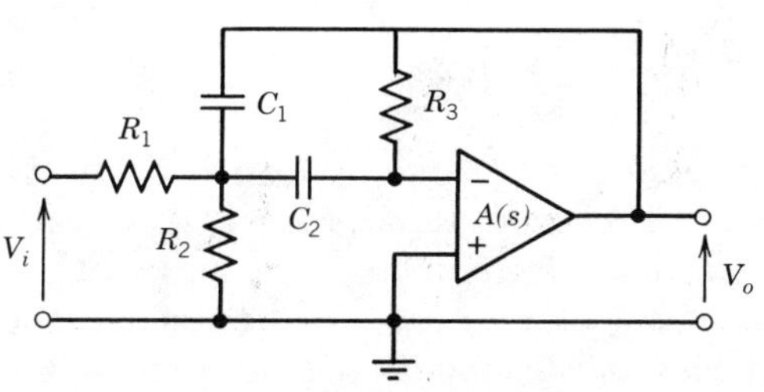

Figure 11.9 Multiple-loop feedback bandpass filter.

simplicity results in easy computation of its element values for the ideal case when the amplifier is assumed to have infinite gain.

When the amplifier is nonideal the ω_0 and Q of the second-order section are modified, and the amplifier's frequency response characteristics increase the order of the circuit to a third (or higher)-order function. In this section the design equations that include the amplifier's frequency response characteristics, the sensitivities of ω_0 and Q to component variations and amplifier parameters, a circuit compensation method, ω_0 and Q tuning methods, and a design example are presented.

MF Transfer Function and Component Sensitivities

The transfer function for the MF circuit given in Fig, 11.9 is

$$\frac{V_0}{V_i} = \frac{\dfrac{-A(s)}{1-A(s)} \cdot \dfrac{-1}{R_1C_1}}{s^2 + \left[\dfrac{1}{R_3\dfrac{C_1C_2}{C_1+C_2}} + \dfrac{1}{\dfrac{R_1R_2}{R_1+R_2}C_1(1-A(s))}\right]s + \dfrac{1}{\dfrac{R_1R_2}{R_1+R_2}R_3C_1C_2}} \tag{11.22}$$

In the special case where $A(s) = -\infty$, Eq. 11.22 reduces to

$$\frac{V_o}{V_i} = \frac{\dfrac{-s}{R_1C_1}}{s^2 + \dfrac{1}{R_3\dfrac{C_1C_2}{C_1+C_2}}s + \dfrac{1}{\dfrac{R_1R_2}{R_1+R_2}R_3C_1C_2}} \tag{11.23}$$

from which

$$\omega_i = \left[\frac{1}{\dfrac{R_1R_2}{R_1+R_2}R_3C_1C_2}\right]^{1/2} \tag{11.24a}$$

$$Q_i = \left[\frac{R_3C_1C_2}{\dfrac{R_1R_2}{R_1+R_2}(C_1+C_2)^2}\right]^{1/2} \tag{11.24b}$$

$$G_i = \frac{V_o}{V_i}(\omega_i) = \frac{-R_3C_2}{R_1(C_1+C_2)} \tag{11.24c}$$

These expressions for the ideal ω_i, Q_i, and G_i can be used to evaluate the classical sensitivities of these parameters to the element values R_1, R_2, R_3, C_1 and C_2 as given in Table 11.5, as well as to calculate the component values required to realize a designed ω_i, Q_i and G_i. The effects of a nonideal $A(s)$ on ω_o, Q, and G are analyzed in what follows, and compensation methods are given to minimize their influence.

Table 11.5 **Classical sensitivity expressions for the multiple-loop feedback bandpass filter of Fig. 11.9**

Element x	$S_x^{\omega_i}$	$S_x^{Q_i}$	$S_x^{G_i}$
R_1	$\frac{-0.5R_2}{(R_1+R_2)}$	$\frac{-0.5R_2}{(R_1+R_2)}$	-1
R_2	$\frac{-0.5R_1}{(R_1+R_2)}$	$\frac{-0.5R_1}{(R_1+R_2)}$	0
R_3	-0.5	0.5	$+1$
C_1	-0.5	$0.5-\frac{C_2}{(C_1+C_2)}$	$\frac{-C_1}{C_1+C_2}$
C_2	-0.5	$0.5-\frac{C_1}{(C_1+C_2)}$	$\frac{+C_1}{C_1+C_2}$

Component Calculations

The numerator and denominator polynomial of Eq. 11.23 can be utilized to calculate the values of R_1, R_2, and R_3 in terms of C_1, C_2, and the desired ω_0, Q_0, and G. It can be shown that the minimum resistor ratio of R_3 to $R_1 \| R_2$ for a given Q can be realized when $C_1 = C_2$. This is also more convenient for component purchase, kitting, and assembly. Therefore, in the following calculations the capacitors are made equal. The design equations are then:

Define:

$$C_1 = C_2 = C \tag{11.25a}$$

$$R = \frac{1}{2f_i C} \tag{11.25b}$$

Then:

$$R_3 = 2Q_i R \tag{11.25c}$$

$$R_1 = \frac{R_3}{2G_i} \tag{11.25d}$$

$$R_2 = \frac{Q_i R}{2Q_i^2 - G_i} \qquad \text{for } G_i < 2Q_i^2 \tag{11.25e}$$

Amplifier Effects on ω_0 and Q

The MF transfer function given by Eq. 11.22 is expressed in terms of the ideal ω_i, Q_i, and G_i of Eqs. 11.25a, 11.25b, and 11.25c and the amplifier's transfer function $A(s)$ is

$$\frac{V_o}{V_i} = G_i \cdot \frac{-A(s)}{1-A(s)} \cdot \frac{(\omega_i/Q_i)s}{s^2 + (\omega_i/Q_i)\{1 + 2Q_i^2/1 - A(s)\} + \omega_i^2} \tag{11.26}$$

The amplifier transfer function $A(s)$ enters into the filter transfer function both as a premultiplier and as a term in the denominator polynomial. When $A(s) = -A/(\tau s + 1)$, Eq. 11.26 becomes

$$\frac{V_o}{V_i} = G_i \cdot \frac{A}{1+A} \cdot \frac{1}{\dfrac{\tau}{1+A}s + 1} \cdot \frac{\dfrac{\omega_i}{Q_i}s}{s^2 + \dfrac{\omega_i}{Q_i}\left[1 + \dfrac{2Q_i^2}{1+A}\left[\dfrac{\tau s + 1}{\dfrac{\tau}{1+A}s + 1}\right]\right]s + \omega_i^2} \tag{11.27}$$

Equation 11.27 can be expanded and rewritten with a third-order denominator, which can then be factored into a real root and a complex root pair whose ω_0 and Q_0 are expressed in terms of ω_i, Q_i, τ, and A. This process results in very complex expressions that are difficult to use. A simpler approach that clearly brings out the practical limitations of the useful range of frequency and Q for this circuit is to consider Eq. 11.27 for three different frequency ranges and use simpler approximations for the amplifier frequency response characteristic in order to examine its effect on ω and Q.

For filters whose frequencies are less than the internal break frequency of the amplifier $(1/\tau)$, the amplifier can be approximated by its open-loop gain. Then $A(s) \approx -A$ and Eq. 11.27 reduces to

$$\frac{V_o}{V_i} = G_i \cdot \frac{A}{1+A} \cdot \frac{\dfrac{\omega_i}{Q_i}s}{s^2 + \dfrac{\omega}{Q_i}\left[1 + \dfrac{2Q_i^2}{1+A}\right]s + \omega_i^2} \tag{11.28}$$

This may be expressed as

$$\frac{V_o}{V_i} = G_o \frac{\dfrac{\omega_0}{Q_0}s}{s^2 + \dfrac{\omega_0}{Q_0}s + \omega_0^2} \tag{11.29}$$

where

$$\omega_0 = \omega_i \tag{11.30a}$$

$$Q_0 = Q_i\left[\frac{1}{1 + \dfrac{2Q_1^2}{1+A}}\right] \tag{11.30b}$$

$$G_o = G_i\left[\frac{A}{A + 1 + 2Q_1^2}\right] \tag{11.30c}$$

The primary effect of the finite gain A of the amplifier is the reduction of Q_0 from the ideal value by the factor $1 + [2Q_i^2/(1 + A)]$.

An amplifier with an open-loop gain of 120 dB or 10^6 will reduce an ideal Q_i of 50 by approximately 0.5%, whereas a gain of 100 dB or 10^5 will reduce it by 5%. Since most integrated operational amplifiers that are available today have open-loop gains in the range of 100 to 120 dB, the MF bandpass filter circuit is limited to realizing Qs in the range of 25 to 50 and then only for frequencies below the internal break frequency of the amplifier.

The gain G_o at ω_0 is also reduced by the finite open-loop gain of the amplifier by an amount comparable to the reduction in Q. Both the gain and Q deviations can be corrected for by calculation or component selection if the value of A is known and is stable.

However, there is a significant variation in A from lot to lot, and within a lot of amplifiers, even the four amplifiers of a quad device will differ from one another. In addition, A varies by as much as 10 dB over the normal operating temperature range of the device. Therefore, it is generally not possible to make a compensation for G and Q that remains stable. Therefore it is generally not desirable to permit the amplifier properties to reduce Q by more than a few percent.

In the frequency range between the internal break frequency of the amplifier but below the unity gain bandwidth frequency $\omega_t = A/\tau$ the amplifier transfer function can be approximated by $A(s) \approx -\omega_t/s \gg 1$. Equation 11.27 then reduces to

$$\frac{V_o}{V_i} \approx \frac{G_i}{\left[1+\dfrac{2Q_i\omega_i}{\omega_t}\right]} \cdot \frac{\dfrac{\omega_i}{Q_i}s}{s^2+\dfrac{\omega_i}{Q_i\left[1+\dfrac{2Q_i\omega_i}{\omega_t}\right]}s+\dfrac{\omega_i^2}{\left[1+\dfrac{2Q_i\omega_i}{\omega_t}\right]}} \tag{11.31}$$

from which

$$\omega_0 = \frac{\omega_i}{\left[1+\dfrac{2Q_i\omega_i}{\omega_t}\right]^{1/2}} \approx \omega_i\left[1+\frac{Q_i\omega_i}{\omega_t}\right]^{-1} \tag{11.32a}$$

$$Q_0 = Q_i\left[1+\frac{2Q_i\omega_i}{\omega_t}\right]^{1/2} \approx Q_i\left[1+\frac{Q_i\omega_i}{\omega_t}\right] \tag{11.32b}$$

$$G_o = G_i \tag{11.32c}$$

Over this frequency range the amplifier affects both the ω and Q of the filter inversely by the same factor. In order that the frequency and Q shifts be less than 1%, the $\omega_i Q_i$ product must be less than $0.01\omega_t$. This poses rather severe restrictions on the useful range of Q_i. To realize a Q of 20 with less than a 1% variation requires $\omega_i < 0.0005\omega_t$. Therefore, an amplifier with an ω_t of 4 MHz could be used for a filter section with a Q of 20 up to only 2 kHz with a 1% decrease in ω and a 1% increase in Q. A 10% deviation in ω and Q would result if the filter were at 20 kHz.

These deviations in ω and Q are not quite as bad as they seem because they can both be corrected by a single trim resistor in parallel with R_2, which increases the ω and decreases the Q to the correct values. In addition the ω_t of an amplifier does not vary as much over temperature as the open-loop gain A; therefore, trimmed filter characteristics will remain reasonably stable.

As a practical guideline, the MF bandpass filter can be built with Qs up to 20 at frequencies up to 1% of ω_t as long as an ω, Q trim is used to correct the deviation caused by the amplifier. For frequencies above $\omega_0 > 0.01\ \omega_t$ the MF bandpass filter has even more restrictions. As ω_t approaches and exceeds ω_t, the Q_0 term is dramatically changed and in the limit as $\omega_0 \gg \omega_t$, $Q_0 = Q_i[1 + 2Q_i^2]^{-1}$. Therefore, a desired $Q_i = 10$ would result in an actual $Q_0 \approx 0.05$. In addition, the first-order factor would already be introducing attenuation and phase shift. As a final point, the model of the amplifier is invalid at these frequencies, so it is not likely that the calculated ω_0 and Q_0 are representative of the real circuit. For these frequency ranges other multiple amplifier circuits must be used.

Compensation Methods

The MF bandpass filter circuit depends heavily upon its amplifier having high enough open-loop gain and gain bandwidth to accurately realize the desired transfer function.

Circuit Compensation The feedback resistor R_3 around the amplifier and the capacitor C_2 into the input of the amplifier are trying to operate as a differentiator with a transfer function of $2Q_i s/\omega_i$, but ideal operation is not achieved because the amplifier runs out of gain and bandwidth. There is no simple circuit modification to compensate for this. The gain of an ideal differentiator approaches infinity as frequency approaches infinity—and practical amplifiers simply do not do that!

The feedback capacitor C_1 and input resistors R_1 and R_2 are attempting to act as an integrator with a transfer function of $\omega_i/2Q_i s$. The amplifier prohibits the R_1, R_2, and C_1 circuit from functioning as an integrator at dc because of its finite open-loop gain and, at high frequencies, because its own internal rolloff introduces an additional pole at ω_t. This pole can be eliminated by introducing a small compensating resistance R_c in series with C_1 such that $R_c C_1 = 1/\omega_t$. This reduces the effect of the amplifier rolloff but does not reduce the dependence of the circuit on the finite open-loop gain. However, it does reduce the ω and Q dependence over the the frequency range $1/\tau < \omega_0 < A/\tau$.

As demonstrated in Section 11.5, in the frequency range $1/\tau < \omega_0 < A/\tau$ where $A(s) \approx -\omega_t/s$, the primary influence of the amplifier is to reduce the frequency and increase the Q by the same factor $[1 + 2Q_i\omega_i/\omega_t]^{1/2}$. Since this quantity can be calculated at least to within the accuracy that ω_t is known, a first-order correction can be made by calculating the reduction in frequency and increase in Q and recomputing R_3, R_1, and R_2 to accommodate these changes.

In this circuit the effect is easy to accommodate because a decrease in R_3 will both increase ω and decrease Q by the same factor. When this is done, R_1 will have to be changed to maintain the correct gain and R_2 will also have to be changed to make $R_1 \| R_2$ remain the same. An alternative to component calculation and kitting is to trim R_3, R_2, and R_1 *in that order* to set frequency, gain, and Q.

Although this method of compensation is effective, examination of the correction term shows that the deviation from ideal of the ω_0 and Q_0 is directly related to Q_i. Even if $\omega_i/\omega_t = 0.01$, a Q_i of 10 results in a 10% deviation in ω_0 and Q_0. Therefore, for accurate realization of a bandpass filter section, it is generally best to use a two or three amplifier realization for bandpass filters when $Q > 20$ and to observe the criterion that the gain of the amplifier at the frequency of the filter section be at least a factor of 10 greater than $4Q^2$.

Amplifier Compensation Some integrated operational amplifiers have external compensation pins and can be compensated by single-pole, two-pole, and feed-forward compensation. Most internally compensated amplifiers use single-pole compensation. The single-pole method provides the amplifier with a first-order internal rolloff, usually via a small capacitor in the differential input stage, that ensures stable operation of the amplifier in the unity gain inverting configuration. With single-pole compensation the amplifier can be modeled quite accurately by the first-order function $A(s) = A/\tau s + 1$ for both the (+) and (−) inputs up to frequencies of $0.1\omega_t$. The model breaks down near ω_t because several other internal time constants come into play, and the amplitude and phase characteristics become higher-order functions. Therefore, single-pole compensation is useful for the amplifier used in a MFBP filter circuit. However, the circuit is limited to the frequency and Q ranges discussed in the previous section.

Two-pole compensation introduces a lead-lag term in the intermediate gain stage of an amplifier that increases the frequency range for which the amplifier operates at full open-loop gain. However, to approach ω_t with a phase margin adequate to maintain stable closed-loop operation, it is necessary to have a second-order rolloff for two or more decades to attenuate the gain, then to go back to first-order rolloff at least a decade before ω_t. When using two-pole compensation, a second-order model should be used for both inputs to the amplifier. The results are much the same as for single-pole compensation. It is still necessary to have adequate gain at $\omega_0(\gg 4Q^2)$ to realize the desired section Q. Since the open-loop-gain frequency range is extended by the two-pole compensation, there are some applications where one can realize a higher Q section at a somewhat higher frequency than can be realized with single-pole compensation. However, most amplifiers that can be two-pole-compensated can also be feed-forward-compensated to extend the open-loop-gain frequency range and the full-signal bandwidth.

Feed-forward compensation bypasses the input stage for high-frequency inputs and directly drives the intermediate gain stage. It provides a substantial increase in the large signal bandwidth of the amplifier. This improvement comes at the expense of other parameters so that care must be taken to accommodate the attendant side effects. In this configuration, only the negative (−) input is bypassed and the positive (+) input must be connected to signal ground. The bandwidth of the amplifier is extended to near the frequency of the poles of the output stage. The phase characteristics of the compensated amplifier leave very little phase margin near ω_t; therefore, when used in the unity gain inverting mode, it is necessary to use a feedback capacitor around the amplifier to ensure stable closed-loop operation. There is also a tendency for the output stage itself to oscillate when driving a capacitive load to ground. The bypass of the input dramatically reduces the ability of the input stage to reject power supply line noise and therefore special consideration must be given to adequately decouple the power supplies. In spite of these apparent drawbacks the feed-forward compensation method works very well in MFBP filter sections, where the positive (+) input is grounded and the capacitive feedback of the circuit ensures its stability.

Trimming Methods for ω_0 and Q

The ω and Q of the MF bandpass filter section are determined by the passive components and the open-loop gain and gain bandwidth frequency of the amplifier. Unfortunately, they are not independently adjustable. In many applications the accuracy of the frequency of the bandpass section is more important than its Q and it is adequate to trim ω and accept the resulting Q. The gain G at ω_0 is relatively independent of ω and Q for high Q circuits and can be trimmed with only a second-order effect on ω_0 and Q.

Examination of the sensitivities of ω, Q, and G to R_1, R_2, and R_3 shows that R_1 changes ω, Q, and G in the same direction; R_2 changes ω and Q in the same direction and has no effect on G; and R_3 changes ω and Q in opposite directions and G in the same direction as Q. Therefore, several trimming sequences are appropriate, depending upon what specifications are most important.

If gain is within tolerance and Q accuracy is not important, R_2 can be trimmed to set ω_0. The gain will not vary in trimming ω_0 but the Q will be increased or decreased in the same amount and direction as ω_0.

If gain and frequency need to be trimmed and Q accuracy is not important, trim R_1 and R_2 in this sequence: R_2 to trim ω_0, then R_1 to trim gain, and repeat until both are within specification. As before, the Q will be varied in the same amount and direction as ω_0.

If gain, frequency, and Q all need to be trimmed, the procedure has to be an iterative one, but there is a sequence that will converge. Begin by trimming R_2 to set ω_0 and R_1 to set gain as in the previous section. Measure Q by determining the -3 dB frequencies f_L, f_H, and calculating $Q = (f_L f_H)^{1/2}/(f_H - f_L)$. If Q is too low by $x\%$, increase R_3 by the square of that percentage, then retrim R_2 and R_1 to set the gain and ω_0. Repeat the Q measurement procedure. This method will usually converge on the first attempt if the amplifier is not contributing second-order terms, and in two iterations even if it is.

EXAMPLE 11.2 Design a one-pole pair bandpass filter with a center frequency of 14.6 kHz, a Q of 7.5, and a gain at the center frequency of 2.5. Assume that it is desirable to maintain an input impedance level of the order of 10.0 kΩ.

PROCEDURE

1 Determine the range of value of C_1 and C_2 using Eq. 11.25.

$$R_1 = \frac{R_3}{2G_i} = \frac{2Q_iR}{2G_i} = \frac{7.5R}{2.5} = 3R = 10\ \text{k}\Omega$$

$$R = 3.333\ \text{k}\Omega$$

$$C_1 = C_2 = \frac{1}{2\pi f_i R} = \frac{1}{2\pi(14{,}600)(3333)} = \frac{10^{-6}}{2\pi(14.6)(3.333)}$$

$$C_1 = C_2 = 0.00327$$

Use standard value capacitors of 0.0033 μF.

2 Determine R_1, R_2, and R_3 for $C_1 = C_2 = 0.0033\ \mu$F from Step 1.

$$R = \frac{1}{2\pi f_1 C} = \frac{1}{2\pi(14{,}600)(0.0033 \times 10^{-6})} = 3{,}303\ \Omega$$

$$R_3 = 2Q_iR = 2 \times 7.5 \times 3303 = 49{,}545\ \Omega$$

$$R_1 = \frac{R_3}{2G_i} = \frac{49{,}545}{2 \times 2.5} = 9909\ \Omega$$

$$R_2 = \frac{RQ_i}{2Q_i^2 - G_i} = \frac{3303 \times 7.5}{2 \times 7.5^2 - 2.5} = 225.2\ \Omega$$

Using nearest standard 1% values, we obtain

$$R_3 = 49.9\ \text{k}\Omega$$

$$R_1 = 10\ \text{k}\Omega$$

$$R_3 = 226\ \Omega$$

If the components were exactly at their nominal values, the f_1, Q_1, and G_1 would be 14,523, 7.513, and 2.495, respectively, all of which are well within 1% of their desired values.

3 Check ω_0 and Q_0 to determine if the amplifier characteristics are substantially changing them from ω_i and Q_i. From Eq. 11.32a and 11.32b:

$$\omega_0 = \omega_i\left(1 + \frac{2Q_i\omega_i}{\omega_t}\right)^{-1/2}$$

$$Q_0 = Q_i\left(1 + \frac{2Q_i\omega_i}{\omega_t}\right)^{1/2}$$

If an internally compensated amplifier such as an LF 351 is used, $f_t = 4$ MHz. Then

$$\left(1 + \frac{2Q_i\omega_i}{\omega_t}\right)^{1/2} = \left(1 + \frac{2 \times 7.5 \times 14{,}600}{4{,}100{,}000}\right)^{1/2} = (1.05475)^{1/2} = 1.027$$

Therefore, using such an amplifier, the frequency will be decreased by 2.7% and the Q will be increased by 2.7% from the ideal value.

4 Compensate the circuit to reduce amplifier effects. If we use the circuit compensation method, this amplifier effect can be compensated for by decreasing R_3 to both increase the frequency to the correct value and decrease Q to the correct value. When this is done, R_1 must also be decreased by the same factor in order to maintain the correct gain, R_2 must then be changed to keep the parallel combination as it was in the original design.

For example,

$$R_{3C} = \frac{R_3}{1.05475} = \frac{49{,}545}{1.05475} = 4697\ \Omega$$

$$R_{1C} = \frac{R_1}{1.05475} = \frac{9909}{1.05475} = 9394\ \Omega$$

$$R_{2C} = \frac{1}{(1/R_1) + (1/R_2) - (1/R_{1C})} = \frac{1}{(1/49{,}545) + (1/225.2) - (1/46973)} = 225.25$$

Using standard values, we obtain $R_{3C} = 47.5$ kΩ $R_{1C} = 9.31$ kΩ $R_{2C} = 226\ \Omega$

5 Consider the effect of an amplifier with $\omega_t = 10$ MHz. An alternative to compensating the circuit is to use a faster amplifier that has less effect on the circuit. For example, if a feed-forward-compensated LM 301AN were used, then $f_c = 10$ MHz and its effect on ω_0 and Q_0 would be reduced to the order of 1% and to the same order of magnitude as the uncertainty in the passive component values.

Based on the result from Step 4 the component values needed for the compensated MF bandpass filter are

$$R_1 = 9.31\ \text{k}\Omega \qquad C_1 = 0.0033\ \mu\text{F}$$
$$R_2 = 226\ \Omega \qquad R_3 = 47.5\ \text{k}\Omega$$

11.6 PRACTICAL CONSIDERATIONS FOR STATE VARIABLE (S-V) FILTER SECTIONS

The three-amplifier biquad (one-pole pair), state-variable active filter structure shown in Fig. 11.10 is often used for low-pass, high-pass, and bandpass filter configurations because of its low sensitivity to both passive and active components. In addition, its capability for independent adjustment of frequency (ω_0) and selectivity (Q) provide for ease of fine tuning and permit this circuit configuration to be used in tunable or programmable applications. The independence of the filter

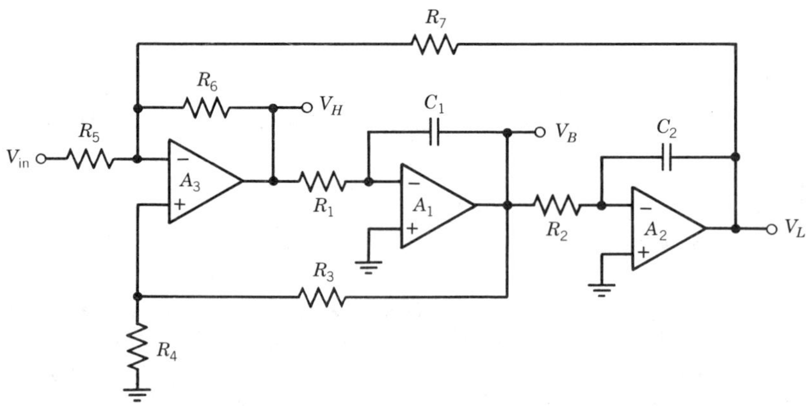

Figure 11.10 Three-amplifier state-variable structure.

parameter variables results in easy computation of its element values for the case where the amplifiers are ideal.

When nonideal amplifiers are used, the ω and Q of the second-order section are modified and the amplifier frequency response characteristics increase the order of the circuit to a third- or higher-order function. In what follows, the design equations for the ideal case, the sensitivities of ω and Q to component variations, circuit compensation methods, and a design example are presented.

S-V Transfer Function and Component Sensitivities

A formal analysis, which includes the three amplifier functions $A_1(s)$, $A_2(s)$, and $A_3(s)$, is formidable. Since this circuit, in various forms, has been analyzed in detail in Chapter 8 and the purpose of this chapter is to focus on practical aspects, we proceed by examining the ideal case and establish how to make the nonideal circuit behave as closely as possible to it.

The transfer function for the S-V circuit of Fig. 11.10 for the low-pass, high-pass, and bandpass outputs for the case of ideal amplifiers are

$$\frac{V_L}{V_{in}} = -\frac{R_7}{R_5} \cdot \frac{\dfrac{R_6}{R_7 R_1 R_2 C_1 C_2}}{s^2 + \dfrac{R_4(R_5 + R_6 + R_7)}{R_1 R_5 (R_3 + R_4) C_1} s + \dfrac{R_6}{R_7 R_1 R_2 C_1 C_2}} \tag{11.33a}$$

$$\frac{V_H}{V_{in}} = -\frac{R_6}{R_5} \cdot \frac{s^2}{s^2 + \dfrac{R_4(R_5 + R_6 + R_7)}{R_1 R_5 (R_3 + R_4) C_1} s + \dfrac{R_6}{R_7 R_1 R_2 C_1 C_2}} \tag{11.33b}$$

$$\frac{V_B}{V_{in}} = \frac{R_6}{R_5} \cdot \frac{\dfrac{s}{R_1 C_1}}{s^2 + \dfrac{R_4(R_5 + R_6 + R_7)}{R_1 R_5 (R_3 + R_4) C_1} s + \dfrac{R_6}{R_7 R_1 R_2 C_1 C_2}} \tag{11.33c}$$

from which:

$$\omega_i = \left[\frac{R_6}{R_7 R_1 R_2 C_1 C_2}\right]^{1/2} \tag{11.34a}$$

$$Q_i = \frac{R_5(R_3 + R_4)}{R_4(R_5 + R_6 + R_7)} \left[\frac{R_1 R_6 C_1}{R_2 R_7 C_2}\right]^{1/2} \tag{11.34b}$$

$$G_{iL} = \left.\frac{V_L}{V_{in}}\right|_{s=0} = -\frac{R_7}{R_5} \tag{11.35a}$$

$$G_{iH} = \left.\frac{V_H}{V_{in}}\right|_{s=0} = -\frac{R_6}{R_5} \tag{11.35b}$$

$$G_{iB} = \left.\frac{V_B}{V_{in}}\right|_{s=jw_i} = \frac{R_6(R_3 + R_4)}{R_4(R_5 + R_6 + R_7)} \tag{11.35c}$$

Table 11.6 **Classical sensitivity expressions for the state-variable filter of Fig. 11.10**

			$\mathcal{S}_x^{G_i}$		
Element	$\mathcal{S}_x^{\omega_i}$	$\mathcal{S}_x^{Q_i}$	*LP*	*HP*	*BP*
R_1	-0.5	0.5	0	0	0
R_2	-0.5	-0.5	0	0	0
R_3	0	$\dfrac{R_3}{R_3 + R_4}$	0	0	$\dfrac{R_3}{R_3 + R_4}$
R_4	0	$\dfrac{-R_3}{R_3 + R_4}$	0	0	$\dfrac{-R_3}{R_3 + R_4}$
R_5	0	$\dfrac{R_6 + R_7}{R_5 + R_6 + R_7}$	-1	-1	$\dfrac{-R_5}{R_5 + R_6 + R_7}$
R_6	0.5	$\dfrac{R_5 - R_6 + R_7}{2(R_5 + R_6 + R_7)}$	0	1	$\dfrac{R_5 + R_7}{R_5 + R_6 + R_7}$
R_7	-0.5	$-\dfrac{R_5 + R_6 + 3R_7}{2(R_5 + R_6 + R_7)}$	1	0	$\dfrac{-R_7}{R_5 + R_6 + R_7}$
C_1	-0.5	0.5	0	0	0
C_2	-0.5	-0.5	0	0	0

The classical sensitivities of these parameters to the element values R_1, R_2, R_3, R_4, R_5, R_6, R_7, C_1, and C_2 are given in Table 11.6.

For most practical applications of the circuit, $R_5 = R_6 = R_7$ and $R_4 < R_3$, which results in all sensitivities being less than unity.

Component Calculations

For this circuit configuration the component values are easy to compute. The frequency, Q, and gain can all be set independently as long as the appropriate sequence is maintained.

Although there are a number of elements that appear to be arbitrarily selectable, there are practical considerations that constrain these choices. In general, it is best to operate both integrators and the feedback amplifier at unity gain at ω_i. This maximizes the dynamic range and minimizes amplitude clipping and saturation. Therefore, it is usually best to select $R_6 = R_7$ and to use equal time constants and time-constant component values for the integrators ($R_1 = R_2$ and $C_1 = C_2$).

This circuit is commonly used for low-pass and high-pass filters with a unity passband gain specification, in which case $R_5 = R_6$ and/or $R_5 = R_7$. When used in the bandpass configuration, the gain resistor R_5 must be selected to insure that the bandpass output has the required gain. When $R_6 = R_7$, $R_1 = R_2$, and $C_1 = C_2$, the parameter equations become

$$\omega_i = \frac{1}{R_1 C_1} \tag{11.36a}$$

$$Q_i = \frac{R_5(R_3 + R_4)}{R_4(R_5 + 2R_6)} \tag{11.36b}$$

$$G_{iL} = -\frac{R_6}{R_5} \tag{11.37a}$$

$$G_{iH} = -\frac{R_6}{R_5} \tag{11.37b}$$

$$G_{iB} = \frac{R_6}{R_5} Q_i \tag{11.37c}$$

Other considerations for component selection are to keep capacitances large enough so that stray capacitance from printed circuit boards, component placement, and encapsulating materials are insignificantly small, and to minimize the influence of input offset voltage and bias current of the amplifiers. Impedance levels of the circuits must be kept high enough to ensure that the amplifiers have sufficient drive current capability to drive them.

The sequence of component calculations that should be used is

1 Select the integrator time constants for the frequency required:

$$R_1 C_1 = \frac{1}{2\pi f}$$

2 Select the value of R_5 to set the gain of the desired output:

$$R_5 = \frac{R_6}{G_{iL}} \quad \text{or} \quad \frac{R_6}{G_{iH}} \quad \text{or} \quad \frac{R_6 Q_i}{G_{iB}}$$

3 Select the values of R_3 and R_4 to realize the desired Q.

$$\frac{R_3 + R_4}{R_4} = \frac{Q_i(R_5 + 2R_6)}{R_5}$$

For bias current compensation, set

$$R_3 \| R_4 = R_5 \| R_6 \| R_7$$

If it is necessary to use large resistors for R_1 and R_2 such that the bias current voltage drop across these elements is of the same order of magnitude or larger than the offset voltage of the amplifier, the effect can be reduced by an order of magnitude by introducing resistors of the same values as R_1 and R_2 in series with the (+) inputs of the two integrator amplifiers. The offset voltage caused by the bias current is then due to the offset current (difference between the + and − input bias currents), which is typically 10 to 20 times less than the bias current. If Q is large, the value of R_3 may become impractically high if the constraint

$R_3 \| R_4 = R_5 \| R_6 \| R_7$ is maintained; in such cases, the values of R_3 and R_4 can be reduced to more practical levels and a resistor can be added in series with the (+) input of A_3 to balance the impedences. This compensation resistor, when added to $R_3 \| R_4$, must equal $R_5 \| R_6 \| R_7$. These bias compensation components do not affect the values of ω, Q, or G and do not change the computed value of any other component.

Amplifier Effects on ω_0 and Q

When the two integrators and the feedback amplifier are considered to have transfer functions of $A_1(s)$, $A_2(s)$, and $A_3(s)$, the second-order transfer function of the S-V filter section increases to a higher-order function that depends upon the model used for the amplifiers. If as in Section 8.5, the assumption can be made that all three amplifiers are identical and can be modeled as a simple first-order function of the form $-A/(\tau s + 1)$ or $-\omega_t/s$, the resulting fifth-order function can be factored down to a third-order function. Even then, the third-order function is very difficult to factor and the effect on the ω and Q of the second-order section is complex. More important, if the filter section is at a frequency that approaches $0.1\omega_t$, the presence of the additional real pole introduced by the amplifier has a dramatic effect on the overall shape of the desired transfer function. When several stages are cascaded to produce a multiple-pole function, these effects compound and result in a considerable deviation from the theoretical function.

From a practical viewpoint we are concerned with how to accommodate or compensate for these amplifier effects. It is easier to understand and correct for them if we consider the nonideal integrators and inverting gain stage individually.

The transfer function of an inverting integrator whose amplifier is modeled as $A(s) = -\omega_t/s$ is

$$\frac{V_o}{V_{\text{in}}} = \frac{-1}{RCs[(s/\omega_t) + 1 + (1/\omega_t)RC]} \tag{11.37}$$

Since the time constant (RC) is generally at a frequency considerably below ω_t, the net effect is the introduction of a first-order lag term (at a frequency very near ω_t) in cascade with the desired integrator function $-1/RCs$.

The transfer function of a unity-gain stage whose amplifier is modeled as $A(s) = -\omega_t/s$ is

$$\frac{V_o}{V_{\text{in}}} = \frac{-1}{[2s/\omega_t + 1]} \tag{11.38}$$

Hence the amplifier introduces a first-order lag term at a frequency of $\omega_t/2$ in cascade with the desired unity-gain inverter.

For both the inverting integrator and the inverting amplifier, an additional lag term is introduced. For frequencies below ω_t, this term reduces the magnitude of, and contributes additional phase lag to, the desired theoretical transfer function. The effect of this lag on the ω and Q of the circuit is essentially the same for all three amplifiers and can be easily understood if they are considered one at a time.

If $\omega < \omega_t$, the inverting integrator transfer function of Eq. 11.37 can be approximated by

$$\frac{V_o}{V_{\text{in}}} = \frac{-1}{RCs}\left[1 - \frac{s}{\omega_t} + \frac{s^2}{\omega_t^2} + \cdots\right] \approx \frac{-1}{RCs}\left[1 - \frac{s}{\omega_t} + \frac{s^2}{\omega_t^2}\right] \tag{11.39}$$

If such a nonideal integrator is used as the second inverting integrator (A_2) of the circuit of Fig. 11.10, the transfer functions become third-order and can be factored into a first-order factor ($s/\omega_t + 1$) and a second-order factor given by

$$\left[R^2C^2 + \frac{1}{\omega_t^2}\right]s^2 + \left[\frac{RC}{Q_i} - \frac{1}{\omega_t}\right]s + 1 \tag{11.40}$$

from which

$$\omega_0 = \omega_i\left[\frac{1}{1 + [\omega_i/\omega_t]^2}\right]^{1/2} \tag{11.41a}$$

$$\approx \omega_i\left[1 - \tfrac{1}{2}[\omega_i/\omega_t]^2\right] \tag{11.41b}$$

and

$$Q_0 = Q_i\frac{\{1 - [\omega_i/\omega_t]^2\}^{1/2}}{1 - Q_i(\omega_i/\omega_t)} \tag{11.42a}$$

$$Q_0 \simeq \frac{Q_i\{1 - \frac{1}{2}(\omega_i/\omega_t)^2\}}{1 - Q_i(\omega_i/\omega_t)} \tag{11.42b}$$

The net effect caused by one of the integrators being nonideal is to decrease the frequency by a second-order term $[-\frac{1}{2}(\omega_i/\omega_t)^2]$ and to increase the Q_0 by a first-order term.

The change in frequency is generally very small. Even if the filter section is operated at $\omega_i = 0.1\omega_t$, the decrease in frequency is only 0.5%. However, the change in Q_0 can be very dramatic. In fact, care must be taken that the denominator term does not go to zero. Even if $\omega_i = 0.01\omega_t$ and $Q_i = 10$, the change in Q_0 is a 10% increase.

When the second integrator and the inverting amplifier also contribute excess phase lag, the effects are substantially the same. Because the frequency of the real pole of the inverting-gain stage is $\omega_t/2$, it contributes twice the excess phase lag of the integrators.

The effect of the phase lag on Q_0 is so large that the circuit must be operated such that $\omega_i \ll \omega_t$ and/or $Q_i\omega_i/\omega_t < 1$. This restriction usually results in the frequency shift being significant.

To derive the best performance from this circuit for high Q requirements or for operation at frequencies approaching ω_t, it is necessary to compensate the circuit to reduce the effect of the phase lag on its amplifiers.

Compensation Methods

If an inverting amplifier with a transfer function of $A(s)$ is compensated with a feedback resistor R_a in series with the feedback integrator capacitor, as shown in

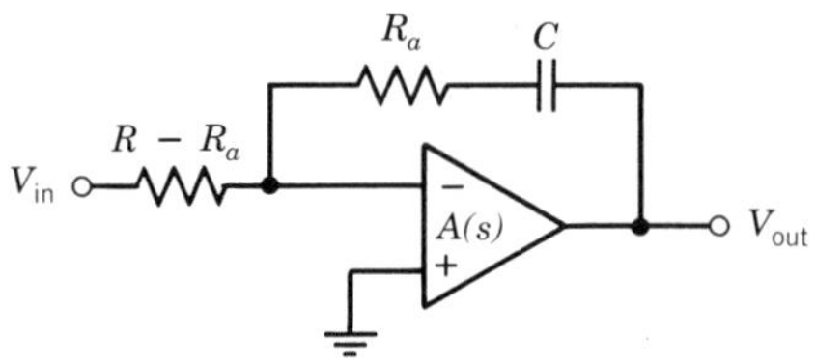

Figure 11.11 Compensated inverting integrator.

Fig. 11.11, its transfer function is given by the equation:

$$\frac{V_{out}}{V_{in}} = \frac{-(R_aCs + 1)}{(R - R_a)Cs - (RCs + 1)/A(s)} \tag{11.43}$$

If the amplifier model $A(s) = -\omega_t/s$ is used, then:

$$\frac{V_{out}}{V_{in}} = \frac{-(R_aCs + 1)}{RCs\,(s/\omega_t) + 1 + (1/\omega_t + RC) - (R_aC/RC)} \tag{11.44}$$

If $R_aC = 1/\omega_t$, this simplifies to

$$\frac{V_{out}}{V_{in}} = \frac{-1}{RCs}$$

Therefore, the placement of a portion (R_a) of the integrator resistor in series with the feedback capacitor, such that the break frequency $1/R_aC$ is the same as ω_t, completely eliminates the effects of the amplifier transfer function.

Unfortunately, the ω_t of a given amplifier is not known precisely and the amplifier does not conform exactly to such a simple model. Hence perfect compensation is not generally possible. However, a significant reduction in phase lag error can be made by calculating R_a for a typical value of ω_t. An even better compensation can be made by trimming the value of R_a to give the integrator 90° of phase shift at ω_0.

The inverting-gain stage shown in Fig. 11.12 has an amplifier that can be modeled as $A(s) = -\omega_t/s$ and $C_a = 0$. Then its transfer function is given by

$$\frac{V_{out}}{V_{in}} = -\frac{R_2}{R_1} \cdot \frac{1}{(R_1 + R_2/R_1\omega_t)s + 1} \tag{11.45a}$$

which, for $R_1 = R_2$ reduces to

$$\frac{V_{out}}{V_{in}} = \frac{-1}{[(2/\omega_t)s + 1]} \tag{11.45b}$$

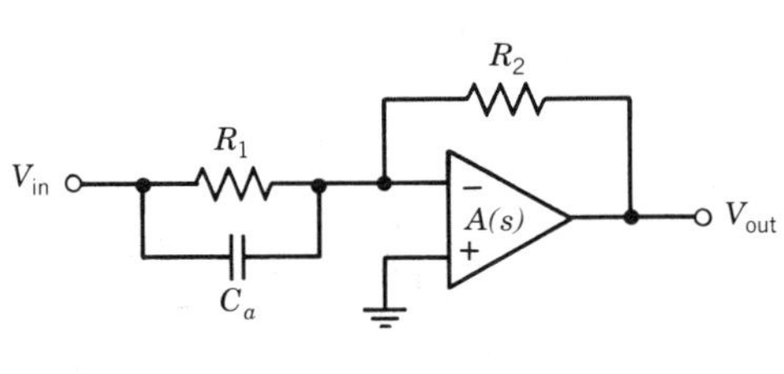

Figure 11.12 Compensated inverting gain stage.

This first-order lag with a corner frequency of $\omega_t/2$ results from the frequency dependence of $A(s)$. If an attempt is made to compensate for this lag term by capacitance C_a in parallel with R_1, the denominator of the transfer function becomes second-order and a perfect cancellation of time constants cannot be achieved.

$$\frac{V_{out}}{V_{in}} = \frac{R_2}{R_1} \cdot \frac{R_1 C_a s + 1}{[R_2 C_a S + R_2/R_1 + 1]s/\omega_t + 1} \tag{11.46}$$

If $R_1 = R_2$ and $\omega_c = 1/R_1 C_a$, then

$$s = \frac{-1}{R_1 C_a}[1 \pm \sqrt{1 - \omega_t R_1 C_a}] = -\omega_c[1 \pm \sqrt{1 - \omega_t/\omega_c}] \tag{11.47}$$

Although a perfect cancellation of numerator and denominator terms cannot be achieved by this form of compensation, several good alternatives do exist.

The first scheme for compensating the inverting gain stage uses C_a to provide a cancellation of the phase shift introduced by the inverting amplifier at the pole frequency of the filter stage. The value of C_a can be computed by selecting ω_c so that the phase shift of the inverting gain stage is zero at ω_0, from which can be derived:

$$\omega_c = \frac{\omega_t}{4}[1 + \sqrt{1 - 8 \cdot \omega_o^2/\omega_t^2}] \tag{11.48}$$

The value of C_a can also be determined by selection of C_a that yields zero phase shift through the inverting amplifier at ω_0.

Several other alternatives exist for use in the S-V design. The integrators and the inverting-gain stage all contribute extra lag terms, all of which can be compensated for in one way or another. However, depending upon which output of the S-V design is to be used, in most cases one or both of the integrators can be overcompensated for in order to correct for the lag of the inverting amplifier. One scheme that works well is to double R_a in each of the integrators in order to make a correction for the $\omega_t/2$ lag term introduced by the inverting amplifier.

By using these compensation methods, moderately high Q (20 to 50) filter sections can be built at frequencies up to 0.02 ω_t.

Trimming Methods for ω and Q

Multistage filters require the cascade of several sections. Their shape is strongly dependent upon the accuracies of the frequencies and Qs of their sections. Therefore, the ability to accurately set or trim these parameters is essential to the precise realization of complex filter functions. The state-variable design is particularly well suited to precision tuning because its amplifiers can be compensated and its ω and Q can be independently adjusted. For example, the G_L, ω, and Q of the low-pass output can be independently trimmed in the following sequence:

1 *Trim* ω. From Eq. 11.33c, the phase shift at the bandpass output should be zero at ω_i. The frequency can be trimmed by varying R_7 (or R_1, R_2, or R_6) to close the loop on a Lissajous pattern using an oscilloscope with V_{in} connected to the horizontal sweep and V_B connected to the vertical sweep.

2 *Trim* G_L. Once the frequency is set, the dc gain can be trimmed by measuring the output/input ratio at $\omega \ll \omega_i$ and adjusting it to the desired value by varying R_5.

3 *Trim* Q. Once ω and G_L have been trimmed, Q can be set by measuring the amplitude ratio V_L/V_{in} at ω_i. From Eq. 11.33a, the gain of the V_L output at ω_i should be Q times the gain at dc. Therefore, Q can be measured directly by normalizing to the gain at dc and measuring the gain at ω_i. The Q can be trimmed to the desired value by varying R_4 or R_3.

When this sequence is followed, each trim is independent of the ones before it, and there is no need to repeat or iterate the sequence. A similar sequence can be used to trim the high-pass output by setting G_H at a frequency above ω_i. Care must be taken not to exceed the slew rate capability of the amplifier A_3 when setting the gain of the high-pass output at high frequency.

The frequency trim of the bandpass output is performed as in the previous example. However, the Q and G_H are not independent of one another and the Q must be calculated by measuring the -3 dB frequencies f_H and f_L and computing $Q = \sqrt{f_L f_H}/(f_H - f_L)$.

EXAMPLE 11.3 Design a two-pole high-pass filter with a frequency of 1.25 kHz, a Q of 2.4 and a high-frequency gain of 1.4. Assume that it is desirable to maintain an input impedance and integrator impedance level of 20 kΩ.

PROCEDURE

1 Determine the value of C_1 and C_2. If $C_1 = C_2$, $R_1 = R_2$, and $R_6 = R_7 = 20$ kΩ,

$$C_1 = \frac{1}{R_1 2\pi f} = \frac{1}{20{,}000(2\pi)(1250)} = \frac{10^{-6}}{20 \times 2\pi \times 1.25}$$

$$C_1 = 0.006366\ \mu\text{F}$$

Use standard value capacitors, $C_1 = C_2 = 0.0068\ \mu$F

2 Compute resistors R_1 and R_2

$$R_1 = R_2 = \frac{1}{2\pi f C} = \frac{1}{2\pi(1250)(0.0068) \times 10^{-6}} = 18{,}724\ \Omega$$

Use standard values $R_1 = R_2 = 18.7$ kΩ.

3 Set high-frequency gain $G_H = 1.4$

$$R_5 = \frac{R_6}{G_{iH}} = \frac{20{,}000}{1.4} = 14{,}285\ \Omega$$

Use a standard value, $R_5 = 14.3$ kΩ

4 Set $Q = 2.4$

$$\frac{R_3 + R_4}{R_4} = \frac{Q(R_5 + R_6 + R_7)}{R_5} = \frac{2.4(14.3 + 20.0 + 20.0)}{14.3} = 9.09$$

To maintain $R_3||R_4 = R_5||R_6||R_7$

$$\frac{R_3R_4}{R_3 + R_4} = \frac{R_3}{9.09} = [20k||20k||14.3k] = 5.88\ \text{k}\Omega$$

$$R_3 = 53.44\ \text{k}\Omega$$

$$R_4 = \frac{R_3}{9.09 - 1} = 6.60\ \text{k}\Omega$$

Use standard values, $R_3 = 53.6\ \text{k}\Omega$, and $R_4 = 6.65\ \text{k}\Omega$

5 Check ω_i, Q_i, and G_{iH} to determine how far from ideal these values will be if the standard value components that have been selected are exact.

a. $f_0 = \dfrac{1}{2\pi R_i C_i} = \dfrac{1}{2\pi(18{,}700)(0.0068) \times 10^{-6}} = 1251.61$

Therefore,

$$\frac{f_a}{f_i} = \frac{1251.61}{1250} = 1.00128$$

b. $G_H = -\dfrac{R_6}{R_5} = \dfrac{20{,}000}{14{,}300} = 1.3986$

Therefore,

$$\frac{G_H}{G_{H_i}} = \frac{1.3986}{1.4} = 0.9990$$

c. $Q = \dfrac{R_3 + R_4}{R_4} \cdot \dfrac{R_5}{R_5 + R_6 + R_7} \cdot \left\{\dfrac{R_1R_6C_1}{R_2R_7C_2}\right\}^{1/2}$

$$= \frac{53.6 + 6.65}{6.65} \cdot \frac{14.3}{14.3 + 20 + 20}[1] = \frac{60.25}{6.65} \cdot \frac{14.3}{54.3}$$

$$= 2.38$$

Hence

$$\frac{Q_0}{Q_i} = \frac{2.38}{2.4} = 0.9916$$

Therefore, if the components were exact, f_0, G_H, and Q would be realized within $+0.128$, -0.1, and -0.84%, respectively. If component tolerances were 1%, these parameters could be further from their ideal values.

If it is necessary for any of these parameters to be realized more precisely, the resistors R_6, R_7, and R_4 can be trimmed using the sequence described in Section 11.6.

6 Determine whether amplifer compensation is needed. This requires knowledge of ω_t for the amplifier used. Assuming the use of a TL081 type amplifier that has an ω_t of 4 MHz, and using Eqs. 11.41b and 11.42b, we obtain

$$\omega_0 = \omega_i\left[1 - \frac{1}{2}\left(\frac{\omega_i}{\omega_t}\right)^2\right] = \omega_i\left[1 - 0.5\left(\frac{1250}{4{,}000{,}000}\right)^2\right]$$

$$= \omega_i[1 - 0.05 \times 10^{-6}]$$

$$Q_0 = \frac{Q_i[1 - \frac{1}{2}(\omega_i/\omega_t)^2]}{1 - Q_i(\omega_i/\omega_t)} = \frac{Q_i(1 - 0.05 \times 10^{-6})}{1 - 10 \times (1.25 \times 10^3/4 \times 10^6)}$$

$$= Q_i\frac{(1 - 0.05 \times 10^{-6}}{1 - 0.0031} = Q_i[1.003]$$

In this case the frequency shift is an insignificant $5 \times 10^{-5}\%$ and the increase in Q is $+0.3\%$, both of which are within the tolerances of the components.

If compensation is necessary, the values of R_a and C_a for the integrators and inverting amplifier can be computed from the discussions of this section.

11.7 SUMMARY

The Sallen–Key low-pass, multiple-feedback bandpass, and state-variable circuits that have been presented represent circuits that are currently used in the manufacture of active filters. They are relatively easy to build and test, and the first-order compensation methods presented help to achieve the best performance that can be realized from the active components. Some of the circuit compensation methods can also be used in other configurations.

With the active components available at this time these circuits can be used to manufacture low-pass and bandpass active filters with Qs up to 50 and $\omega_0 Q$ products up to 1,000,000.

Other practical considerations such as resistor and capacitor ratios, amplifier offset and noise, capacitive coupling from printed circuit boards and encapsulation materials must also be considered. These properties become especially important in high Q circuits and at frequencies approaching the ω_t of the amplifiers.

11.8 REFERENCE

D. Hilberman, "An approach to the sensitivity and statistical variability of biquadratic filters," IEEE Trans. CT-20, pp. 382–390, 1973.

CHAPTER 12

Computer Programs

G. Stirk
General Electric
Charlottesville, Virginia

12.1 INTRODUCTION

The preceding chapters have given valuable information about the design and performance characteristics of various active filter networks. Detailed filter theory, as well as practical design considerations, have been presented.

This chapter gives the reader a set of computer programs (tools) to aid the analysis and design of active filter networks. The programs are written in structured BASIC, which is explained in the following paragraphs of this section. The computer system used is an IBM PC (Personal Computer).* Every attempt is made to use all the available features of the IBM PC, especially in the area of screen graphics. Although much of the BASIC code is transportable from one system to another, limiting statements to only standard BASIC would have greatly reduced the capabilities of this program.†

For those not familiar with BASIC computer language, the first section of this chapter presents a brief description of BASIC command types and their required syntax or instruction format. The intent is not to teach the reader BASIC programming but to give enough knowledge of the language to enable understanding and efficient use of the programs provided. As a further aid to understanding,

* IBM is a trademark of International Business Machines.

† Further information on the programs described in this chapter may be obtained from Applied Computers, P.O. Box 6541, Charlottesville, Va. 22906.

the programs are presented in structured form using RATBAS*, a structured BASIC language. A brief description of RATBAS is also included in this section.

The remaining four sections of the chapter are devoted to the computer programs. Section 12.2 provides programs for the calculation of coefficients, roots, minimum order, and biquadratic Qs and ω_0s for the Chebyshev and Butterworth approximating functions. A limited set of information is also provided for the Thomson (Bessel) approximation.

Section 12.3 presents a program that, following entry of specific filter requirements (i.e., high-pass, bandpass, bandstop, etc.), transforms the roots and coefficients to their normalized values. These normalized values can then be used for other filter parameters such as the calculation of Qs and ω_0s of Section 12.2.

Frequency-response plots of a specified filter may be obtained by the use of the program shown in Section 12.4. This program prompts the user for the filter specifications that may be in either biquad or coefficient form. The resulting frequency analysis yields a tabulation of gain, phase shift, and delay versus frequency for the specified filter. This data is stored on disk for immediate or future plotting.

All of the programs presented in Sections 12.2 through 12.4 are tied together in an interactive filter analysis program. This program allows the user, from a menu-driven environment, to select any of the filter analyses given in the other sections. Hence it is possible to design a filter, including component calculations, by running this single program. A design example is presented to illustrate the use of the program package as a complete filter design tool.

BASIC (Beginner All Symbolic Instruction Code) was developed at Dartmouth College in 1964 with the aim of providing a language that could be learned by a beginner in just a few hours. For the simplest of programs this is still true. However, the language was so easy to use that it was soon expanded and may now take several days to learn. Also, instructions have been added that are unique to a particular machine so as to utilize the full power of the system. This is the case with the IBM PC. Many powerful BASIC statements have been added to the standard BASIC to allow easy use of the system graphics.

The most common program structure of BASIC is the interpreter. The BASIC interpreter is a machine-dependent software block that reads sequentially down a program and executes line by line what the programmer has requested in his or her instructions. Each program line, therefore, must have a line number. It is good programming practice to make each line number at least 5 or 10 greater than the previous one. Since the BASIC interpreter executes sequentially, lines may then be inserted without renumbering the program.

Figure 12.1 shows that BASIC instructions may be grouped into one of three catagories: *commands*, *statements*, and *functions*.

Instructions may be entered in two modes, direct and indirect. Direct instructions usually do not appear in programs because they are file and program management operations such as listing the program or saving and loading the files.

* RATBAS is a structured BASIC translator produced by Wells Fargo Investment Advisors.

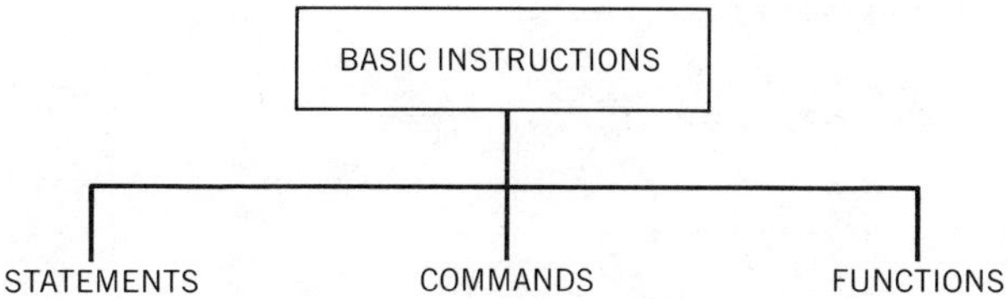

Figure 12.1 Breakdown of "Basic" instruction types.

These direct mode instructions have traditionally been classified as "command types."

Indirect mode instructions are those which are usually executed by the program; they control program flow or change variable values. Those which control the program flow are classified as "statements," while those controlling variable values are "functions."

Let us now examine what a typical program looks like and how it runs. The listing in Fig. 12.2 shows a simple program for calculating the equivalent resistance of a resistor network. It prompts the user for the type of circuit, the number of resistors in the network, and their values.

Lines 10 through 280 are REMark statements. They are completely ignored by the program at the time of execution. Their purpose is for documentation and information. In nonstructured programming they are as important as the instructions themselves.

Line 300 prompts the user for the number of resistors in the network and line 310 asks for the type of resistor network to be calculated. After the program has this information, it will check to see what type of calculation is required (i.e., series or parallel). This is done in line 320 with the IF statement shown. This statement will branch instruction flow to line 400 if a parallel analysis is requested (type 1) and to line 500 for a series analysis (type 2). Since a GOSUB branch was used, the instruction execution will return to the line following the IF statement (line 330) if a return is encountered. Lines 400 to 490 and lines 500 to 580 define the "subroutines" that calculate equivalent resistance for parallel and series networks respectively. Subroutines are ended with a RETURN statement thereby returning program execution to the main program. Line 330 PRINTs the requested equivalent resistance as

THE TOTAL EQUIVALENT RESISTANCE IS 3.45 K ohms

for an example that would result in an equivalent resistance of 3.45 kΩ.

A FOR/NEXT loop as seen in lines 440 to 470 and 540 to 570 is a means of repeating a set of instructions a specific number of times. The FOR statement assigns an initial value to a variable ($n = 1$ for this case) and the NEXT statement increments its value (i.e., $n = n + 1$). If the value of the variable is less than or equal to the value after the TO in the FOR statement, then lines after the FOR statement up to the NEXT statement are executed once again. The END statement will cause the program to stop execution.

```
10 REM                   ********************
20 REM                   *                  *
30 REM                   *     NETWORK      *
40 REM                   *    RESISTANCE    *
50 REM                   *    CALCULATION   *
60 REM                   *                  *
70 REM                   ********************
80 REM
90 REM
100 REM *******************************************************
110 REM * THIS PROGRAM WILL CALCULATE THE EQUIVALENT RESISTANCE
120 REM * OF EITHER A SERIES OR PARALLEL RESISTANCE STRING OF
130 REM * LENGTH N.
140 REM *
150 REM * THE TOTAL NUMBER OF RESISTORS (NUM.OF.RES) IS
160 REM * ENTERED FIRST WHEN PROMPTED. THEN THE TYPE OF NETWORK.
170 REM * THE USER SHOULD RESPOND WITH A "1" FOR A PARALLEL
180 REM * NETWORK AND A "2" FOR A SERIES NETWORK FOLLOWED BY A
190 REM * CARRAGE RETURN.
200 REM *
210 REM * THE PROGRAM WILL PROMPT YOU FOR THE RESISTANCE VALUES
220 REM * IN K ohms FOR THE SPECIFIED NUMBER OF RESISTORS.
230 REM * THE FINAL RESULT IS EXPRESSED IN K ohms
240 REM *
250 REM ********************************************************
260 REM
270 REM ++++++++++++++++++ BEGIN PROGRAM ++++++++++++++++++++
280 REM
300 INPUT "ENTER NUMBER OF RESISTORS IN NETWORK : ";NUM.OF.RES
310 INPUT "ENTER TYPE OF NETWORK , PARALLEL/SERIES (1)/(2)  : ";NET.TYPE
320 IF NET.TYPE = 1 THEN GOSUB 400 ELSE GOSUB 500
330 PRINT "THE TOTAL EQUIVALENT RESISTANCE IS : ";EQUIV.RES;" K ohms"
340 END
400 REM ***************************************
410 REM *  PARALLEL RESISTOR NETWORK SUBROUTINE
420 REM ***************************************
440 FOR N = 1 TO NUM.OF.RES
450 PRINT " RESISTOR #";N;" VALUE IN K ohms : ";:INPUT "",RESISTOR(N)
460 INV.OF.RES = 1/RESISTOR(N)+INV.OF.RES
470 NEXT N
480 EQUIV.RES = 1/INV.OF.RES
490 RETURN
500 REM ***************************************
510 REM *  SERIES RESISTOR NETWORK SUBROUTINE
520 REM ***************************************
540 FOR N = 1 TO NUM.OF.RES
550 PRINT " RESISTOR #";N;" VALUE IN K ohms : ";:INPUT "",RESISTOR(N)
560 EQUIV.RES = EQUIV.RES + RESISTOR(N)
570 NEXT N
580 RETURN
```

Figure 12.2 Example of a "Basic" program.

One can easily see that with a very long program and many GOTO, GOSUB, and IF THEN ELSE statements, the program would be extremely difficult to follow. For this reason a structured BASIC language called RATBAS was developed. RATBAS allows all BASIC statements except for GOTOs and GOSUBs. The reason for this is that no line numbers are used, which makes for much easier program reading. Subroutines are defined with procedure names and then the procedure name may be used as a "new structured" BASIC statement. Any special structured statement is marked with a "]" symbol at the end of the line to distinguish it from a standard BASIC statement.

The RATBAS language has only two structures or instructions that are unique to RATBAS (versus BASIC). They are the CASE variable OF] statement and a structured IF THEN ELSE statement. The standard BASIC IF THEN ELSE statement allows only one BASIC statement to follow the THEN or the ELSE qualifiers. For example,

```
IF a = b THEN c = d :e = f :g = h ELSE d = c :f = e :h = g
```

would not be correct, since $c = d$, $e = f$, and so on are multiple statements. This statement would have to be written as

```
10    IF a = b THEN GOTO 60
20    d = c
30    f = e
40    h = g
50    GOTO 90
60    c = d
70    e = f
80    g = h
90    ........ more program
```

In structured BASIC this would be written as:

```
IF a = b THEN]

      c = d
      e = f
      g = h

   ELSE]

      d = c
      f = e
      h = g

IFEND]
```

which is certainly clearer than the previous example.

The CASE variable OF] statement was created to avoid looping IFs inside IFs. This would be the case if there were more than two responses to a program question or prompt. Such is often the case in menu-driven programs where the user can select from 1 to 10 items on a list by pressing a single key. An example of the structured CASE statement follows:

```
CASE choice OF]

1:]
      STATEMENT
          |
      STATEMENT
2:]
      STATEMENT
          |
      STATEMENT
|
|
10:]
      STATEMENT
          |
      STATEMENT
CEND]
```

This example will execute the statements following the"1:]" and preceding the "2:]" if the value of choice is 1. It will execute the statements following the 2:] and preceding the implied 3:] if the choice is 2, and so on, up to a value of 10. An OTHERWISE] statement might be added after the 10:] statements with a list of statements to be executed, should the choice contain a value not in the range 1 to 10.

As mentioned previously, a BASIC subroutine in RATBAS may be defined as a PROCEDURE]. The parallel resistance calculation subroutine of the pro-

```
PROCEDURE GET.NUM.OF.RES.AND.TYPE.OF.NETWORK
          ' THIS  PROCEDURE  PROMPTS THE USER FOR THE NUMBER
          ' RESISTORS AND NETWORK TYPE. TYPE 1 = PARALLEL
          ' TYPE 1 = PARALLEL; TYPE 2 = SERIES
       INPUT "ENTER NUMBER OF RESISTORS IN NETWORK : ";NUM.OF.RES
       INPUT"ENTER TYPE OF NETWORK, PARALLEL/SERIES (1)/(2): ";NET.TYPE
PEND]

PROCEDURE  CALCULATE.PARALLEL.RESISTANCE (X,V:Y)]
          ' This procedure calculates the parallel resistance of
          ' a parallel resistor network of X resistors
          '. and assigns the result to the variable Y
      FOR N = 1 TO X
        PRINT "RESISTOR #";N;" VALUE IN K ohms  : ";
        INPUT RESISTOR(N)
        INV.OF.RES = INV.OF.RES + 1/REAISTOR(I)
      NEXT N
      Y = 1/INV.OF.RES
PEND]

PROCEDURE CALCULATE.SERIES.RESISTANCE (X,V:Y)]
                                   :
                                   :
PEND]                              :

PROGRAM]
     ' THIS PROGRAM WILL CALCULATE THE EQUIVALENT RESISTANCE
     ' OF EITHER A SERIES OR PARALLEL RESISTANCE STRING OF
     ' LENGTH N.
     '
     ' THE TOTAL NUMBER OF RESISTORS (NUM.OF.RES) IS
     ' ENTER FIRST WHEN PROMPTED. THEN THE TYPE OF NETWORK
     ' THE USER SHOULD RESPOND WITH A "1" FOR A PARALLEL
     ' NETWORK AND A "2" FOR A SERIES NETWORK FOLLOWED BY A
     ' CARRIAGE RETURN.
     '
     ' THE PROGRAM WILL PROMPT YOU FOR THE RESISTANCE VALUES
     ' IN K ohms FOR THE SPECIFIED NUMBER OF RESISTORS.
     ' THE FINAL RESULT IS EXPRESSED IN K ohms
     '
     '    GET.NUM.OF.RES.AND.FILTER.TYPE]
          IF TYPE = 1 THEN]
                  CALCULATE.PARALLEL.RESISTANCE (NUM.OF.RES,EQUIV.RES)]
               ELSE]
                  CALCULATE.SERIES.RESISTANCE (NUM.OF.RES,EQUIV.RES)]
          IFEND]
          PRINT"THE TOTAL EQUIVALENT RESISTANCE IS:";EQUIV.RES;"K ohms"
END 'PROGRAM
```

Figure 12.3 Example of a "Ratbas Basic" program.

gram in Fig. 12.2 could be written as a procedure in RATBAS:

```
PROCEDURE CALCULATE.PARALLEL.RESISTANCE (X,V:Y)]

' This procedure calculates the parallel resistance of
' a parallel resistor network of X resistors
' and assigns the result to the variable Y

FOR N = 1 TO X
  PRINT "RESISTOR #";N;" VALUE IN K ohms : ";
  INPUT RESISTOR(N)
  INV.OF.RES = INV.OF.RES + 1/RESISTOR(I)
NEXT N
Y = 1/INV.OF.RES

PEND]
```

Executing a similar procedure for the series case would result in a structured RATBAS program (equivalent to that of Fig. 12.2) shown in Fig. 12.3. Notice that our previous names NUM. OF. RES and EQUIV.RES were substituted for the variables *X* and *Y*, respectively, as the procedure arguments. This allows the procedure to be used by different sections of the program (this was not the case, however, for our simple example). The "V:" preceding the *Y* parameter tells the RATBAS translator that *Y* is to be redefined at the exit of the procedure as the variable name specified (EQUIV.RES in our case). If the "V:" does not precede the variable, then the assignment of the dummy variable (*X*) is made only at the entry of the subroutine.

After a program has been written in RATBAS form, it must be translated to standard BASIC form with line numbers before it can be run. This is done with a translator program. The programs in the following sections, as mentioned, will be written and shown in the RATBAS form for ease of reading.

12.2 BUTTERWORTH, CHEBYSHEV, AND BESSEL APPROXIMATION FUNCTIONS

This section explains the programs that are used to calculate the minimum order, coefficients, roots, and biquadratic *Q*s and ω_0s for Butterworth and Chebyshev approximation functions. Limited data on Bessel functions are also presented.

Butterworth Approximation

The Butterworth minimum order is determined by[1]

```
                   GamaMin/10           GamaMax/10
          Log[(10            -1)/(10            -1)]
NB  =  ------------------------------------------------        (12.1)
                          Log(W2/W1)
```

The values of GamaMin and GamaMax* must be entered before the minimum order of the Butterworth can be determined. The procedure is as follows.

* These are equivalent to the A_{min} and A_{max} of Chapters 2 and 13.

```
PROCEDURE GET.FILTER.CONSTANTS]
    ' PROCEDURE PROMPTS THE USER
    ' FOR THE FILTER PARAMETERS:
    '                        GAMA.MAX
    '                        GAMA.MIN
    '                        W1,W2
    ' WHERE W1 IS THE START OF THE STOPBAND
    ' AND  W2 IS THE END
   COLOR 7,0
   CLS
   ANS$ = " "
   WHILE ANS$ = " "
     IF LAST.KEY$ <> CHR$(59) THEN]
       LOCATE 5,5,1
       COLOR 0,7
       PRINT "DO YOU WISH TO USE EXISTING
               MIN ORDER FILTER OF ";MIN.ORDER;
       COLOR 7,0
       PRINT "(Y/N)"
       GET.KEY(ANS$)]
     ELSE]
       ANS$ = "N"
     IFEND]
     CASE ANS$ OF]
       "N":]
       COLOR 7,0
       CLS
       LOCATE 10,1,0
       INPUT "ENTER VALUE OF GAMA.MIN ";GAMA.MIN
       INPUT "ENTER VALUE OF GAMA.MAX ";GAMA.MAX
       INPUT "ENTER VALUE OF W1,W2 ";W1,W2
       "n":]

       COLOR 7,0
       CLS
       LOCATE 10,1,0
       INPUT "ENTER VALUE OF GAMA.MIN ";GAMA.MIN
       INPUT "ENTER VALUE OF GAMA.MAX ";GAMA.MAX
       INPUT "ENTER VALUE OF W1,W2 "; W1,W2
       "Y":]
       "y":]
     OTHERWISE]
       LOCATE 24,1,0
       COLOR 0,7
       PRINT "ENTER 'Y' OR 'N' PLEASE!!";
       ANS$ = " "
     CEND]
   WEND
   PEND]
```

This procedure is used whenever a calculation of minimum order is requested. For example, if the Butterworth coefficients were requested from the main program and the minimum order had not yet been calculated, this procedure would be called as the first part of the minimum order calculation. For this reason the above procedure will test to see whether the minimum order was requested directly or if it was needed as a result of another request such as root calculations. The test for Last.key < > chr$(59) does this testing (Last.Key = chr$(59) indicates a direct request for a minimum order calculation).

If the procedure is called indirectly, then the program displays the current value of the minimum order and gives the user a choice of using this value or of entering new values for GamaMin, GamaMax, ω_1, and ω_2.* The procedure

* These values are equivalent to the ω_p and ω_s of Chapter 2.

handles screen prompting of the variables. The procedure for calculating the Butterworth minimum order is

```
PROCEDURE BTR.MIN.ORDER]
' CALCULATES THE MINIMUM ORDER OF
' BUTTERWORTH FILTER REALIZATION
' GAMA.MIN = STOBAND ATTENUATION IN DB
' W1,W2 ARE THE FREQUENCIES DEFINING THE STOPBAND
' NB IS PASSED BACK TO THE MAIN PROGRAM AS
' THE MINIMUM ORDER REQUIRED

GET.FILTER.CONSTANTS]
 NUM.ARG=(((10^(GAMA.MIN/10))-1)/((10^(GAMA.MAX/10))-1))
 LOG10(DENOM,W2/W1)]
 LOG10(NUMER,NUM.ARG)]
 NB = NUMER/(2*DENOM)
DET.NEXT.LARGEST.INT(NB)]
BTR.EPSILON = SQR((10^(0.1*GAMA.MAX))-1)
BTR.CONVERSION.FACTOR = BTR.EPSILON^(1/NB)
PEND]
```

Notice that after evaluation of Eq. 12.1, the procedure CALCULATE.NEXT. LARGEST.INT] is called. This procedure, which is listed below, substitutes an integer value for the minimum order NB, which is the calculated value of NB rounded up to the nearest integer (e.g., 3.02 would round up to NB = 4). Taking the INTeger function of a value rounds the value to a less positive value. For example, INT(3.02) = 3 and INT(−3.02) = −4. Thus, in order to round the minimum order upward, it is necessary to take the negative of the INT(−MIN.ORDER), as shown in the DET.NEXT.LARGEST.INT] procedure.

```
PROCEDURE DET.NEXT.LARGEST.INT(V:VALUE)]
' FINDS THE NEXT LARGES INTERGER OF VALUE
  VALUE = -INT(-VALUE)
pend]
```

The coefficients and roots of the Butterworth polynomials must be adjusted when GAMA.MAX is not specified as 3 dB. This procedure was explained in Chapter 2 and is reflected in the calculations of BTR.EPSILON and BTR.CORRECTION.FACTOR shown at the end of the BTR.MIN.ORDER] procedure above. The correction factor is then used to readjust the roots and coefficients of the specified Butterworth polynomial. The Butterworth roots are calculated from the equation[1]:

```
Root(n) = -SIN(PI*(2K-1)/(2*N) + J COS(PI*(2K-1)/(2N))
```
(12.2)

The Butterworth root calculation procedure loops through the above equation NB times as follows:

```
PROCEDURE BTR.ROOTS]
' CALCULATES NB ROOTS (NB = MINIMUM ORDER BUTTERWORTH)
  FOR I = 1 TO NB
      BTR.REAL.ROOT(I) = -SIN(((2*I-1)/NB)*(PI/2))
      BTR.IMAG.ROOT(I) =  COS(((2*I-1)/NB)*(PI/2))
      R(I) = BTR.REAL.ROOT(I)/BTR.CONVERSION.FACTOR
      I!(I) = BTR.IMAG.ROOT(I)/BTR.CONVERSION.FACTOR
  NEXT I
PEND]
```

The Butterworth coefficients are a function of the previous coefficient where the coefficient of the s^0 term is 1. This expression is given by Eq. 12.3:

```
                     COS((PI*(K1)/2NB)
COEFF(S^K)  =  ---------------------------          (12.3)
                    SIN((PI*K/(2NB)
```

The coefficients specified by Eq. 12.3 are evaluated by the procedure shown:

```
PROCEDURE BTR.COEF]

' CALCULATE THE COEFFICIENTS OF S^N TERMS
  COEF(0) = 1
  FOR I = 1 TO NB
     BTR.COEF(0) = 1.0
     NUMER.BTR(I) = COS((I-1)*PI/(NB*2))
     DENOM.BTR(I) = SIN(I*PI/(NB*2))
     BTR.COEF(I) = (NUMER.BTR(I)/DENOM.BTR(I))
                   *BTR.COEF(I-1)
                   *BTR.CONVERSION.FACTOR
     COEF(I) = BTR.COEF(I)
  NEXT I
PEND]
```

Observe that this procedure also adjusts to the minimum order previously calculated.

The Q and ω_0 of each biquadratic is determined by examining the real and imaginary parts of the calculated roots. The ω_0 values are given by

$$\omega_0 \text{ (biquad } n) = \text{square root } (R(n)\hat{\ }2 + I(n)\hat{\ }2) \qquad (12.4)$$

where $R(n)$ and $I(n)$ are the real and imaginary parts of the complex pole pairs found by the root procedure shown above. The Qs of each biquad are then found using Eq. 12.4 in the expression

$$Q \text{ (biquad } n) = \omega_0(n)/(2*R(n)) \qquad (12.5)$$

The calculation of the Butterworth Q and ω_0 for each biquad are performed in a single procedure called BTR.Q.W]:

```
PROCEDURE BTR.Q.W]
  BTR.ROOTS]
  FOR I = 1 TO NB/2
    IF ABS(I!(I)) > .0001 THEN]
       Wo(I) = SQR( (R(I)^2) + (I!(I)^2) )
       Q(I)    =Wo(I)/(2*R(I))
    IFEND]
  NEXT I
PEND]
```

This procedure checks for an imaginary part of the root close to 0 (i.e., less than 0.0001). If the imaginary part is found to be approximately 0, then Q and ω_0 are not calculated, since this is not a true biquad.

Chebyshev Approximation

The routine for calculating the minimum order of a Chebyshev network is identical to that for the Butterworth except the expression for NC (the minimum order of the Chebyshev network), which is given by Eq. 12.6.

```
          -1       GamaMin/10              GamaMax/10         0.5
      COSH    [(10             - 1)/(10               - 1)]
NC = ----------------------------------------------------------------   (12.6)
                                  -1
                              COSH   (W2/W1)
```

Where GamaMin, GamaMax, ω_1 and ω_2 are supplied by the user when the GET. FILTER.CONSTANTS] procedure is executed. The Chebyshev minimum order procedure is

```
PROCEDURE CBY.MIN.ORDER]
' CALCULATES THE MINIMUM ORDER OF
' CHEBYSHEV FILTER REALIZATION
' GAMA.MIN = STOPBAND ATENUATION IN DB
' W1,W2 ARE THE FREQUENCIES DEFINING THE STOPBAND
' NC IS PASSED BACK TO THE MAIN PROGRAM AS
' THE MINIMUM ORDER REQUIRED
    GET.FILTER.CONSTANTS]
    ARCCOSH(DENOM,W2/W1)]
    NUM.ARG=(((10^(GAMA.MIN/10))-1)/((10^(GAMA.MAX/10))-1))^.5
    ARCCOSH(NUMER,NUM.ARG)]
    NC = NUMER/DENOM
    DET.NEXT.LARGEST.INT(NC)]
PEND]
```

The procedure DET.NEXT.LARGEST.INT] was used in both the Butterworth and Chebyshev minimum order procedures simply by changing the argument. The procedure for ARCCOSH(V:ARG1,ARG2) may be seen. There is no longer an appendix in the listing of the program. This procedure takes the ARCCOSH of ARG2 and places the result in ARG1. Therefore, the ARCCOSH function call in the above procedure is equivalent to the statement "DENOM = ARCCOSH(ω_2/ω_1)".

Calculation of the Chebyshev coefficients as well as the biquad coefficients is simplified if the root locations are determined first. The left-hand plane pole locations of the Chebyshev function can be found by evaluating the following expression [1]:

```
ROOT(K) = -SINH(A)*SIN[(2K-1)*PI/(2N)] +
           j COSH(A)*COS[(2K-1)*PI/(2N)]          (12.7)
```

where k is incremented from 1 to the minimum order N, and A is given by the expression

```
                      -1
A = 1/N * SINH   [1/EPSILON]                      (12.8)
```

EPSILON is a real constant determined by the procedure

```
PROCEDURE CALC.EPSILON]
      EPSILON = SQR((10^(GAMA.MAX/10))-1)
PEND]
```

The CBY.ROOTS procedure that follows separates the calculation into two parts, real and imaginary, representing the first two terms of Eq. 12.7.

```
PROCEDURE CBY.ROOTS]
' CALCULATES NC ROOTS (NC = MINIMUM ORDER CHEBYSHEV)
' AND PASSES THEM AS CHBY.REAL.ROOT(N) - REAL PART OF ROOT N
' AND                 CHBY.IMAG.ROOT(N) - IMAG PART OF ROOT N
' WHERE N IS FROM 1 TO NC
  CALC.EPSILON]
  E = EPSILON
  ARCSINH(PARAM1,1/E)]
  A = (1/NC)*PARAM1
  'SET PARAM2 = SINH(A)
  'SET PARAM3 = COSH(A)
  SINH(PARAM2,A)]
  COSH(PARAM3,A)]

  FOR I = 1 TO NC
      CHBY.REAL.ROOT(I) = -(PARAM2*SIN((2*I-1)*PI/(NC*2)))
      CHBY.IMAG.ROOT(I) =  (PARAM3*COS((2*I-1)*PI/(NC*2)))
      R(I) = CHBY.REAL.ROOT(I)
      I!(I) = CHBY.IMAG.ROOT(I)
  NEXT I
PEND]
```

The Chebyshev coefficients may be calculated by multiplying out all of the roots produced by the CHBY.ROOTS] procedure above. Each pole is first multiplied together to give a set of biquad terms. This will eliminate any imaginary terms from further calculations. Next, each of the biquads is multiplied by the resulting polynomial of the previous multiplication. The procedure thus determines NC + 1 coefficients. The index of the coefficient in the procedure below represents the power of *s* for which it is the coefficient.

```
PROCEDURE CBY.COEF]
    CBY.ROOTS]
    FOR I = 0 TO 10: COEF(I) = 0: NEXT I
    PCOEF(2,0) = 0
    PCOEF(1,0) = 0
    PCOEF(0,0) = 1
    FOR I = 1 TO NC\2
       PCOEF(2,I) = 1
       PCOEF(1,I) = -2*R(I)
       PCOEF(0,I) = R(I)^2 + I!(I)^2
    NEXT I
    J = NC MOD 2
    IF J = 1 THEN]
        PCOEF(0,0) = -R(NC\2+1)
        PCOEF(1,0) = 1
        PCOEF(2,0) = 0
    IFEND]
    FOR I = 1 TO NC\2
      FOR J = 0 TO 2
        FOR K = 0 TO I*2
          COEF(K + J) = PCOEF(K,0) * PCOEF(J,I) + COEF(J+K)
        NEXT K
      NEXT J
      FOR K = 0 TO I*2+2
        PCOEF(K,0) = COEF(K)
        IF I < NC\2 THEN COEF(K) = 0.0
      NEXT K
    NEXT I
PEND]
```

The Q and ω_0 of each biquad of the Chebyshev network is calculated in the same manner as for the Butterworth networks (i.e., by using the coefficients of the biquads) and is given by Eqs. 12.4 and 12.5. Since the biquads are calculated by multiplying pole-pairs, the root solver procedure must be evaluated before the following CBY.Q.W procedure can evaluate these equations.

```
PROCEDURE CBY.Q.W]
    CBY.ROOTS]
    FOR I = 1 TO INT(NC/2)
      IF ABS(I!(I)) >.0001 THEN]
        Wo(I) = SQR( (R(I)^2) + (I!(I)^2) )
        Q(I)    =Wo(I)/(2*R(I))
      IFEND]
    NEXT
PEND]
```

Bessel Approximation

There is no simple expression for determining the minimum order of a Bessel network from a set of filter specifications. Therefore, the procedure simply requests from the user the required filter minimum order to be used and performs no actual calculation of this parameter. The procedure will, however, check to see if the minimum order was requested directly by the user. This is done by checking to see if the menu key F1 (chr$(59)) was the last key to be pressed before entering the BSL.MIN.ORDER procedure (shown below). If the user selects this function, then the program simply asks for the desired minimum order. If the minimum order routine is entered indirectly through another procedure requiring a minimum order (such as a root or coefficient calculation), then the program tells the user what the present specified minimum order is and allows it to be changed at this time.

```
PROCEDURE BSL.MIN.ORDER]
     'NONE AVAILABLE - ENTERED MANUALLY
  COLOR 7,0
  CLS
  ANS$ = " "
  WHILE ANS$ = " "
  IF LAST.KEY$ <> CHR$(59) THEN]
    LOCATE 5,5,1
    COLOR 0,7
    PRINT "DO YOU WISH TO USE EXISTING
           MIN ORDER FILTER OF ";MIN.ORDER;
    COLOR 7,0
    PRINT "(Y/N)"
    GET.KEY(ANS$)]
  ELSE]
    ANS$ = "N"
  IFEND]
  CASE ANS$ OF]
    "N":]
      COLOR 7,0
      CLS
      LOCATE 10,1,0
      INPUT "ENTER DESIRED ORDER OF BESSEL FILTER : ";NBSL
    "n":]
      COLOR 7,0
      CLS
```

```
        LOCATE 10,1,0
        INPUT "ENTER DESIRED ORDER OF BESSEL FILTER : ";NBSL

    "Y":]
     "Y":]
   OTHERWISE]
        LOCATE 24,1,0
        COLOR 0,7
        PRINT "ENTER 'Y' OR 'N' PLEASE!!";
        ANS$ = " "
   CEND]
   WEND
  MIN.ORDER = NBSL
PEND]
```

As in the other procedures, the LOCATE, CLS, and COLOR statements are used for screen formatting.

The coefficients of an nth order Bessel network may be found via the following relationship:

```
                                  ORDER = C
     COEF(S^ORDER) = SUMMATION(2*ORDER-1)*COEF(S^ORDER-1))          (12.9)
                                  ORDER = MIN.ORDER
```

The coefficients as determined by Eq. 12.9 are evaluated by the procedure BSL.COEF] as shown:

```
PROCEDURE BSL.COEF]
    'LATER
    DIM BSL(30,15):BSL(0,0)=1:BSL(1,0)=1:BSL(1,1)=1
    IF NBSL => 2 THEN]
       FOR ORDR=2 TO MIN.ORDER
         FOR S=0 TO ORDR-1
           IF S<2 THEN
               BSL(ORDR,S)=(2*ORDR-1)*BSL(ORDR-1,S)
           ELSE
               BSL(ORDR,S)=(2*ORDR-1)*BSL(ORDR-1,S)
               +BSL(ORDR-2,S-2)
         NEXT S
       BSL(ORDR,ORDR)=1
       NEXT ORDR
    IFEND]
    FOR I=MIN.ORDER TO 0 STEP -1
         COEF(I) = BSL(MIN.ORDER,I)
    NEXT I
   ERASE BSL ' undimensions to avoid redimension error
PEND]
```

There is no closed formula for determining the Bessel roots from a given set of filter specifications as in the Butterworth and Chebyshev filters. Thus a root-finder program is required. The procedure begins by entering the coefficients of the polynomial and its degree. The program first tests for zero roots by checking for a 0 constant value (i.e., coef (s ^ 0) = 0). If there is a zero root, the program will readjust all of the coefficients and decrease the order by one. Then it will repeat the process until no zero roots are found. The program next determines the degree of the polynomial. If the degree is 1 or 2, the roots are calculated using standard methods (SINGLE.ROOT], DOUBLE.ROOT]). If the degree is greater than 2, the program extracts a quadratic factor using the Lin-Bairstow method (ROOT.FINDER] AND ITERATE]) for finding the quadratic factor of a given polynomial. Once a quadratic factor is extracted, the roots can be calculated using the quadratic formula. The remaining polynomial is then checked again for the resulting order after the quadratic extraction, and the process is repeated until all roots have been calculated. This root finding (ROOTS]) procedure is listed here along with the required support routines.

```
PROCEDURE ENTER.COEFS]

     COLOR 7,0: CLS: COLOR 0,7: LOCATE 3,10
     PRINT " DO YOU WISH TO USE EXISTING DEGREE AND ROOTS : ";
     COLOR 7,0:PRINT "(Y/N)"
     GET.KEY(ROOT$)]
     IF ROOT$ = "n" OR ROOT$ = "N" THEN]
       INPUT " ENTER DEGREE ";M3
       FOR I = 0 TO M3
         PRINT "ENTER S^";I;" TERM: ";:INPUT " ",C(I)
       NEXT I

     ELSE]
       M3 = MIN.ORDER
       FOR I = 0 TO M3
         C(I) = COEF(I)
       NEXT I
     IFEND]
       COMPLEX.ROOT.N = 0.0:REAL.ROOT.N = INT ((M3/2)+1)+M3
       RMAX = REAL.ROOT.N
       NO = M3
PEND]

PROCEDURE SINGLE.ROOT]
     IF L1<= 0 THEN R=1/R
       FLAG = 1
       PRINT R
       ROOT.REAL(REAL.ROOT.N) = R:ROOT.IMAG(REAL.ROOT.N) = 0.0
       REAL.ROOT.N = REAL.ROOT.N-1
       N=N-1
       FOR I=0 TO N
         C(I)=D(I)
       NEXT I
PEND]

PROCEDURE DOUBLE.ROOT]
   FLAG = 1
   IF L1<=0 THEN]
     P=P/Q
     Q=1/Q
   IFEND]
   T=P*P/4-Q
   IF T<0 THEN]
     T1=-P/2
     T=SQR(-T)
     S=SQR(ABS(Q))
     COMPLEX.ROOT.N = COMPLEX.ROOT.N + 1
     ROOT.REAL(COMPLEX.ROOT.N) = T1:ROOT.IMAG(COMPLEX.ROOT.N) = T
   ELSE]
     T = SQR(T) + ABS (P)/2
     IF P >=0 THEN T=-T
     ROOT.REAL(REAL.ROOT.N) = Q/T
     REAL.ROOT.N = REAL.ROOT.N -1
     ROOT.REAL(REAL.ROOT.N) = T
     REAL.ROOT.N = REAL.ROOT.N -1
   IFEND]
     N=N-2
     FOR I=0 TO N
       C(I)=B(I)
     NEXT I
PEND]
```

```
PROCEDURE ITERATE]
' THIS PROCEDURE IS USED AS PART OF THE LIN-BAIRSTOW QUADRATIC
' EXTRACTION PROCESS. A QUADRATIC FACTOR IS ESTIMATED THEN TESTED
' AND THEN RE-ESTIMATED. THIS PROCESS CONTINUES UNTIL THE
' ITERATION YEILDS A VALID QUADRATIC FACTOR.
         FOR L2=1 TO 20
             B(0)=C(0)
             C1=B(0)
             D(0)=C(0)
             B(1)=C(1)-B(0)*P
             C2=B(1)-C1*P
             D(1)=C(1)+R*D(0)
             E1=D(1)+R*D(0)
             FOR I=2 TO N-1
               B(I)=C(I)-B(I-1)*P-B(I-2)*Q
               C0=C1
               C1=C2
               C2=B(I)=C1*P-C0*Q
               D(I)=C(I)+R*D(I-1)
               E0=E1
               E1=D(I)+R*E0
             NEXT I
             T1=C(N)-P*B(N-1)-Q*B(N-2)
             C2=-P*C1-Q*C0
             T=C1*C1-C2*C0
             P3=B(N-1)*C1-T1*C0
             Q3=T1*C1-B(N-1)*C2
             A1 = 0
             IF ABS(P3)<=X*ABS(P*T) THEN IF ABS(Q3)<=X*ABS(Q*T) THEN
             RETURN
             IF T = 0 THEN]
               P=P-2
               Q=Q*(Q+1)
             ELSE]
               P=P+P3/T
               Q=Q+Q3/T
             IFEND]
             T=C(N)+R*D(N-1)
             A1 = 1
             IF ABS(T)<=X*ABS(R*E1) THEN RETURN
             A1 = 2
             IF E1 = 0 THEN R = R-1 ELSE R = R-T/E1
             NEXT L2
             P1=P2
             Q1=Q2
             P2=P
             Q2=Q
             X=10*X
PEND]

PROCEDURE ROOT.FINDER]
' THIS IS THE ACTUAL ROOT SOLVING ALGORITHM AS DESCRIBED IN THE
' PARAGRAPH ABOVE
    FLAG = 0
    XXX = 0
    N1 = N0
    FOR I = 0 TO N1
      G(I) = C(I)
    NEXT I
    N = N0
    L1 = -1
    WHILE C(N) = 0
      N = N-1
    WEND
    IF N= 0 THEN]
      PRINT "DEGREE IS ZERO, NO ROOTS"
    RETURN
```

```
 IFEND]
WHILE N > 0
 IF C(0)=0 THEN]
     PRINT 0
     N = N-1
     FOR I = 0 TO N
       C(I) = C(I+1)
     NEXT I
   WEND
   IF N <= 0 THEN RETURN
 IFEND]
 WHILE N > 0
 LOCATE 10,25,0:COLOR 0,7:PRINT " CALCULATING ROOTS ":FOR I=1 TO 500
 NEXT I
 COLOR 7,0:CLS
   CASE N OF]
     1:]
        R = C(1)/C(0)
        SINGLE.ROOT]
     2:]
        P = C(1)/C(0)
        Q=C(2)/C(0)
        DOUBLE.ROOT]
     >3:]
        P1 = 0
        P2 = 0
        Q1 = 0
        Q2 = 0
        X = .0000001
        S = SQR(ABS(C(0)*C(N)))
        FOR I = 0 TO N
          C(I) = C(I)/S
        NEXT I
        WHILE FLAG = 0
 IF (XXX=1) OR ((XXX = 0) AND (ABS(C(1)/C(0))
    < ABS(C(N-1)/C(N)))) THEN]
   XXX = 1
   L1 = -L1
   FOR I = 0 TO (N-1)/2+.1
     J = N-I
     T = C(I)
     C(I) = C(J)
     C(J) = T
   NEXT I
 IFEND]
 XXX = 1
 IF Q1 <> 0 THEN]
   P = P1
   Q = Q1
 ELSE]
   P = C(N-2)
   IF P = 0 THEN]
     Q = 1
     P = -2
   ELSE]
     Q = C(N-1)/P
     P = Q-C(N-3)*Q/P
   IFEND]
   R=0
IFEND]
ITERATE]
 CASE A1 OF]
   0:]
      DOUBLE.ROOT]
   1:]
      SINGLE.ROOT]
```

```
              OTHERWISE]
                FLAG = 0
            CEND]
            WEND
            FLAG = 0
        CEND]
    IF N <= 0 THEN RETURN
    WEND
PEND]


PROCEDURE ROOT]
  DIM ROOT.REAL(20),ROOT.IMAG(20)
  CLS
  LOCATE 12,29,0
  COLOR 0,7
  PRINT " ROOT SOLVING PROGRAM "
  ENTER.COEFS]
  ROOT.FINDER]
  COLOR 7,0
  CLS
  FOR I = 1 TO COMPLEX.ROOT.N
    ROOT.REAL(M3-I+1) = ROOT.REAL(I):ROOT.IMAG(M3-I+1) = -ROOT.IMAG(I)
    R(I)=ROOT.REAL(I):I!(I)=ROOT.IMAG(I)
        NEXT I
  TEMP.VAR = RMAX
  FOR I = COMPLEX.ROOT.N+ 1 TO M3-COMPLEX.ROOT.N
    ROOT.REAL(I) = ROOT.REAL(TEMP.VAR)
    R(I) = ROOT.REAL(I)
    TEMP.VAR = TEMP.VAR - 1
  NEXT I
  FOR I = M3-COMPLEX.ROOT.N+1 TO M3
    R(I) = ROOT.REAL(I):I!(I) = ROOT.IMAG(I)
  NEXT I
' END OF ROOT FINDING ROUTINE
PEND]
```

The procedure ENTER.COEFS] allows the user to enter a polynomial or use the last set of coefficients that were computed by the BSL.COEF] procedure. The root-finding procedure will also order the roots in such a manner that the PRINT.ROOT] procedure shown needs only to read one-half of the complex conjugate pole pairs.

```
PROCEDURE PRINT.ROOTS]
' PRINTS ROOTS IN FORM OF "REAL +/- J IMAG"
     COLOR 7,0
     CLS
     COLOR 0,7
     LOCATE 5,1,0
       FOR I = 1 TO MIN.ORDER
          IF I!(I) = > 0 THEN SIGN$ = "+" ELSE SIGN$ = "-"
          PRINT " ROOT S(";I;") = ";
          PRINT USING "##.####";R(I);
          PRINT " "; SIGN$;" J";
          PRINT USING "##.####";ABS(I!(I))
       NEXT I
PEND]
```

As before, the ω_0 and Q are determined from the Eqs. 12.4 and 12.5, respectively. The Bessel procedure for evaluating these variables is shown in the following procedure BSL.Q.ω_0].

```
PROCEDURE BSL.Q.W]
    BSL.ROOTS]
    FOR I = 1 TO INT(NBSL/2)
      IF ABS(I!(I)) > .0001 THEN]
      Wo(I) = SQR( (R(I)^2) + (I!(I)^2) )
      Q(I)    =Wo(I)/(2*R(I))
      IFEND]
    NEXT I
PEND]
```

12.3 FREQUENCY TRANSFORMATION PROGRAMS

This section explains the method and use of the frequency transformation programs introduced in Section 2.7. These programs allow the transformation of a normalized low-pass filter into a specified high-pass, bandpass, or bandstop filter. Transformations are also possible in the reverse direction. That is, a filter network may be specified and then transformed into a normalized low-pass filter. After normalization, the minimum order required to realize the specified filter can be calculated by using the minimum order routine described in Section 12.2.

The other programs described in Section 12.2 may also be selected. Routines that allow plotting of the gain and phase frequency responses for the transformed or normalized filter networks are also available and will be explained in Section 12.4.

The transform program is selected from the main menu. A new menu will display a choice of either transforming an “Existing Low-pass Transformed to Specified Filter Type” or transforming a “Specified Filter to a Normalized Low-pass” filter.

If a filter type is to be transformed to a normalized low-pass filter, then a menu appears that displays the three possible filter transforms. The program then prompts for the appropriate filter parameters as shown in Fig. 12.4. By definition, ω_1 of the normalized low pass will be 1; ω_2 of the normalized low-pass networks is shown in Fig. 12.4*a*.

The procedure CALCULATE.LOW.PASS.PROTOTYPE] shown calculates the low-pass filter frequencies ω_1 and ω_2:

```
PROCEDURE CALCULATE.LOW.PASS.PROTOTYPE
'
   W1.LP = 1
   CASE LAST.KEY$ OF]
     CHR$(59):]
        W2.LP = W2/W1
     CHR$(60):]
        W2.LP = Q.FLTR*(W4-W3)/WO
     CHR$(61):]
        W2.LP = WO/(Q.FLTR*(W4-W3))
     OTHERWISE]
        CLS:PRINT "KEY ERROR"
   CEND]
PEND]
```

Filter type	ω_2 OF LOWPASS NETWORK
High pass	ω_2/ω_1
Bandpass	$Q(\omega_4 - \omega_3)/\omega_0$
Band stop	$\dfrac{\omega_0}{[Q(\omega_4 - \omega_3)]}$

(a)

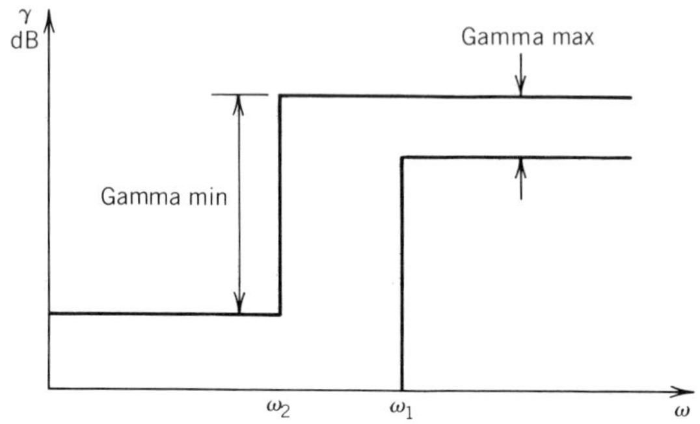

(b)

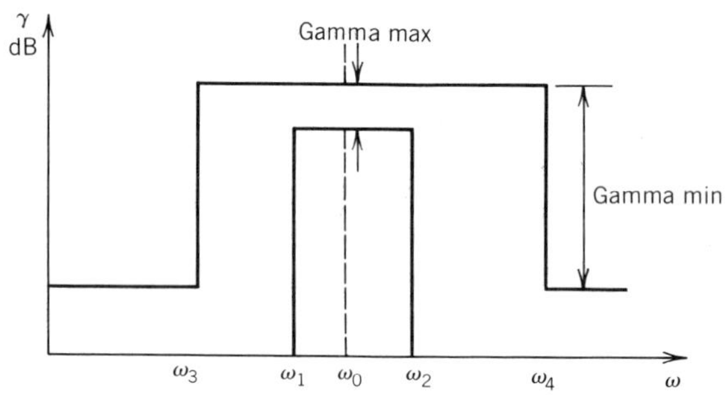

(c)

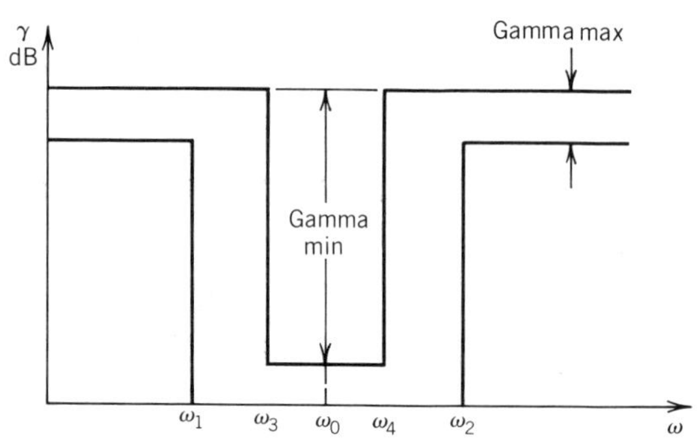

(d)

Figure 12.4 Frequency transformations and filter specifications: (a) Frequency transformations; (b) high-pass specification; (c) bandpass specification; (d) bandstop specification.

The frequencies ω_1 and ω_2 for the bandpass and bandstop filters need not be specified. Also, ω_0 and Q, which are prompted by the program, define ω_1 and ω_2 by the relationships

$$\omega_0^2 = \omega_1\omega_2 \tag{12.10}$$

and

$$Q = \omega_0/(\omega_2 - \omega_1) \tag{12.11}$$

ω_3, ω_4, GamaMin, and GamaMax are values that are also prompted by the program. After the filter parameters are entered and the transform frequencies ω_1 and ω_2 for the normalized low-pass filter are calculated, the program automatically links back to the appropriate Chebyshev or Butterworth analysis programs.

When a previously defined low-pass filter is to be transformed to a high-pass filter, only the new passband frequency ω_2 and high-frequency gain need to be specified. The biquadratic coefficients must have been calculated before selection of the transform routine. A new set of biquadratic coefficients is calculated by substituting ω_2/s for s in the low-pass biquadratic expression. If the biquadratic terms were not previously entered, the program asks for the number of biquads and then prompts for all the required coefficients of both numerator and denominator terms. The DO.LOW.PASS.TO.HI.PASS.TRANSFORM] procedure that follows is a listing of the procedure that calculates the new biquadratic coefficients.

```
PROCEDURE DO.LOW.PASS.TO.HI.PASS.TRANSFORM]
        CLS
        LOCATE 10,3,0
        INPUT "W2 = ",W2
        W2 = W2 * SCALE:FACTOR
        LOCATE 11,3,0
        INPUT "H (GAIN AT W >> W2) = ",H
        FOR I = 1 TO NUM.OF.BI.QUADS
           IF BI.QUAD.S2(I) <> 0 THEN]
                BI.QUAD.S2(I) = BI.QUAD.S0(I)
                BI.QUAD.S1(I) = BI.QUAD.S1(I)*W2
                BI.QUAD.S0(I) = BI.QUAD.S2(I)*W2^2
                IF I = 1 THEN]
                   N.BI.QUAD.S2(I) = H
                ELSE]
                   N.BI.QUAD.S2(I) = 1
                IFEND]
                N.BI.QUAD.S1(I) = 0
                N.BI.QUAD.S0(I) = 0
           ELSE]
                BI.QUAD.S2(I) = 0
                BI.QUAD.S1(I) = BI.QUAD.S0(I)
                BI.QUAD.S0(I) = BI.QUAD.S1(I)*W2
                IF I = 1 THEN]
                   N.BI.QUAD.S1(I) = H
                ELSE]
                   N.BI.QUAD.S1(I) = 1
                IFEND]
                N.BI.QUAD.S2(I) = 0
                N.BI.QUAD.S0(I) = 0
           IFEND]
        NEXT I
PEND]
```

When a low-pass filter is to be transformed to a bandpass or bandstop, then ω_0, Q, ω_3, ω_4, and H must be entered. H is the midband gain and dc gain for the bandpass and bandstop filters, respectively. Selection of ω_2 and ω_3 should be as shown in Figs. 12.4*c* and *d*. Specifically, $(\omega_3 < \omega_1 < \omega_0)$ and $(\omega_0 < \omega_2 < \omega_4)$ for bandpass and $(\omega_1 < \omega_3 < \omega_0)$ and $(\omega_0 < \omega_4 < \omega_2)$ for the bandstop. If incorrectly specified, the program adjusts the values of ω_3 and ω_4 so that the value of $[\omega_0/(\omega_4 - \omega_3)]$ is half that of the passband Q for a bandpass filter and twice that of the stopband Q. When ω_3 and ω_4 are specified correctly but are not symmetric $(\omega_3\omega_4 \neq \omega_0^2)$, they are then adjusted for symmetry in a direction that tends to increase the order of the required filter.

The low-pass filter is transformed into a bandpass filter by substituting $Q[s/\omega_0 + \omega_0/s]$ for s and transformed into a bandstop by the substitution of $1/Q[1/(s/\omega_0 + \omega_0/s)]$ for s in the biquadratic expressions.

These procedures are shown in the following listings. The low-pass to bandpass transform procedure is

```
PROCEDURE DO.LOW.PASS.TO.BAND.PASS.TRANSFORM]
          CLS
          LOCATE 10,3,0
          INPUT "WO = ",WO
          WO = WO * SCALE.FACTOR
          LOCATE 11,3,0
          INPUT "H (MID BAND GAIN) = ",H
          LOCATE 12,3,0
          INPUT "Q = ",Q.FLTR
          FOR I = 1 TO NUM.OF.BI.QUADS
               A.S2(I) = BI.QUAD.S2(I)
               B.S1(I) = BI.QUAD.S1(I)
               C.SO(I) = BI.QUAD.SO(I)
          NEXT I
          BI.QUAD.COUNTER = 0
          FOR II = 1 TO NUM.OF.BI.QUADS
              IF A.S2(II) <> 0 THEN
                    DEGREE = 4
                    COEF = A.S2(II)*Q.FLTR*Q.FLTR
                    COEF(4) = 1
                    COEF(3) = (B.S1(II)*Q.FLTR*WO)/COEF
                    COEF(2) = ((C.SO(II)+2*A.S2(II)*Q.FLTR^2)*WO^2)/COEF
                    COEF(1) = (B.S1(II)*Q.FLTR*WO^3)/COEF
                    COEF(0) = WO^4
                    ROOT]
                    BI.QUAD.S2(2*II-1) = 1
                    BI.QUAD.S1(2*II-1) = -(R(1) + R(4))
                    BI.QUAD.SO(2*II-1) = R(1)*R(4) + I!(1)^2
                    BI.QUAD.S2(2*II) = 1
                    BI.QUAD.S1(2*II) = -(R(2) + R(3))
                    BI.QUAD.SO(2*II) = R(2)*R(3) + I!(2)^2
                    IF II = 1 THEN]
                         N.BI.QUAD.S1(2*II-1) = (WO/SQR(COEF))*H
                         N.BI.QUAD.S1(2*II)    = (WO/SQR(COEF))
                         N.BI.QUAD.SO(2*II-1) = 0
                         N.BI.QUAD.SO(2*II)    = 0
                    ELSE]
                         N.BI.QUAD.S1(2*II-1) = (WO/SQR(COEF))
                         N.BI.QUAD.S1(2*II) = (WO/SQR(COEF))
                         N.BI.QUAD.SO(2*II-1) = 0
                         N.BI.QUAD.SO(2*II)    = 0
                    IFEND]
                    N.BI.QUAD.S2(2*II-1) = 0
                    N.BI.QUAD.S2(2*II)    = 0
                    BI.QUAD.COUNTER = BI.QUAD.COUNTER + 1
```

```
        ELSE]
              COEF  = B.S1(II)*Q.FLTR
              COEF(2)  = 1
              COEF(1)  = C.SO(II)*WO/COEF
              COEF(O)  = WO*WO
              BI.QUAD.S2(2*II-1)  = 1
              BI.QUAD.S1(2*II-1)  = COEF(1)
              BI.QUAD.SO(2*II-1)  = COEF(O)
              IF II = 1 THEN]
                   N.BI.QUAD.S1(2*II-1)  = (WO/COEF)*H
              ELSE]
                   N.BI.QUAD.S1(2*II-1)  = WO/COEF
              IFEND]
              N.BI.QUAD.SO(2*II-1)  = 0
              N.BI.QUAD.S2(2*II-1)  = 0
        IFEND]
      NEXT II
      NUM.OF.BI.QUADS = NUM.OF.BI.QUADS + BI.QUAD.COUNTER
PEND]
```

The low-pass to bandstop transform procedure is

```
PROCEDURE DO.LOW.PASS.TO.BAND.STOP.TRANSFORM]
      CLS
      LOCATE 10,3,0
      INPUT "WO = ",WO
      WO = WO * SCALE.FACTOR
      LOCATE 11,3,0
      INPUT "H (DC GAIN) = ",H
      LOCATE 12,3,0
      INPUT "Q = ",Q.FLTR
      FOR I = 1 TO NUM.OF.BI.QUADS
           A.S2(I) = BI.QUAD.S2(I)
           B.S1(I) = BI.QUAD.S1(I)
           C.SO(I) = BI.QUAD.SO(I)
      NEXT I
      BI.QUAD.COUNTER = 0
      FOR II = 1 TO NUM.OF.BI.QUADS
          IF A.S2(II) <> 0 THEN]
                DEGREE  = 4
                COEF  = C.SO(II)*Q.FLTR*Q.FLTR
                COEF(4)  = 1
                COEF(3)  = (B.S1(II)*Q.FLTR*WO)/COEF
                COEF(2)=A.S2(II)+2*C.SO(II)*Q.FLTR*Q.FLTR*WO^2/COEF
                COEF(1)  = (B.S1(II)*Q.FLTR*WO^3)/COEF
                COEF(O)  = WO^4
                ROOT]
                BI.QUAD.S2(2*II-1)  = 1
                BI.QUAD.S1(2*II-1)  = -(R(1) + (R(4))
                BI.QUAD.SO(2*II-1)  = R(1)*R(4) + I!(1)^2
                BI.QUAD.S2(2*II)  = 1
                BI.QUAD.S1(2*II)  = -(R(2) + R(3))
                BI.QUAD.SO(2*II)  = R(2)*R(3) + I!(2)^2
                IF II = 1 THEN]
                     N.BI.QUAD.S2(2*II-1)  = (Q.FLTR/SQR(COEF))*H
                     N.BI.QUAD.S2(2*II)    = (Q.FLTR/SQR(COEF))
                     N.BI.QUAD.SO(2*II-1)  = (Q.FLTR/SQR(COEF))*H*WO*WO
                     N.BI.QUAD.SO(2*II)    = (Q.FLTR/SQR(COEF))*WO*WO
                ELSE]
                     N.BI.QUAD.S2(2*II-1)  = (Q.FLTR/SQR(COEF))
                     N.BI.QUAD.S2(2*II)  = (Q.FLTR/SQR(COEF))
                     N.BI.QUAD.SO(2*II-1)  = (Q.FLTR/SQR(COEF))*WO*WO
                     N.BI.QUAD.SO(2*II)    = (Q.FLTR/SQR(COEF))*WO*WO
                IFEND]
                N.BI.QUAD.S1(2*II-1)  = 0
                N.BI.QUAD.S1(2*II)    = 0
                BI.QUAD.COUNTER = BI.QUAD.COUNTER + 1
```

```
            ELSE]
                COEF = C.SO(II)*Q.FLTR
                COEF(2) = 1
                COEF(1) = B.S1(II)*WO/COEF
                COEF(0) = WO*WO
                BI.QUAD.S2(2*II-1) = 1
                BI.QUAD.S1(2*II-1) = COEF(1)
                BI.QUAD.SO(2*II-1) = COEF(0)
                IF II = 1 THEN]
                    N.BI.QUAD.S2(2*II-1) = (Q.FLTR/COEF)*H
                    N.BI.QUAD.SO(2*II-1)   = (Q.FLTR/COEF)*H
                ELSE]
                    N.BI.QUAD.S2(2*II-1) = Q.FLTR/COEF
                    N.BI.QUAD.SO(2*II-1)   = Q.FLTR/COEF
                IFEND]
                N.BI.QUAD.S1(2*II-1) = 0
            IFEND]
        NEXT II
        NUM.OF.BI.QUADS = NUM.OF.BI.QUADS + BI.QUAD.COUNTER
PEND]
```

For the last two transforms shown, the order of each biquadratic section is doubled. Thus the transformation will create a fourth-order denominator for each second-order biquadratic section. This will increase the number of biquadratic sections by 1 for each second-order low-pass biquadratic section. The root finder routine is called for this case to separate the transformed fourth-order expression into two biquadratic sections. The factor H that is entered is a multiplier on the normalized gain. If the low-pass filter being transformed is a Butterworth filter, then the final gain is H. However, if a Chebyshev filter has been transformed, the gain varies from H by a factor of one over the product of the s^0 terms of each biquad.

12.4 GAIN AND PHASE FREQUENCY RESPONSES

This section describes two programs. The first calculates the gain and phase frequency responses of a network function. The results are tabulated on the screen and stored on a disk. The second program reads these data from the disk, scales the data, and plots either the gain or phase response on the screen.

The data required for the gain and phase frequency response program are:

1 Number of biquadratic sections.

2 Coefficients of the numerator biquads.

3 Coefficients of the denominator biquads.

4 Starting and ending frequencies.

5 Frequency units, that is, rads/s or hertz.

The procedure GET.XFER.FUNCT] (get transfer function) shown below collects these data for use by the program. If a filter has been analyzed using the programs of Section 12.2, the value of NUM.OF.BI.QUADS is the number of biquadratic sections for the analyzed filter. The program uses all previously calculated biquad coefficients if the value of NUM.OF.BI.QUADS is not zero. If the value is zero, then the program requires manual entry of the number of biquadratic sections and all the necessary coefficients. The start and stop frequencies are always entered manually as well as their scaling (i.e., Hz or rad/s).

```
PROCEDURE GET.XFER.FUNCT]
   CLS
   PRINT "WILL FREQUENCIES BE ENTERED IN HZ OR RADIANS - (H/R)? : ";
   ANS$ = ""
   WHILE ANS$ = ""
      ANS$ = INKEY$
      IF ANS$ = "R" OR ANS$ = "r" THEN]
         SCALE.FACTOR = 1
         PRINT ANS$
      ELSE]
         IF ANS$ = "H" OR ANS$ = "h" THEN]
           SCALE.FACTOR = (2*3.141592654)
           PRINT ANS$
         ELSE]
           ANS$ = ""
         IFEND]
      IFEND]
   WEND
   PRINT:PRINT
   IF NUM.OF.BI.QUADS = 0 THEN]
      INPUT "ENTER NUMBER OF BI-QUADRATIC SECTIONS ",NUM.OF.BI.QUADS
      FOR I = 1 TO NUM.OF.BI.QUADS
          CLS
          PRINT "BIQUAD #";I;" COEFFICIENTS....":PRINT
          PRINT "ENTER S^2 COEF OF NUMER : ";:INPUT "",N.BI.QUAD.S2(I)
          PRINT "ENTER S^1 COEF OF NUMER : ";:INPUT "",N.BI.QUAD.S1(I)
          PRINT "ENTER S^0 COEF OF NUMER : ";:INPUT "",N.BI.QUAD.SO(I)
          PRINT
          PRINT "ENTER S^2 COEF OF DENOM : ";:INPUT "",BI.QUAD.S2(I)
          PRINT "ENTER S^1 COEF OF DENOM : ";:INPUT "",BI.QUAD.S1(I)
          PRINT "ENTER S^0 COEF OF DENOM : ";:INPUT "",BI.QUAD.SO(I)
      NEXT I
   IFEND]
   PRINT:INPUT"ENTER START FREQUENCY :",FS
   PRINT:INPUT"ENTER STOP FREQUENCY :",FF
   DATA.COUNTER = 0
   MIN.FREQ = FS
   MAX.FREQ = FF
   FREQ = FS
   FT = 10^(.05)          '20 EQUALLY SPACED SAMPLES PER DECADE
   WHILE FREQ <= FF+FF/10
     DATA.COUNTER = DATA.COUNTER+1
     FREQ = FREQ*FI
   WEND
   MIN.GAIN=1E10
   MAX.GAIN =-1E10
   CLS
   PRINT "Freq (HZ)","     Gain (dB)","PHASE (DEG)"
   PRINT
PEND]
```

The frequency analysis program increments ω in 20 exponential steps per decade and then substitutes $j\omega$ for s in each of the biquads. This yields an expression for the transfer function with a real and imaginary part. The gain is calculated by taking the magnitude of this complex expression. The phase shift is calculated by finding the arc tangent of the quotient of the imaginary and real parts. The phase is automatically limited to + or − 180°. The program also stores the values of the maximum and minimum gain for use by the plotting routine. The procedure DO.FREQ.ANALYSIS], which follows, calculates the gain and phase frequency responses described above.

```
PROCEDURE DO.FREQ.ANALYSIS]
   FREQ = FS
   FI = 10^(.05)        '20 EQUALLY SPACED SAMPLES PER DECADE
   WHILE FREQ <= FF+FF/10
    W = SCALE.FACTOR*FREQ
    S = W
    OLD.X = 1
    DB = 0!:PHASE = 0!:DELAY = 0!
    FOR I = 1 TO NUM.OF.BI.QUADS
    NUMR.REAL = -N.BI.QUAD.S2(I)*S^2 + N.BI.QUAD.S0(I)
    NUMR.IMAG = S*N.BI.QUAD.S1(I)
    DEMR.REAL = -BI.QUAD.S2(I)*S^2 + BI.QUAD.S0(I)
    DEMR.IMAG = S*BI.QUAD.S1(I)
    DIVISOR = DEMR.REAL*DEMR.REAL + DEMR.IMAG*DEMR.IMAG
    NEW.NUMR.REAL =  NUMR.REAL*DEMR.REAL + NUMR.IMAG*DEMR.IMAG
    NEW.NUMR.IMAG = -NUMR.REAL*DEMR.IMAG + NUMR.IMAG*DEMR.REAL
    NUMR.DIV.DEMR.REAL = NEW.NUMR.REAL/DIVISOR
    NUMR.DIV.DEMR.IMAG = NEW.NUMR.IMAG/DIVISOR
    X = SQR(NUMR.DIV.DEMR.REAL*NUMR.DIV.DEMR.REAL +
        NUMR.DIV.DEMR.IMAG*NUMR.DIV.DEMR.IMAG)
    IF X < 1E-10 THEN X = OLD.X
    OLD.X = X
    DB = DB + 20! * LOG(X)/LOG(10)
    RL = NUMR.DIV.DEMR.REAL
    XIMG = NUMR.DIV.DEMR.IMAG
    ATAN2(ATAN.ANGLE,RL,XIMG)]
    PHASE = PHASE + ATAN.ANGLE
    NEXT I
    IF DB > MAX.GAIN THEN MAX.GAIN = DB
    IF DB < MIN.GAIN THEN MIN.GAIN = DB
    PHASE= PHASE*57.295577951#
    WHILE PHASE => 360.
       PHASE = PHASE -360
    WEND
    WHILE PHASE <= -360
       PHASE = PHASE + 360
    WEND
    IF PHASE > 180 THEN]
       PHASE = PHASE -360
    IFEND]
    IF PHASE < -180 THEN]
       PHASE = PHASE + 360
    IFEND]
    PRINT USING "####.###              ";FREQ;
    PRINT USING "####.###          ";DB;
    PRINT USING "####.###          ";PHASE
    WRITE #1,FREQ,DB
    WRITE #2,FREQ,PHASE
    FREQ = FREQ*FI
    WEND
PEND]
```

The following RATBAS program performs the complete gain and phase frequency analysis using the two procedures described above.

```
PROGRAM]
        CLS
        PRINT "ENTER FILENAME FOR GAIN PLOT DATA (8 CHARACTERS MAX) : ";
        INPUT "",GAINFILE$
        PRINT:PRINT
        PRINT "ENTER FILENAME FOR PHASE PLOT DATA (8 CHARACTERS MAX) : ";
        INPUT "",PHASFILE$
        OPEN GAINFILE$ FOR OUTPUT AS #1
        OPEN PHASFILE$ FOR OUTPUT AS #2
        GET.XFER.FUNCT]
        WRITE #1,SCALE.FACTOR,MIN.FREQ,MAX.FREQ,DATA.COUNTER
        WRITE #2,SCALE.FACTOR,MIN.FREQ,MAX.FREQ,DATA.COUNTER
```

```
        DO.FREQ.ANALYSIS
        WRITE #1,MIN.GAIN,MAX.GAIN
        CLOSE #1
        CLOSE #2
        CLS
        LOCATE 24,1,0
        COLOR 0,7
        PRINT "HIT ANY KEY TO RETURN TO MAIN MENU";
        COLOR 7,0
        Z$ = ""
        WHILE Z$ = ""
        Z$ = INKEY$
        WEND
        CHAIN "FILTER",,ALL
END
```

The plotting routine plots the gain or phase response of any data file that has been stored on the disk from a previous gain and phase analysis. A menu lets the user select either a gain plot or a phase plot. If a gain plot is selected, then a program called "GAINPLOT" will be run. This program starts by prompting for the name of the gain data file stored on a disk. The procedure GET.GRAPH. PARAMETERS], which follows, reads in all required data from the disk.

```
PROCEDURE GET.GRAPH.PARAMETERS]
        DEFSNG A-H,L-Z
        CLS
        PRINT "ENTER FILE NAME TO BE PLOTTED : ";:INPUT FILE$
        OPEN FILE$ FOR INPUT AS #1
        INPUT #1,SCALE.FACTOR,MIN.FREQ,MAX.FREQ,DATA.COUNTER
        DIM FREQ(DATA.COUNTER),GAIN(DATA.COUNTER)
        FOR I = 1 TO DATA.COUNTER
          INPUT #1,FREQ(I),GAIN(I)
        NEXT I
        INPUT #1,MIN.GAIN,MAX.GAIN
        CLOSE #1
        TITLE$ = " GAIN VS. FREQ"
PEND]
```

After the horizontal and vertical scaling has been set up and the graph, with proper labeling, has been constructed (procedures not shown), the data are scaled and plotted by the procedure listed below.

```
PROCEDURE PLOT.DATA]
        FOR I = 2 TO DATA.COUNTER
           X.F1 = ((LOG(FREQ(I-1)/LEFT.AXIS)/LOG(10))*H.DECADE.INCR)
                  + H.OFFSET
           X.F2 = ((LOG(FREQ(I-0)/LEFT.AXIS)/LOG(10))*H.DECADE.INCR)
                  + H.OFFSET
           Y.G1 = BOTTOM-(((GAIN(I-1)-LOWER.AXIS)/20)*V.DECADE.INCR)
           Y.G2 = BOTTOM-(((GAIN(I-0)-LOWER.AXIS)/20)*V.DECADE.INCR)
           LINE (X.F1,Y.G1) - (X.F2,Y.G2)
        NEXT I
PEND]
```

LEFT.AXIS, H.DECADE.INCR, H.OFFSET, and so on, are scale factors that scale the data so that all plots have the same size. The phase plotting program, which is not shown, is done in a similar manner. After completion of a plot, any key press displays the plot menu for selection of another plot or return to the main menu. The following program listing is the RATBAS program for the gain plot routine described above.

```
PROGRAM]
        GET.GRAPHIC.NUMBERS.UP]
        GET.GRAPHIC.ASCII.ROT]
        GET.GRAPH.PARAMETERS]
        CLS
        DETERMINE.H.AXIS]
        DETERMINE.V.AXIS]
        MAKE.GRAPH]
        PLOT.DATA]
        LOCATE 23,1
        LOCATE 25,1,0
        PRINT " PRESS ANY KEY TO RETURN TO PLOT MENU";
        Z$ = ""
        WHILE Z$ = ""
        Z$ = INKEY$
        WEND
        CLS
        RUN "PLOTS
END
```

12.5 A DESIGN EXAMPLE

All of the programs presented in Sections 12.2 through 12.4 are linked together in an interactive filter analysis program. This program allows the user (from a menu-driven environment) to select any of the filter analyses given in the other sections. Hence it is possible to design a filter by running this single program. A design example is presented in this section to illustrate the use of the program package as a complete filter design tool. A calculator program has also been included in this section. It provides the user with an additional filter design tool that may be used to calculate component values using methods and equations presented in other chapters.

The BASIC interpreter limits the maximum program size to 64K bytes. Most of the programs described in the previous sections were themselves close to this limit. Thus each of the program selections of the main menu is chained. Chaining is a method by which variables may be passed from one program to another. Each of the program selections also chains back to the main menu selection. This allows variables calculated in one program (e.g., biquad coefficients in the Butterworth program) to be passed to another program (e.g., transformation program). There is one exception to this rule. When the plot routine returns to the main menu, it will return using a RUN instruction rather than a CHAIN. This allows all of the variables to be reset and a new design analysis can be started.

Figure 12.5 shows the main menu program selection, which is the menu that appears when the program is started. A Butterworth response with the following characteristics will illustrate the use of this program.

```
FILTER TYPE:         BUTTERWORTH
FREQUENCY TYPE:      BANDPASS
CENTER FREQUENCY:    Wo = 1.5 KHz
FILTER Q:            5
STOPBAND FREQ'S:     W3 = 550 Hz, W4 = 3 KHz
GAMA MIN:            10 dB
GAMA MAX:            3 dB
MID BAND GAIN:       H = 10 (20 dB)
```

```
                BUTTERWORTH FILTER ANALYSIS PROGRAM

F1  REQUIRED MINIMUM ORDER          F2  ROOT LOCATIONS

F3  COEFFICIENTS                    F4  Q & W OF EACH 2nd ORDER FACT



F9  HELP                            F10  RETURN TO MAIN MENU


PRESS F1 - F10 TO SELECT DESIRED PROGRAM
```

Figure 12.5 Main Menu Program Section.

When the program is started, the main program selection menu is displayed as shown in Fig. 12.5. Since none of the low-pass filter specifications were given, the transformation program must first be selected by pressing the "F5" key. A new menu (Fig. 12.6) now appears, which lets the designer select the type of transformation to be performed. The selection of "F1" starts the analysis, which transforms

```
SELECT TRANSFORMATION TYPE

     F1    - SPECIFIED FILTER TO NORMALIZED LOW PASS

     F2    - EXISTING LOWPASS TRANSFORMED TO SPECIFIED FILTER TYPE



     F10   RETURN TO MAIN MENU
```

Figure 12.6 Transformation Selection Menu.

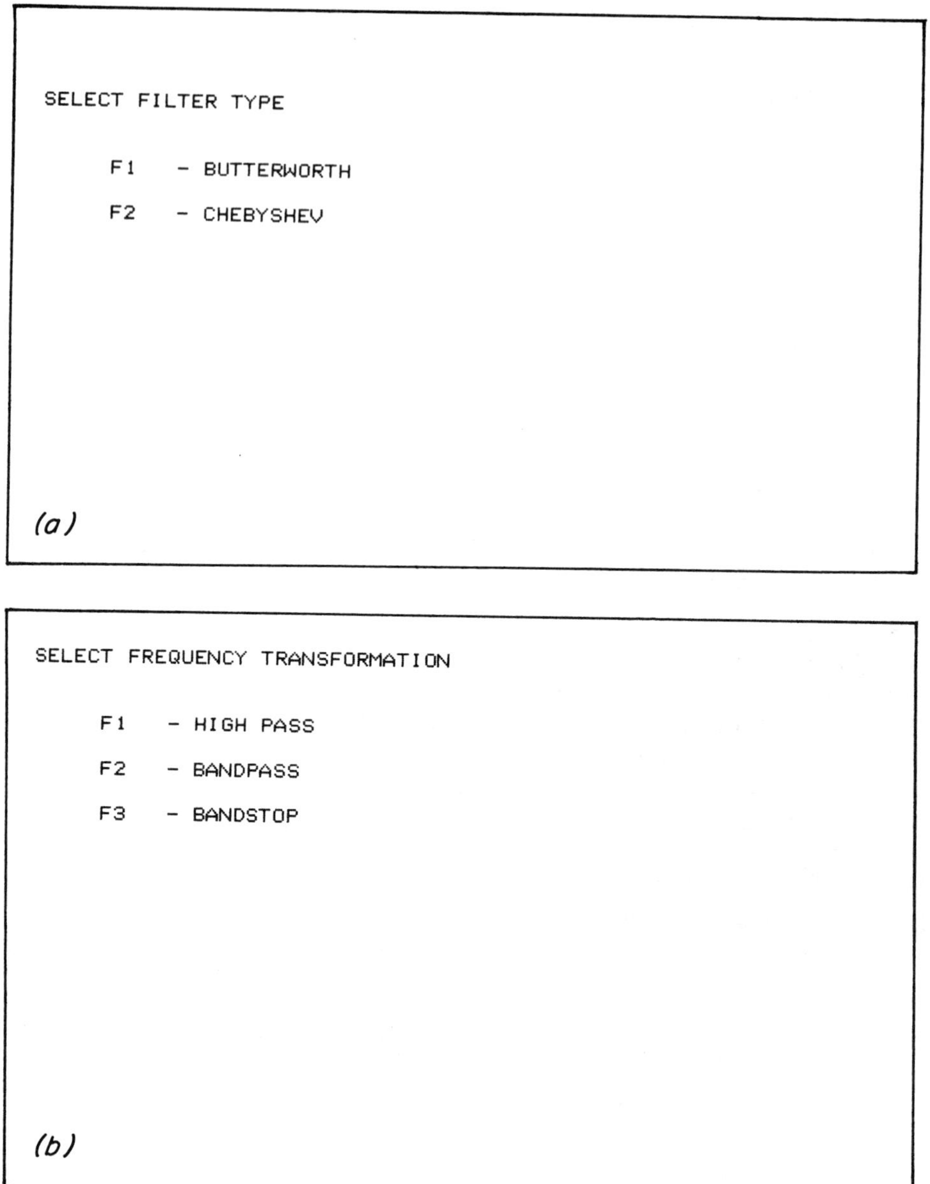

Figure 12.7 *(a) Selection of filter type; (b) selection of frequency transformation.*

a specified filter to a normalized low-pass form. The program then prompts data that are required to perform the analysis and transformation. This includes a filter type and transformation selection shown by Figs. 12.7*a* and (*b*), respectively. The filter specification prompts are shown in Fig. 12.8.

```
WILL FREQ BE ENTERED IN 'HZ' OR 'RAD'  (ENTER H OR R ) H



CENTER FREQ - W0 = 1.5E+3
Q = 5
STOPBAND FREQ - W3 = 2E+3
STOPBAND FREQ - W4 = 900
GAMA MIN = 15
GAMA MAX = 3
```

Figure 12.8 Specification prompts and data entry.

As soon as the last of the data has been entered the program returns directly to the Butterworth filter analysis menu shown in Fig. 12.9. The minimum order will automatically be calculated with no data entry, since the transform program has passed all of the necessary data to the Butterworth analysis program. Selection of "F2"—"ROOT LOCATIONS" yields a display of the root locations (Fig. 12.10)

```
                    FILTER ANALYSIS PROGRAM


F1  BUTTERWORTH ANALYSIS          F2  CHEBYSHEV ANALYSIS

F3  BESSEL ANALYSIS               F4  ROOT SOLVING PROGRAM

F5  NETWORK TRANSFORMATIONS       F6  GAIN AND PHASE CALC'S

F7  FREQUENCY PLOTS               F8  CALCULATOR MODE

F9  HELP                          F10  END PROGRAM (ie. QUIT)


PRESS F1 - F10 TO SELECT DESIRED PROGRAM
```

Figure 12.9 Butterworth analysis menu.

```
ROOT S( 1 ) =   -0.7071 + J  0.7071
ROOT S( 2 ) =   -0.7071 - J  0.7071






PRESS ANY KEY TO DISPLAY BI-QUADS
```

Figure 12.10 Butterworth root locations.

and low-pass biquads (Fig. 12.11). After display of the biquads any key will return the program to the Butterworth selection menu.

For our example, the order is determined to be 2, with root locations of $-0.707 +/-j.707$. This produces the denominator biquad of $s^2 + (1.414)s + 1$ and a numerator of 1. Since the transformation program requires only the coefficents of the biquads, no further Butterworth analyses are required. Selection of

```
BI.QUAD S^2 ( 1 ) =    1.0000
BI.QUAD S^1 ( 1 ) =    1.4142
BI.QUAD S^0 ( 1 ) =    1.0000






PRESS ANY KEY TO RETURN TO MENU
```

Figure 12.11 Biquad coefficients.

```
W0 = 1.5E3
H (MID BAND GAIN) = 10
Q = 5
```

Figure 12.12 Filter specification.

the "F10" key redisplays the main menu and the transform program "F5" is selected once more.

This time, however, we wish to transform an already existing low-pass filter to the required bandpass. Selection of "F2" prompts for the data as shown in Fig. 12.12. When all the data have been entered, the program will calculate the new set of transformed biquads and display them on the screen as shown in Fig. 12.13.

```
BI.QUAD S^2 ( 1 ) =  1
BI.QUAD S^1 ( 1 ) =  1238.619
BI.QUAD S^0 ( 1 ) =  7.709428E+07

BI.QUAD S^2 ( 2 ) =  1
BI.QUAD S^1 ( 2 ) =  1427.111
BI.QUAD S^0 ( 2 ) =  1.023439E+08

        PRESS ANY KEY TO RETURN TO MAIN MENU
```

Figure 12.13 Biquad coefficients.

For our case, two biquads are displayed that, from a design standpoint, represent a two-stage filter. Notice that the specified gain of 10 has been lumped into the first-stage specification. This is done to reduce the noise output of the filter.

The coefficients of these biquads are necessary for component calculations. Therefore, these values should be written down for later use. After display of the biquads, any keypress will return the program back to the main menu. The gain and phase program and plot program may be run by following the procedure outlined in Section 12.4. This produces the plots shown in Figs. 12.14*a* and 12.14*b*. Returning to the main menu again allows selection of the calculator mode.

The calculator mode provides a tool for calculation of component values using formulas presented in other chapters. The calculator is set up to operate in reverse polish notation. That is, numbers are entered first and then operated upon. For example, if 3 is to be multiplied by 2, then 3 and 2 are both entered before the "*" is pressed. Function labels on the displayed calculator keys may be changed

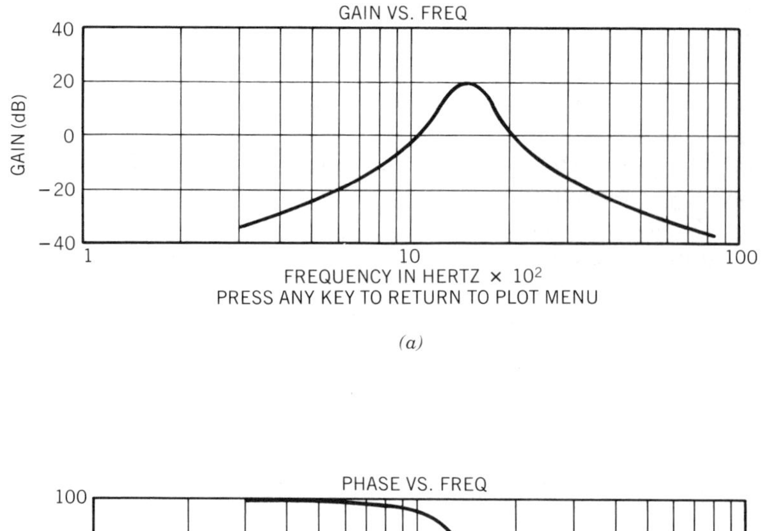

(a)

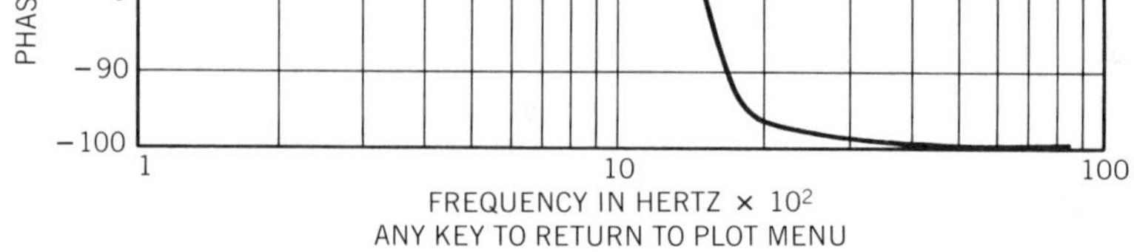

(b)

Figure 12.14 *(a) Magnitude response plot; (b) phase response plot.*

between three menus by repeated pressing of the "F9" menu key. The "F10" key, labeled "EXIT", returns the program to the main menu.

By using the design equations of Chapter 7, the values of R1, R2, and R3 may be calculated for each stage after choosing values for C1 and C2.

12.6 REFERENCES

[1] P. Bowron and F. W. Stephenson, *Active Filters for Communications and Instrumentation*, London: McGraw-Hill, 1979.

CHAPTER 13

Design Aids

David Baez–Lopez

Instituto Naçional de Astrofisica
Puebla, Mexico

13.1 INTRODUCTION

This chapter presents information of value in the design process. Because it is frequently necessary, at the outset of a design, to analyze alternative structures, a basic introduction to signal-flow graphs is provided. This material is meant to supplement the discussion in Chapter 5 and to complement the remaining sections of this chapter, most of which are associated with ladder realizations.

Section 13.3 contains tabulations of approximating functions. Of particular value are the listings of Cauer (or elliptic) functions. Coupled with the various nomographs of Section 13.4, this material allows the designer to specify a wide variety of network functions in terms of their poles and zeros or as rational functions of s.

The LC realization tables of Sections 13.5–13.6 are invaluable aids to the design of high-order multiloop structures as well as direct replacement methods, all of which are presented in Chapter 10. Although not as exhaustive a collection as those appearing elsewhere [4, 5], these tabulations allow one to satisfy a wide range of design specifications.

The chapter concludes with a description of components suitable for use in RC-active filters.

13.2 SIGNAL-FLOW GRAPH FUNDAMENTALS

Signal-flow graphs (SFG), introduced by S. J. Mason [1] in 1953, have been used to solve graphically sets of linear equations characterizing electrical networks.

Their usefulness stems from the fact that once a network has been reduced to a flow graph, there are well-defined rules to obtain the network transfer functions.

A signal-flow graph is a graphical representation of a set of linear equations in terms of *nodes* and *branches*. Before discussing the rules for solving the signal-flow graph, we define various terms related to SFG.

Node. Represents a specified quantity or variable.

Branch. Represents the functional dependence between two variables (nodes). With each branch there is associated a direction (the branch direction) represented by an arrow, and the signal travels along the branch only in the direction indicated by its associated arrow.

Source Node. A node with no branches entering, that is, representing an independent variable.

Sink node. A node with no branches leaving. It represents a variable upon which no other variable in the system is dependent, for example, an output variable.

Intermediate Node. A node with branches entering and leaving. This will be referred to simply as a *node*.

Path. A continuous unidirectional succession of branches having the same direction.

Open path. Any path along which each node is encountered only once.

Forward path. An open path between a source node and a sink node.

Loop. A path in which the starting and ending nodes are the same. A loop is also called a closed path.

Self-loop. A loop consisting of a single branch and a single node.

Branch transmission. The quantity relating the nodes of a branch. This quantity is also called the gain of the branch. Thus if x is the variable at the starting node and t is the gain of the branch, the value of the variable delivered to the ending node is $y = tx$.

Rules for Reducing Signal-Flow Graphs

The signal-flow graph of a network presents the information contained in the descriptive equations from which it is derived and it permits the visualization of the signal transmission paths in the network. Moreover, it indicates the interaction between variables within the network and points out existing feedback loops. Most important, a set of simple rules permits the graph reduction to simpler and more tractable forms by successive elimination of dependent variables, that is, by elimination of graph nodes. In other words, the reduction of signal-flow graphs is equivalent to solving a corresponding set of network equations; but, because of the existence of certain systematic rules, the reduction of the graph is often more convenient. The set of rules for reducing (solving) a signal-flow graph follows.

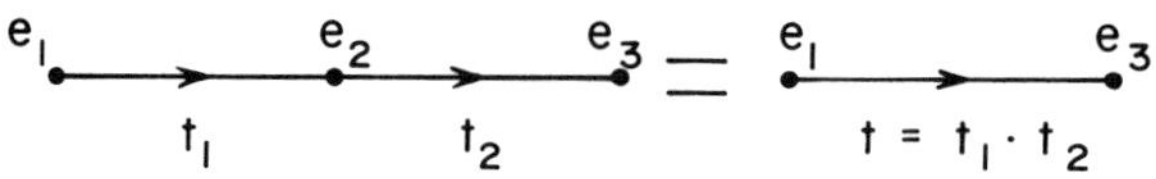

Figure 13.1 *Cascade transformation of a SFG.*

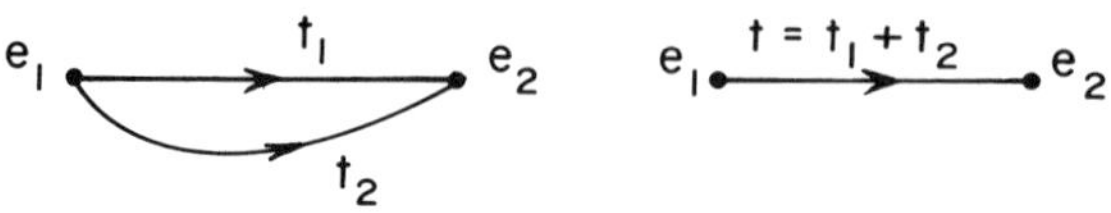

Figure 13.2 *Parallel transformation of a SFG.*

Rule 1. Cascade Transformation

The total gain of a path, that is, of a series of branches in cascade, is equal to the product of the individual branch gains, as shown in Fig. 13.1.

Rule 2. Parallel Transformation

The total gain of parallel branches equals the sum of the individual branch transmissions. The example of Fig. 13.2 shows this transformation.

Rule 3. Node Elimination

A node of origin or termination of a branch may be removed, as long as the gain between the nodes of interest remains unchanged, as shown in Fig. 13.3.

Rule 4. Removal of a Self-Loop

A self-loop whose loop gain is L is removed from a node by dividing the gain of all other branches entering that node by $1 - L$. An example is given in Fig. 13.4.

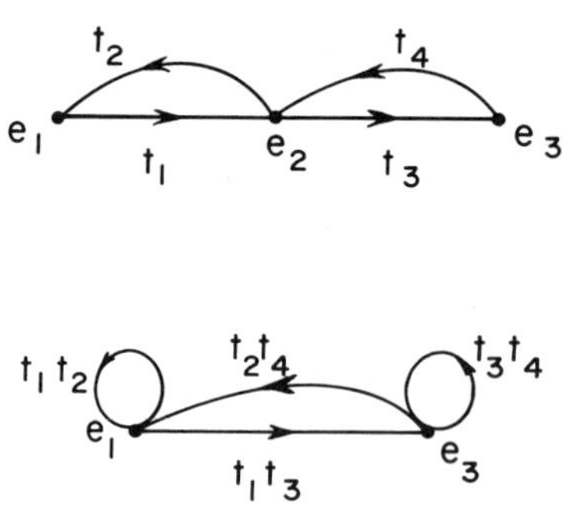

Figure 13.3 *Node elimination in a SFG.*

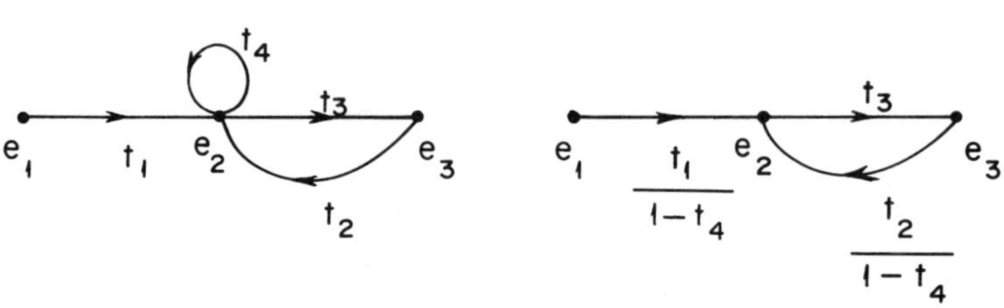

Figure 13.4 *Removal of a self-loop in a SFG.*

Rule 5. General Multiloop Reduction Rule

Consider the case in which there is more than one forward path from a source node to a sink node. If we assume that the source node variable is x and the output node variable is y, the general formula for determining the transfer function T between nodes x and y, using SFG is given by

$$T = \frac{y}{x} = \frac{\sum_{i=1}^{n} \mathrm{FP}_i(\delta_i)}{\delta} \tag{13.1}$$

where

1. FP_i is the ith forward path from the source node to the sink node.
2. δ_i is given as follows: $\delta_i = 1 -$ (sum of loops not touching FP_i taken one at a time) + (sum of loops not touching FP_i taken two at a time) − (sum of loops not touching FP_i taken three at a time) + $\cdots$.
3. $\delta = 1 -$ (sum of loops taken one at a time) + (sum of products of non touching loops taken two a time) − (sum of products of non touching loops taken three at a time) + $\cdots$. Equation 13.1 is known as the Mason–Zimmerman formula. A few examples will demonstrate the use of Eq. 13.1.

EXAMPLE 13.1 Determine the voltage transfer function of the state-variable filter shown in Fig. 13.5*a*.

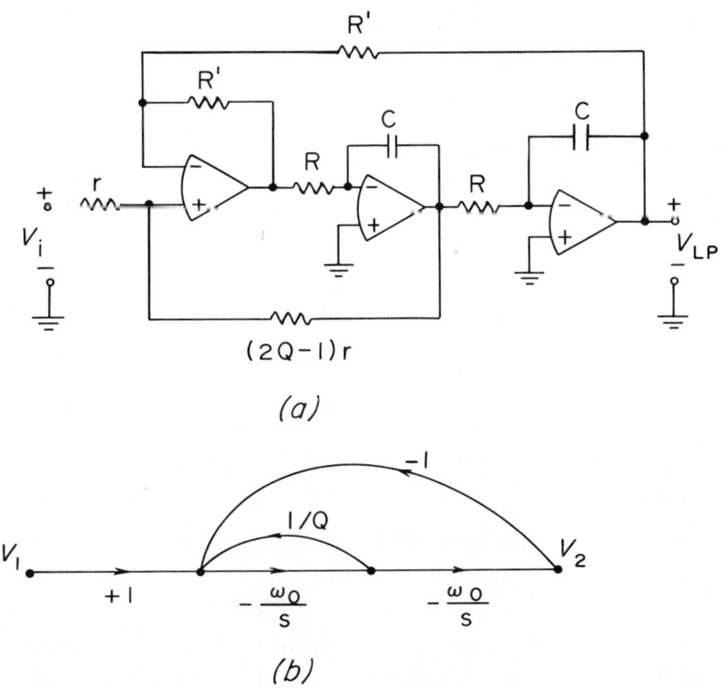

Figure 13.5 (a) Active-RC state-variable filter; (b) corresponding SFG.

PROCEDURE

The SFG of Fig. 13.5*b* corresponds to the active-RC state-variable filter [2] of Fig. 13.5*a*. We have only one FP and two loops denoted by L_1 and L_2. That is,

$$\mathrm{FP} = \frac{\omega_0^2}{s^2}$$

$$L_1 = \frac{-\omega_0}{sQ}, \qquad L_2 = -\frac{\omega_0^2}{s^2}$$

If the output node is V_{LP}, the transfer function from input to output is given by

$$T = \frac{V_{LP}}{V_{\text{in}}} = \frac{\mathrm{FP}}{1-(L_1+L_2)} = \frac{\omega_0^2/s^2}{1+\omega_0/sQ+\omega_0^2/s^2} = \frac{\omega_0^2}{s^2+s\omega_0/Q+\omega_0^2}$$

EXAMPLE 13.2 Determine E_o/E_{in} for the flow graph of Fig. 13.6.

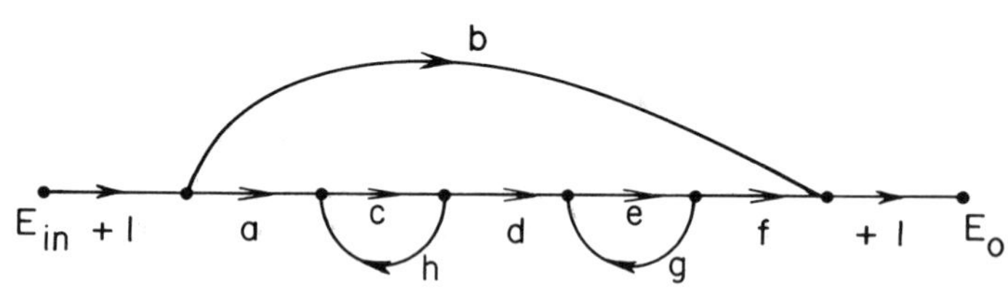

Figure 13.6 SFG for Example 13-2.

PROCEDURE

The SFG of Fig. 13.6 can be solved as follows:

$$\mathrm{FP}_1 = 1\cdot b\cdot 1 = b$$

$$\mathrm{FP}_2 = 1\cdot a\cdot c\cdot d\cdot e\cdot f\cdot 1 = acdef$$

$$L_1 = ch$$

$$L_2 = eg$$

Both loops, L_1 and L_2, are nontouching loops. Thus their product will be used in δ_1 and δ. Using Eq. 13.1, we have that δ, δ_1 and δ_2 are given by

$$\delta = 1-(L_1+L_2)+(L_1L_2)$$

$$\delta_1 = 1-(ch+eg)+(cheg)$$

$$\delta_2 = 1-0 = 1$$

Thus $T = E_0/E_{\text{in}}$ is given by

$$T = \frac{E_0}{E_{\text{in}}} = \frac{b[1-(ch+eg)+cheg]+acdef\,[1-0]}{1-(ch+eg)+(cheg)}$$

So far, in our study of SFG we have defined the transfer function between a source node and a sink node. If we wish to find the transfer function between

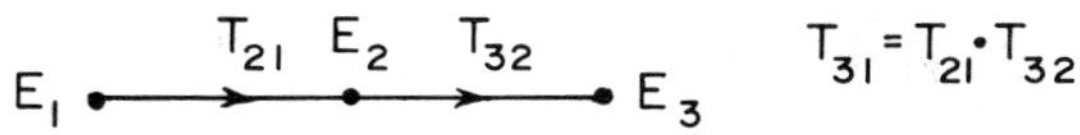

Figure 13.7 SFG showing how to obtain a transfer function between a nonsource and a sink node.

a node that is not a source node and another node, we proceed as follows. Consider the SFG of Fig. 13.7. To find $T_{32} = E_3/E_2$, note that E_2 does not correspond to a source node variable. Thus we evaluate

$$T_{32} = \frac{E_3}{E_2} = \frac{E_3/E_1}{E_2/E_1} = \frac{T_{31}}{T_{21}} \tag{13.2}$$

and note that in T_{32}, the denominators of T_{31} and T_{21} always cancel.

13.3 TABLES OF APPROXIMATING FUNCTIONS

In this section we present tables of some of the more important low-pass approximation functions. Tables of Butterworth, Chebyshev, elliptic, and Thomson low-pass functions are included. Chebyshev and elliptic functions have been selected for different values of passband ripple and will enable the designer to choose from a wide variety of functions and passband ripples. Tables 13.1 to 13.11 include the order of the function, poles, quadratic factor coefficients a_0 and a_1, which are of the form $a_0 + a_1 s + s^2$, and -3dB passband cutoff frequency ω_c* (in the case of Butterworth functions, ω_c is unity for all orders and is not given). All functions are normalized to a passband of 0 to 1 rad/s. An example follows.

EXAMPLE 13.3 Obtain a low-pass network function with the following characteristics:

a. Passband ripple of 0.5 dB extending from 0 to 1 rad/s.

b. The function should be of fifth order.

PROCEDURE

1 From Table 13.6 we find that for $n = 5$ the poles and quadratic factors are

$$P_{1,2} = -0.11196294 \pm j1.01155746 \qquad s^2 + 0.22392587s + 1.03578424$$
$$P_{3,4} = -0.29312274 \pm j0.62517685 \qquad s^2 + 0.58624548s + 0.47676703$$
$$P_5 = -0.36231962 \qquad s + 0.36231962$$

2 Thus the transfer function will be

$$N(s) = \frac{K}{(s^2 + 0.22392587s + 1.03578424)(s^2 + 0.58624548s + 0.47676703)(s + 0.36231962)}$$

where K is a constant.

* Note that ω_p in Fig. 2.2 denotes the edge of the passband and is equal to ω_c if 3dB of passband ripple is specified.

3 By multiplying out the factors in the denominator, $N(s)$ becomes

$$N(s)=\frac{K}{s^5+1.172490970s^4+2.161293654s^3+1.015422787s^2+0.752518296s+0.178923492}$$

4 This $N(s)$ corresponds to a transfer function of fifth order with a 0 to 1 rad/s passband and with a passband ripple of 0.5 dB. For this function the -3 dB cutoff frequency, ω_c, is equal to 1.05925918 rad/s. A plot of the magnitude of this function is shown in Fig. 13.8 for $K = 0.178923492$.

Table 13.1 **Pole locations and a_1 coefficient of the quadratic factors $1 + a_1s + s^2$ of Butterworth low-pass functions**

n	poles	a_1
2	−0.70710683 + J 0.70710677	1.41421366
3	−0.50000000 + J 0.86602539	1.00000000
	−1.0	
4	−0.38268346 + J 0.92387950	0.76536691
	−0.92387956 + J 0.38268343	1.84775913
5	−0.30901703 + J 0.95105654	0.61803406
	−0.80901700 + J 0.58778524	1.61803401
	−1.0	
6	−0.25881904 + J 0.96592581	0.51763809
	−0.70710683 + J 0.70710677	1.41421366
	−0.96592587 + J 0.25881895	1.93185174
7	−0.22252093 + J 0.97492796	0.44504187
	−0.62348980 + J 0.78183144	1.24697959
	−0.90096885 + J 0.43388367	1.80193770
	−1.0	
8	−0.19509032 + J 0.98078525	0.39018065
	−0.55557024 + J 0.83146960	1.11114049
	−0.83146966 + J 0.55557019	1.66293931
	−0.98078525 + J 0.19509023	1.96157050
9	−0.17364819 + J 0.98480767	0.34729639
	−0.50000000 + J 0.86602539	1.00000000
	−0.76604450 + J 0.64278752	1.53208899
	−0.93969268 + J 0.34202006	1.87938535
	−1.0	
10	−0.15643446 + J 0.98768836	0.31286892
	−0.45399049 + J 0.89100653	0.90798098
	−0.70710683 + J 0.70710677	1.41421366
	−0.89100659 + J 0.45399040	1.78201318
	−0.98768836 + J 0.15643448	1.97537673

Table 13.2 **Pole locations, coefficients a_0 and a_1 of the quadratic factors of $a_0 + a_1 s + s^2$ of 0.01 dB ripple Chebyshev low-pass functions, and half power frequencies ω_c**

n	poles			a_1	a_0	ω_c
2	-2.22778678 +	J	2.33731318	4.45557356	10.42606735	3.30364513
3	-0.79469121 +	J	1.62622344	1.58938241	3.27613688	1.87719059
	-1.58938253					
4	-0.41086909 +	J	1.35553312	0.82173818	2.00628352	1.46690917
	-0.99192572 +	J	0.56148022	1.98385143	1.29917669	
5	-0.25251383 +	J	1.22820294	0.50502765	1.57224572	1.29122055
	-0.66108972 +	J	0.75907117	1.32217944	1.01322865	
	-0.81715184					
6	-0.17146651 +	J	1.15866840	0.34293303	1.37191331	1.19941461
	-0.46845526 +	J	0.84820414	0.93691051	0.93890059	
	-0.63992178 +	J	0.31046414	1.27984357	0.50588787	
7	-0.12426424 +	J	1.11664510	0.24852848	1.26233792	1.14526975
	-0.34818065 +	J	0.89547974	0.69636130	0.92311370	
	-0.50313562 +	J	0.49695373	1.00627124	0.50010848	
	-0.55843842					
8	-0.09429328 +	J	1.08933818	0.18858656	1.19554889	1.11061001
	-0.26852456 +	J	0.92349637	0.53704911	0.92495102	
	-0.40187541 +	J	0.61706048	0.80375081	0.54226750	
	-0.47404432 +	J	0.21668275	0.94808865	0.27166945	
9	-0.07404611 +	J	1.07060397	0.14809223	1.15167570	1.08706522
	-0.21320726 +	J	0.94147336	0.42641452	0.93182945	
	-0.32665253 +	J	0.69878703	0.65330505	0.59500515	
	-0.40069863 +	J	0.37181678	0.80139726	0.29880714	
	-0.42641455					
10	-0.05971299 +	J	1.05719769	0.11942599	1.12123263	1.07033193
	-0.17329386 +	J	0.95371181	0.34658772	0.93959701	
	-0.26991153 +	J	0.75686997	0.53982306	0.64570439	
	-0.34010036 +	J	0.48594034	0.68021673	0.35181171	
	-0.37701303 +	J	0.16744368	0.75402606	0.17017621	

Table 13.3 **Pole locations, coefficients a_0 and a_1 of the quadratic factors of $a_0 + a_1 s + s^2$ of 0.05 dB ripple Chebyshev low-pass functions, and half power frequencies ω_c**

n	poles	a_1	a_0	ω_c
2	-1.44453073 + J 1.60831225	2.88906145	4.67333746	2.26859260
3	-0.56856781 + J 1.31141460	1.13713562	2.04307771	1.51209927
	-1.13713574			
4	-0.30548799 + J 1.18215024	0.61097598	1.49080205	1.27839601
	-0.73751318 + J 0.48966265	1.47502637	0.78369522	
5	-0.19128513 + J 1.11852312	0.38257027	1.28768396	1.17536879
	-0.50079095 + J 0.69128525	1.00158191	0.72866690	
	-0.61901164			
6	-0.13124590 + J 1.08302021	0.26249179	1.19015825	1.12073624
	-0.35857049 + J 0.79282576	0.71714097	0.75714552	
	-0.48981640 + J 0.29019427	0.97963279	0.32413283	
7	-0.09572170 + J 1.06130481	0.19144340	1.13553059	1.08824265
	-0.26820624 + J 0.85110027	0.53641248	0.79630625	
	-0.38756922 + J 0.47232494	0.77513844	0.37330076	
	-0.43016940			
8	-0.07293745 + J 1.04708874	0.14587490	1.10171461	1.06733227
	-0.20770828 + J 0.88767904	0.41541657	0.83111680	
	-0.31085745 + J 0.59312814	0.62171489	0.44843337	
	-0.36668131 + J 0.20827882	0.73336262	0.17783524	
9	-0.05744039 + J 1.03728783	0.11488078	1.07926536	1.05307722
	-0.16539298 + J 0.91217566	0.33078596	0.85941929	
	-0.25339678 + J 0.67704153	0.50679356	0.52259517	
	-0.31083718 + J 0.36024624	0.62167436	0.22639711	
	-0.33078599			
10	-0.04641714 + J 1.03025055	0.09283427	1.06357074	1.04292107
	-0.13470778 + J 0.92940241	0.26941556	0.88193500	
	-0.20981230 + J 0.73757792	0.41962460	0.58804238	
	-0.26437891 + J 0.47355407	0.52875781	0.29414967	
	-0.29306626 + J 0.16317566	0.58613253	0.11251413	

Table 13.4 **Pole locations, coefficients a_0 and a_1 of the quadratic factors of $a_0 + a_1 s + s^2$ of 0.1 dB ripple Chebyshev low-pass functions, and half power frequencies ω_c**

n	poles			a_1	a_0	ω_c
2	-1.18617785 +	J	1.38094819	2.37235570	3.31403589	1.94321907
3	-0.48470268 +	J	1.20615506	0.96940535	1.68974674	1.38899481
	-0.96940541					
4	-0.26415631 +	J	1.12260973	0.52831262	1.33003116	1.21309912
	-0.63772976 +	J	0.46500021	1.27545953	0.62292445	
5	-0.16653369 +	J	1.08037198	0.33306739	1.19493711	1.13471794
	-0.43599081 +	J	0.66770661	0.87198162	0.63592011	
	-0.53891426					
6	-0.11469334 +	J	1.05651891	0.22938669	1.12938666	1.09293067
	-0.31334805 +	J	0.77342552	0.62669611	0.69637406	
	-0.42804140 +	J	0.28309330	0.85608280	0.26336125	
7	-0.08384094 +	J	1.04183340	0.16768189	1.09244609	1.06800056
	-0.23491712 +	J	0.83548546	0.46983424	0.75322199	
	-0.33946505 +	J	0.46365938	0.67893010	0.33021656	
	-0.37677780					
8	-0.06398011 +	J	1.03218126	0.12796022	1.06949162	1.05192661
	-0.18219994 +	J	0.87504107	0.36439988	0.79889369	
	-0.27268150 +	J	0.58468372	0.54536301	0.41621026	
	-0.32164973 +	J	0.20531355	0.64329946	0.14561220	
9	-0.05043805 +	J	1.02550948	0.10087611	1.05421376	1.04095471
	-0.14523056 +	J	0.90181798	0.29046112	0.83436757	
	-0.22250617 +	J	0.66935378	0.44501233	0.49754348	
	-0.27294421 +	J	0.35615566	0.54588842	0.20134540	
	-0.29046115					
10	-0.04078866 +	J	1.02071047	0.08157732	1.04351354	1.03313065
	-0.11837330 +	J	0.92079616	0.23674659	0.86187780	
	-0.18437074 +	J	0.73074794	0.36874148	0.56798512	
	0.23232070 +	J	0.46916896	0.46464139	0.27409244	
	-0.25752947 +	J	0.16166465	0.51505893	0.09245688	

Table 13.5 **Pole locations, coefficients a_0 and a_1 of the quadratic factors of $a_0 + a_1 s + s^2$ of 0.25 dB ripple Chebyshev low-pass functions, and half power frequencies ω_c**

n	poles		a_1	a_0	ω_c
2	-0.89834177 + J	1.14324880	1.79668355	2.11403584	1.59814143
3	-0.38361135 + J	1.09154618	0.76722270	1.33863068	1.25288808
	-0.76722276				
4	-0.21251822 + J	1.05678213	0.42503643	1.16195250	1.13976789
	-0.51306432 + J	0.43773347	1.02612865	0.45484561	
5	-0.13502531 + J	1.03788376	0.27005062	1.09543455	1.08872390
	-0.35350078 + J	0.64144742	0.70700157	0.53641760	
	-0.43695101				
6	-0.09339190 + J	1.02688622	0.18678381	1.06321740	1.06134069
	-0.25515142 + J	0.75173289	0.51030284	0.63020462	
	-0.34854335 + J	0.27515322	0.69708669	0.19719176	
7	-0.06844716 + J	1.02000809	0.13689432	1.04510152	1.04494607
	-0.19178468 + J	0.81798291	0.38356936	0.70587742	
	-0.27713689 + J	0.45394623	0.55427378	0.28287205	
	-0.30759874				
8	-0.05232144 + J	1.01544499	0.10464288	1.03386605	1.03435194
	-0.14899886 + J	0.86085272	0.29799771	0.76326805	
	-0.22299257 + J	0.57520336	0.44598514	0.38058460	
	-0.26303765 + J	0.20198448	0.52607530	0.10998654	
9	-0.04129513 + J	1.01227176	0.08259025	1.02639937	1.02710986
	-0.11890455 + J	0.89017695	0.23780911	0.80655330	
	-0.18217237 + J	0.66071343	0.36434475	0.46972901	
	-0.22346748 + J	0.35155827	0.44693497	0.17353094	
	-0.23780912				
10	-0.03342277 + J	1.00997984	0.06684554	1.02117634	1.02194023
	-0.09699666 + J	0.91111594	0.19399332	0.83954060	
	-0.15107585 + J	0.72306567	0.30215171	0.54564792	
	-0.19036669 + J	0.46423665	0.38073337	0.25175515	
	-0.21102308 + J	0.15996510	0.42204615	0.07011957	

Table 13.6 **Pole locations, coefficients a_0 and a_1 of the quadratic factors of $a_0 + a_1 s + s^2$ of 0.5 dB ripple Chebyshev low-pass functions, and half power frequencies ω_c**

n	poles			a₁	a₀	ω_c
2	−0.71281230 +	J	1.00404239	1.42562461	1.51620257	1.38974369
3	−0.31322822 +	J	1.02192748	0.62645644	1.14244759	1.16748524
	−0.62645650					
4	−0.17535307 +	J	1.01625288	0.35070613	1.06351864	1.09310198
	−0.42333975 +	J	0.42094570	0.84667951	0.35641184	
5	−0.11196294 +	J	1.01155746	0.22392587	1.03578424	1.05925918
	−0.29312274 +	J	0.62517685	0.58624548	0.47676703	
	−0.36231962					
6	−0.07765008 +	J	1.00846076	0.15530016	1.02302253	1.04102957
	−0.21214396 +	J	0.73824453	0.42428792	0.59001005	
	−0.28979403 +	J	0.27021617	0.57958806	0.15699735	
7	−0.05700320 +	J	1.00640857	0.11400639	1.01610768	1.03008997
	−0.15971941 +	J	0.80707693	0.31943882	0.67688346	
	−0.23080122 +	J	0.44789386	0.46160245	0.25387812	
	−0.25617003					
8	−0.04362008 +	J	1.00500202	0.08724016	1.01193178	1.02301073
	−0.12421948 +	J	0.85199958	0.24843895	0.74133372	
	−0.18590760 +	J	0.56928790	0.37181520	0.35865036	
	−0.21929294 +	J	0.19990726	0.43858588	0.08805230	
9	−0.03445272 +	J	1.00400388	0.06890544	1.00921071	1.01816678
	−0.09920264 +	J	0.88290632	0.19840528	0.78936470	
	−0.15198728 +	J	0.65531701	0.30397457	0.45254052	
	−0.18644001 +	J	0.34868684	0.37288001	0.15634239	
	−0.19840530					
10	−0.02789941 +	J	1.00327325	0.05579883	1.00733566	1.01470673
	−0.08096725 +	J	0.90506583	0.16193451	0.82569987	
	−0.12610947 +	J	0.71826428	0.25221893	0.53180718	
	−0.15070720 +	J	0.46115398	0.31781441	0.23791450	
	−0.17614998 +	J	0.15890288	0.35229996	0.05627894	

Table 13.7 **Pole locations, coefficients a_0 and a_1 of the quadratic factors of $a_0 + a_1 s + s^2$ of 1.0 dB ripple Chebyshev low-pass functions, and half power frequencies ω_c**

n	poles	a_1	a_0	ω_c
2	-0.54886711 + J 0.89512849	1.09773421	1.10251009	1.21762609
3	-0.24708526 + J 0.96599865	0.49417052	0.99420452	1.09486806
	-0.49417055			
4	-0.13953598 + J 0.98337907	0.27907196	0.98650467	1.05300188
	-0.33686966 + J 0.40732896	0.67373931	0.27939805	
5	-0.08945837 + J 0.99010712	0.17891674	0.98831493	1.03381455
	-0.23420502 + J 0.61191982	0.46841004	0.42929786	
	-0.28949332			
6	-0.06218101 + J 0.99341124	0.12436202	0.99073237	1.02344227
	-0.16988169 + J 0.72722751	0.33976337	0.55771965	
	-0.23206270 + J 0.26618364	0.46412539	0.12470683	
7	-0.04570897 + J 0.99528396	0.09141795	0.99267948	1.01720512
	-0.12807369 + J 0.79815573	0.25614738	0.65345544	
	-0.18507186 + J 0.44294295	0.37014371	0.23045005	
	-0.20541427			
8	-0.03500823 + J 0.99645132	0.07001647	0.99414080	1.01316381
	-0.09969501 + J 0.84475064	0.19939002	0.72354275	
	-0.14920413 + J 0.56444430	0.29840827	0.34085926	
	-0.17599827 + J 0.19820641	0.35199654	0.07026117	
9	-0.02766745 + J 0.99722958	0.05533489	0.99523234	1.01039636
	-0.07966522 + J 0.87694901	0.15933044	0.77538610	
	-0.12205422 + J 0.65089536	0.24410844	0.43856201	
	-0.14972167 + J 0.34633416	0.29944333	0.14236394	
	-0.15933046			
10	-0.02241445 + J 0.99777550	0.04482890	0.99605834	1.00841820
	-0.06504928 + J 0.90010631	0.13009855	0.81442279	
	-0.10131662 + J 0.71432841	0.20263325	0.52053010	
	-0.12766640 + J 0.45862696	0.25533280	0.22663739	
	-0.14151928 + J 0.15803212	0.28303856	0.04500186	

Table 13.8 **Pole locations, coefficients a_0 and a_1 of the quadratic factors of the form $a_0 + a_1 s + s^2$ and half power frequencies ω_c of Thomson (maximally flat delay) low-pass functions**

n	poles	a_1	a_0	ω_c
2	-1.5000000000 + J 0.8660253882	3.0000000000	3.0000000000	1.36165
3	-1.8389073610 + J 1.7543808222	3.6778147221	6.4594326019	1.75567
	-2.3221852779			
4	-2.1037893295 + J 2.6574182510	4.2075786591	11.4878015518	2.11392
	-2.8962106705 + J 0.8672341108	5.7924213409	9.1401309967	
5	-2.3246741295 + J 3.5710229874	4.6493482590	18.1563148499	2.42741
	-3.3519563675 + J 1.7426613569	6.7039127350	14.2724809647	
	-3.6467385292			
6	-3.7357084751 + J 2.6262722015	7.4714169502	20.8528232574	2.70339
	-2.5159323215 + J 4.4926729202	5.0318646431	26.5140266418	
	-4.2483592033 + J 0.8675096631	8.4967184067	18.8011283875	
7	-2.6856770515 + J 5.4206938744	5.3713541031	36.5967826843	2.95172
	-4.7582902908 + J 1.7392860651	9.5165805817	25.6664428711	
	-4.0701389313 + J 3.5171740055	8.1402778625	28.9365444183	
	-4.9717869759			
8	-2.8389837742 + J 6.3539113998	5.6779675484	48.4320182800	3.17961
	-5.2048406601 + J 2.6161751747	10.4096813202	33.9347381592	
	-5.5878858566 + J 0.8676144481	11.1757717133	31.9772224426	
	-4.3682889938 + J 4.4144425392	8.7365779877	38.5692520142	
9	-2.9792609215 + J 7.2914628983	5.9585218430	62.0414276123	3.39169
	-6.1293168068 + J 1.7378656864	12.2586336136	40.5886993408	
	-5.6044416428 + J 3.4981408119	11.2088832855	43.6467552185	
	-4.6384367943 + J 5.3172774315	9.2768735886	49.7885360718	
	-6.2970876694			
10	-3.1089136600 + J 8.2326974869	6.2178273201	77.4426498413	3.59097
	-5.9674363136 + J 4.3847947121	11.9348726273	54.8367195129	
	-6.9219722748 + J 0.8669555187	13.8439445496	48.6653137207	
	-6.6154313087 + J 2.6120018959	13.2308626175	50.5864868164	
	-4.8862466812 + J 6.2250113487	9.7724933624	62.6261749268	

Table 13.9 **Pole and zero locations of normalized elliptic low-pass functions of 0.1 dB passband ripple and 1 rad/sec 0.1 dB passband. (*a*) Odd case and even-order class A**

n	Ω_s	Amin	poles			zeros
3	1.50	14.85	-0.289646178	+J	1.212427735	1.675116181
			-1.298181772			
	2.00	24.01	-0.381858438	+J	1.217904687	2.270067930
			-1.116764903			
	2.50	30.52	-0.420815110	+J	1.215538621	2.856308699
			-1.055745363			
	3.00	35.62	-0.441039622	+J	1.213293552	3.439158440
			-1.026754260			
	3.50	39.85	-0.452928662	+J	1.211657643	4.020282745
			-1.010467172			
4	1.50	29.06	-0.173602447	+J	1.108107805	3.477878332
			-0.698749244	+J	0.617000997	1.592148542
	2.00	41.45	-0.216253519	+J	1.116819739	4.922069550
			-0.670443475	+J	0.535640061	2.143172264
	2.50	50.14	-0.234273940	+J	1.119468212	6.297524452
			-0.658091068	+J	0.506945729	2.689261198
	3.00	56.95	-0.243679449	+J	1.120622039	7.646634102
			-0.651650012	+J	0.493044019	3.233480930
	3.50	62.59	-0.249229774	+J	1.121231198	8.982359886
			-0.647859991	+J	0.485154450	3.776745319
5	1.10	20.05	-0.329691559	+J	0.953298509	1.480909348
			-0.049533281	+J	1.039346099	1.122194409
			-0.932112157			
	1.20	28.30	-0.379155248	+J	0.875398099	1.722895145
			-0.075429901	+J	1.051645041	1.233339787
			-0.782857478			
	1.50	43.42	-0.417037278	+J	0.775766015	2.331875801
			-0.114129379	+J	1.066150665	1.557406425
			-0.649752915			
	2.00	58.90	-0.429091632	+J	0.721329212	3.250804663
			-0.138912484	+J	1.073567510	2.089246511
			-0.590933323			
	2.50	69.77	-0.432354957	+J	0.700317919	4.135804653
			-0.149322003	+J	1.076304317	2.618067980
			-0.569953918			
6	1.10	29.68	-0.036955025	+J	1.025835752	2.970661402
			-0.194426298	+J	0.951625824	1.309153557
			-0.599798739	+J	0.517634988	1.115020156
	1.20	39.63	-0.054593772	+J	1.034293532	3.598949194
			-0.235427678	+J	0.907697976	1.495312572
			-0.547630370	+J	0.429689229	1.222710013

Table 13.9 (a) (*continued*)

n	Ω_s	Amin	poles			zeros
6	1.50	57.77	-0.080384858	+J	1.044896960	5.082049847
			-0.277387530	+J	0.846777678	1.981678486
			-0.487831414	+J	0.349732071	1.539249182
	2.00	76.35	-0.096676044	+J	1.050733447	7.235802650
			-0.296649605	+J	0.810842872	2.732050657
			-0.457234770	+J	0.314305097	2.061105251
	2.50	89.40	-0.103480890	+J	1.052997589	9.278976440
			-0.303458214	+J	0.796428084	3.460955381
			-0.445700765	+J	0.301697373	2.580876827
7	1.05	30.47	-0.362386286	+J	0.791218579	1.647534966
			-0.097929962	+J	0.979496002	1.143827200
			-0.018274479	+J	1.012906075	1.057128668
			-0.697913051			
	1.10	39.36	-0.372605830	+J	0.706869304	1.874771833
			-0.129117623	+J	0.957427263	1.234481096
			-0.028279115	+J	1.018274426	1.110913038
			-0.599629283			
	1.20	50.96	-0.371194631	+J	0.627190828	2.228609085
			-0.161447927	+J	0.928551912	1.393320203
			-0.041080363	+J	1.024497986	1.216500044
			-0.519720137			
	1.50	72.13	-0.359473556	+J	0.543128490	3.087082624
			-0.198305324	+J	0.887344599	1.820436954
			-0.059559058	+J	1.032552958	1.528568745
			-0.443710178			
	2.00	93.81	-0.350169420	+J	0.501993001	4.354434013
			-0.217125237	+J	0.862266958	2.490336657
			-0.071123950	+J	1.037140250	2.044515371
			-0.408602297			
8	1.05	38.87	-0.014530487	+J	1.009558916	3.323612928
			-0.068626054	+J	0.979963779	1.380621672
			-0.223627791	+J	0.856770933	1.114948153
			-0.518430352	+J	0.408224314	1.055395246
	1.10	49.03	-0.022205081	+J	1.013627529	3.886669874
			-0.092340037	+J	0.964023352	1.542857409
			-0.246415690	+J	0.795957565	1.194613814
			-0.466762960	+J	0.344817638	1.108280182
	1.20	62.30	-0.031932183	+J	1.018410802	4.731130600
			-0.118065245	+J	0.943251848	1.803560734
			-0.261711299	+J	0.732750654	1.337725043
			-0.419006765	+J	0.293877214	1.212539792
	1.50	86.49	-0.045849293	+J	1.024717331	6.712673187
			-0.149083614	+J	0.913365483	2.453940630
			-0.270991623	+J	0.659774721	1.730797648
			-0.369041562	+J	0.246296585	1.521741509
	1.70	97.75	-0.050403163	+J	1.026665926	7.890916348
			-0.158082053	+J	0.903648019	2.852371693
			-0.272201091	+J	0.639137149	1.982722640
			-0.355670929	+J	0.234367803	1.726844311

Table 13.9 (a) (continued)

n	Ω_s	A_{min}	poles			zeros
9	1.01	29.30	-0.353795081	+J	0.790357351	1.592390895
			-0.092967801	+J	0.965875864	1.113389254
			-0.021403983	+J	0.996712208	1.028781176
			-0.003812146	+J	1.002768278	1.011454225
			-0.702470243			
	1.05	47.28	-0.355256587	+J	0.616991401	1.998417497
			-0.149551108	+J	0.893073559	1.262644172
			-0.050862618	+J	0.982010305	1.097911716
			-0.011772206	+J	1.007371068	1.054232359
			-0.514178574			
	1.10	58.71	-0.341730624	+J	0.544812560	2.302009583
			-0.173148543	+J	0.847266614	1.392019749
			-0.069512889	+J	0.969793320	1.170603991
			-0.017843811	+J	1.010566831	1.106502414
			-0.448274761			
	1.20	73.63	-0.323598087	+J	0.480739743	2.766295910
			-0.192807883	+J	0.797069132	1.605959058
			-0.090331338	+J	0.953983843	1.303693652
			-0.025489906	+J	1.014360309	1.209858060
			-0.392972082			
	1.50	100.84	-0.299691409	+J	0.415878683	3.874834776
			-0.210182026	+J	0.735885024	2.153215408
			-0.116302155	+J	0.931223392	1.675116301
			-0.036362100	+J	1.019421220	1.517108083
			-0.339005381			
10	1.01	35.71	-0.003196034	+J	1.002181530	3.115820408
			-0.016121050	+J	0.996274948	1.317109585
			-0.060997620	+J	0.971309543	1.078554869
			-0.210897222	+J	0.865315557	1.024101377
			-0.530286551	+J	0.428810209	1.011166215
	1.05	55.68	-0.009703588	+J	1.005862594	4.106043816
			-0.039276320	+J	0.984256327	1.591743231
			-0.106316864	+J	0.916451216	1.199442983
			-0.246341065	+J	0.730421901	1.086950660
			-0.422208637	+J	0.302646190	1.053409457
	1.10	68.38	-0.014626367	+J	1.008441925	4.814956188
			-0.054334003	+J	0.974525154	1.806084871
			-0.127784356	+J	0.880488157	1.309230089
			-0.251250118	+J	0.665084004	1.154917240
			-0.377369165	+J	0.259862483	1.105243802
	1.20	84.96	-0.020797068	+J	1.011523962	5.874712467
			-0.071471266	+J	0.962020278	2.141376019
			-0.147515953	+J	0.839831948	1.495322585
			-0.250235796	+J	0.602564454	1.281183481
			-0.337237298	+J	0.224733755	1.207956076
	1.30	97.00	-0.024719415	+J	1.013413906	6.765391827
			-0.081594326	+J	0.953976989	2.431232452
			-0.157343462	+J	0.816009700	1.664289832
			-0.247724935	+J	0.569882154	1.402236938
			-0.317077488	+J	0.208022341	1.310135007

Table 13.9 **Pole and zero locations of normalized elliptic low-pass functions of 0.1 dB passband ripple and 1 rad/sec 0.1 dB passband. (*b*) Even-order class B**

n	Ω_s	A_{min}	poles			zeros
4	1.50	26.31	-0.183945701	+J	1.112633824	
			-0.693676710	+J	0.581059575	1.610689640
	2.00	38.70	-0.223113567	+J	1.118519902	
			-0.665767848	+J	0.521239638	2.171726942
	2.50	47.39	-0.238886833	+J	1.120301127	
			-0.654896796	+J	0.498926878	2.726861715
	3.00	54.20	-0.246954784	+J	1.121105433	
			-0.649388373	+J	0.487861186	3.279782772
	3.50	59.84	-0.251665473	+J	1.121544719	
			-0.646185219	+J	0.481503576	3.831585169
6	1.10	27.88	-0.038727645	+J	1.027325392	
			-0.206641793	+J	0.947162211	1.341571569
			-0.588098049	+J	0.479475766	1.116277099
	1.20	37.83	-0.056663863	+J	1.035617232	
			-0.243665650	+J	0.901849866	1.537823915
			-0.538162768	+J	0.410730362	1.224532127
	1.50	55.97	-0.082130797	+J	1.045701146	
			-0.280867875	+J	0.842354238	2.046826124
			-0.483133197	+J	0.343213022	1.542280197
	2.00	74.55	-0.097774304	+J	1.051149249	
			-0.298090816	+J	0.808339894	2.828690052
			-0.455027610	+J	0.311686665	2.065770388
	2.50	87.59	-0.104210675	+J	1.053251863	
			-0.304260731	+J	0.794840932	3.586820841
			-0.444390535	+J	0.300228029	2.587035179
8	1.05	37.53	-0.015009743	+J	1.009939194	
			-0.071700841	+J	0.979388893	1.426355481
			-0.232469738	+J	0.847645164	1.119366884
			-0.506417751	+J	0.386317492	1.055655718
	1.10	47.69	-0.022813516	+J	1.014036894	
			-0.095249183	+J	0.962910533	1.599105358
			-0.251502782	+J	0.787764251	1.200483561
			-0.458373487	+J	0.332940996	1.108653545
	1.20	60.95	-0.032591268	+J	1.018790245	
			-0.120409369	+J	0.941863298	1.875204206
			-0.264031768	+J	0.726644397	1.345701218
			-0.413886666	+J	0.287902713	1.213077664
	1.50	85.14	-0.046373866	+J	1.024969339	
			-0.150388762	+J	0.912223399	2.560610533
			-0.271505862	+J	0.656662405	1.743459344
			-0.366857499	+J	0.244168863	1.522641420
	1.80	101.27	-0.052425016	+J	1.027525902	
			-0.162025824	+J	0.899337053	3.183829784
			-0.272661984	+J	0.630117297	2.123939037
			-0.349862099	+J	0.229222983	1.830455422

Table 13.9 (b) **(*continued*)**

n	Ω_s	A_{min}	poles			zeros
10	1.01	33.65	−0.003378154	+J	1.002347589	
			−0.017580895	+J	0.996338308	1.383392453
			−0.069287360	+J	0.969653368	1.087567091
			−0.247064307	+J	0.849348485	1.025384307
			−0.614560962	+J	0.384771526	1.011247277
	1.05	53.58	−0.010169439	+J	1.006193519	
			−0.041772861	+J	0.983714461	1.674769163
			−0.115388557	+J	0.911607325	1.212528586
			−0.272544265	+J	0.710688055	1.089310288
			−0.486271679	+J	0.273775965	1.053591609
	1.10	66.26	−0.015255612	+J	1.008850694	
			−0.057219815	+J	0.973594427	1.900638700
			−0.136496127	+J	0.874512315	1.324606180
			−0.273555130	+J	0.645766199	1.157894492
			−0.434145987	+J	0.234898359	1.105487347
	1.20	82.83	−0.021583991	+J	1.011997819	
			−0.074619800	+J	0.960706890	2.253465652
			−0.155606613	+J	0.833127797	1.513745785
			−0.269400507	+J	0.584238052	1.284918666
			−0.387617171	+J	0.202683076	1.208274961
	1.30	94.86	−0.025582504	+J	1.013914943	
			−0.084826596	+J	0.952478290	2.558379650
			−0.165043339	+J	0.809077621	1.685148358
			−0.265446752	+J	0.552204788	1.406527042
			−0.364257693	+J	0.187298477	1.310506225

Table 13.9 **Pole and zero locations of normalized elliptic low-pass functions of 0.1 dB passband ripple and 1 rad/sec 0.1 dB passband. (*c*) Even-order class C**

n	Ω_s	A_{min}	poles			zeros
4	1.50	23.72	−0.206200436	+J	1.136413813	
			−0.863118589	+J	0.598000824	1.615388989
	2.00	36.02	−0.253433555	+J	1.142939806	
			−0.817437947	+J	0.520717561	2.175621748
	2.50	44.69	−0.272470564	+J	1.144693732	
			−0.799947441	+J	0.492344260	2.730037689
	3.00	51.48	−0.282209396	+J	1.145434737	
			−0.791140556	+J	0.478363305	3.282447100
	3.50	57.11	−0.287895858	+J	1.145823479	
			−0.786035538	+J	0.470358163	3.833875418
6	1.10	26.22	−0.040909626	+J	1.029744744	
			−0.230054840	+J	0.952150941	1.355422258
			−0.715427041	+J	0.476982594	1.117069840
	1.20	36.11	−0.060074959	+J	1.038655877	
			−0.269706607	+J	0.902241409	1.550623417
			−0.646193027	+J	0.398587525	1.225421190
	1.50	54.20	−0.087368153	+J	1.049424529	
			−0.308433712	+J	0.836973429	2.056805611
			−0.572095394	+J	0.324067682	1.543098807

***Table 13.9 (c)* (*continued*)**

n	Ω_s	A_{min}	poles			zeros
	2.00	72.76	-0.104186811	+J	1.055208564	
			-0.325870037	+J	0.799822450	2.835983992
			-0.534997463	+J	0.290135562	2.066415310
	2.50	85.80	-0.111118577	+J	1.057433009	
			-0.332019418	+J	0.785111964	3.592585802
			-0.521076024	+J	0.277943462	2.587558508
8	1.05	36.27	-0.015567600	+J	1.010441065	
			-0.076086313	+J	0.979592800	1.443968773
			-0.257533282	+J	0.843667030	1.122271299
			-0.604072690	+J	0.371389717	1.055855036
	1.10	46.40	-0.023658860	+J	1.014702678	
			-0.100633107	+J	0.962425888	1.615201116
			-0.276092917	+J	0.780152142	1.203595161
			-0.541763008	+J	0.314616710	1.108895659
	1.20	59.64	-0.033798341	+J	1.019636512	
			-0.126735583	+J	0.940478444	1.889040351
			-0.287400216	+J	0.715924621	1.348769069
			-0.485226631	+J	0.267901152	1.213341594
	1.50	83.81	-0.048106883	+J	1.026041865	
			-0.157688260	+J	0.909547865	2.570681334
			-0.292845398	+J	0.643036246	1.746017575
			-0.426488370	+J	0.223590448	1.522883773
	1.80	99.93	-0.054397356	+J	1.028689981	
			-0.169659168	+J	0.896093667	3.191883326
			-0.293096393	+J	0.615555823	2.126084328
			-0.405493110	+J	0.208686978	1.830666423
10	1.01	34.64	-0.003287109	+J	1.002262592	
			-0.016824901	+J	0.996276021	1.362523079
			-0.064640462	+J	0.970131516	1.083684206
			-0.221774966	+J	0.854235947	1.024785757
			-0.515442252	+J	0.400070041	1.011208892
	1.05	54.61	-0.009913792	+J	1.006005287	
			-0.040313896	+J	0.983945370	1.656818986
			-0.109291062	+J	0.913996100	1.207940578
			-0.249668628	+J	0.721758723	1.088336229
			-0.414730877	+J	0.293586969	1.053512218
	1.10	67.31	-0.014883441	+J	1.008598804	
			-0.055378195	+J	0.974066019	1.884909391
			-0.130001873	+J	0.877950430	1.320070982
			-0.252591819	+J	0.658709645	1.156807542
			-0.372588634	+J	0.254859596	1.105391502
	1.20	83.89	-0.021067692	+J	1.011673331	
			-0.072379157	+J	0.961489737	2.240415335
			-0.148914903	+J	0.837635636	1.509570479
			-0.250516325	+J	0.598449945	1.283807039
			-0.334462613	+J	0.222156540	1.208169460
	1.30	95.93	-0.024974072	+J	1.013546944	
			-0.082363322	+J	0.953463435	2.547032118
			-0.158323646	+J	0.814167321	1.681328416
			-0.247721523	+J	0.566895843	1.405456901
			-0.315155238	+J	0.206336021	1.310401797

Table 13.10 **Pole and zero locations of normalized elliptic low-pass functions of 0.5 dB passband ripple and 1 rad/sec 0.5 dB passband. (*a*) Odd case and even-order class A**

n	Ω_s	A_{min}	poles			zeros
3	1.50	21.92	-0.226428181	+J	1.047806740	1.675116181
			-0.766952217			
	2.00	31.19	-0.268935859	+J	1.037382841	2.270067930
			-0.692124724			
	2.50	37.71	-0.286000371	+J	1.031974792	2.856308699
			-0.665551603			
	3.00	42.82	-0.294709712	+J	1.028949857	3.439158440
			-0.652630627			
	3.50	47.04	-0.299788266	+J	1.027104020	4.020282745
			-0.645282507			
4	1.50	36.25	-0.127478063	+J	1.021854520	3.478300095
			-0.460004836	+J	0.510127485	1.592303157
	2.00	48.64	-0.150577784	+J	1.019755960	4.922104359
			-0.442278832	+J	0.463364929	2.143185854
	2.50	57.33	-0.160027534	+J	1.018568993	6.297529221
			-0.435015976	+J	0.446363479	2.689263105
	3.00	64.15	-0.164895341	+J	1.017885327	7.646635532
			-0.431290299	+J	0.438017249	3.233481407
	3.50	69.78	-0.167748392	+J	1.017462134	8.982360840
			-0.429113388	+J	0.433246970	3.776745796
5	1.10	27.21	-0.246152118	+J	0.839608252	1.480909348
			-0.040820617	+J	1.010403275	1.122194409
			-0.573107302			
	1.20	35.49	-0.268527925	+J	0.777460277	1.722895145
			-0.057874668	+J	1.011845827	1.233339787
			-0.497944862			
	1.50	50.61	-0.285117894	+J	0.703363180	2.331875801
			-0.081731014	+J	1.012528539	1.557406425
			-0.425970703			
	2.00	66.09	-0.290271938	+J	0.663882077	3.250804663
			-0.096276127	+J	1.012299538	2.089246511
			-0.392612040			
	2.50	76.96	-0.291640013	+J	0.648710430	4.135804653
			-0.102247529	+J	1.012075901	2.618067980
			-0.380483717			
6	1.10	36.87	-0.028678060	+J	1.006121874	2.970880508
			-0.144375697	+J	0.893164337	1.309214830
			-0.396521926	+J	0.432554632	1.115053177
	1.20	46.82	-0.040425483	+J	1.007236362	3.598975420
			-0.167971954	+J	0.851722121	1.495320678
			-0.363651812	+J	0.375259697	1.222715020
	1.50	64.96	-0.056836016	+J	1.008161426	5.082051277
			-0.191730320	+J	0.798694193	1.981678843
			-0.326750726	+J	0.319660932	1.539249420

***Table 13.10* (*a*) (*continued*)**

n	Ω_s	A_{min}	poles			zeros
6	2.00	83.55	−0.066844299	+J	1.008418679	7.235803127
			−0.202626169	+J	0.768760920	2.732050896
			−0.307877123	+J	0.293759823	2.061105490
	2.50	96.59	−0.070955805	+J	1.008462667	9.278977394
			−0.206494525	+J	0.756939054	3.460955620
			−0.300743669	+J	0.284334362	2.580877066
7	1.05	37.66	−0.254999757	+J	0.706704438	1.647534966
			−0.074674204	+J	0.948049486	1.143827200
			−0.014257696	+J	1.003106117	1.057128668
			−0.447036952			
	1.10	46.55	−0.256961346	+J	0.640873969	1.874771833
			−0.094671115	+J	0.922896922	1.234481096
			−0.021209056	+J	1.004004717	1.110913038
			−0.391507298			
	1.20	58.15	−0.253327876	+J	0.578669488	2.228609085
			−0.114863433	+J	0.893110156	1.393320203
			−0.029804889	+J	1.004879236	1.216500044
			−0.344432056			
	1.50	79.32	−0.244232759	+J	0.512168288	3.087082624
			−0.137465000	+J	0.853735268	1.820436954
			−0.041798126	+J	1.005760312	1.528568745
			−0.298117280			
	2.00	101.00	−0.237867668	+J	0.479100466	4.354434013
			−0.148912162	+J	0.830870390	2.490336657
			−0.049109086	+J	1.006132722	2.044515371
			−0.276245415			
8	1.05	46.06	−0.010987154	+J	1.002143025	3.323636770
			−0.051230084	+J	0.960376620	1.380628228
			−0.160151854	+J	0.805507839	1.114951253
			−0.344304860	+J	0.355920434	1.055397153
	1.10	56.22	−0.016304918	+J	1.002820492	3.886672020
			−0.066938467	+J	0.941568673	1.542858005
			−0.172667474	+J	0.751468837	1.194614172
			−0.311794907	+J	0.310342401	1.108280420
	1.20	69.49	−0.022871571	+J	1.003518701	4.731131077
			−0.083614260	+J	0.919011593	1.803560853
			−0.180838242	+J	0.696560860	1.337725163
			−0.281639963	+J	0.271752924	1.212539792
	1.50	93.68	−0.032024018	+J	1.004293799	6.712673187
			−0.103376850	+J	0.888692558	2.453940630
			−0.185527384	+J	0.633813381	1.730797648
			−0.249813974	+J	0.233944818	1.521741629
	1.70	104.94	−0.034968201	+J	1.004500866	7.890916824
			−0.109059945	+J	0.879199922	2.852371931
			−0.186054796	+J	0.616099358	1.982722759
			−0.241238013	+J	0.224182129	1.726844311

Table 13.10 (a) (continued)

n	Ω_s	A_{min}	poles			zeros
9	1.01	36.49	-0.250002414	+J	0.704686224	1.592390895
			-0.071396887	+J	0.935939729	1.113389254
			-0.016836684	+J	0.989262640	1.028781176
			-0.003015001	+J	1.000687361	1.011454225
			-0.448483646			
	1.05	54.47	-0.243096888	+J	0.566801012	1.998417497
			-0.107365131	+J	0.860746086	1.262644172
			-0.037377570	+J	0.968819141	1.097911716
			-0.008718332	+J	1.001565337	1.054232359
			-0.339524418			
	1.10	65.90	-0.232773095	+J	0.508299708	2.302009583
			-0.121904768	+J	0.817044795	1.392019749
			-0.049931690	+J	0.954144239	1.170603991
			-0.012918365	+J	1.002095222	1.106502414
			-0.299159199			
	1.20	80.82	-0.220165625	+J	0.455314308	2.766295910
			-0.133898884	+J	0.770558834	1.605959058
			-0.063695803	+J	0.936420918	1.303693652
			-0.018099779	+J	1.002662897	1.209858060
			-0.264472127			
	1.50	108.03	-0.204129696	+J	0.400447488	3.874834776
			-0.144461200	+J	0.715010941	2.153215408
			-0.080604456	+J	0.912372530	1.675116301
			-0.025315816	+J	1.003329873	1.517108083
			-0.229960799			
10	1.01	42.90	-0.002457376	+J	1.000509858	3.115864515
			-0.012352766	+J	0.991221726	1.317120910
			-0.046117201	+J	0.953534722	1.078559160
			-0.152606726	+J	0.813044369	1.024103165
			-0.351563126	+J	0.369148165	1.011167288
	1.05	62.87	-0.007082723	+J	1.001193523	4.106044292
			-0.028526202	+J	0.974839032	1.591743350
			-0.076137170	+J	0.894984484	1.199442983
			-0.171244860	+J	0.692414463	1.086950660
			-0.283034176	+J	0.276667804	1.053409457
	1.10	75.57	-0.010484145	+J	1.001619935	4.814956665
			-0.038749516	+J	0.963037372	1.806084991
			-0.089996040	+J	0.859260261	1.309230089
			-0.173116162	+J	0.634984910	1.154917359
			-0.254322529	+J	0.242670834	1.105243802
	1.20	92.15	-0.014677346	+J	1.002089024	5.874712467
			-0.050211381	+J	0.948723316	2.141376019
			-0.102617256	+J	0.820084214	1.495322704
			-0.171575695	+J	0.579955280	1.281183481
			-0.228386447	+J	0.213642523	1.207956076
	1.30	104.19	-0.017310886	+J	1.002357721	6.765391827
			-0.056915149	+J	0.939854383	2.431232691
			-0.108879037	+J	0.797527492	1.664289832
			-0.169602588	+J	0.551064670	1.402236938
			-0.215266407	+J	0.199469090	1.310135007

Table 13.10 **Pole and zero locations of normalized elliptic low-pass functions of 0.5 dB passband ripple and 1 rad/sec 0.5 dB passband. (*b*) Even-order class B**

n	Ω_s	Amin	poles			zeros
4	1.50	33.50	−0.133921266	+J	1.022083759	
			−0.455254227	+J	0.490987211	1.610986590
	2.00	45.89	−0.154379696	+J	1.019501209	
			−0.439317852	+J	0.455193222	2.171753168
	2.50	54.58	−0.162488982	+J	1.018308043	
			−0.433108032	+J	0.441678554	2.726866245
	3.00	61.40	−0.166612238	+J	1.017669678	
			−0.429967403	+J	0.434941500	3.279783964
	3.50	67.03	−0.169012532	+J	1.017288923	
			−0.428143591	+J	0.431059986	3.831585407
6	1.10	35.07	−0.030164016	+J	1.006509185	
			−0.151943401	+J	0.886143088	1.341669798
			−0.388065726	+J	0.410574138	1.116327405
	1.20	45.02	−0.041951813	+J	1.007500529	
			−0.172766045	+J	0.845392287	1.537836909
			−0.357647359	+J	0.363459498	1.224539638
	1.50	63.16	−0.057981759	+J	1.008255959	
			−0.193690538	+J	0.794757426	2.046826363
			−0.323865741	+J	0.315189481	1.542280436
	2.00	81.74	−0.067527585	+J	1.008443952	
			−0.203438342	+J	0.766658902	2.828690052
			−0.306518614	+J	0.291863471	2.065770388
	2.50	94.78	−0.071401142	+J	1.008471489	
			−0.206948772	+J	0.755628288	3.586821079
			−0.299934864	+J	0.283247054	2.587035418
8	1.05	44.72	−0.011380302	+J	1.002241969	
			−0.053492796	+J	0.958816111	1.426365018
			−0.165243849	+J	0.798207011	1.119371057
			−0.336671203	+J	0.341947705	1.055658340
	1.10	54.88	−0.016767273	+J	1.002911806	
			−0.068956666	+J	0.939779520	1.599106431
			−0.175505355	+J	0.743908167	1.200484157
			−0.306564391	+J	0.302224159	1.108653903
	1.20	68.14	−0.023341592	+J	1.003589153	
			−0.085169129	+J	0.917280912	1.875204325
			−0.182102188	+J	0.691202402	1.345701337
			−0.278438151	+J	0.267387658	1.213077664
	1.50	92.33	−0.032375205	+J	1.004328728	
			−0.104211807	+J	0.887495756	2.560610771
			−0.185794383	+J	0.631141663	1.743459344
			−0.248428181	+J	0.232271507	1.522641420
	1.80	108.46	−0.036273103	+J	1.004590750	
			−0.111548409	+J	0.875004768	3.183830023
			−0.186247110	+J	0.608357787	2.123939037
			−0.237509146	+J	0.219957262	1.830455422

Table 13.10 (b) *(continued)*

n	Ω_s	A_{min}	poles			zeros
10	1.01	41.83	-0.002536832	+J	1.000533223	
			-0.012927027	+J	0.990930974	1.362538695
			-0.048812773	+J	0.951209188	1.083689928
			-0.158924609	+J	0.801499963	1.024788260
			-0.341997683	+J	0.351040989	1.011210322
	1.05	61.80	-0.007243554	+J	1.001226306	
			-0.029285938	+J	0.974240839	1.656819105
			-0.078135781	+J	0.892071128	1.207940578
			-0.173042312	+J	0.684676528	1.088336349
			-0.278347492	+J	0.270135134	1.053512216
	1.10	74.50	-0.010672162	+J	1.001651764	
			-0.039485496	+J	0.962351382	1.884909511
			-0.091445126	+J	0.856550872	1.320071101
			-0.173801199	+J	0.629402995	1.156807542
			-0.251298875	+J	0.238852173	1.105391502
	1.20	91.08	-0.014867720	+J	1.002115488	
			-0.050831519	+J	0.948048353	2.240415335
			-0.103515029	+J	0.817894876	1.509570479
			-0.171680078	+J	0.576362133	1.283807039
			-0.226609364	+J	0.211571023	1.208169460
	1.30	103.12	-0.017486572	+J	1.002379537	
			-0.057432778	+J	0.939243138	2.547032118
			-0.109504312	+J	0.795739710	1.681328416
			-0.169557393	+J	0.548447907	1.405456901
			-0.214026168	+J	0.198077410	1.310401797

Table 13.10 **Pole and zero locations of normalized elliptic low-pass functions of 0.5 dB passband ripple and 1 rad/sec 0.5 dB passband. (*c*) Even-order class C**

n	Ω_s	A_{min}	poles			zeros
4	1.50	30.91	-0.155360967	+J	1.029769540	
			-0.619333148	+J	0.456954092	1.615933061
	2.00	43.21	-0.179572806	+J	1.026269197	
			-0.589134157	+J	0.413667589	2.175670385
	2.50	51.88	-0.189149916	+J	1.024636865	
			-0.577568352	+J	0.397558153	2.730046749
	3.00	58.67	-0.194015220	+J	1.023762226	
			-0.571759641	+J	0.389575154	3.282449007
	3.50	64.30	-0.196846187	+J	1.023240209	
			-0.568398654	+J	0.384990156	3.833875895
6	1.10	33.41	-0.032326829	+J	1.007311821	
			-0.172980562	+J	0.879145741	1.355566740
			-0.515852034	+J	0.370770156	1.117144227
	1.20	43.30	-0.044930097	+J	1.008378863	
			-0.194498286	+J	0.834957838	1.550642610
			-0.468594640	+J	0.321314603	1.225432396
	1.50	61.39	-0.062077891	+J	1.009157658	
			-0.215431452	+J	0.780447602	2.056806087
			-0.417828918	+J	0.272261649	1.543099165

***Table 13.10* (*c*) (*continued*)**

n	Ω_s	A_{min}	poles			zeros
	2.00	79.95	-0.072301865	+J	1.009323239	
			-0.224887490	+J	0.750359178	2.835983992
			-0.392356187	+J	0.249126643	2.066415310
	2.50	92.99	-0.076453969	+J	1.009334683	
			-0.228235692	+J	0.738575280	3.592586040
			-0.382783800	+J	0.240673453	2.587558746
8	1.05	43.46	-0.011870511	+J	1.002388716	
			-0.057185087	+J	0.957101226	1.443981647
			-0.185891867	+J	0.782527566	1.122276902
			-0.438275158	+J	0.299842536	1.055858493
	1.10	53.59	-0.017461129	+J	1.003087878	
			-0.073258363	+J	0.937281668	1.615202665
			-0.195164800	+J	0.728015602	1.203595996
			-0.394980371	+J	0.261085331	1.108896136
	1.20	66.83	-0.024279799	+J	1.003791094	
			-0.090014584	+J	0.913894713	1.889040351
			-0.200427935	+J	0.673573971	1.348769069
			-0.355480075	+J	0.227940798	1.213341594
	1.50	91.00	-0.033652429	+J	1.004553556	
			-0.109588064	+J	0.882964790	2.570681810
			-0.202318177	+J	0.612037003	1.746017814
			-0.314181536	+J	0.195288807	1.522884011
	1.80	107.12	-0.037699729	+J	1.004821301	
			-0.117099173	+J	0.869999170	3.191883564
			-0.202027783	+J	0.588816464	2.126084328
			-0.299346387	+J	0.184012115	1.830666542
10	1.01	40.84	-0.002618504	+J	1.000558734	
			-0.013574126	+J	0.990649521	1.383412004
			-0.052680869	+J	0.948802531	1.087574363
			-0.179800510	+J	0.786860228	1.025387406
			-0.445251584	+J	0.307980388	1.011248946
	1.05	60.77	0.007416811	+J	1.001273274	
			-0.030426014	+J	0.973508179	1.674769282
			-0.082830571	+J	0.887772858	1.212528706
			-0.191074446	+J	0.666665018	1.089310408
			-0.355579555	+J	0.230681226	1.053591609
	1.10	73.45	-0.010957055	+J	1.001709938	
			-0.040879793	+J	0.961325586	1.900638819
			-0.096320786	+J	0.851344109	1.324606180
			-0.190164194	+J	0.610652447	1.157894611
			-0.318761587	+J	0.201898217	1.105487466
	1.20	90.02	-0.015250636	+J	1.002184749	
			-0.052482873	+J	0.946680427	2.253465652
			-0.108441867	+J	0.811815798	1.513745904
			-0.186347649	+J	0.557301581	1.284918666
			-0.285680979	+J	0.177257538	1.208274961
	1.30	102.06	-0.017930826	+J	1.002454519	
			-0.059224501	+J	0.937666535	2.558379650
			-0.114409313	+J	0.789203167	1.685148358
			-0.183309615	+J	0.529363275	1.406527042
			-0.268991679	+J	0.165220425	1.310506225

Table 13.11 **Pole and zero locations of normalized elliptic low-pass functions of 1.0 dB passband ripple and 1 rad/sec 1.0 dB passband. (*a*) Odd case and even-order class A**

n	Ω_s	Amin	poles			zeros
3	1.50	25.18	-0.187698022	+J	0.994225144	1.675116181
			-0.591015339			
	2.00	34.45	-0.217033699	+J	0.981575310	2.270067930
			-0.539958417			
	2.50	40.97	-0.228663549	+J	0.975860238	2.856308699
			-0.521547496			
	3.00	46.08	-0.234572858	+J	0.972804964	3.439158440
			-0.512538731			
	3.50	50.31	-0.238011077	+J	0.970980406	4.020282745
			-0.507398248			
4	1.50	39.52	-0.104408421	+J	0.993936777	3.478352785
			-0.364989161	+J	0.480694294	1.592322469
	2.00	51.91	-0.121478364	+J	0.989176095	4.922109604
			-0.351272941	+J	0.442497790	2.143187761
	2.50	60.60	-0.128394186	+J	0.987046897	6.297531128
			-0.345728755	+J	0.428465307	2.689263821
	3.00	67.41	-0.131942734	+J	0.985910416	7.646635532
			-0.342895806	+J	0.421546191	3.233481646
	3.50	73.05	-0.134018332	+J	0.985231936	8.982360840
			-0.341243416	+J	0.417582363	3.776745558
5	1.10	30.47	-0.202144533	+J	0.804784656	1.480909348
			-0.034620721	+J	1.000220776	1.122194409
			-0.446561664			
	1.20	38.76	-0.217567787	+J	0.748166680	1.722895145
			-0.048083685	+J	0.998478413	1.233339787
			-0.391578615			
	1.50	53.87	-0.228874624	+J	0.681678116	2.331875801
			-0.066540629	+J	0.995253742	1.557406425
			-0.337846249			
	2.00	69.36	-0.232338041	+J	0.646439850	3.250804663
			-0.077624545	+J	0.992913544	2.089246511
			-0.312598795			
	2.50	80.23	-0.233244047	+J	0.632909298	4.135804653
			-0.082143128	+J	0.991878688	2.618067980
			-0.303368032			
6	1.10	40.14	-0.023926357	+J	0.999415636	2.970907927
			-0.118729413	+J	0.874516606	1.309222460
			-0.315090895	+J	0.409247220	1.115057349
	1.20	50.09	-0.033261102	+J	0.998304248	3.598978519
			-0.136580199	+J	0.834255815	1.495321631
			-0.289466858	+J	0.359828532	1.222715616
	1.50	68.23	-0.046116482	+J	0.996373713	5.082051277
			-0.154480219	+J	0.783830702	1.981678963
			-0.260803789	+J	0.310775340	1.539249539

Table 13.11 (a) (continued)

n	Ω_s	A_{min}	poles			zeros
	2.00	86.81	-0.053871013	+J	0.995015264	7.235802650
			-0.162691265	+J	0.755715072	2.732050657
			-0.246135786	+J	0.287533075	2.061105251
	2.50	99.86	-0.057040170	+J	0.994421184	9.278974533
			-0.165610760	+J	0.744660974	3.460954905
			-0.240586907	+J	0.279011726	2.580876589
7	1.05	40.93	-0.206293359	+J	0.681552768	1.647534966
			-0.061952632	+J	0.937640369	1.143827200
			-0.011920066	+J	0.999752045	1.057128668
			-0.352248251			
	1.10	49.82	-0.206797123	+J	0.621264279	1.874771833
			-0.077645950	+J	0.911761940	1.234481096
			-0.017523870	+J	0.999243736	1.110913038
			-0.310174763			
	1.20	61.42	-0.203316048	+J	0.564153254	2.228609085
			-0.093370825	+J	0.881885648	1.393320203
			-0.024379862	+J	0.998471022	1.216500044
			-0.274068803			
	1.50	82.59	-0.195790157	+J	0.502754033	3.087082624
			-0.110878393	+J	0.843206465	1.820436954
			-0.033844158	+J	0.997188687	1.528568745
			-0.238188341			
	2.00	104.27	-0.190684393	+J	0.472046733	4.354434013
			-0.119724914	+J	0.821043849	2.490336657
			-0.039565977	+J	0.996308744	2.044515371
			-0.221130684			
8	1.05	49.33	-0.009102376	+J	0.999652207	3.323639870
			-0.042261697	+J	0.953923225	1.380629063
			-0.130371213	+J	0.789532244	1.114951611
			-0.274078667	+J	0.341142416	1.055397391
	1.10	59.49	-0.013388376	+J	0.999258935	3.886672497
			0.051735798	+J	0.934314847	1.542858124
			-0.139688283	+J	0.737744629	1.194614172
			-0.248664156	+J	0.300352275	1.108280420
	1.20	72.76	-0.018637845	+J	0.998688698	4.731130600
			-0.067891054	+J	0.911293209	1.803560734
			-0.145709708	+J	0.685428560	1.337725163
			-0.225049093	+J	0.265193075	1.212539792
	1.50	96.94	-0.025894416	+J	0.997774124	6.712672234
			-0.083399802	+J	0.880915821	2.453940153
			-0.149084762	+J	0.625788748	1.730797410
			-0.200052753	+J	0.230187804	1.521741390
	1.70	108.21	-0.028216215	+J	0.997454524	7.890916348
			-0.087847725	+J	0.871505678	2.852371693
			-0.149437129	+J	0.608955741	1.982722640
			-0.193302348	+J	0.221059754	1.726844311

Table 13.11 (a) (continued)

n	Ω_s	Amin	poles			zeros
9	1.01	39.76	-0.202470899	+J	0.679164052	1.592390895
			-0.059357684	+J	0.925989509	1.113389254
			-0.014111457	+J	0.986705244	1.028781176
			-0.002531712	+J	0.999967396	1.011454225
			-0.353065878			
	1.05	57.74	-0.195247293	+J	0.551838040	1.998417497
			-0.087514065	+J	0.850481927	1.262644172
			-0.030696824	+J	0.964496195	1.097911716
			-0.007178310	+J	0.999639809	1.054232359
			-0.269877911			
	1.10	69.17	-0.186736435	+J	0.497328341	2.302009583
			-0.098791964	+J	0.807558239	1.392019749
			-0.040723436	+J	0.949093997	1.170603991
			-0.010562799	+J	0.999326646	1.106502414
			-0.238541424			
	1.20	84.09	-0.176577941	+J	0.447596312	2.766295910
			-0.108068891	+J	0.762285829	1.605959058
			-0.051655497	+J	0.930818737	1.303693652
			-0.014711155	+J	0.998887777	1.209858060
			-0.211419314			
	1.50	111.30	-0.163781390	+J	0.395695418	3.874834776
			-0.116230488	+J	0.708497882	2.153215408
			-0.065021351	+J	0.906413257	1.675116301
			-0.020450752	+J	0.998201430	1.517108083
			-0.184274971			
10	1.01	46.16	-0.002046414	+J	0.999941647	3.115870237
			-0.010275022	+J	0.989512861	1.317122340
			-0.038188517	+J	0.947631478	1.078559637
			-0.124594852	+J	0.796663284	1.024103403
			-0.279687315	+J	0.352456123	1.011167407
	1.05	66.14	-0.005806164	+J	0.999658942	4.106044292
			-0.023346879	+J	0.971767843	1.591743350
			-0.062030755	+J	0.888117850	1.199442983
			-0.138222516	+J	0.680725634	1.086950660
			-0.225977108	+J	0.269047499	1.053409457
	1.10	78.84	-0.008546646	+J	0.999403894	4.814956665
			-0.031536885	+J	0.959336698	1.806084991
			-0.072952271	+J	0.852541566	1.309230208
			-0.139375925	+J	0.625732958	1.154917359
			-0.203387380	+J	0.237537503	1.105243802
	1.20	95.42	-0.011907362	+J	0.999054909	5.874712944
			-0.040676214	+J	0.944480538	2.141376257
			-0.082871281	+J	0.813874006	1.495322704
			-0.137939602	+J	0.572984338	1.281183481
			-0.182920709	+J	0.210278720	1.207956195
	1.30	107.46	-0.014010065	+J	0.998820007	6.765391827
			-0.046005063	+J	0.935366750	2.431232452
			-0.087786913	+J	0.791725755	1.664289832
			-0.136294544	+J	0.545245588	1.402236938
			-0.172544688	+J	0.196854144	1.310135007

Table 13.11 **Pole and zero locations of normalized elliptic low-pass functions of 1.0 dB passband ripple and 1 rad/sec 1.0 dB passband. (*b*) Even-order class B**

n	Ω_s	A_{min}	poles			zeros
4	1.50	36.77	-0.109354325	+J	0.993062377	
			-0.361064196	+J	0.465490818	1.611023545
	2.00	49.16	-0.124305300	+J	0.988442123	
			-0.348974675	+J	0.435852677	2.171756506
	2.50	57.85	-0.130204633	+J	0.986518681	
			-0.344266623	+J	0.424616963	2.726866961
	3.00	64.66	-0.133199021	+J	0.985523462	
			-0.341887027	+J	0.419006258	3.279783964
	3.50	70.30	-0.134940609	+J	0.984939158	
			-0.340505749	+J	0.415770710	3.831585407
6	1.10	38.34	-0.025184665	+J	0.999425113	
			-0.124561764	+J	0.866978347	1.341681957
			-0.308338463	+J	0.391155481	1.116333723
	1.20	48.29	-0.034509271	+J	0.998231232	
			-0.140215605	+J	0.827910662	1.537838578
			-0.284782559	+J	0.349826604	1.224540591
	1.50	66.43	-0.047022097	+J	0.996256411	
			-0.155954644	+J	0.780065238	2.046826363
			-0.258565992	+J	0.306850940	1.542280436
	2.00	85.01	-0.054402445	+J	0.994926989	
			-0.163302585	+J	0.753736019	2.828690052
			-0.245080680	+J	0.285837084	2.065770149
	2.50	98.05	-0.057384480	+J	0.994359136	
			-0.165953189	+J	0.743432760	3.586821318
			-0.239958137	+J	0.278032154	2.587035418
8	1.05	47.99	-0.009435039	+J	0.999653339	
			-0.044113997	+J	0.952060640	1.426366210
			-0.134218544	+J	0.780319393	1.119371533
			-0.268114090	+J	0.329175204	1.055658579
	1.10	58.15	-0.013771215	+J	0.999244571	
			-0.056359999	+J	0.932328641	1.599106669
			-0.141813129	+J	0.730432332	1.200484157
			-0.244588822	+J	0.293222040	1.108654022
	1.20	71.41	-0.019019803	+J	0.998660266	
			-0.069126002	+J	0.909468055	1.875204444
			-0.146648765	+J	0.680311441	1.345701337
			-0.222549587	+J	0.261267513	1.213077664
	1.50	95.60	-0.026174368	+J	0.997742176	
			-0.084055848	+J	0.879706323	2.560610533
			-0.149279311	+J	0.623250902	1.743459225
			-0.198965654	+J	0.228644341	1.522641420
	1.80	111.73	-0.029244691	+J	0.997311652	
			-0.089794792	+J	0.867351532	3.183829308
			-0.149562344	+J	0.601599157	2.123938799
			-0.190366462	+J	0.217105150	1.830455184

Table 13.11 (b) *(continued)*

n	Ω_s	A_{min}	poles			zeros
10	1.01	45.10	-0.002114868	+J	0.999944091	
			-0.010760629	+J	0.989122033	1.362540603
			-0.040399298	+J	0.944951117	1.083690643
			-0.129378185	+J	0.785164118	1.024788499
			-0.272192091	+J	0.337012112	1.011210442
	1.05	65.07	-0.005939762	+J	0.999654591	
			-0.023969498	+J	0.971078157	1.656819224
			-0.063624419	+J	0.885077596	1.207940578
			-0.139552757	+J	0.673297226	1.088336349
			-0.222316340	+J	0.263196558	1.053512216
	1.10	77.77	-0.008700744	+J	0.999394894	
			-0.032133155	+J	0.958580434	1.884909272
			-0.074098229	+J	0.849789739	1.320070982
			-0.139871508	+J	0.620398462	1.156807423
			-0.201017827	+J	0.234048694	1.105391383
	1.20	94.35	-0.012061563	+J	0.999042273	
			-0.041173875	+J	0.943761885	2.240415573
			-0.083577491	+J	0.811692059	1.509570599
			-0.138002902	+J	0.569550037	1.283807039
			-0.181522384	+J	0.208352208	1.208169580
	1.30	106.39	-0.014151519	+J	0.998806655	
			-0.046418600	+J	0.934726298	2.547032118
			-0.088277914	+J	0.789957702	1.681328297
			-0.136248171	+J	0.542740941	1.405456901
			-0.171566516	+J	0.195548207	1.310401797

Table 13.11 **Pole and zero locations of normalized elliptic low-pass functions of 1.0 dB passband ripple and 1 rad/sec 1.0 dB passband. (*c*) Even-order class C**

n	Ω_s	A_{min}	poles			zeros
4	1.50	34.18	-0.128323227	+J	0.994626820	
			-0.524312377	+J	0.404482543	1.616000652
	2.00	46.48	-0.145844191	+J	0.988870800	
			-0.499020368	+J	0.370742828	2.175676584
	2.50	55.14	-0.152737617	+J	0.986460328	
			-0.489348292	+J	0.358129859	2.730047703
	3.00	61.94	-0.156232506	+J	0.985211194	
			-0.484495997	+J	0.351867437	3.282449484
	3.50	67.57	-0.158263981	+J	0.984477341	
			-0.481690317	+J	0.348266900	3.833876133
6	1.10	36.68	-0.027108757	+J	0.999598563	
			-0.142907187	+J	0.855557084	1.355584741
			-0.437217206	+J	0.330312908	1.117153525
	1.20	46.57	-0.037076995	+J	0.998324156	
			-0.158827782	+J	0.813560903	1.550644994
			-0.397575766	+J	0.289707810	1.225433826
	1.50	64.66	-0.050449759	+J	0.996191382	
			-0.174301863	+J	0.762524605	2.056806087
			-0.355020493	+J	0.248803988	1.543099165

Table 13.11 **(*c*)** **(*continued*)**

n	Ω_s	A_{min}	poles			zeros
	2.00	83.22	−0.058342103	+J	0.994744897	
			−0.181298807	+J	0.734590113	2.835984468
			−0.333675981	+J	0.229268044	2.066415548
	2.50	96.25	−0.061532702	+J	0.994125009	
			−0.183779076	+J	0.723683953	3.592586279
			−0.325655252	+J	0.222087532	2.587558985
8	1.05	46.73	−0.009858536	+J	0.999670088	
			−0.047268417	+J	0.949664712	1.443983197
			−0.151847437	+J	0.763118982	1.122277617
			−0.371900916	+J	0.270505220	1.055858970
	1.10	56.86	−0.014359698	+J	0.999248207	
			−0.059982117	+J	0.929136455	1.615202904
			−0.158468083	+J	0.711532295	1.203595996
			−0.335498929	+J	0.237676173	1.108896255
	1.20	70.10	−0.019803321	+J	0.998642445	
			−0.073157266	+J	0.905405879	1.889040351
			−0.162094772	+J	0.660151780	1.348769069
			−0.302266985	+J	0.209204972	1.213341594
	1.50	94.27	−0.027224494	+J	0.997686565	
			−0.088478148	+J	0.874557674	2.570681334
			−0.163159296	+J	0.602102876	1.746017575
			−0.267487705	+J	0.180774808	1.522883773
	1.80	110.39	−0.030411119	+J	0.997237027	
			−0.094343446	+J	0.861760676	3.191883564
			−0.162808657	+J	0.580184102	2.126084566
			−0.254983127	+J	0.170864329	1.830666661
10	1.01	44.11	−0.002185799	+J	0.999947488	
			−0.011316013	+J	0.988717258	1.383414507
			−0.043694600	+J	0.941860676	1.087575197
			−0.147211850	+J	0.766972780	1.025387764
			−0.377716303	+J	0.277059674	1.011249185
	1.05	64.04	−0.006110627	+J	0.999652803	
			−0.024923222	+J	0.970174491	1.674769402
			−0.067536555	+J	0.880131483	1.212528706
			−0.154773578	+J	0.652742267	1.089310408
			−0.302220272	+J	0.211052373	1.053591609
	1.10	76.72	−0.008937627	+J	0.999387503	
			−0.033288799	+J	0.957369149	1.900638700
			−0.078131482	+J	0.843989789	1.324606180
			−0.153647870	+J	0.599498212	1.157894492
			−0.271183014	+J	0.185925081	1.105487347
	1.20	93.29	−0.012376973	+J	0.999026716	
			−0.042531654	+J	0.942201078	2.253465652
			−0.087629832	+J	0.805089056	1.513745785
			−0.150336385	+J	0.548672616	1.284918666
			−0.243253693	+J	0.164163828	1.208274961
	1.30	105.32	−0.014515738	+J	0.998785317	
			−0.047885992	+J	0.932955921	2.558379650
			−0.092302039	+J	0.782939136	1.685148358
			−0.147809148	+J	0.521998286	1.406527042
			−0.229149014	+J	0.153448477	1.310506225

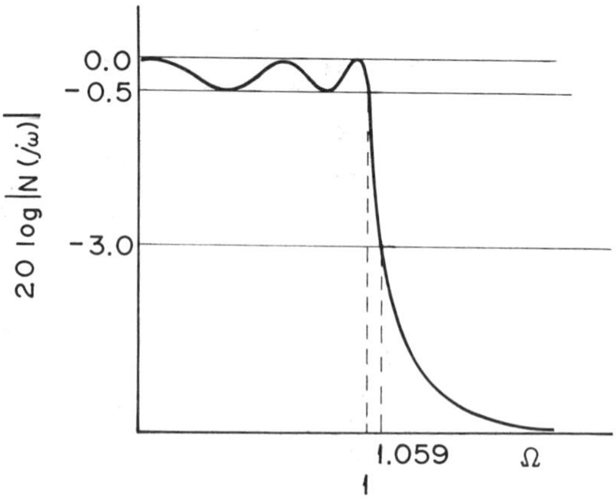

Figure 13.8 *Magnitude of Chebyshev fifth-order function.*

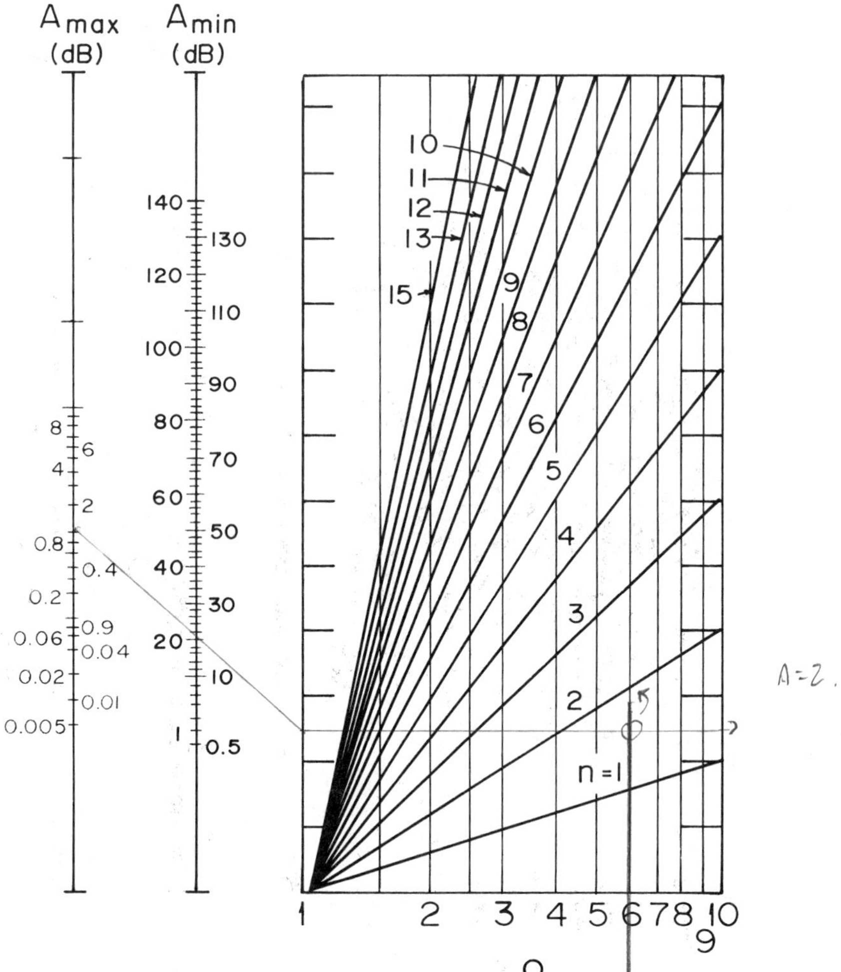

Figure 13.9 *Nomograph for Butterworth (maximally flat magnitude) normalized low-pass functions.*

Nomographs for Approximating Functions

In addition to the tables, we have at our disposal nomographs that are used when the order of the function is unknown. That is, we know the passband cutoff frequency and passband ripple and a prescribed attenuation at a given stopband frequency. Since the tables give information only about poles and zeros, we cannot obtain information about the magnitude of the transfer function at some given frequency. Thus we must use an alternative process to obtain the order of the function. For this purpose, nomographs are presented (Figs. 13.9 to 13.11) to obtain the order of the function given passband ripple $A_{\max}$ and stopband attenuation $A_{\min}$, at a given frequency. The nomographs given [3] are for Butterworth, Chebyshev, and elliptic functions. Their use is illustrated in Fig. 13.12. An example follows.

EXAMPLE 13.4 Determine a low-pass elliptic filter function with 1 dB passband ripple extending from 0 to 1 rad/s and at least 40 dB of attenuation at $\omega = 2$ rad/s.

PROCEDURE

1 Use the nomograph of Fig. 13.11 to determine n as lying between 3 and 4.

2 Since the specifications may be exceeded, choose $n = 4$ for the design.

3 Determine the network function from Table 13.11, as described in the following section.

13.4 TABLES OF ELEMENT VALUES FOR ELLIPTIC NETWORKS

One major approach to the design of active filters is covered in Chapter 10 and is based on the simulation of an LC ladder realization of the filter. That is, the starting point for active-RC filter realization is a passive-LC filter. The main advantage of this approach is to obtain active-RC filters with the same low sensitivities as the passive LC double-resistance terminated filter. The element values of the passive filter can be obtained by the use of closed-form expressions or filter tables [4, 5]. As indicated in Chapter 2, one of the most useful passive-LC filter realizations is given by the elliptic filters. However, to obtain the network element values for elliptic filters requires lengthy calculations that are best done by a digital computer. For this reason we include in this section tables of element values of elliptic networks (Tables 13.12 to 13.17). The associated networks are presented in Fig. 13.13.

As was seen in Chapter 2, the elliptic filter can be specified by its order n, the ripple passband in dB, $A_{\max}$, and the normalized stopband frequency Ω_s, as shown in Fig. 13.14. In addition, for even-order filters, we have to specify which of the three classes is to be chosen (A, B, or C), where class A may have negative element values resulting from its finite value at infinite frequency; that is, does not have a zero at infinity. Class B has positive element values but unequal terminating resistances, and class C has positive element values and equal terminating resistances.

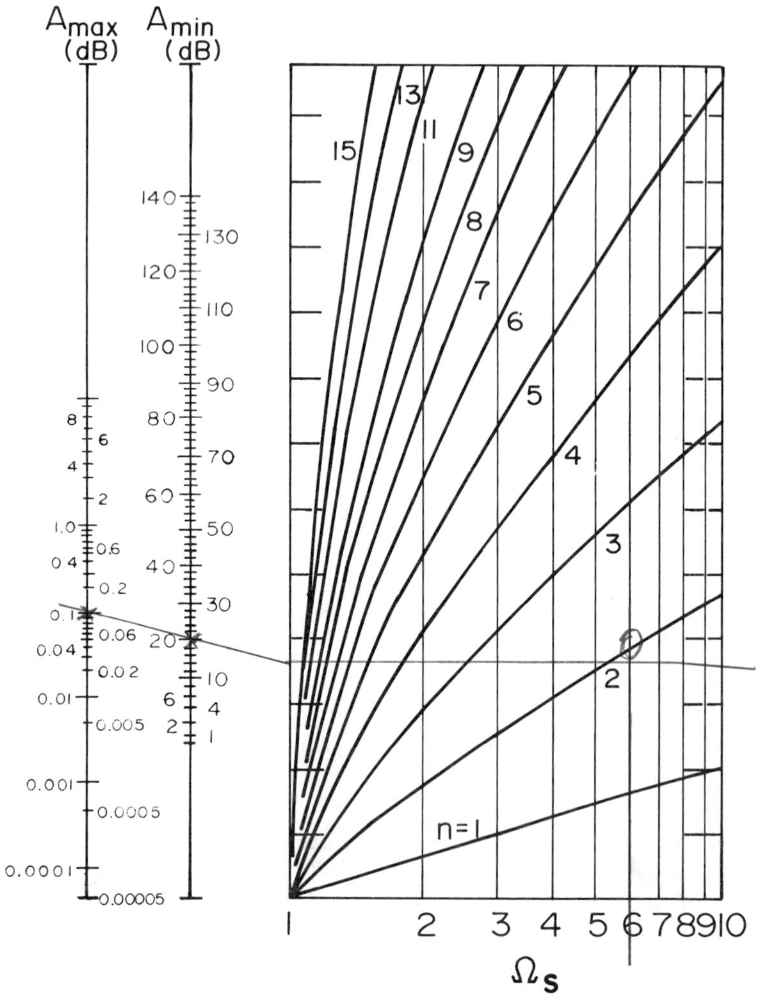

Figure 13.10 Nomograph for Chebyshev normalized low-pass functions.

In this section, we present tables of element values of elliptic networks for three different values of passband ripple A_{max}, namely 0.1 dB, 0.5 dB, and 1.0 dB, and different values of Ω_s. Since for even-order class A functions, negative element values may result, we have omitted these networks from the tables. In addition, since odd-order networks and class C even-order networks have equal terminating resistances, they are listed in the same tables, leaving even-order class B in other tables; in this case the terminating resistance value depends upon the passband ripple. In all cases, the input resistance is normalized to unity.

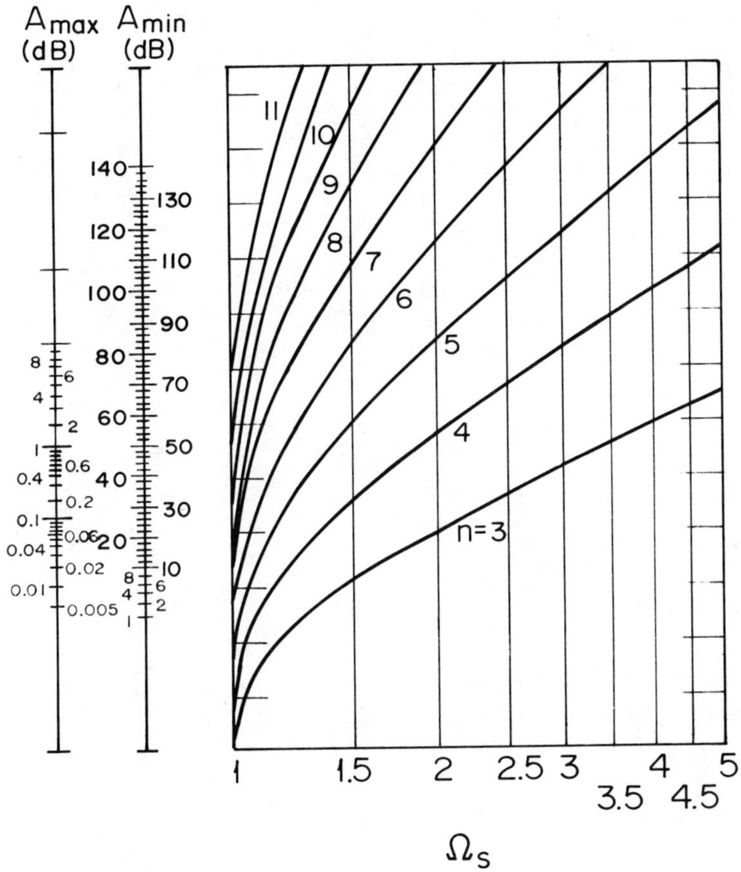

Figure 13.11 Nomograph for elliptic normalized low-pass functions.

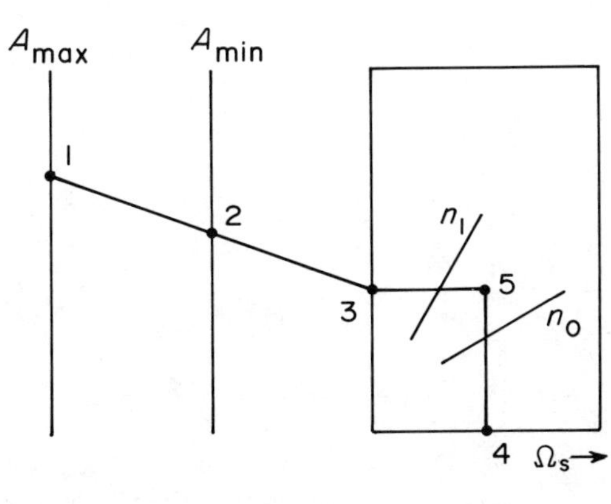

Figure 13.12 How to use the nomographs of this section. The order chosen should be n_1.

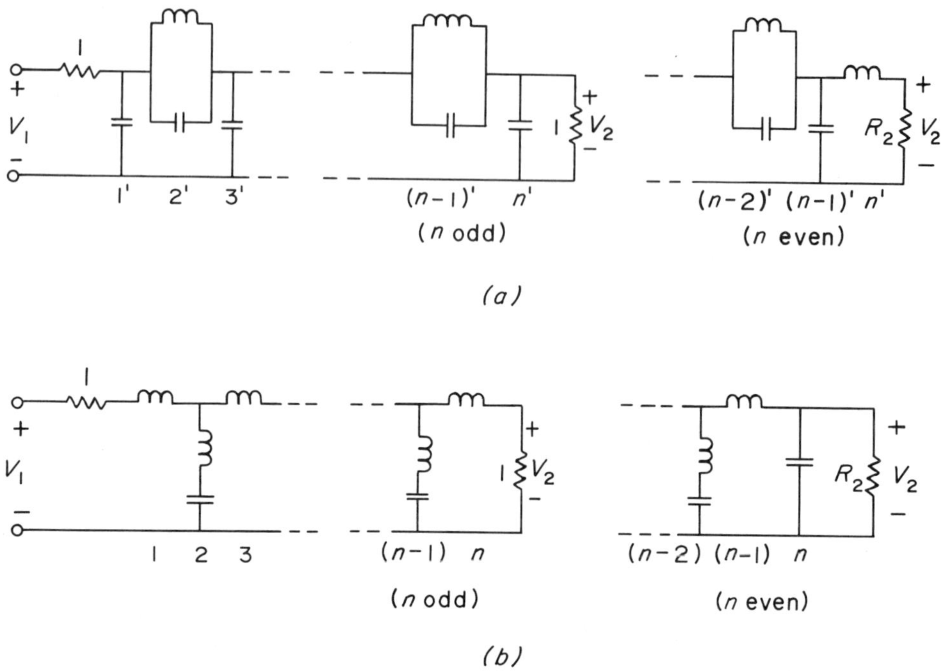

Figure 13.13 Network configuration for Tables 13.12 to 13.17. (a) Minimum inductance realization, (b) Dual of realization in part (a).

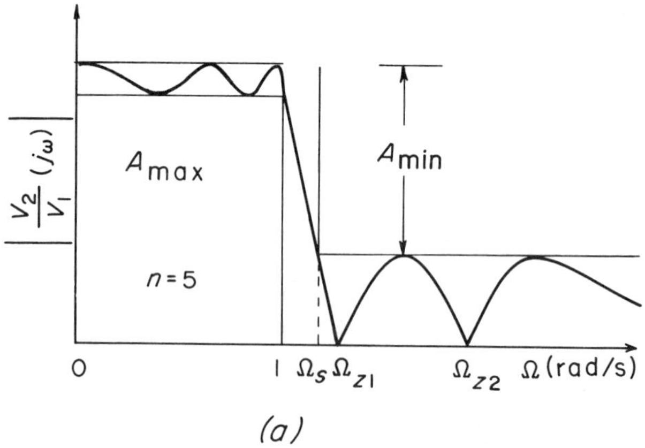

Figure 13.14 Transfer function magnitude plot of elliptic networks, (a) odd order, (b) even-order class A, (c) even-order class B, and (d) even-order class C.

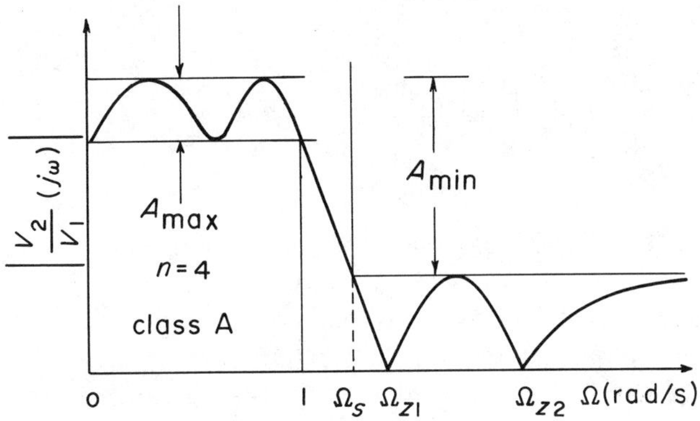

(b)

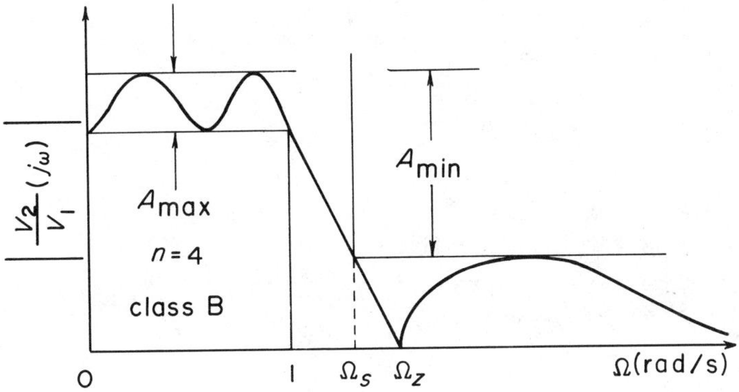

(c)

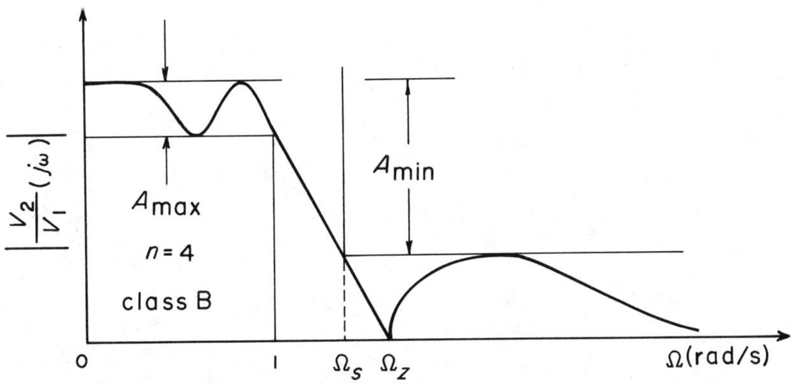

(d)

Figure 13.14 (continued)

EXAMPLE 13.5 Obtain a normalized passive elliptic filter that has a 0.5 dB passband ripple and an attenuation of at least 55 dB at a frequency greater than or equal to 2.0 rad/s. The terminating resistance should be unity, and the realization, minimum inductance.

PROCEDURE

1 Use the nomograph of Fig. 13.12 to determine the need for a fifth-order filter.

2 Use the tables for elliptic filter realizations. From Table 13.10, with $n = 5$ and $\Omega_s = 2$, the value of A_{min} satisfying the specifications is equal to 66.09 dB.

3 The final minimum inductance realization is obtained from Table 13.13 and is shown in Fig. 13.15. Note that the specification has been satisfied by a 0.5 dB passband ripple filter.

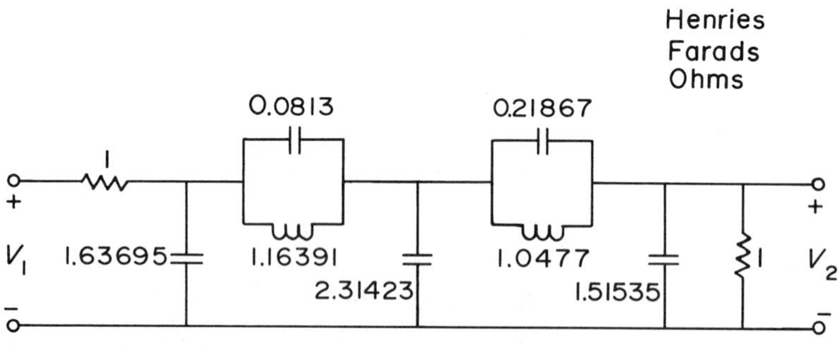

Figure 13.15 Network realization for Example 13.5.

Table 13.12 **Element values for normalized low-pass elliptic double-terminated lossless-ladder realizations of 0.1 dB passband ripple and $R_L = 1\ \Omega$ (class C for even-order realizations)**

n	Ω_s	A_{min}	C_1'	L_2'	C_2'	C_3'	L_4'	C_4'	C_5'	L_6'	C_6'	C_7'	L_8'	C_8'	C_9'	L_{10}'
3	1.50	14.85	0.77031	0.74561	0.47797	0.77031										
	2.00	24.01	0.89544	0.93759	0.20697	0.89544										
	2.50	30.52	0.94720	1.01731	0.12049	0.94720										
	3.00	35.62	0.97394	1.05854	0.07987	0.97394										
	3.50	39.85	0.98964	1.08275	0.05714	0.98964										
4	1.50	23.72	0.62785	0.93962	0.40784	1.24694	0.93517									
	2.00	36.02	0.77553	1.17644	0.17958	1.33473	0.93382									
	2.50	44.69	0.83469	1.27440	0.10528	1.37218	0.93247									
	3.00	51.48	0.86498	1.32510	0.07004	1.39179	0.93166									
	3.50	57.11	0.88268	1.35491	0.05021	1.40338	0.93116									
5	1.10	20.05	0.81296	0.92418	0.49338	1.22445	0.37193	2.13500	0.29125							
	1.20	28.30	0.91441	1.06516	0.31628	1.38201	0.60131	1.09329	0.52974							
	1.50	43.42	1.0278[illegible]	1.21517	0.15134	1.63179	0.93525	0.44083	0.81549							
	2.00	58.90	1.08758	1.29322	0.07317	1.79387	1.14330	0.20038	0.97720							
	2.50	69.77	1.11066	1.32351	0.04417	1.86209	1.22961	0.11865	1.04217							
6	1.10	26.22	0.57614	0.88777	0.61313	0.97284	0.59029	1.35761	0.94287	1.01380						
	1.20	36.11	0.70983	1.06264	0.39138	1.15973	0.87405	0.76189	1.09175	1.02462						
	1.50	54.20	0.86595	1.27403	0.18554	1.43106	1.27235	0.33007	1.28253	1.03317						
	1.80	66.28	0.92764	1.35976	0.11502	1.55265	1.44841	0.20010	1.36335	1.03548						
	2.00	72.76	0.95131	1.39297	0.08926	1.60132	1.51866	0.15421	1.39521	1.03621						
7	1.05	30.47	0.91937	1.07659	0.34220	1.09623	0.40518	2.20850	0.84335	0.50342	1.51827	0.41098				
	1.10	39.36	0.98821	1.16726	0.24374	1.27743	0.59720	1.35681	1.04029	0.67881	0.96669	0.58282				
	1.20	50.96	1.05029	1.24872	0.16124	1.48377	0.82869	0.81542	1.28723	0.87428	0.58918	0.75395				
	1.50	72.13	1.11593	1.33554	0.07857	1.75687	1.15174	0.37160	1.63827	1.12502	0.26822	0.95588				
	2.00	93.81	1.14910	1.37979	0.03822	1.92026	1.35221	0.17692	1.85664	1.27023	0.12694	1.06720				
8	1.05	36.27	0.68104	1.01038	0.47468	0.91101	0.48836	1.83677	0.67610	0.70497	1.12624	0.98460	1.04903			
	1.10	46.40	0.77921	1.13273	0.33839	1.08668	0.70666	1.15081	0.86562	0.92698	0.74468	1.09627	1.06140			
	1.15	53.70	0.83495	1.20281	0.26881	1.20269	0.85707	0.86518	0.99654	1.07113	0.57169	1.16453	1.06769			
	1.20	59.64	0.87250	1.25079	0.22404	1.28897	0.97154	0.69915	1.09601	1.17713	0.46698	1.21321	1.07163			
	1.40	77.09	0.95513	1.35543	0.13310	1.49755	1.25576	0.39492	1.34163	1.42924	0.26770	1.32525	1.07913			
9	1.01	29.30	0.90421	1.05676	0.37319	0.98992	0.30110	3.13798	0.40697	0.17277	5.65758	0.61326	0.44284	1.82164	0.34286	
	1.05	47.28	1.02597	1.21654	0.20583	1.29803	0.60675	1.36728	0.76114	0.44745	2.01086	0.94134	0.74311	0.84408	0.63918	
	1.10	58.71	1.07226	1.27741	0.14773	1.46403	0.79046	0.92320	1.00154	0.63575	1.28473	1.14956	0.89544	0.57633	0.77015	
	1.20	73.63	1.11295	1.33139	0.09815	1.64257	0.99964	0.58858	1.29050	0.86538	0.78945	1.39230	1.05143	0.36877	0.89697	
	1.50	100.84	1.15493	1.38761	0.04800	1.86765	1.27611	0.27927	1.69055	1.18960	0.36523	1.72037	1.23694	0.17437	1.04130	
10	1.01	33.65	0.64267	0.95901	0.54486	0.79305	0.33016	2.88076	0.33299	0.22076	4.42964	0.46579	0.59773	1.41443	0.92372	1.05056
	1.05	53.58	0.82096	1.17319	0.30389	1.09174	0.66563	1.26609	0.64759	0.54491	1.65323	0.76830	0.97997	0.69407	1.11484	1.07974
	1.10	66.26	0.89544	1.26213	0.21933	1.25856	0.87195	0.85540	0.86008	0.76734	1.06636	0.95503	1.18233	0.48204	1.20555	1.09090
	1.15	75.40	0.93632	1.31138	0.17518	1.36465	1.00811	0.66381	1.00574	0.92160	0.81059	1.07841	1.30718	0.37802	1.25932	1.09666
	1.20	82.83	0.96431	1.34441	0.14648	1.44189	1.10918	0.54607	1.11622	1.03951	0.65893	1.17011	1.39630	0.31255	1.29689	1.10032
			L_1	C_2	L_2	L_3	C_4	L_4	L_5	C_6	L_6	L_7	C_8	L_8	L_9	C_{10}

$A_{max} = 0.1$

Table 13.13 **Element values for normalized low-pass elliptic double-terminated lossless-ladder realizations of 0.5 dB passband ripple and $R_L = 1\ \Omega$ (class C for even-order realizations)**

n	Ω_s	A_{min}	C_1'	L_2'	C_2'	C_3'	L_4'	C_4'	C_5'	L_6'	C_6'	C_7'	L_8'	C_8'	C_9'	L_{10}'
3	1.50	21.92	1.30386	0.78815	0.45217	1.30386										
	2.00	31.19	1.44483	0.93666	0.20718	1.44483										
	2.50	37.71	1.50251	0.99759	0.12287	1.50251										
	3.00	42.82	1.53226	1.02902	0.08216	1.53226										
	3.50	47.04	1.54971	1.04746	0.05907	1.54971										
4	1.50	30.91	1.01065	1.08666	0.35242	1.36684	1.29083									
	2.00	43.21	1.15514	1.30375	0.16204	1.44950	1.30089									
	2.50	51.88	1.21386	1.39376	0.09627	1.48416	1.30426									
	3.00	58.67	1.24408	1.44044	0.06443	1.50222	1.30587									
	3.50	64.30	1.26180	1.46790	0.04635	1.51287	1.30677									
5	1.10	27.21	1.32222	0.86396	0.52777	1.53073	0.42221	1.88076	0.79647							
	1.20	35.49	1.43762	0.97486	0.34557	1.77043	0.61277	1.07284	1.03384							
	1.50	50.61	1.56789	1.09837	0.16743	2.10688	0.88157	0.46767	1.33720							
	2.00	66.09	1.63695	1.16391	0.08130	2.31423	1.04770	0.21867	1.51535							
	2.50	76.96	1.66374	1.18943	0.04915	2.40008	1.11657	0.13066	1.58791							
6	1.10	33.41	0.96572	0.95223	0.57150	1.06188	0.71212	1.12520	1.02341	1.38666						
	1.20	43.30	1.10523	1.10809	0.37532	1.27023	1.00008	0.66586	1.14509	1.41238						
	1.50	61.39	1.27248	1.29576	0.18243	1.56001	1.40484	0.29894	1.30626	1.43815						
	1.80	73.47	1.33960	1.37134	0.11405	1.68750	1.58446	0.18292	1.37572	1.44702						
	2.00	79.95	1.36548	1.40053	0.08878	1.73829	1.65626	0.14140	1.40325	1.45023						
7	1.05	37.66	1.43532	0.97157	0.37919	1.41773	0.38282	2.33747	1.05428	0.50932	1.50069	0.89685				
	1.10	46.55	1.51353	1.04502	0.27226	1.64070	0.54082	1.49825	1.31959	0.65422	1.00301	1.07461				
	1.20	58.15	1.58480	1.11227	0.18102	1.89238	0.73120	0.92414	1.63753	0.81390	0.63289	1.25859				
	1.50	79.32	1.66092	1.18476	0.08857	2.22472	0.99714	0.42922	2.07634	1.01737	0.29660	1.48213				
	2.00	101.00	1.69967	1.22195	0.04316	2.42356	1.16221	0.20584	2.34510	1.13487	0.14208	1.60758				
8	1.05	43.46	1.07834	1.03844	0.46185	1.02277	0.53574	1.67431	0.70181	0.81447	0.97482	1.02826	1.44254			
	1.10	53.59	1.18566	1.14206	0.33563	1.21424	0.75038	1.08377	0.89548	1.05025	0.65727	1.11603	1.46872			
	1.15	60.89	1.24769	1.20088	0.26925	1.34046	0.89761	0.82610	1.02671	1.20419	0.50851	1.17068	1.48285			
	1.20	66.83	1.29034	1.24093	0.22582	1.43433	1.00942	0.67292	1.12554	1.31779	0.41714	1.21002	1.49209			
	1.40	84.29	1.38377	1.32767	0.13588	1.66140	1.28630	0.38554	1.36747	1.58922	0.24075	1.30145	1.51089			
9	1.01	36.49	1.41628	0.95299	0.41382	1.28234	0.28118	3.36024	0.50622	0.16972	5.75943	0.76684	0.45357	1.77854	0.82397	
	1.05	54.47	1.55374	1.08136	0.23156	1.65545	0.53336	1.55541	0.95971	0.40011	2.24876	1.19765	0.70072	0.89515	1.12958	
	1.10	65.90	1.60673	1.13128	0.16681	1.85562	0.68515	1.06511	1.25763	0.55654	1.46757	1.45985	0.82482	0.62568	1.27156	
	1.20	80.82	1.65370	1.17591	0.11113	2.07133	0.85794	0.68579	1.61197	0.74678	0.91483	1.76092	0.95158	0.40746	1.41192	
	1.50	108.03	1.70259	1.22278	0.05447	2.34392	1.08621	0.32809	2.09932	1.01481	0.42814	2.16347	1.10215	0.19570	1.57432	
10	1.01	40.84	1.03783	0.98884	0.52841	0.89772	0.36126	2.63272	0.35004	0.25044	3.90456	0.48094	0.69813	1.21100	0.97324	1.44108
	1.05	60.77	1.23526	1.16413	0.30626	1.22823	0.68563	1.22916	0.68170	0.58237	1.54688	0.77852	1.10993	0.61280	1.11680	1.49845
	1.10	73.45	1.32081	1.23484	0.22418	1.41405	0.88202	0.84564	0.90141	0.80955	1.01076	0.95729	1.33194	0.42790	1.18666	1.52188
	1.15	82.59	1.36904	1.27348	0.18039	1.53276	1.01070	0.66211	1.05128	0.96707	0.77248	1.07433	1.47011	0.33612	1.22845	1.53448
	1.20	90.02	1.40170	1.29922	0.15157	1.61943	1.10583	0.54772	1.16472	1.08750	0.62985	1.16090	1.56926	0.27810	1.25780	1.54272
			L_1	C_2	L_2	L_3	C_4	L_4	L_5	C_6	L_6	L_7	C_8	L_8	L_9	C_{10}

$A_{max} = 0.5$

Table 13.14 Element values for normalized low-pass elliptic double-terminated lossless-ladder realizations of 1.0 dB passband ripple and $R_L = 1\ \Omega$ (class C for even-order realizations)

n	Ω_s	A_{min}	C'_1	L'_2	C'_2	C'_3	L'_4	C'_4	C'_5	L'_6	C'_6	C'_7	L'_8	C'_8	C'_9	L'_{10}
3	1.50	25.18	1.69200	0.73340	0.48592	1.69200										
	2.00	34.45	1.85199	0.85903	0.22590	1.85199										
	2.50	40.97	1.91737	0.91048	0.13462	1.91737										
	3.00	46.08	1.95107	0.93700	0.09023	1.95107										
	3.50	50.31	1.97084	0.95257	0.06495	1.97084										
4	1.50	34.18	1.25672	1.11427	0.34366	1.38980	1.53225									
	2.00	46.48	1.40677	1.32367	0.15960	1.46762	1.55071									
	2.50	55.14	1.46798	1.41068	0.09511	1.50012	1.55742									
	3.00	61.94	1.49954	1.45585	0.06375	1.51703	1.56072									
	3.50	67.57	1.51806	1.48244	0.04589	1.52699	1.56261									
5	1.10	30.47	1.69[illegible]91	0.77511	0.58827	1.79892	0.39922	1.98907	1.12109							
	1.20	38.76	1.82812	0.87005	0.38720	2.09095	0.56347	1.16672	1.38094							
	1.50	53.87	1.97[illegible]87	0.97694	0.18824	2.49161	0.79362	0.51950	1.71889							
	2.00	69.36	2.05595	1.03392	0.09152	2.73567	0.93561	0.24486	1.91939							
	2.50	80.23	2.08666	1.05613	0.05536	2.83631	0.99447	0.14670	2.00138							
6	1.10	36.68	1.22057	0.94233	0.57749	1.10898	0.75715	1.05826	1.01675	1.64681						
	1.20	46.57	1.37146	1.08633	0.38284	1.32610	1.05109	0.63355	1.12484	1.68498						
	1.50	64.66	1.55426	1.25876	0.18779	1.62529	1.46558	0.28655	1.26961	1.72482						
	1.80	76.74	1.62308	1.32787	0.11778	1.75642	1.64998	0.17566	1.33237	1.73902						
	2.00	83.22	1.65561	1.35450	0.09179	1.80860	1.72376	0.13586	1.35729	1.74424						
7	1.05	40.93	1.82156	0.86343	0.42668	1.67632	0.34381	2.60271	1.23696	0.46779	1.63392	1.22362				
	1.10	49.82	1.91040	0.92662	0.30705	1.93579	0.48016	1.68753	1.55276	0.59277	1.10699	1.41994				
	1.20	61.42	1.99168	0.98474	0.20446	2.22804	0.64444	1.04856	1.92724	0.73012	0.70551	1.62539				
	1.50	82.59	2.07382	1.04761	0.10016	2.61372	0.87393	0.48973	2.44021	0.90483	0.33349	1.87717				
	2.00	104.27	2.12329	1.07993	0.04884	2.84446	1.01638	0.23538	2.75306	1.00567	0.16034	2.01924				
8	1.05	46.73	1.34673	1.00921	0.47522	1.08540	0.54692	1.64008	0.70772	0.86532	0.91753	1.00315	1.72154			
	1.10	56.86	1.46597	1.10092	0.34817	1.28842	0.75883	1.07170	0.90030	1.11345	0.61997	1.07811	1.75961			
	1.15	64.16	1.53547	1.15253	0.28054	1.42245	0.90386	0.82039	1.03007	1.27620	0.47982	1.12510	1.78049			
	1.20	70.10	1.58346	1.18748	0.23599	1.52224	1.01386	0.66997	1.12753	1.39664	0.39359	1.15904	1.79429			
	1.40	87.55	1.68918	1.26273	0.14287	1.76399	1.28582	0.38569	1.36541	1.68543	0.22701	1.23825	1.82284			
9	1.01	39.76	1.79904	0.84667	0.46579	1.51674	0.25172	3.75347	0.59291	0.15398	6.34818	0.89985	0.41808	1.92952	1.14201	
	1.05	57.74	1.95471	0.95672	0.26172	1.94887	0.46951	1.76694	1.12605	0.35392	2.54224	1.40979	0.63099	0.99407	1.47897	
	1.10	69.17	2.01503	0.99976	0.18875	2.18048	0.60062	1.21501	1.47339	0.48929	1.66927	1.71691	0.73764	0.69962	1.63779	
	1.20	84.09	2.06867	1.03833	0.12585	2.43022	0.74985	0.78464	1.88531	0.65376	1.04499	2.06810	0.84652	0.45803	1.79580	
	1.50	111.30	2.12470	1.07895	0.06173	2.74601	0.94696	0.37634	2.45079	0.88531	0.49076	2.53625	0.97584	0.22103	1.97958	
10	1.01	44.11	1.30310	0.96163	0.54336	0.95685	0.36714	2.59058	0.35794	0.25998	3.76133	0.48206	0.74463	1.13538	0.95128	1.71785
	1.05	64.04	1.52461	1.11270	0.32041	1.31260	0.68239	1.23499	0.69671	0.59634	1.51065	0.77126	1.18327	0.57482	1.06945	1.79995
	1.10	76.72	1.62236	1.17191	0.23621	1.51411	0.87139	0.85595	0.92029	0.82679	0.98969	0.94303	1.42284	0.40056	1.12735	1.83423
	1.15	85.86	1.67789	1.20387	0.19083	1.64333	0.99465	0.67280	1.07272	0.98672	0.75710	1.05498	1.57287	0.31416	1.16210	1.85291
	1.20	93.29	1.71564	1.22500	0.16075	1.73787	1.08551	0.55798	1.18808	1.10906	0.61761	1.13757	1.68092	0.25962	1.18655	1.86522
			L_1	C_2	L_2	L_3	C_4	L_4	L_5	C_6	L_6	L_7	C_8	L_8	L_9	C_{10}

$A_{max} = 1.0$

Table 13.15 **Element values for even-order class B normalized elliptic low-pass double-terminated lossless-ladder realizations of 0.1 dB passband ripple and $R_L = 0.73781\ \Omega$**

n	Ω_s	A_{min}	C'_1	L'_2	C'_2	C'_3	L'_4	C'_4	C'_5	L'_6	C'_6	C'_7	L'_8	C'_8	C'_9	L'_{10}
4	1.50	26.31	0.79945	0.88288	0.43659	1.53661	0.84069									
	2.00	38.70	0.95051	1.08630	0.19518	1.64683	0.83004									
	2.50	47.39	1.01120	1.16994	0.11495	1.69333	0.82549									
	3.00	54.20	1.04229	1.21313	0.07663	1.71758	0.82313									
	3.50	59.84	1.06046	1.23848	0.05500	1.73189	0.82175									
6	1.10	27.88	0.70771	0.81689	0.68016	1.10468	0.51872	1.54712	1.19765	0.88523						
	1.20	37.83	0.84243	0.98081	0.43112	1.32790	0.75657	0.88147	1.37707	0.87993						
	1.50	55.97	0.99836	1.17887	0.20248	1.64500	1.08849	0.38623	1.61158	0.87198						
	1.80	68.06	1.05945	1.25881	0.12505	1.78539	1.23473	0.23491	1.71193	0.86849						
	2.00	74.55	1.08280	1.28970	0.09690	1.84134	1.29301	0.18123	1.75160	0.86710						
8	1.05	37.53	0.79899	0.92995	0.52855	1.02241	0.42916	2.09093	0.78904	0.59417	1.34322	1.24820	0.89367			
	1.10	47.69	0.89538	1.04944	0.37264	1.22051	0.61878	1.31484	1.01703	0.77697	0.89306	1.38901	0.89114			
	1.15	55.00	0.94948	1.11808	0.29413	1.35015	0.74974	0.98947	1.17408	0.89490	0.68762	1.47614	0.88933			
	1.20	60.95	0.98606	1.16509	0.24409	1.44606	0.84957	0.79988	1.29323	0.98124	0.56277	1.53864	0.88795			
	1.40	78.42	1.06463	1.26758	0.14367	1.67650	1.09789	0.45187	1.58704	1.18530	0.32389	1.68326	0.88452			
10	1.01	34.64	0.75550	0.87751	0.61385	0.88074	0.28825	3.30346	0.38136	0.19170	5.10145	0.54288	0.49767	1.71100	1.17048	0.E9761
	1.05	54.61	0.92833	1.09111	0.33387	1.21270	0.58502	1.44313	0.74643	0.46759	1.92689	0.91278	0.80946	0.84666	1.41610	0.E9549
	1.10	67.31	0.99839	1.18097	0.23833	1.39449	0.76963	0.97094	0.99183	0.65631	1.24697	1.14180	0.97201	0.59039	1.53570	0.E9383
	1.15	76.45	1.03678	1.23092	0.18920	1.50909	0.89208	0.75152	1.15971	0.78698	0.94942	1.29335	1.07139	0.46406	1.60715	0.89267
	1.20	83.89	1.06237	1.26448	0.15755	1.59212	0.98322	0.61709	1.28687	0.88676	0.77257	1.40608	1.14190	0.38429	1.65727	0.89179
			L_1	C_2	L_2	L_3	C_4	L_4	L_5	C_6	L_6	L_7	C_8	L_8	L_9	C_{10}

$A_{max} = 0.1$

Table 13.16 **Element values for even-order class B normalized elliptic low-pass double-terminated lossless-ladder realization of 0.5 dB passband ripple and $R_L = 0.5042\ \Omega$**

n	Ω_s	A_{min}	C'_1	L'_2	C'_2	C'_3	L'_4	C'_4	C'_5	L'_6	C'_6	C'_7	L'_8	C'_8	C'_9	L'_{10}
4	1.50	33.50	1.32922	0.86136	0.44734	2.07704	0.85546									
	2.00	45.89	1.49441	1.02046	0.20777	2.21460	0.84895									
	2.50	54.58	1.56154	1.08588	0.12385	2.27188	0.84624									
	3.00	61.40	1.59609	1.11967	0.08303	2.30163	0.84485									
	3.50	67.03	1.61633	1.13952	0.05973	2.31914	0.84403									
6	1.10	35.07	1.20822	0.76855	0.72283	1.41366	0.50363	1.59334	1.65603	0.88397						
	1.20	45.02	1.35679	0.90225	0.46865	1.70456	0.69697	0.95684	1.85874	0.88059						
	1.50	63.16	1.53194	1.06425	0.22428	2.10240	0.96588	0.43526	2.13106	0.87573						
	1.80	75.25	1.60126	1.12955	0.13936	2.27548	1.08443	0.26747	2.24922	0.87362						
	2.00	81.74	1.62785	1.15475	0.10823	2.34412	1.13171	0.20706	2.29614	0.87278						
8	1.05	44.72	1.30232	0.85196	0.57692	1.31699	0.39273	2.28488	0.99998	0.54511	1.46408	1.69610	0.88925			
	1.10	54.88	1.40963	0.94936	0.41192	1.56108	0.54975	1.47993	1.28762	0.69491	0.99853	1.85594	0.88768			
	1.15	62.19	1.47049	1.00545	0.32708	1.71987	0.65793	1.12755	1.48259	0.79143	0.77752	1.95650	0.88657			
	1.20	68.14	1.51188	1.04390	0.27242	1.83705	0.74029	0.91795	1.62945	0.86206	0.64057	2.02923	0.88572			
	1.40	85.61	1.60132	1.12781	0.16147	2.11787	0.94494	0.52500	1.98896	1.02898	0.37309	2.19902	0.88362			
10	1.01	41.83	1.25270	0.80710	0.66739	1.13880	0.26545	3.58712	0.47992	0.17857	5.47657	0.69043	0.46116	1.84644	1.60471	0.89175
	1.05	61.80	1.44420	0.98015	0.37167	1.54496	0.51242	1.64757	0.93864	0.40939	2.20079	1.15212	0.71584	0.95740	1.88055	0.89040
	1.10	74.50	1.52324	1.05330	0.26722	1.76590	0.66529	1.12322	1.24059	0.56618	1.44549	1.43297	0.84868	0.67618	2.01883	0.88939
	1.15	83.64	1.56685	1.09406	0.21287	1.90503	0.76651	0.87463	1.44587	0.67447	1.10780	1.61789	0.92995	0.53463	2.10229	0.88867
	1.20	91.08	1.59602	1.12148	0.17764	2.00578	0.84180	0.72076	1.60090	0.75706	0.90493	1.75512	0.98766	0.44431	2.16115	0.88813
			L_1	C_2	L_2	L_3	C_4	L_4	L_5	C_6	L_6	L_7	C_8	L_8	L_9	C_{10}

$A_{max} = 0.5$

Table 13.17 **Element values for even-order class B normalized elliptic low-pass double-terminated lossless-ladder realization of 1.0 dB passband ripple and $R_L = 0.50402\ \Omega$**

n	Ω_s	A_{min}	C'_1	L'_2	C'_2	C'_3	L'_4	C'_4	C'_5	L'_6	C'_6	C'_7	L'_8	C'_8	C'_9	L'_{10}
4	1.50	36.77	1.71481	0.78305	0.49204	2.49366	0.79924									
	2.00	49.16	1.90048	0.91820	0.23091	2.65459	0.79441									
	2.50	57.85	1.97617	0.97377	0.13811	2.72140	0.79242									
	3.00	64.66	2.01517	1.00249	0.09273	2.75606	0.79139									
	3.50	70.30	2.03804	1.01936	0.06682	2.77645	0.79079									
6	1.10	38.34	1.56905	0.69148	0.80338	1.66829	0.45608	1.75944	1.99785	0.82076						
	1.20	48.29	1.73631	0.80659	0.52424	2.01190	0.62218	1.07185	2.22816	0.81822						
	1.50	66.43	1.93461	0.94611	0.25229	2.47740	0.85305	0.49283	2.53990	0.81461						
	1.80	78.52	2.01336	1.00233	0.15705	2.67899	0.95486	0.30376	2.67569	0.81305						
	2.00	85.01	2.04359	1.02402	0.12205	2.75884	0.99547	0.23540	2.72966	0.81243						
8	1.05	47.99	1.67197	0.76069	0.64615	1.55530	0.34969	2.56606	1.17579	0.48610	1.64182	2.03669	0.82473			
	1.10	58.15	1.79301	0.84446	0.46309	1.83957	0.48544	1.67598	1.51273	0.61510	1.12808	2.21830	0.82355			
	1.15	65.46	1.86191	0.89274	0.36838	2.02418	0.57888	1.28152	1.74015	0.69821	0.88133	2.33313	0.82272			
	1.20	71.41	1.90884	0.92585	0.30716	2.16030	0.65000	1.04547	1.91110	0.75902	0.72753	2.41639	0.82209			
	1.40	88.88	2.01048	0.99813	0.18245	2.48626	0.82663	0.60014	2.32877	0.90273	0.42527	2.61124	0.82054			
10	1.01	45.10	1.61537	0.72146	0.74660	1.34627	0.23682	4.02081	0.56337	0.15973	6.12248	0.81260	0.41243	2.06460	1.93150	0.82661
	1.05	65.07	1.83096	0.86999	0.41873	1.81829	0.45061	1.87359	1.10022	0.35991	2.50339	1.35198	0.63152	1.08523	2.24375	0.82560
	1.10	77.77	1.92046	0.93288	0.30171	2.07454	0.58273	1.28236	1.45156	0.49556	1.65147	1.67847	0.74584	0.76942	2.40166	0.82484
	1.15	86.91	1.96996	0.96794	0.24060	2.23584	0.67017	1.00038	1.69000	0.58918	1.26816	1.89313	0.81580	0.60944	2.49728	0.82431
	1.20	94.35	2.00313	0.99155	0.20092	2.35262	0.73518	0.82529	1.86993	0.66055	1.03714	2.05233	0.86549	0.50703	2.56480	0.82390
			L_1	C_2	L_2	L_3	C_4	L_4	L_5	C_6	L_6	L_7	C_8	L_8	L_9	C_{10}

$A_{max} = 1.0$

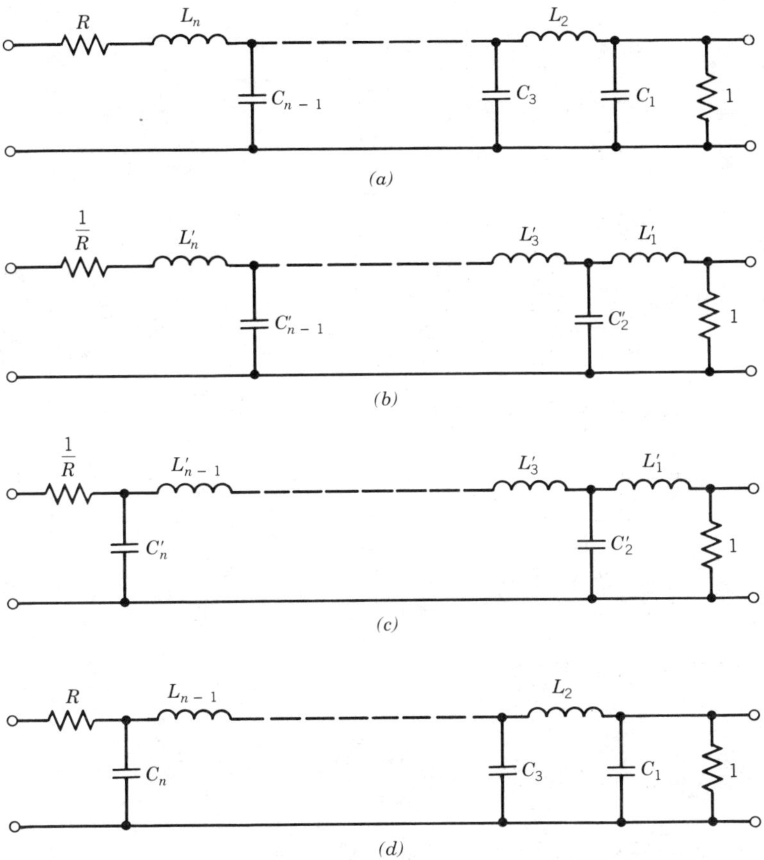

Figure 13.16 *Low-pass LC structures: (a) Even-order structure, (b) odd-order structure, (c) alternative even-order structure, (d) alternative odd-order structure.*

13.5 TABLES OF ELEMENT VALUES FOR BUTTERWORTH, CHEBYSHEV, AND THOMSON (BESSEL) DOUBLY TERMINATED NETWORKS

The doubly terminated low-pass structures of Fig. 13.16 represent the conventional LC realizations of all-pole approximating functions. Figures 13.16*a* to *b* show the even-order and odd-order structures, respectively. An alternative form is presented in Figs. 13.16*c* to *d*.

The normalized element values for Butterworth, Chebyshev (0.5 dB and 1 dB ripple), and Thomson (Bessel) functions are listed in Tables 13.18 (equally terminated) and 13.19 (source and load resistors related by a factor of two). It should be noted that there is no even-order Chebyshev solution for 1 dB in either table and for 0.5 dB only in Table 13.19.

Table 13.18 **Element values for low-pass double-resistance-terminated lossless-ladder realizations with $R = 1\ \Omega$[a]**

Elements in Figs. 13.16*a* (*even*) *and* 13.16*d* (*odd*)

n	C_1	L_2	C_3	L_4	C_5	L_6	C_7	L_8	C_9	L_{10}
2	1.4142	1.4142				Butterworth				
3	1.0000	2.0000	1.0000			(1 rad/s bandwidth)				
4	0.7654	1.8478	1.8478	0.7654						
5	0.6180	1.6180	2.0000	1.6180	0.6180					
6	0.5176	1.4142	1.9319	1.9319	1.4142	0.5176				
7	0.4450	1.2470	1.8019	2.0000	1.8019	1.2470	0.4450			
8	0.3902	1.1111	1.6629	1.9616	1.9616	1.6629	1.1111	0.3902		
9	0.3473	1.0000	1.5321	1.8794	2.0000	1.8794	1.5321	1.0000	0.3473	
0	0.3129	0.9080	1.4142	1.7820	1.9754	1.9754	1.7820	1.4142	0.9090	0.3129
3	1.5963	1.0967	1.5963			$\frac{1}{2}$-dB ripple Chebyshev				
5	1.7058	1.2296	2.5408	1.2296	1.7058	(1 rad/s bandwidth)				
7	1.7373	1.2582	2.6383	1.3443	2.6383	1.2582	1.7373			
9	1.7504	1.2690	2.6678	1.3673	2.7239	1.3673	2.6678	1.2690	1.7504	
3	2.0236	0.9941	2.0236			1-dB ripple Chebyshev				
5	2.1349	1.0911	3.0009	1.0911	2.1349	(1 rad/s bandwidth)				
7	2.1666	1.1115	3.0936	1.1735	3.0936	1.1115	2.1666			
9	2.1797	1.1192	3.1214	1.1897	3.1746	1.1897	3.1214	1.1192	2.1797	
2	1.5774	0.4226				Thomson (1-s delay at dc)				
3	1.2550	0.5528	0.1922							
4	1.0598	0.5116	0.3181	0.1104						
5	0.9303	0.4577	0.3312	0.2090	0.0718					
6	0.8377	0.4116	0.3158	0.2364	0.1480	0.0505				
7	0.7677	0.3744	0.2944	0.2378	0.1778	0.1104	0.0375			
8	0.7125	0.3446	0.2735	0.2297	0.1867	0.1387	0.0855	0.0289		
9	0.6678	0.3203	0.2547	0.2184	0.1859	0.1506	0.1111	0.0682	0.0230	
10	0.6305	0.3002	0.2384	0.2066	0.1808	0.1539	0.1240	0.0911	0.0557	0.0187
n	L'_1	C'_2	L'_3	C'_4	L'_5	C'_6	L'_7	C'_8	L'_9	C'_{10}

Elements in Figs. 13.16*b* (*odd*) *and* 13.16*c* (*even*)

[a] Reprinted by permission from L. P. Huelsman and P. E. Allen, *Introduction to the Theory and Design of Active Filters*, New York: McGraw-Hill, 1980.

Table 13.19 **Element values for low-pass double-resistance-terminated lossless-ladder realizations with $R = \frac{1}{2}\ \Omega^a$**

Elements in Figs. 13.16a *(even) and* 13.16*d (odd)*

n	C_1	L_2	C_3	L_4	C_5	L_6	C_7	L_8	C_9	L_{10}
						Butterworth (1 rad/s bandwidth)				
2	3.3461	0.4433								
3	3.2612	0.7789	1.1811							
4	3.1868	0.8826	2.4524	0.2175						
5	3.1331	0.9237	3.0510	0.4955	0.6857					
6	3.0938	0.9423	3.3687	0.6542	1.6531	0.1412				
7	3.0640	0.9513	3.5532	0.7512	2.2726	0.3536	0.4799			
8	3.0408	0.9558	3.6678	0.8139	2.6863	0.5003	1.2341	0.1042		
9	3.0223	0.9579	3.7426	0.8565	2.9734	0.6046	1.7846	0.2735	0.3685	
10	3.0072	0.9588	3.7934	0.8864	3.1795	0.6808	2.1943	0.4021	0.9818	0.0825
						$\frac{1}{2}$-dB ripple Chebyshev (1 rad/s bandwidth)				
2	1.5132	0.6538								
3	2.9431	0.6503	2.1903							
4	1.8158	1.1328	2.4881	0.7732						
5	3.2228	0.7645	4.1228	0.7116	2.3197					
6	1.8786	1.1884	2.7589	1.2403	2.5976	0.7976				
7	3.3055	0.7399	4.3575	0.8132	4.2419	0.7252	2.3566			
8	1.9012	1.2053	2.8152	1.2864	2.8479	1.2628	2.6310	0.8063		
9	3.3403	0.7395	4.4283	0.8341	4.4546	0.8235	4.2795	0.7304	2.3719	
10	1.9117	1.2127	2.8366	1.2999	2.8964	1.3054	2.8744	1.2714	2.6456	0.8104
n	L'_1	C'_2	L'_3	C'_4	L'_5	C'_6	L'_7	C'_8	L'_9	C'_{10}

Elements in Figs. 13.16*b (odd) and* 13.16*c (even)*

Table 13.19 **(*continued*)**

Elements in Figs. 13.16*a* (*even*) *and* 13.16*d* (*odd*)

n	C_1	L_2	C_3	L_4	C_5	L_6	C_7	L_8	C_9	L_{10}
3	3.4774	0.6153	2.8540				1-dB ripple Chebyshev			
5	3.7211	0.6949	4.7448	0.6650	2.9936		(1 rad/s bandwidth)			
7	3.7916	0.7118	4.9425	0.7348	4.8636	0.6757	3.0331			
9	3.8210	0.7182	5.0013	0.7485	5.0412	0.7429	4.9004	0.6797	3.0495	
2	2.6180	0.1910					Thomson			
3	2.1156	0.2613	0.3618				(1 s delay at dc)			
4	1.7893	0.2461	0.6127	0.0530						
5	1.5686	0.2217	0.6456	0.1015	0.1393					
6	1.4102	0.1999	0.6196	0.1158	0.2894	0.0246				
7	1.2904	0.1821	0.5797	0.1171	0.3497	0.0542	0.0735			
8	1.1964	0.1676	0.5395	0.1135	0.3685	0.0683	0.1684	0.0142		
9	1.1202	0.1558	0.5030	0.1081	0.3580	0.0744	0.2195	0.0336	0.0453	
10	1.0569	0.1460	0.4710	0.1024	0.3586	0.0763	0.2456	0.0450	0.1100	0.0925
n	L'_1	C'_2	L'_3	C'_4	L'_5	C'_6	L'_7	C'_8	L'_9	C'_{10}

Elements in Figs. 13.16*b* (*odd*) *and* 13.16*c* (*even*)

[a] Reprinted by permission from L. P. Huelsman and P. E. Allen, *Introduction to the Theory and Design of Active Filters*, New York: McGraw-Hill, 1980.

Table 13.20 **Typical characteristics of active-RC filter resistors**

	Thick-film resistor	*Thin-film resistors*		
Characteristic	*Cermet*	*Nichrome*	*Cermet*	*Tantalum oxynitride*
TCR (ppm/°K)	±100	±100	−150, −300	−100, −250
Resistance range (Ω)	10 Ω–5 MΩ	15 Ω–15 kΩ	20 Ω–100 kΩ	10 Ω–1 MΩ
Maximum sheet resistivity (Ω/□)	10^6	100–250	3000	100
Operating temperature range (°C)	−55 to +175	−55 to +125	−55 to +125	−55 to +125
Practical (trimmed) tolerance (%)	0.5	0.5	0.5	0.1
	Discrete resistors			
	Wirewound	*Carbon film*	*Metal film*	*Carbon composition*
TRC (ppm/°K)	±10	+200, −500	±25	±500
Resistance range (Ω)	0.1 Ω–1.2 MΩ	10 Ω–5 MΩ	10 Ω–5 MΩ	10 Ω–10 MΩ
Maximum sheet resistivity (Ω/□)	—	—	—	—
Operating temperature range (°C)	−55 to +145	−55 to +165	−55 to +125	−55 to +130
Practical tolerance (%)	0.01	5.0	0.5	1.0

Table 13.21 **Typical characteristics of active-RC filter capacitors**

Characteristics	*Mica*	*Polystyrene*	*Mylar*	*NPO ceramic chip*	*High-K ceramic*	*Thin-film tantalum*	*Polycarbonate*
Q (1 KHz)	1000	5000	200	1000	40	400	500
TCC (ppm/°C)	±200	−120 ± 30	+600	±30	−5000 to +20,000	+200 ± 25	+40
Operating temperature range (°C)	−55 to +125	−40 to +85	−55 to +125	−40 to +85	−55 to +125	−55 to +85	−40 to +85
Capacitance range (pF–μF)	1–0.1	500–10	5000–1	10–0.05	100–20	1000–200	1000–10
Minimum tolerance (%)	±1	±0.5	±20	±0.25	±10	±5	±0.25

13.6 RESISTORS AND CAPACITORS FOR USE IN ACTIVE FILTERS

The RC-active filter designer must take into account several factors that influence the choice of the filter components to be used. Among these factors are power consumption, size, weight, number of units to be manufactured, and so on. Passive components are usually manufactured using thin- or thick-film technologies, or discrete components, whereas the active elements are monolithic operational amplifiers (see Chapter 5). In this section, data on passive components used in the manufacture of active-RC filters are presented. This discussion supplements the information presented in Chapter 11.

In the manufacture of active-RC filters, three approaches are most commonly used. Two of them make use of passive thick- and thin-film elements and operational amplifiers. One important advantage of these approaches is the automation of the filter tuning [6]. The remaining approach makes use of discrete passive elements, together with operational amplifiers, to realize the desired transfer function. In this case, we do not have automated tuning of the filter. In addition, we do not have the advantage of batch processing and of temperature tracking elements. However, this approach is useful when we need to manufacture a small number of active-RC filters.

Thin-film elements have a thickness varying from 50 to 20,000 Å. This thickness range is obtained through deposition by vacuum or vapor methods onto a glass or ceramic substrate. Thick-film elements are deposited by means of screen-and-fire methods and the thickness varies between 125,000 and 625,000 Å.

A more basic distinction between thin and thick films lies in the manner by which they are formed. A thin film is one that is grown in situ from a vapor or liquid (sputtering, anodization, etc.) In this case the growth process determines the properties of the film. Thick films, in contrast, are formed using a material that is applied by some means (screening, spraying, etc.) onto a substrate. The properties of thick films are dependent on the properties of the starting material and not on the way the material is applied. Both technologies are used to manufacture the passive elements of active-RC filters.

Among the materials used for thin-film resistors are Nichrome (nickel–chromium), tantalum compounds, Cermet (mixture of SiO and Cr); for capacitors, SiO, SiO_2, Ta_2O_5, TiO_2, and Al_2O_3 (among others) are used. For thick-film resistors, the material used is usually thick-film Cermet.

Data on resistors and capacitors used in the manufacture of active-RC filters are presented in Tables 13.20 and 13.21.

13.7 REFERENCES

[1] S. J. Mason, "Feedback Theory—Some Properties of Signal Flow Graphs," *Proc. IRE*, vol. 41, pp. 1144–1156, 1953.

[2] A. Sedra and P. O. Bracket, *Filter Theory and Design: Active and Passive*, Champaign, Ill.: Matrix Publishers, Inc., 1978, Chapter 9.

[3] M. Kawakami, "Nomographs for Butterworth and Chebyshev Filters." *IEEE Trans. Circuit Theory*, vol. CT-10, June 1963, pp. 288–298.

[4] R. Saal and W. Entenmann, *Handbook of Filter Design*, AEG-Telefunken Berlin, 1979.

[5] A. I. Zverev, *Handbook of Filter Synthesis*, New York: Wiley, 1967.

[6] P. V. Lopresti, "Optimum Design of Linear Tuning Algorithms," *IEEE Trans. on Circuits and Systems*, vol. CAS-24, pp. 144–151, March 1977.

INDEX